Protocols for Macroalgae Research

Protocols for Macroalgae Research

Edited by
Bénédicte Charrier
Thomas Wichard
C.R.K. Reddy

CRC Press
Taylor & Francis Group
Boca Raton London New York

CRC Press is an imprint of the
Taylor & Francis Group, an **informa** business

Central cover photo courtesy of (c) Station Biologique de Roscoff, Yann Fontana.

CRC Press
Taylor & Francis Group
6000 Broken Sound Parkway NW, Suite 300
Boca Raton, FL 33487-2742

Printed on acid-free paper

International Standard Book Number-13: 978-1-4987-9642-2 (Hardback)

Library of Congress Cataloging-in-Publication Data

Names: Charrier, Bénédicte, author.
Title: Protocols for macroalgae research / Bénédicte Charrier, Thomas Wichard, and CRK Reddy.
Description: Boca Raton : Taylor & Francis, 2018. | Includes bibliographical references.
Identifiers: LCCN 2017048183 | ISBN 9781498796422 (hardback : alk. paper) | ISBN 9781351132954 (ebook)
Subjects: LCSH: Marine algae--Research--Popular works.
Classification: LCC QK570.2 .C49 2017 | DDC 579.8/177--dc23
LC record available at https://lccn.loc.gov/2017048183

Visit the Taylor & Francis Web site at
http://www.taylorandfrancis.com

and the CRC Press Web site at
http://www.crcpress.com

Contents

Preface

Macroalgae are multicellular, (usually) macroscopic organisms that inhabit either marine or freshwater environments. They are vital organisms on our coastlines, where they are the major primary producers and an essential link in the food chain. However, macroalgae are not only of high ecological importance, they also hold great economic value, being rich in vitamins, nutrients, and other organic substances making them useful for humans in many ways. Their direct use as food, feed, and fertilizers, or as sources of polysaccharides, minerals, and proteins for the food, pharmaceutical, and textile industries, together promote seaweeds as a promising resource for diverse human activities.

Macroalgae are attractive study organisms owing to their unique biological and evolutionary features. They have evolved independently of terrestrial plants and have hence developed some inherently specific cellular features. For example, their chlorophyll pigments are partly different from those found in land plants and genome sequence analyses have revealed the acquisition of novel genes, often by horizontal transfers that have resulted in the selection of distinct biosynthesis and regulation pathways. Furthermore, the marine environment imposes several specific conditions: mechanical forces are much higher than on land, light intensities are reduced, and high salinity promotes chemical modifications of cell components. Nevertheless, despite significant achievements in fundamental research over the past decades, macroalgae remain largely under-investigated. The stumbling blocks to work with macroalgae are practical ones: growing them and getting them to reproduce in controlled conditions are difficult at best. Macroalgae are highly sensitive to seawater composition, and this seawater is often artificially reconstituted in labs distant from the shore. Fertility and reproduction can be intractable in artificial conditions—as in animals—making genetic approaches sometimes close to impossible in the laboratory. Macroalgal growth and development can be further complicated by reciprocally beneficial interactions with bacteria. These few examples illustrate that working with these organisms poses many challenges, requiring intellectual flexibility and the development of specific, innovative tools. Opportunely, seaweed research is now experiencing an enormous impetus, driven by their potential applications.

This book is a collection of well-established methodologies, including current and updated protocols for research on macroalgae. It is not only intended for neophytes starting out in the study of this exciting group of organisms, but also for confirmed phycologists who are interested in the extraordinary possibilities offered by these organisms to address essential questions in algal physiology, developmental biology, and sustainable aquaculture. How do specific environments affect their growth and development? To what extent have they explored evolutionary alternatives, in comparison with their terrestrial counterparts? Answering these questions requires the development of a full range of cutting-edge techniques dedicated explicitly to macroalgae, the focus and the feature of this book. Progress in technological developments has now reached a stage where many techniques initially developed in and for other organisms can finally be transferred to macroalgae; genomics is one of the most convincing examples. Nevertheless, even straightforward extraction techniques require adjustment to be effective on macroalgae.

This book contains 31 chapters written by experts and reflects the variety of studies conducted recently on brown, red, and green macroalgae, both in the seaweed aquaculture sector and academic research. It is divided into three sections dedicated to *Cultivating and preserving seaweeds* (Section I), *Chemical composition* (Section II), and *Cellular and molecular characterization* (Section III), and each chapter offers the necessary *Tips and Tricks* to carry out the protocols successfully.

Section I (Chapters 1 through 9) deals with growing macroalgae in their natural environment. Standardizing material preparation and monitoring algal growth are a prerequisite to any attempt to study their biology in this changing environment. This first section covers protocols describing the production of standardized juveniles in the laboratory (Chapter 2 by Forbord et al., Chapter 3 by I. Bartsch, and Chapter 9 by Califano and Wichard), their transfer offshore, and the monitoring of their growth, sometimes in hostile environments (Chapter 1 by Buck and Grote and Chapter 6 by Leal and Roleda). Each step in the sustainable cultivation requires good knowledge on algal biology throughout its life cycle, starting with the early stages (Chapters 2, 3, and 6). Deciphering regulatory *checkpoints* in the *life cycle* is also of prime importance to improve aquaculture practices, but difficult to investigate at the molecular level (Chapter 8 by Kessler et al., and Chapter 9).

In Section II (Chapters 10 through 20), protocols provide access to the chemical substances found in seaweeds, which can be recognized as chemical factories. Approaches have been adapted to extract substances of various chemical composition, such as charcoal as a soil amendment (biochar; Chapter 10 by Contreras-Porcia et al.), and to identify and quantify a series of specific types of molecules, such as sugars (Chapter 11 by Graiff et al., and Chapter 12 by Huijgen et al.), proteins (Chapter 13 by Safi et al.), phytohormones (Chapter 14 by Matsuura et al.), antioxidants (Chapter 15 by Hou et al.), pigments (Chapter 16 by Beattie et al.), and oxygen species (Chapter 17 by Kumar et al.). Often inspired by research on microalgae, mass spectrometry techniques now offer a view of the general landscape of the metabolic activities occurring in macroalgae (Chapter 18 by Kuhlisch et al.). Most important, together with genomics-mediated identification of the holobiome (Chapter 5 by Aires et al.), surface metabolic profiling contributes to a better understanding of the cross talk between macroalgae and their biotic environment (Chapter 19 by F. Weinberger, and Chapter 20 by Cirri and Pohnert).

In Section III (Chapters 21 through 31), the experimenter will find approaches for the acquisition of data at the cellular level. Special attention is given to studies carried out at the surface of individual cells including *in situ* mapping of polysaccharides (Chapter 21 by Siméon et al.), elasticity measurements (Chapter 22 by Torode et al.), and growth patterning (Chapter 23 by Rabillé et al.) of the cell wall. These studies are extended further to other, additional levels, that is, the subcellular (e.g., cytoskeleton organization; Chapter 24 by Rabillé et al.) and tissue levels, with a fine-scale description of the spatial organization of tissues. Work on these levels requires adjustments of traditional microscopy techniques due to the special biological (e.g., for calcareous seaweeds; Chapter 27 by Kaleb et al.) and chemical (e.g., the high polysaccharide content; Chapter 25 by Nagasato et al., and Chapter 26 by Raimundo et al.) features of algal cells. New technologies, such as laser capture microdissection, shed light on the gene expression at the single-cell level (Chapter 29 by Saint-Marcoux et al.) and will be an important complementary technique in the transcriptomics of macroalgae (Chapter 28 by S. Heinrich). All these approaches aim to put together an instructive and coherent atlas of the features at the tissue, cellular, and subcellular levels. Furthermore, the determined differences between the various tissues, organs, or cells of the thallus will pave the way to improved knowledge on how macroalgae develop into multicellular organisms and will allow us to address new questions that may be relevant for the potential applications of macroalgae.

The last chapters raise prospects as to developing genetic engineering alternatives to the exploitation of natural, wild populations. *In vitro* production of native proteins in heterologous organisms is now possible for some species (Chapter 30 by A. Groisillier), and genetic transformation (Chapter 31 by Boesger et al.) offers the possibility to modify the activity of macroalgal cells in a targeted and controlled approach (e.g., in *Ulva).* Cross-breeding (Chapter 3) and protoplast fusions (Chapter 7 by Gupta and Reddy) are alternative approaches to modifying the genetic traits of seaweeds. The modification of the macroalgal genetic heritage should be accompanied by enhanced development of cryopreservation protocols, still too limited for these species (Chapter 4 by J. Day). Further developments in algal genetics (Chapters 30 and 31) in an ethical, regulated, and optimally legislated framework will enlighten fundamental research and increase the economic value of macroalgae as food sources or possible alternatives for the production of renewable energy.

This book is a production of the COST Action FA1406 *Phycomorph* (2015–2019), which endeavors to acquire knowledge on the biological mechanisms that control macroalgal development, growth, and reproduction (http://www.cost.eu/COST_Actions/fa/FA1406). The COST Action workshops and training schools carried out as part of the *Phycomorph* inspired us to compile the collective experience and knowledge of the participating phycologists to provide a sorely needed manual on the well-established and novel methodologies in *Protocols for Macroalgae Research.* We thank all the authors for their enthusiastic collaboration in the development and constitution of this book that, we hope, will inspire readers and researchers.

Bénédicte Charrier
Roscoff, France

Thomas Wichard
Jena, Germany

C.R.K. Reddy
Bhavnagar, India

Editors

Dr. Bénédicte Charrier is a senior scientist from Centre National de la Recherche Scientifique (CNRS) working at Station Biologique de Roscoff, Roscoff, France. She is a specialist in macroalgae morphogenesis and development, with an expertise on the filamentous alga *Ectocarpus siliculosus*. She is currently the Chair of the European network *Phycomorph*, supported by the Cooperation in Science & Technology (COST) Association (2015–2019), which aims to coordinate research in growth, reproduction, and morphogenesis of macroalgae in Europe and in association with Asian and North American laboratories.

Dr. Thomas Wichard is a research group leader at the Institute for Inorganic and Analytical Chemistry of the Friedrich Schiller University Jena, Jena, Germany. The main focus of his research group is to elucidate the mutualistic interactions between bacteria and the marine macroalga *Ulva* (*cross-kingdomcross-talk*). The group applies various methodologies in analytical chemistry, chemical ecology, and molecular biology to understand the basis of ecophysiological processes.

Dr. C.R.K. Reddy is a chief scientist in Seaweed Biology and Biotechnology Cultivation at CSIR–Central Salt and Marine Chemicals Research Institute in Bhavnagar, India. He is an expert in marine macroalgae, making significant contributions to macroalgae tissue culture and protoplast techniques, including genetic improvement of seaweeds, seaweed biomass for chemicals and biofuels, seaweed biodiversity, and seaweed biology and cultivation.

Contributors

Tânia Aires*
Centro de Ciencias do Mar (CCMAR)
CIMAR, Universidade do Algarve
Faro, Portugal

G. Alongi
Department of Biological, Geological and Environmental Sciences
University of Catania
Catania, Italy

Taghreed Alsufyani
Institute for Inorganic and Analytical Chemistry (IAAC)
Friedrich Schiller University Jena
Jena, Germany

and

Algal Research Laboratory
Taif University
Taif, Saudi Arabia

Matías Araya
Departamento de Ecología y Biodiversidad, Facultad de Ecología y Recursos NaturalesUniversidad Andres Bello
Santiago, Chile

and

Centro de Investigación Marina Quintay (CIMARQ), Facultad de Ecología y Recursos Naturales
Universidad Andres Bello
Quintay, Chile

Inka Bartsch*
Alfred-Wegener-Institute
Helmholtz Center for Polar and Marine Research
Bremerhaven, Germany

Stewart William Beattie
Mer, Molécule, Santé
Université de Nantes
Nantes, France

Bernard Billoud
Morphogenesis of Macroalgae
Station Biologique de Roscoff
Roscoff, France

Anne-Belinda Bjerre
Center of Bioresource and Biorefinery
Danish Technological Institute
Taastrup, Denmark

Jens Boesger
Institute for Inorganic and Analytical Chemistry
Friedrich Schiller University Jena
Jena, Germany

and

Leibniz Institute on Aging–Fritz Lipmann Institute (FLI)
Jena, Germany

Kristine Braaten Steinhovden
SINTEF Ocean
Trondheim, Norway

Siobhan A. Braybrook*
The Sainsbury Laboratory
University of Cambridge
Cambridge, United Kingdom

and

Department of Molecular, Cell, and Developmental Biology
University of California at Los Angeles
Los Angeles, CA

Lambertus van den Broek
Wageningen Food & Biobased Research
Wageningen, the Netherlands

Bela H. Buck*
Alfred-Wegener Institute Helmholtz Centre for Polar and Marine Research (AWI)
Bremerhaven, Germany

and

Applied Marine Biology
Bremerhaven University of Applied Sciences
Bremerhaven, Germany

Cristian Bulboa
Departamento de Ecología y Biodiversidad, Facultad de Ecología y Recursos Naturales
Universidad Andres Bello
Santiago, Chile

Gianmaria Califano
Institute for Inorganic and Analytical Chemistry
Friedrich Schiller University Jena
Jena, Germany

Bénédicte Charrier*
Morphogenesis of Macroalgae
Station Biologique de Roscoff
Roscoff, France
E-mail: Benedicte.Charrier@sb-roscoff.fr

Emilio Cirri
Bioorganic Analytics, Institute for Inorganic and Analytical Chemistry
Friedrich Schiller University
Jena, Germany

E.M. Cobussen-Pool
Biomass & Energy Efficiency
Energy Research Centre of the Netherlands (ECN)
Petten, the Netherlands

Loretto Contreras-Porcia*
Departamento de Ecología y Biodiversidad, Facultad de Ecología y Recursos Naturales
Universidad Andres Bello
Santiago, Chile

and

Center of Applied Ecology & Sustainability (CAPES)
Pontificia Universidad Católica de Chile, Avda
Santiago, Chile

and

Centro de Investigación Marina Quintay (CIMARQ), Facultad de Ecología y Recursos Naturales
Universidad Andres Bello
Quintay, Chile

John G. Day*
Culture Collection of Algae and Protozoa, Scottish Association for Marine Science
Scottish Marine Institute
Oban, United Kingdom

Paul Déléris
Mer, Molécule, Santé
Université de Nantes
Nantes, France

David S. Domozych*
Department of Biology and Skidmore Microscopy Imaging Center
Skidmore College
Saratoga Springs, New York

Delphine Duffieux
Marine Biology
Station Biologique de Roscoff
Roscoff, France

Justine Dumay*
Mer, Molécule, Santé
Université de Nantes
Nantes, France

B.F. van Egmond
Biomass & Energy Efficiency
Energy Research Centre of the Netherlands (ECN)
Petten, the Netherlands

Aschwin H. Engelen
Centro de Ciencias do Mar (CCMAR)
CIMAR, Universidade do Algarve
Faro, Portugal

Nicole Engelen-Smit
Wageningen Food & Biobased Research
Wageningen, the Netherlands

Camila Espinoza
Departamento de Ecología y Biodiversidad, Facultad de Ecología y Recursos Naturales
Universidad Andres Bello
Santiago, Chile

and

Centro de Investigación Marina Quintay (CIMARQ), Facultad de Ecología y Recursos Naturales
Universidad Andres Bello
Quintay, Chile

A. Falace*
Department of Life Sciences
University of Trieste
Italy

Joël Fleurence
Mer, Molécule, Santé
Université de Nantes
Nantes, France

Silje Forbord*
SINTEF Ocean
Trondheim, Norway

and

Department of Biology
Norwegian University of Science and Technology (NTNU)
Trondheim, Norway

Elizabeth Garrido-Ramírez
Centro de Investigación para la Sustentabilidad, Facultad de Ecología y Recursos Naturales
Universidad Andres Bello
Santiago, Chile

Angelika Graiff*
Applied Ecology and Phycology, Institute of Biological Sciences
University of Rostock
Rostock, Germany

Agnès Groisillier*
Algal Biology and Interactions with the Environment
Station Biologique de Roscoff
Roscoff, France

Britta Grote
Alfred-Wegener Institute Helmholtz Centre for Polar and Marine Research (AWI)
Bremerhaven, Germany

Vishal Gupta
Biological Oceanography Division
CSIR–National Institute of Oceanography
Goa, India

J.W. van Hal
Biomass & Energy Efficiency
Energy Research Centre of the Netherlands (ECN)
Petten, the Netherlands

Aleksander Handå
SINTEF Ocean
Trondheim, Norway

Paulien Harmsen
Wageningen Food & Biobased Research
Wageningen, the Netherlands

Sandra Heinrich*
Biozentrum Klein Flottbek
University of Hamburg
Hamburg, Germany

Cécile Hervé*
Marine Biology
Station Biologique de Roscoff
Roscoff, France

Takashi Hirayama
Institute of Plant Science and Resources
Okayama University
Kurashiki, Japan

Xiaoru Hou*
Center of Bioresource and Biorefinery
Danish Technological Institute
Taastrup, Denmark

Wouter J.J. Huijgen*
Biomass & Energy Efficiency
Energy Research Centre of the Netherlands (ECN)
Petten, the Netherlands

Yoko Ikeda
Institute of Plant Science and Resources
Okayama University
Kurashiki, Japan

S. Kaleb
Department of Life Sciences
University of Trieste
Italy

Ulf Karsten
Applied Ecology and Phycology, Institute of Biological Sciences
University of Rostock
Rostock, Germany

Christos Katsaros*
Section of Botany, Department of Biology
National and Kapodistrian University of Athens
Athens, Greece

Ralf W. Kessler
Institute for Inorganic and Analytical Chemistry
Friedrich Schiller University Jena
Jena, Germany

Kaia Kjølbo Rød
Seaweed Energy Solutions
Trondheim, Norway

Paul Knox
Centre for Plant Sciences, Faculty of Biological Sciences
University of Leeds
Leeds, United Kingdom

Maria Koutalianou
Section of Botany, Department of Biology
National and Kapodistrian University of Athens
Athens, Greece

Constanze Kuhlisch
Bioorganic Analytics, Institute for Inorganic and Analytical Chemistry
Friedrich Schiller University Jena
Jena, Germany

Manoj Kumar*
Climate Change Cluster (C3), Faculty of Science
University of Technology Sydney (UTS)
Sydney, Australia

Nirali M. Kumar
Faculty of Science, Medicine, and Health (SMAH), School of Chemistry
University of Wollongong
Wollongong, Australia

Michiel Kwantes
Institute for Inorganic and Analytical Chemistry
Friedrich Schiller University Jena
Jena, Germany

Jane A. Langdale
Department of Plant Sciences
University of Oxford
Oxford, United Kingdom

Sophie Le Panse
Merimage Platform
Station Biologique de Roscoff
Roscoff, France

Pablo P. Leal*
Instituto de Fomento Pesquero (IFOP)
Puerto Montt, Chile

Jelle van Leeuwen
Wageningen Food & Biobased Research
Wageningen, the Netherlands

Marina Linardic
The Sainsbury Laboratory
University of Cambridge
Cambridge, United Kingdom

J. Louis Kaplan
The Sainsbury Laboratory
University of Cambridge
Cambridge, United Kingdom

Takakazu Matsuura
Institute of Plant Science and Resources
Okayama University
Kurashiki, Japan

Koji Mikami*
Faculty of Fisheries Sciences
Hokkaido University
Hakodate, Japan

College of Fisheries and Life Science
Shanghai Ocean University
Shanghai, China

Michèle Morançais
Mer, Molécule, Santé
Université de Nantes
Nantes, France

Izumi C. Mori
Institute of Plant Science and Resources
Okayama University
Kurashiki, Japan

Taizo Motomura
Muroran Marine Station, Field Science Center for Northern Biosphere
Hokkaido University
Muroran, Japan

Gerard Muyzer
Microbial Systems Ecology, Department of Freshwater and Marine Ecology, Institute for Biodiversity and Ecosystem Dynamics
University of Amsterdam
Amsterdam, the Netherlands

Chikako Nagasato*
Muroran Marine Station, Field Science Center for Northern Biosphere
Hokkaido University
Muroran, Japan

Randi Neerup
Center of Bioresource and Biorefinery
Danish Technological Institute
Taastrup, Denmark

Georg Pohnert*
Bioorganic Analytics, Institute for Inorganic and Analytical Chemistry
Friedrich Schiller University Jena
Jena, Germany

Hervé Rabillé
Morphogenesis of Macroalgae
Station Biologique de Roscoff
Roscoff, France

Sandra C. Raimundo
Department of Biology and Skidmore Microscopy Imaging Center
Skidmore College
Saratoga Springs, New York

Peter J. Ralph
Climate Change Cluster (C3), Faculty of Science
University of Technology Sydney (UTS)
Sydney, Australia

C.R.K. Reddy*
Discipline of Marine Biotechnology and Ecology
CSIR–Central Salt and Marine Chemicals Research Institute
Bhavnagar, India
E-mail: crk@csmcri.res.in

Jean Pierre Remonsellez
Departamento de Ecología y Biodiversidad, Facultad de Ecología y Recursos Naturales
Universidad Andres Bello
Santiago, Chile

and

Facultad de Administración y Economía
Universidad de Santiago de Chile, Avda
Santiago, Chile

Jorge Rivas
Departamento de Ciencias Químicas, Facultad de Ciencias Básicas
Universidad de Chile
Santiago, Chile

and

Instituto de Ciencias Químicas Aplicadas, Facultad de Ingeniería
Universidad Autónoma de Chile
San Miguel, Chile

Michael Y. Roleda
Norwegian Institute of Bioeconomy Research
Bodø, Norway

Elodie Rolland
Morphogenesis of Macroalgae
Station Biologique de Roscoff
Roscoff, France

Wolfgang Ruth
Analytical and Technical Chemistry, Institute of Chemistry
University of Rostock
Rostock, Germany

Carl Safi*
Wageningen Food & Biobased Research
Wageningen, the Netherlands

Denis Saint-Marcoux*
Université de Lyon
UJM-Saint-Étienne
Saint-Étienne, France

and

Department of Plant Sciences
University of Oxford
Oxford, United Kingdom

Ester A. Serrão
Centro de Ciencias do Mar (CCMAR)
CIMAR, Universidade do Algarve
Faro, Portugal

Amandine Siméon
Marine Glycobiology
Station Biologique de Roscoff
Roscoff, France

Jorunn Skjermo
SINTEF Ocean
Trondheim, Norway

Yvette Telleman
Wageningen Food & Biobased Research
Wageningen, the Netherlands

Thomas Torode
Centre for Plant Sciences, Faculty of Biological Sciences
University of Leeds
Leeds, United Kingdom

and

The Sainsbury Laboratory
University of Cambridge
Cambridge, United Kingdom

Florian Weinberger*
GEOMAR Helmholtz-Zentrum für Ozeanforschung Kiel
Kiel, Germany

Thomas Wichard*
Institute for Inorganic and Analytical Chemistry
Friedrich Schiller University Jena
Jena, Germany
E-mail: Thomas.Wichard@uni-jena.de

Javier Zapata
Departamento de Ecología y Biodiversidad, Facultad de Ecología y Recursos Naturales
Universidad Andres Bello
Santiago, Chile

and

Centro de Investigación Marina Quintay (CIMARQ), Facultad de Ecología y Recursos Naturales
Universidad Andres Bello
Quintay, Chile

* Corresponding author.

section one

Cultivating and preserving seaweeds

chapter one

Seaweed in high-energy environments

Protocol to move Saccharina *cultivation offshore*

Bela H. Buck and Britta Grote

Contents

1.1 Introduction to offshore aquaculture

In recent years, the development of offshore aquaculture to increase global aquaculture production has become a new challenging frontier of the *Blue Revolution*. Driven by the stagnating or in parts even decreasing capture fisheries, a strong shift towards the growth in aquaculture production to fulfill the increasing global food demand could be observed in recent years (FAO 2016). As a significant expansion of coastal aquaculture is constrained in most areas by the fact that suitable areas are already used for aquaculture and competition for space between different stakeholders thus limits the aquaculture expansion, substantial research and development to explore the possibilities of aquaculture in the open sea is in progress. *Offshore aquaculture* refers to the cultivation of marine organisms not only far from coastal areas but also mainly at the sites exposed to the high-energy environment of the open ocean (Ryan 2005). However, somewhat in contrast to aquaculture development in coastal areas, offshore aquaculture faces numerous additional technological, operational, economic, and political challenges (Langan 2009). In most cases, offshore operations are set in the exclusive economic zone (EEZ), which is defined by the United Nations Convention on the Law of the Sea (UNCLOS) as a 200-nautical mile zone beyond the 12 nautical mile coastal sea of a state. The EEZ is a special zone, either controlled or allocated in a number of different ways under various legal systems, such as national and international law, depending on the respective country. Besides the heavy institutional setup for marine spatial planning, offshore operations also face challenges pertaining to the technology of aquaculture structures, operation, and maintenance. All these need to be able to withstand the harsh environment under high-energy conditions. Over the past decade, considerable international research efforts have fostered major technological advances, and new offshore use concepts and several have been successfully tested at small scale. However, currently only a few offshore aquaculture businesses are in service, and the implementation of the research results is hence limited.

In addition, the installation of offshore aquaculture systems needs to be accompanied by the development of sustainable solutions for commercial aquaculture to minimize adverse effects on the ecosystem (Buck and Langan 2017a). To achieve this goal, the waste management of most aquaculture operations can be improved by using extractive species to take up access nutrients, and therefore mitigate environmental impacts (Neori et al. 2004, Corey et al. 2012). Thus, seaweed is an ideal extractive component for eco-intensification of aquaculture operations, leading to an integrated multitrophic aquaculture (IMTA). However, for offshore cultivation, candidate species must tolerate the harsh and exposed conditions and concurrently provide enough economic potential to counterbalance the high costs involved in offshore cultivation.

1.2 State of the art of offshore seaweed cultivation

In 2012, 95.6% of seaweed production was derived from mariculture (FAO 2014, Chopin 2014, Clements and Chopin 2016). The Asian countries, such as China, Indonesia, the Philippines, the Republic of Korea, Japan, and Malaysia host the highest seaweed production of about 96.3% of the total yield (Clements and Chopin 2016). The central focus of the production is hereby placed on the cultivation of the brown seaweed *Saccharina japonica* and *Undaria pinnatifida* as well as on the red seaweed *Porphyra tenera* and Eucheumatoids (species of *Kappaphycus* and *Eucheuma*) (McHugh 2003). In many Asian countries, people are growing, harvesting, and cooking temperate seaweed species in large quantities, whereas cultivation in the western countries reached only 54,000 tons with an annual value of U.S.$51 million in 2013 (FAO 2016). Interestingly, despite that seaweed cultivation can be viewed as an emerging new industry in western countries, the western market request is increasing due to a rising demand for the pharmaceutical and cosmetics production, the health food industry, for animal feed, or newly as a substrate for biofuel production (Edwards and Dring 2011).

Moving the cultivation of seaweed from coastal areas into the offshore realm could solve currently existing spatial and environmental constraints. New technological approaches to move aquaculture offshore were investigated in the Americas and Europe in the course of the past 15 years. However, depending on species, technology, and selected sites, there are still problems to be solved before upscaling to a commercial level can take place. In the early 1990s, predominantly Laminarian seaweed was used for first offshore cultivation trials in Europe (Lüning and Buchholz 1996). Subsequently, the first trials of offshore seaweed cultivation in France and Germany were followed about 10 years later by UK and Ireland. Nowadays, countries such as Germany, Belgium, Norway, the Netherlands, France, the United Kingdom, Denmark, Spain, Portugal, the United States, the Republic of Korea, and China, are all carrying out research and/or seaweed farming trials further off the coast, most of which are still in pilot scale. Only a few offshore operations for seaweed cultivation are commercial-scale businesses, such as enterprises in Germany (CRM 2001), Norway (SES 2015a, b), and the Netherlands (Hortimare 2016).

One of the leading institutes in offshore seaweed research is the Alfred Wegener Institute (AWI) Helmholtz Centre for Polar and Marine Research in Bremerhaven (Germany) (Figure 1.1), where trials of offshore seaweed cultivation started already in the early 1990s with focus on the system design that investigates the suitability of longline, ladder, grid, and ring constructions (Lüning and Buchholz 1996). Subsequent studies concentrated on reproduction, seeding strategies, growth, and site selection (Lüning

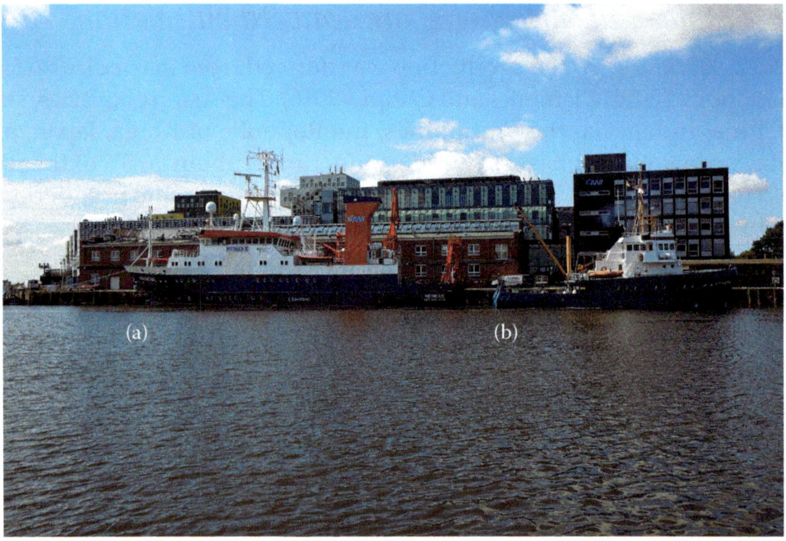

Figure 1.1 Alfred Wegener Institute Helmholtz Centre for Polar and Marine Research (AWI) in Bremerhaven (Germany). The image shows the main building with two research vessels in front. (a) *RV Heincke* vessel was used to carry the equipment for the offshore farm and for deploying the moorings, and (b) *RV Uthörn* vessel was used to install the complete harness, including system design, buoyancy, seeded Laminarian ropes, and for O&M and harvest. (Courtesy of Müller-Elsner and AWI.)

and Buchholz 1996; Buck and Buchholz 2004a, 2005). The technology used in earlier studies was then revised with a focus on the easy deployment, maintenance, harvesting, strength and connectedness to the foundations of offshore wind turbines (Lüning and Buchholz 1996, Buck 2001; Buck 2004; Buck and Buchholz 2004a, b). For this purpose, the first co-use studies for the combination of aquaculture and offshore wind farms emerged (Buck 2004). One result of these studies was the patented ring construction for the offshore cultivation of seaweed on an industrial scale (Buck and Buchholz 2004a, b).

Similar to the research in Germany, the offshore seaweed cultivation is discussed in co-use with existing offshore structures in Norway (Skjermo et al. 2014). The main reason to relocate aquaculture offshore is the negative impact of nutrient-rich effluents from the Norwegian salmon farms on marine ecosystems in coastal areas. In 2011, the project SWEEDTECH investigated the development of an economic system for commercial-scale offshore seaweed cultivation, including the development of seeding strategies of carrier material, design, and improvement for a structural rig, and of innovative deployment and harvesting methods (SES 2015a, b). As a result of this project, a new seaweed cultivation development in Norway, *AquaCulture Engineering* (ACE) *AS* and *SINTEF Fisheries and Aquaculture* will start Laminarian seaweed cultivation in a 3 ha site with an anticipated production of 1500 tons per year (ACE 2015).

Furthermore, the enterprise *Seaweed Energy Solutions* (SES) AS patented (SES 2015a) the first modern structure for industrial-scale seaweed cultivation in Norway. This sheet-like structure, named *Seaweed Carrier*, copies a huge seaweed thallus, moving freely in the sea connected to a single mooring on the seabed. So far, two major tangible prototypes of the carrier (SES 2015b) were tested with promising results. In the long run, this structure should be deployed in deeper and more exposed waters for offshore seaweed cultivation.

In the Netherlands, Ecofys and Hortimare undertook one of the first co-use trials of the seaweed cultivation in an offshore wind farm in 2012 to test whether cultivation at this site is feasible. The 400 m^2 large cultivation site comprises a set of steel cables, set up at 2 m below the sea surface attached to anchors and floating buoys. Connected to these cables are horizontal nets of 100 m^2 seeded with Laminarian sporophytes. In June 2015, Hortimare successfully cultivated brown seaweed at a small scale 10 km north–east from the Island of Texel in the Dutch North Sea. More seaweed cultivation trials are planned in an offshore area off the coast of Scheveningen (The Hague).

In some offshore areas, where nutrients are limiting factor for optimal seaweed growth, an offshore IMTA is recommendable (Troell et al. 2009; Buchholz et al. 2012). The current technological possibilities make kelp species ideal candidates for such type of cultivation as they have low maintenance requirements and are relatively easy to harvest. However, recent research with a focus on the strain development could lead to seaweed cultures with desired morphological and physiological traits accommodating site-specific aquaculture demands while providing high-production output. This would also ease the pressure on wild stocks used for seed collection.

In the Americas and Europe, mainly kelp species (primarily the sugar kelp, *Saccharina latissima*) have been cultivated during the past two decades (Buck and Buchholz 2004a, 2005; Barrington et al. 2009; Broch et al. 2013; Kraemer et al. 2014; Kim et al. 2015; Marinho et al. 2015). Laminarian species are some of the most robust seaweed with high potential for offshore aquaculture because of their morphology, followed by the red seaweed *Palmaria* sp. and the green seaweed *Ulva* sp. Laminarian species can adapt to high currents and reach high growth rates if deployed offshore as young individuals (Buck and Buchholz 2004a, 2005).

In the following chapter, we present a protocol for offshore seaweed aquaculture with a focus on the sugar kelp *S. latissima*, including different aspects and challenges of offshore cultivation, also in multi-use settings. The described protocol is based on research efforts and experience since

2000 until today (Buck et al. 2008; Buck et al. 2017a, b). Several types of seaweed cultivation in the offshore realm were tested, such as longline or ring constructions (Buck and Buchholz 2004a). The protocol briefly describes the land-based preparation, including a list of equipment needed, the seeding procedure, and the preparations to move offshore. However, as this chapter is focusing on the offshore cultivation of seaweed, for example, *S. latissima*, we do not go into detail regarding the complete life cycle and reproduction of this species and will describe the cultivation mainly about the offshore aspect, including the technical design. For detailed descriptions of the life cycle of *S. latissima*, the seedling protocol and hybridization of kelp also refer Chapter 2 by Forbord et al. (2017) and Chapter 3 by Bartsch within this compilation.

1.3 *Land-based preparation and site selection*

In the following, the different offshore seaweed technologies and system design will be described. Procedures before the transfer at sea are in most parts similar to the process described by Bartsch (Chapter 3) and Forbord et al. (Chapter 2). Only the seeding procedure at a certain stage slightly differs from common technologies and will be presented as follows.

In the late 1980s and early 1990s, macroalgal farming of Laminarian species was tried out at some places further off the coast. For instance, Neushul and coworkers installed a pilot farm in a hostile environment off the Californian coast (Neushul and Harger 1985). Here, longline technologies were deployed to test the biomass yield from a nearshore test farm of approximately 0.5 ha in size. Further, Neushul et al. (1992) measured water flow and nitrate uptake to show that flow rates over a crop in the sea can be controlled. Similar studies were done in Europe: at the Isle of Man (Kain 1991) and the Ile d'Ouessant (Perez et al. 1992). However, all these studies used equipment adapted to nearshore, mainly sheltered conditions. The move into a high-energy environment, however, needs equipment that withstands the forces induced by strong current velocities and high waves (Buck et al. 2017b).

In 1993, another feasibility study on the offshore algal culture in the North Sea was launched. One major component of this research was to develop an appropriate technical device to grow macroalgae, which withstands the harsh environmental conditions of the German North Sea shelf with maximum wind speeds of 150–180 km/h, a wave height of up to 5–8 m during storms. This was the first study to move seaweed farming into severe sea states and was funded by the German government and conducted at the marine station of the *Biologische Anstalt Helgoland* (BAH) off the island of Helgoland (North Sea). Because of the extreme drawbacks caused by too strong wave climate conditions and resulting loss and damage of material, the research was conducted again starting in the early 2000 (Buck 2004). As a result, a ring design was developed that resisted all sea conditions, whereas at the same time producing approximately one ton of seaweed biomass (wet weight) on roughly 20 m^2 (Buck and Buchholz 2004a,b). Since then, several system designs were developed offshore from different institutions and withstood storm conditions.

In the following section, we summarize the major design criteria when moving from nearshore sheltered conditions in offshore environments:

1.3.1 Site selection

Locating an offshore site involves much prior research, integrating scientific, technological, biological, socioeconomic, and legislative issues. The decision to build an offshore installation, therefore, requires not only finance but also a dedicated interdisciplinary team, the right equipment, and the correct biological parameters that endorse all profits and growth performance of the crop (McElwee 1997).

Before a seaweed farm is deployed at a respective area, a meticulous site assessment must be conducted to gather a variety of parameters. These parameters can be subdivided into three parameter categories, which will be carefully evaluated: (1) advance information of the local area ashore and at sea; (2) site-specific, oceanographic, and water quality parameter; and (3) information on the economic/technical/expansion feasibility. A complete list of parameters and a progression of site visits and investigation is shown in Figure 1.2. The actual values from the selected site and the target value for the seaweed enterprise depends on the species, the size of the farm, economic potential, etc.; a generalization of parameters for all seaweed species does not exist and is not feasible, because every species has certain requirements, which are reflected in different max and min values compared to other selected species, and every site is different. However, important is the fact that some parameters need special attention when moving into the open ocean. These parameters are designated with an exclamation mark (!) in Figure 1.2.

To find the perfect site is not easy and requires time, patience, and staying power. It is important to take this into account as some potential *young seaweed growers* would like to start soon and may ignore that the processes of finding the farm site, selecting the appropriate design of the farm that is tailored to the respective site, and the permitting procedures at the authorities and finances tend to last long and must be viewed very strategically (because of local regulations/stakeholders/permits/banks, etc.). Following Benetti et al. (2010), there are two primary issues during site selection: (1) it is essential to understand that site selection finally is a compromise and a process of elimination in which the majority of areas initially identified as suitable are eliminated because of user conflicts and (2) depending on the project size, the level of investment, and the

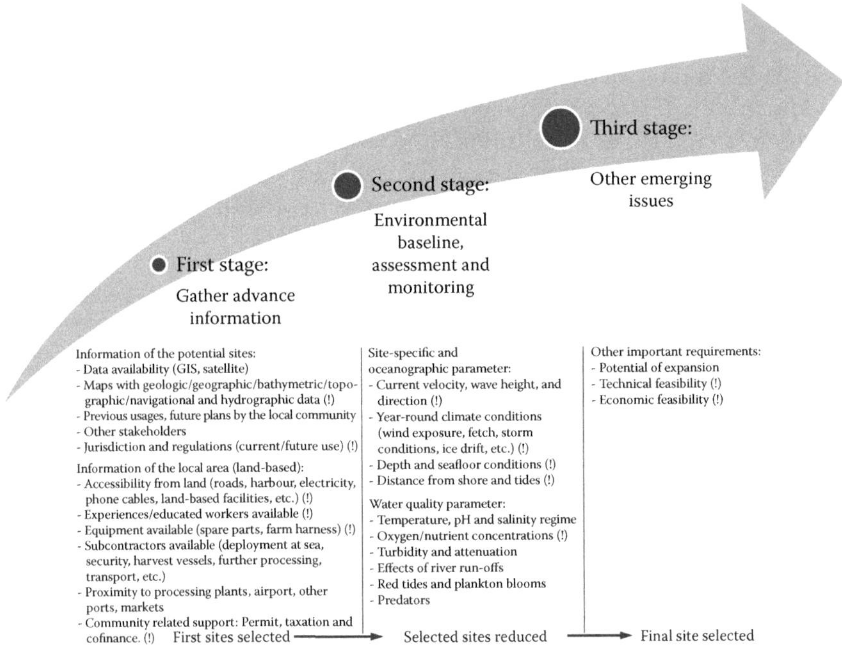

Figure 1.2 Site-selection process for a seaweed farm, including typical seaweed site-selection criteria. *First stage* is the start of the selection process and will end through the *second stage* at the end of the *third stage*. At the end of each stage, several sites are identified but will be reduced to the specification process (grey arrows). Parameters with an exclamation mark need special attention when moving off the coast into a hostile environment.

timeframe for development, it may be necessary to create infrastructure. However, the list of issues is long and at various planning levels to be conducted in the planning process (see list in Section 1.3.1.1).

1.3.1.1 *Site selection: Advance information of the local area ashore and at sea*

What is the coastal population/community size and setup next to your site?

- This will influence the acceptance of the local community, which, in turn, could provoke stakeholder conflicts.
- This issue is also relevant to have access to skilled workers, subcontractors, and equipment.
- Size/organizational level of the local community could influence the developmental stage of roads, harbors, and electricity, and the proximity to other important infrastructure facilities (airport, highways, etc.)

Is there an aquaculture act, any legal framework, an experienced person (authority delegate) in charge, and information on taxation and finance for that respective site?

- It could be relevant for the permit, and any other emerging issues related to regulations and jurisdictions. It could support or hinder the issue of a lease and can be influenced if there are already reserved areas, which, in turn, would have the marginal potential for stakeholder conflicts.
- It could also be important for potential cofinance, any other financial support, and taxation.

Is information on the selected site (geology, sediment, topography, and depth) available?

- The condition of the seafloor (soft bottom, hard substrate, and depth) will reduce the number of selected sites, and therefore ease the site selection process.
- Preapproved areas for aquaculture would support the selection of good sites.

1.3.1.2 Site selection: Site-specific, oceanographic, and water quality parameter

Is reliable information available on physical, biological, geographical, and other local parameters of the selected site?

- This could influence the growth, morphology, and health, and the biofouling on the seaweed (nutrient concentration, light, water quality, etc., on short-term and an annual level). Most of the conditions should be stable year round.
- The offshore conditions (wave/swell, current velocity, wind/fetch, etc.), and the depth/topography, will affect the technology needed, as the construction should, of course, withstand all potential worst-case conditions while also having a long, expected lifetime.
- It is urgent to have a land-based facility close to the shore/harbor/pier to allow quick and logistically easy handling.

Is there any information on potential local users or other factors, which could influence the water quality?

- This information is needed to avoid areas with harmful river run-offs or other issues influencing the seaweed production (predator, sediment loads/suspended matter/turbidity/light attenuation).

- Further, freshwater or brackish water from rivers/creeks/estuarine areas cannot be tolerated as these conditions would also harm the plant and the production results.

1.3.1.3 Site selection: Economic/technical/expansion feasibility

Is the selected site feasible taking the current technology development and system design into account?

- There is plenty of information that is available to design a seaweed farm from a technological and O&M perspective. However, depending on the local parameters, the site may be unfeasible because of the prevailing wave climate, current conditions, and etc., as this could lead to the loss of biomass/parts of the construction, permanently harm or destroy the construction, and increase the level of blade tip loss.
- The technology can be floating, semi-submerged, or fully submerged. This could affect the growth of the seaweed as some species would extremely reduce their growth rate in submerged conditions/deeper environments.

Is the selected site perfect regarding economic feasibility?

- With regard to the selected site, it is also important to calculate the distance to shore, the access to the farm depending on weather conditions (how many days at sea per month/year?), and the equipment needed to run an offshore business (adapted system design, vessels adapted to severe weather conditions, etc.) as these parts are different as compared to inshore or sheltered farms and farm equipment.
- The production costs per ton of biomass could be more expensive for offshore cultivated seaweed. This could be slightly reduced depending on the complete production volume of the entire farm. Even then, does the local market balance these higher production costs? Is the selected site too small, and therefore the produced biomass too low for a break even?

Is the selected site restricted to a certain production amount?

- In the case of a *perfect site* for seaweed cultivation, this would reduce the potential to expand the business in the future.
- The selected site could be too small to run it economically feasible.

The success of the planning procedure of the enterprise and the site selection will depend to a large extent on a thorough site survey and proper assessment about the project development.

1.3.2 Equipment and system design

For the farming of Laminarian species there are plenty of systems used, either in sheltered or in hostile conditions: (1) longlines, (2) ladder, and (3) grid systems (Figure 1.3a–g), ring devices (Figure 1.4a–d) and technologies mounted to the seafloor (e.g., extensive fixed off-bottom longline technique in the Baltic Sea—Kiel/Germany; CRM 2001). In direct comparison, installations in the open ocean need modifications in system design to cope with the hostile conditions offshore. Therefore, to establish the

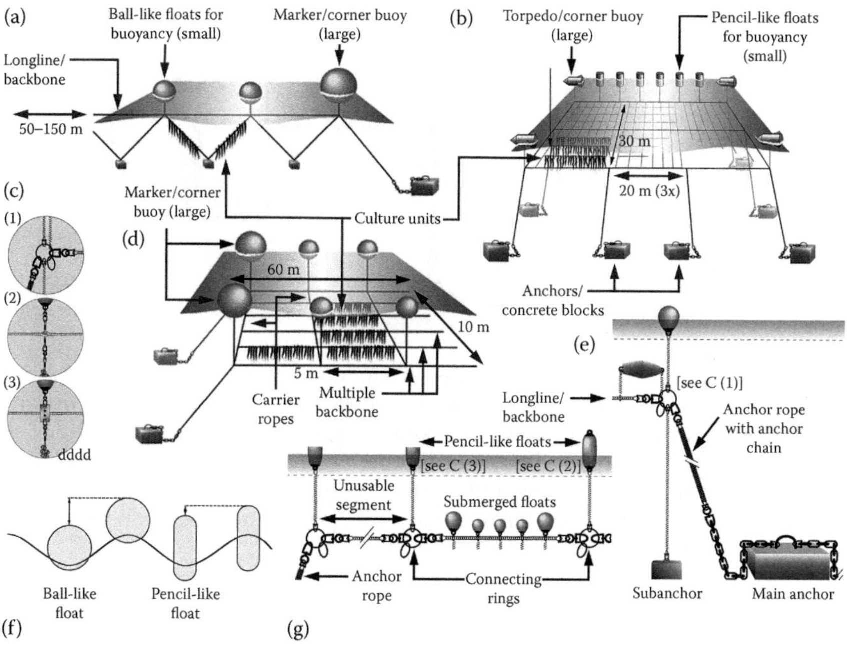

Figure 1.3 (a–g): System design and the concept of seaweed cultivation devices. (a) Longline (backbone) system with floating buoyancy and the seaweed culture unit hanging perpendicular in the water column; (b) large grid design with floating buoyancy and rectangular culture units; (c) connection devices with (C 1) rings as coupling center piece, connection of floats to the backbone within the strands of the rope (C 2), or by using metal tiles (C 3); (d) ladder construction with culture units attached to the multiple backbone of the system; (e) unit at one end of a backbone with different anchors, holding devices (chains, ropes), and flotation; (f) different buoy shapes and dipping depths when riding the swell; and (g) backbone with floating and submerged flotation and the *unusable segment*. (Modified after Buck 2007 and Buck & Buchholz 2004.)

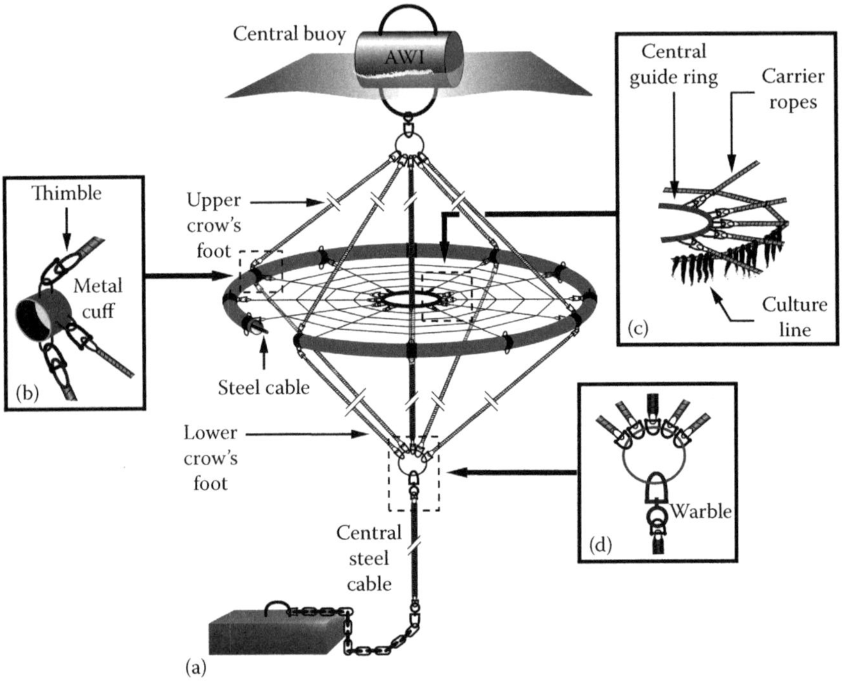

Figure 1.4 (a–d): Offshore-ring device for the cultivation of seaweed. (a) Ring construction for the culture of Laminarian species at offshore locations, including major elements magnified in; (b) metal cuffs, to which the upper and lower crow's feet, the ring tube, and the carrier ropes are attached; (c) the central guide ring with attached carrier ropes and culture lines; and (d) transition between the central steel cable of the mooring and that of the lower crow's foot. (Modified after Buck and Buchholz 2004a.)

farm in such a hostile environment will cost more in capital equipment and running costs. Further, no matter what system design one chooses, it must be durable, flexible, sturdy, noncorroding, hardy, workable, and accessible to work on (McElwee 1997). All parts of the installation, such as anchors, concrete blocks, wires, ropes, and chains, all parts for coupling, buoys, and floats and pipes (ring device) and other parts, are important; however, special attention should be given to the mooring design (Figures 1.5 and 1.6).

In the following section, we give some important hints when choosing the equipment and designing the installation.

Figure 1.5 (a–f): Different development steps of the offshore-ring device. (a) Shows the center piece in the first stage, which was because of the severe conditions in the open ocean exchanged by the central guide ring; (c) connection of the carrier ropes to the ring and the modified coupling devices (d); (e) small ring device with 2.5–3.0 m in diameter and the version with 5 m in diameter during the harvest (f). (Modified after Buck et al. 2017b.)

Figure 1.6 (a–e): Farm equipment stored on land/board before deployment. (a) Wires of different length according to the depth of the mooring site, (b) typical corner/marker buoy with radar reflectors for later identification of the mooring positions, (c) concrete blocks at the peer before transfer on the deck of the service vessel (during winter time), (d) concrete blocks lashed on the board of the service vessel *RV Heincke*, and (e) mooring chains of various length depending on the depth of the farm to anchor the system to the concrete blocks.

1.3.2.1 System design for the offshore seaweed farm

Vessel

- It would be most beneficial to use those vessels that are appropriate for offshore aquaculture with a size of approximately 15–25 m in length (minimum) to allow enough space aboard to prepare the harness, moorings, flotation, etc.
- The vessel should be equipped with a crane having enough power to easily lift the moorings and all the other equipment.
- In addition, the vessel should have a slip hook at midship and a large freeboard to assure easy handling of all floating devices from the entire installation.
- A vessel, which could maneuver accurate to a meter at the spot of deployment to guarantee the right placement of the anchor stones/concrete blocks, is necessary.

Ropes and connecting pieces

- Ropes and lines will have different diameters and are made of different materials as well. For the seeding ropes polypropylene (PP) lines of 6, 8, and 10 mm proved well (Buck and Buchholz 2004). Ropes used as backbones are mostly made of either PP or polyethylene (PE) and are usually much thicker from 15 to 32 mm depending on the length and local conditions.
- The backbone holds the seeded line coiled around the backbone or hanging down from the backbone in single droppers or in loops. As current velocity and waves are strong and quite frequent, offshore spacing between loops and droppers is important to avoid the entanglement of the seeded ropes, which would lead to abrasion and detachment or loss of the plants (see Figures 1.3, 1.4, and 1.6).
- Seeded ropes could either be coiled around the backbone in a continuous version or lashed to the backbone where the loops and droppers meet the backbone.

Flotation

- All floats are normally lashed to the backbone or other parts of the system. If the farm site is extremely hostile, it may be possible to use shackles.
- Corner/marker buoys are normally at a single chain or mooring rope/wire and do not have a direct contact to the backbone (see Figure 1.3).
- The spacing of floats depends on the various parameters and the size of the floats. Larger floats, of course, increase the buoyancy per length of rope and decrease the number of floats to be used. On the other hand, large floats experience stress as due to the large surface the load forces increase.

- Further, different types of floats are available: ball-like floats ride the swell, and therefore directly impact everything hanging below the floats, whereas pencil-like floats partly dip into the impinging wave, and therefore decrease the stress on the system (Figure 1.3).

Mooring

- There are different mooring devices available on the market: concrete blocks, metal, tension, and drill anchors. These various mooring parts can be used in combinations but mainly depend on the condition and depth of the seafloor.
- Concrete blocks are usually used if the bottom is hard substrate. In softer conditions, anchors of all types can be used at the seafloor.

1.3.3 Seeding procedure

To prevent the loss of seaweed in high-energy environments, the holdfasts need to adapt to offshore expected forces. Buck and Buchholz (2005) investigated the sensibility of macroalgal aquaculture in the offshore realm of the North Sea. They measured and compared the drag forces on Laminarian blades originating from sheltered sites (undulated/ruffled margin and wide blade) and offshore adapted species (elongated/streamlined and flat/narrow blade). As the drag varied according to (1) frond size, (2) current velocity, and (3) acceleration reaction, dislodgement of Laminarian holdfasts and the forces necessary to break the stipe depend on the blade length and the surface area. Both measured and our calculated values of drag did not exceed those forces occurring offshore. Even during storm conditions with maximum current velocities of 1.52 m s^{-1} and wave heights of up to 6.4 m *L. saccharina* (*S. latissima*) withstood the high-energy environment.

To support the adaptation to offshore stress conditions, indoor seeding strategies should be conducted by using *Automated Laminaria–Saccharina Seeding Drums* (Figure 1.7) coiled with ropes, to be seeded by the spores (Buck and Buchholz 2004a; Buck et al. 2017a). First, these *seeding bodies* allow a several 100 m long seeding rope as substrate (instead of *curtain* systems) for the sporophytes to grow. Further, the drums rotate slowly at early stages (up to 20 days) and then exceed their speed. The advantage is that because of this induced current at the drum/rope to the water surface, which can be increased by accelerating the rotation velocity of the drums, that this velocity impinging the plant stipes and holdfasts stresses the sporophytes leading to an adaptation to more hostile environments (Buck et al. 2017a, b). When transferring the plants at sea in high-energy environments, plants will not detach from its

Figure 1.7 (a–h): Preparation of seeded ropes for the transfer to the offshore farm. (a) Seaweed production site with greenhouse and seaweed production tanks at the AWI in List (Sylt, North-Germany); (b) adaptation of seaweed blades (*Saccharina latissima*) originating from offshore and sheltered environments; (c–e) tanks with seeding drums and its motorization; (f–h) seeding rope preparation from empty curtain devices to young sporophytes ready for transfer at sea. Red arrows display the process from prepared empty curtain-like ropes (f) to seeded ropes coiled on drums (c), through young sporophytes (g) to storage for the transfer at sea (h). (Modified after Buck & Buchholz 2005; Buck et al. 2017a.)

substrate (Buck and Buchholz 2005). Major growth of young sporophytes first takes place in forming a withstanding holdfast and later followed by a blade length increase.

1.4 Transfer at sea

1.4.1 Deployment of the offshore farm

As a first step, the installation of aquaculture devices offshore needs a full suite of complementary oceanographic, environmental, topographic, and site-specific dataset, a complete understanding of the water motions on the farm structure, and its associated candidates. Even if most of these data were already gathered during the site-selection process, it is of prime importance to get the data updated. Depending on current weather conditions, the identification of the right time slot for the entire installation is crucial because this step may take several days in a row.

By and large, one of the major features for the seaweed farm should be the ability to survive the described conditions, where it will be deployed, while at the same time allow suitable growth performance. To avoid major stress on the installation induced by prevailing and worst-case conditions, the major part of the farm installations should be submergible. That means that a significant part of the entire culture unit is located beneath the surface, whereas the bulk of the structure's mass is mounted on the surface to allow buoyancy and stabilization. Further, the entire mooring devices should be below the zone of turbulence and wave action (Starchild 1980). None of these technologies and system designs are available off the shelf on the market as every site is different. A *One size fits all* way to solve this is not existing and a far cry from being a solution.

One very important lesson is as follows: any preparation, which can be conducted on land, should be done on land! The farm could be in an extreme location subject to full rigors of the environment at all times. That said, everything prepared offshore is not only more expensive about the vessel time, it is also not convenient because of offshore and weather conditions, which will even complicate the work and endanger the work staff (e.g., Figure 1.12). Only the installation of the farm should be done offshore.

The farm cannot be deployed without having experts calculating the various forces using the oceanographic, topographic, and climatic parameters to conduct the numerical modeling. Even if these calculations have been done and a farm design was suggested, the deployment still can only be done by experienced experts/crews. Therefore, we will not explain in detail how to maneuver the vessel, how the various pieces of the system design will be carried with the cranes to be deployed at the site. In the following section, we will provide some considerations for the farm deployment, which support the planning process.

1.4.1.1 Deployment of the farm

Vessel

- It would be most beneficial to have somebody/a crew member on the service vessel with mooring experience.
- Check the equipment of the vessel to be appropriate to lift and lower heavy stones, anchors, chains, etc., with enough service loads (e.g., cranes, winches, slip hooks, and line hauler). It is also important to use vessels that can not only operate in sheltered environments, but also navigate through narrow passages (e.g., between other farms) and maneuver on the spot (Figures 1.8 and 1.9).
- The vessel should have enough freight hold to allow the transfer of the complete farm harness.

Figure 1.8 (a–d): Deployment at sea under harsh conditions. (a) Short time slots are not rare and should be used if necessary to deploy heavy equipment (e.g., concrete blocks); (b) the vessel needs to maneuver on the spot; (c) equipment can be extremely heavy and needs the entire crew to handle, also at night times; (d) typical harsh conditions at the seaweed farm site at *Roter Sand*, 17 nautical miles off the coast of the city of Bremerhaven (Germany).

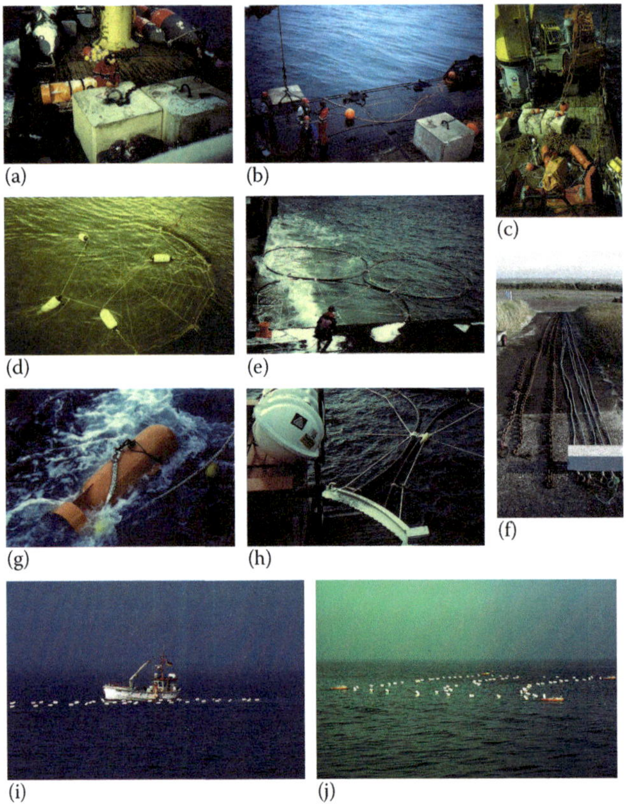

Figure 1.9 (a–j): Deployment of the seaweed farm at the offshore site. (a)–(c) concrete anchor clocks (2 tons) lashed on board of the service vessel *RV Heincke* and deployment of the concrete blocks at the farm site; (d) floating-ring device (first generation) after deployment; (e) coupling of various ring devices before towing them to the farm site; (f) mooring line length adjusting and preparation depending on different depths; (g) torpedo-like sensor for towing tests to examine force loads resulting from and on farm devices; (h) deployment of farm designs in a connected mode; (i) O&M vessel *Aade* at the farm site; (j) longlines after deployment. (Modified after Buck et al. 2017a,b.)

Ropes, equipment connections, and buoyancy

- Every possible connection, which does not complicate, block, or hinder the deployment, should be done land-based (Figure 1.10). The remaining work will be done onsite.
- It is important that previous to the deployment all calculations with regard to buoyancy force, flotation, and gravity force on the various bodies used, load and resistance forces, and rope sizing (mooring, backbone, connections, buoys, and anchors) had been done to complete satisfaction.

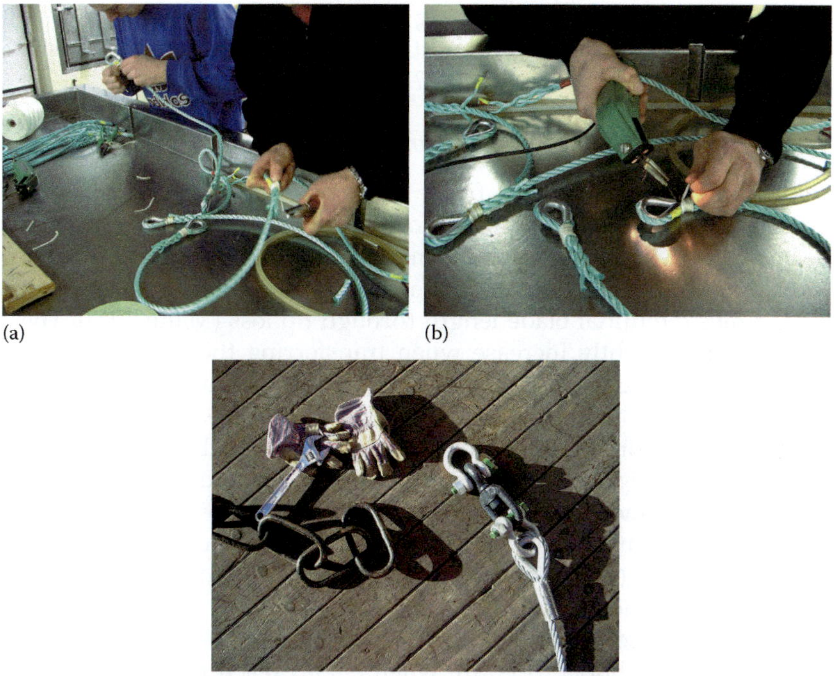

(a) (b) (c)

Figure 1.10 (a–c): Land-based work previously conducted before the system will be transferred at sea. (a) Preparation and arrangement of swivels, thimbles, and shackles with the Laminarian seeding line; (b) heat preparation of rope ends by using an electric heat cutter to avoid unraveling and opening of rope ends; (c) image shows the size of chains, shackles, swivels, and thimbles to connect the buoyancy to the backbone.

Mooring

- Care and consideration should be taken for the moorings procedure to deploy the backbone and the entire seaweed harness.
- Topography conditions and sediment type is an important information to ease the mooring procedure.
- The mooring device is only effective if it remains in its position under the farm load at all times unless the farm should be retrieved for any reason.

Diving at the farm site

- Diving in the open ocean is another challenge and needs experienced divers with the right certificate while also following all potential safety regulations.
- However, diving still needs to be done as the farmer from time to time needs to know what is happening with the system under

the surface: Is the mooring still in place, are the ropes, chains connected, how is the condition of the connecting material (swivels, thimbles, shackles, etc.)?

1.4.2 *Operation, maintenance, and harvest*

It is well known that blades from, for example, *S. latissima* and other Laminarian species behave similar to budging belts of tissue, abrading at the blade tips while growing at the meristem next to the stipe/blade transition zone. This leads to the fact that a total year's growth may amount to 1–5 times the initial blade length through tip loss (Mann 1972). These characteristics slightly increase when transferring the plants at an offshore farm because of the physical stress resulting from stronger currents and wave action (Buck and Buchholz 2005). However, when following the seeding process for offshore cultivated seaweed (see subchapter "Seeding procedure" and Figure 1.7), the biomass loss will be reduced.

After deployment of the farm and the installation of the seeded ropes, the work at sea can be divided into two major work areas: (1) operation and maintenance (O&M) and (2) harvest. O&M mainly focuses on biological and technical issues, whereas harvesting the biomass depends on species, distance from shore, skill level of workers, and equipment. Normally, it is a short procedure when blades reach market size at the end of the growing season. For the complete work at the farm site next to the skilled workforce, a suitable sturdy work boat with the equipment needed to conduct O&M (crane, low deck, etc.) and harvest (strong crane, storage space, etc.) is required. The O&M vessel does not necessarily have to be the same vessel as it is used for harvest. Harvest vessels typically are larger in size as these vehicles need to have large space to store the harvested biomass.

1.4.2.1 *Operation and maintenance at sea*

Biological issues

- Observe the length increment of the plant during the grow-out. At the beginning of the cultivation process, seeding density and early growth should be determined. In the further grow-out weekly/two-weekly visits are sufficient during normal weather conditions.
- In the case of prolonging severe conditions or storm events, a site visit after that is of prime importance as seaweed detachment or complete loss could occur.

Technical issues

- The system design and the entire farm equipment also need a detailed inspection at any time during the visits to prevent failures or the damage/loss of equipment.

- Moorings (still at the same place?), connecting pieces (shackles, thimbles, swivels, rings, etc., are ok and service loads not exceeded?), backbones or other ropes, and chains of the harness (no unraveled or entangled parts?), and the complete buoyancy (too deep, fully submerged, loss?) need to be scrutinized.
- The system has to be controlled for fouling organisms settling potentially on all hard substrates of the harness. If the fouling increases a certain biomass, the entire system will experience too much stress because of the increase in volume and the resulting surface. Due to the increasing surface the drag coefficient will change resulting into higher loads.
- As already mentioned earlier, after prolonging severe conditions, a farm visit is mandatory.

1.4.2.2 Harvest

Identification of market size

- To find the right time of harvest, a certain length of the cultivated plants should be gained. Market size depends on species, seeding density, time of the year, and site conditions (nutrients, physical stress, etc.). For example, *S. latissima* grown in the North Sea (East Atlantic) normally reach a market size of approximately 200 cm between May–June (Figure 1.11).
- If the harvest will be conducted too early, it is important to plan the further process. As open ocean locations could be subject to swift weather condition change, it might be necessary to harvest the crop, even if the plant lamina did not reach the planned market size. Here, it is essential to check the local weather forecast.
- Further, if the harvest will be planned too late, an increased tip loss could result as fouling organisms settle on the blades and

Figure 1.11 Single *Saccharina latissima* plant at the harvest size.

induce a different less streamlined morphology in the flow further resulting into increased load forces and bending.

Harvest procedure

- The service vessel conducting the harvest need to have a certain time at sea. Therefore, it is of prime importance that the weather conditions are appropriate to fulfill safety regulations while also allowing an easy navigation and maneuvering at the farm site.
- The service vessel and its equipment can only be handled by experienced personnel as offshore conditions can change quickly and waves, swell, and wind could hamper the operation. If the wave height exceeds a certain altitude, the harvest cannot be conducted. Otherwise, next to safety regulations embed for the crew and the operation procedure, system damage, and loss of harvested biomass could result.
- If weather conditions are acceptable, each backbone or seeded line will be detached from the farm harness, transferred to the deck of the boat, and the seaweed cutoff the lines. After finalizing this process, the harvested biomass will be transferred to the land-based facility for further processing (washing in seawater to eliminate fouling, transport to the processing company, drying, etc.).
- Once the seaweed is harvested the farming process starts again from the seeding procedure.

1.4.3 *Problems and failures*

In general, living, working, and operation conditions offshore, and technical requirements, are in stark contrast to any other aquaculture environment (Figure 1.12). Materials have to be stronger in terms of service loads and expected live time, and technical constructions have to resist extreme hostile conditions to avoid breaking free traveling their own way through the offshore realm. Heavy sea states influence the operation at the farm and increase the risk for workers while also reducing working days per year. However, the major problem is that some of these hostile conditions may last for a longer period, with the waves impinging the farm every 5–8 s the whole day year round. In the case of severe problems at the farm, it might be possible that a service vessel cannot sail out because of the prevailing weather conditions, which in turn could even worsen the problem at the farm site. Therefore, all potential worst-case scenarios have to be considered thoroughly in the first place, calculations have to be done with enough room concerning safety, robustness, and life expectancy, and the entire farm and operation need to have worst-case contingency plans in place. To the contrary, the suggested culture candidates can quickly adapt to these offshore conditions and would therefore not pose a problem (Buck & Buchholz 2005).

(a) (b)

Figure 1.12 (a–b): Conditions in a high-energy environment. (a) Shows a slightly stronger but typical wave climate at the offshore farm in the German Bight during winds. The device with the white table top is normally used to prepare the sensor equipment for the seaweed farm; and (b) shows the same table during hostile environments.

In the following section, some of the problems and failures are listed, which should be taken into account before a seaweed enterprise would be installed in the open ocean. Offshore aquaculture constructions were engineered to withstand tremendous forces for many years, but can be damaged or destroyed during heavy storms. Staying ahead of the O&M required on the ring device proved problematic during the project, and it was a constant challenge for personnel to maintain the integrity of the installation and mooring. Entanglement, damage, and loss occurred during continuing storm and heavy seas conditions and even exacerbated in the next round of damage if repairs could not be organized quickly between weather events (Figure 1.13). Other problems accumulated in designing incorrect dimensions in size, connectedness, attachment, coupling, and space (Figure 1.13), and resulted in further entanglement, increased loads, and, corresponding to the service load, in too much biomass per size of construction. Escaping the wave and current energy at the water surface into a submerged mode is critical, as the seaweed need a certain amount of irradiance to allow photosynthesis. Learning from these miscalculations and setbacks, the following issues have to be taken into account.

1.4.3.1 *Prevention of mistakes*

Operation

- Eliminate modes of failure in all parts of the production system, starting from land-based precultivation to transfer at sea and deployment. Keeping the complete operation and the entire farm design, including the harness, the buoyancy, and the mooring, strong, but as simple as possible to prevent the reveal of component weaknesses by the waves and currents, which could often result in a catastrophe.

(a) (b) (c) (d) (e) (f)

Figure 1.13 (a–f): Failures and miscalculations from first offshore seaweed farms at Helgoland and at the offshore light house *Roter Sand* (a,b) Moreover, displays submerged buoys in full and collapsed shape; (c) shows a ravel of buoys, ropes, seeding lines, and connecting pieces after a heavy storm event; (d) shows a broken buoy rope leading to the loss of equipment; (e) is an image of an unstable ring device (First generation in 1994) with a too large biomass for harvest; and (f) shows the entanglement of little ring devices resulting from a wrong mooring and harness calculation. (Modified after Kaiser and Chambers 2017; Buck and Langan 2017b.)

Planning

- Ocean conditions and weather determine everything offshore, so plans must be made and adjusted accordingly. Nothing can be forced if conditions do not allow it as the ocean will always win in that situation.
- Redundancy and backup systems where applicable, especially with regards to moorings, which are arguably the most critical components of the farm system.
- Prepare as much equipment as possible while the system is onshore. It becomes exponentially harder to work on ropes, chains, anchors, backbones, shackles, etc., the moment it enters the water.

1.5 Multi-use of offshore installations

As the marine aquaculture sector is proliferating, competition for space by different coastal stakeholders and emerging environmental problems in coastal ecosystems caused by intensification of aquaculture operations, drive the development of offshore cultivation as a solution. Nevertheless, moving the aquaculture offshore is related to economic drawbacks because of higher costs for operation and maintenance and an increased risk of loss of cultures because of a high-energy environment. As a result, the idea of a co-use of offshore structures by different marine sectors emerged (Figure 1.14a–d). However, also further development of offshore aquaculture activities often overlaps with other marine interests, leading to competing claims for the ocean space

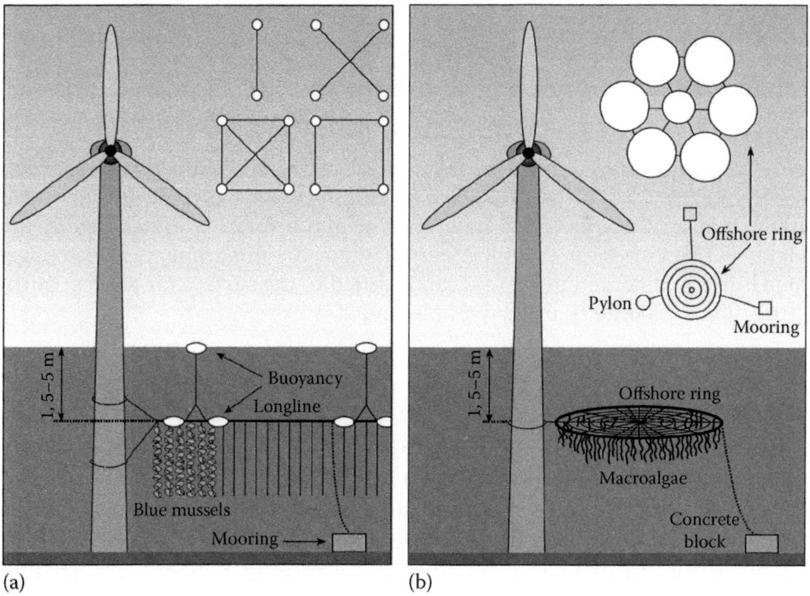

Figure 1.14 (a–b): First drawings to show the Multi-use concept combining aquaculture with wind farms. (a) Top a birds-eye view of longline constructions. Special pylon anchorages ensure access and maintenance of the wind energy plants by ship; down side view of a submerged mussel and seaweed longline culture. The wind energy plant pylon is used as the anchorage for one end of the long line. The line can reach a length of 100–300 m. Longer lines may be mounted to the next pylon; (b) top A bird's-eye view of the described ring construction. The combined rings are fixed around the pylon and a pylon with a single offshore ring construction using two anchorages; down side view of a pylon with offshore ring construction used for cultivation of algae and mussels. An anchor stone is placed at the outermost point of the ring construction; *(Continued)*

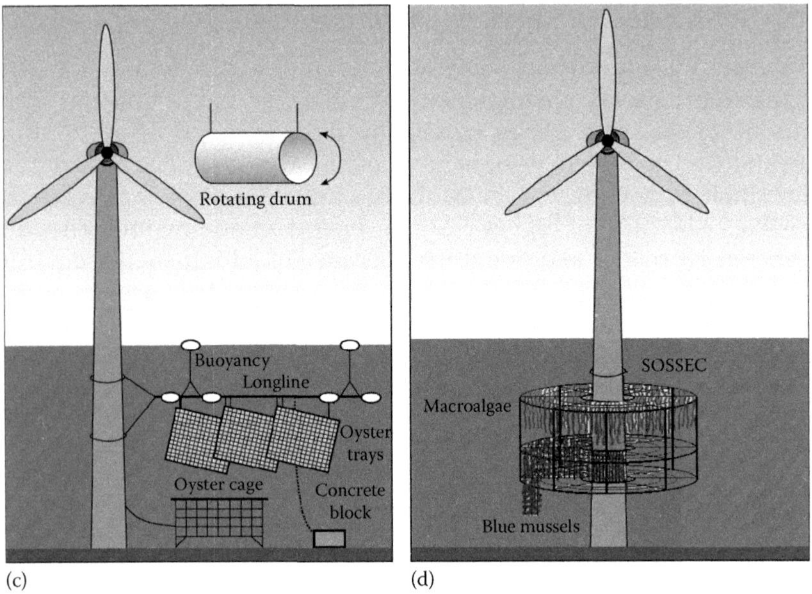

Figure 1.14 (Continued) (c–d): First drawings to show the Multi-use concept combining aquaculture with wind farms. (c) side view of a turbine–oyster culture combination (oyster trays fixed to a longline and a rotating oyster cage/drum); and (d) SOSSEC design (Submersible Offshore Shellfish and Seaweed Cage), which is in a submerged mode and can be lifted to the surface for harvest during culture. (Buck 2000, 2004; Buck and Langan 2017.)

(Holm et al. 2017). These claims can not only lead to user conflicts but also to smart combinations of multiple uses of marine space with economic and ecological benefits (Michler-Cieluch et al. 2009; Krause et al. 2011; Van Hoof et al. 2014). Such multi-use concepts are not new ideas as diverse marine activities already coexisted for decades, such as fishing and shipping, nature conservation, and a mussel fishery, and artificial reefs and angling.

Resulting from the Blue Growth strategy of the European Commission, the importance of the perspective of offshore aquaculture is increasing. Moving aquaculture offshore has the potential for further growth of the industry in many European offshore areas (Troell et al. 2009; Rosenthal et al. 2012a, b). In recent years, many studies investigating the co-use of offshore structures and areas have been conducted or are still ongoing (i.e., MERMAID, COEXIST, TROPOS, ICES WGAQUA, and MUSES).

The North Sea is, for example, a highly used marine space with fisheries, tourism, shipping, oil and gas extraction, cables and pipelines, gravel extraction and wind farms, and marine protected areas (MPAs) such as Natura 2000 areas (Jongbloed et al. 2014). Some of these activities and MPAs may limit the development of aquaculture, some combinations of marine activities with offshore aquaculture were identified as beneficial, and others are economical or for safety reasons not feasible (COEXIST project: Poelman and Bolman 2010; Stelzenmüller et al. 2013). Most promising in this regard is the co-use of offshore wind farms, certain types of fishery and nature conservation in combination with aquaculture (Buck et al. 2004; Benassai et al. 2014; Lagerveld et al. 2014; Gimpel et al. 2015). However, the multi-use concepts can be highly site-specific dependent on environmental requirements, and address legal and logistic issues (Naylor et al. 2000; Stickney and McVey 2002; Stuiver et al. 2012). Overall, the feasibility of a combined seaweed aquaculture enterprise with offshore wind farming depends on the biological feasibility, the technological implementation, the environmental sustainability of farming aquatic organisms, and the economic revenue of this commercial operation (Buck and Krause 2012).

There are quite a few studies available on the feasibility of offshore seaweed culture in combination with offshore structures such as wind farms. For instance, German scientists are involved in the cooperation for a project planning to integrate seaweed cultivation within a projected wind farm in Nantucket Sound (Massachusetts, USA) (Buck et al. 2011). However, progress on how to appropriately connect the seaweed culture devices to offshore foundations of wind turbines is still limited (Buck and Krause 2012).

Results from stakeholder analyses point out that economic and technical feasibility are prerequirements to convince wind farm operators and mariculturists to support the multi-use concept (Michler-Cieluch et al. 2009; Krause et al. 2011; Buck and Krause 2012). Stakeholders need to be engaged at an early stage, and site-specific data on risk assessment, production rate, and product quality should be available to convince stakeholders to participate.

More often than not, the economic revenues are only based on the market value of the aquaculture production, whereas the value of ecosystem services is neglected (Costanza and Folke 1997, Chopin 2014). Seaweeds provide ecosystem services as they can be used for bioremediation of the fed aquaculture in IMTAs or of nutrients from other sources (Chopin 2014, Grote 2016). They can form new habitats for fish or even whole communities and can be used for habitat restoration. Furthermore, as photosynthetic organisms, they produce oxygen and take up carbon dioxide, thereby locally increasing oxygen and pH of the seawater and similarly acting locally against oxygen depletion and ocean acidification (Clements and Chopin 2016).

References

ACE 2015. ACE expands with seaweed site. Accessed February 20, 2017, at http://aceaqua.no/en/2015/09/ace-expands-with-seaweed-site

Barrington, K., Chopin, T., and Robinson, S. 2009. Integrated multi-trophic aquaculture (IMTA) in marine temperate waters. In: D. Soto (Ed.), *Integrated Mariculture: A Global Review. FAO Fisheries and Aquaculture Technical Paper* No. 529, Rome, Italy: FAO, pp. 7–46.

Bartsch, I. 2017. Derivation of clonal stock cultures and hybridisation of kelps: Tool for strain preservation and breeding programs. In: B. Charrier, T. Wichard, and C.R.K. Reddy (Eds.), *Protocols for Macroalgae Research*, Chapter 3. Boca Raton, FL: CRC Press, Taylor & Francis Group.

Benassai, G., Mariani, P., Stenberg, C., and Christoffersen, M. 2014. A sustainability index of potential co-location of offshore wind farms and open water aquaculture. *Ocean Coast. Manag.* 95: 213–218.

Benetti, D., Benetti, G.I., Rivera, J.A., and O'Hanlon, B. 2010. Site selection criteria for open ocean aquaculture. *Mar. Tech. Soc. J.* 44(3): 22–35.

Broch, O.J., Ellingsen, I.H., Forbord, S., Wang, X., Volent, Z., Alver, M.O., Handå, A. et al. 2013. Modelling the cultivation and bioremediation potential of the kelp *Saccharina latissima* in close proximity to an exposed salmon farm in Norway. *Aquac. Environ. Interact.* 4: 186–206.

Buchholz, C.M., Krause, G., and Buck, B.H. 2012. Seaweed and man. In: C. Wiencke and C. K. Bischof (Eds.), *Seaweed Biology, Ecological Studies*. Berlin, Germany: Springer-Verlag, pp. 471–493.

Buck, B.H. 2000. Open ocean aquaculture und offshore-windparks: Eine machbarkeitsstudie über die multifunktionale nutzung von offshore-windparks und offshore-marikultur im raum nordsee. Reports on Polar and Marine Research, 412. Alfred Wegener Institute for Polar and Marine Research, Bremerhaven, Germany.

Buck, B.H. 2001. Combined utilisation of wind farming and mariculture in the North Sea. Biennial Report of Science Highlights at the Alfred Wegener Institute Helmholtz Centre for Polar and Marine Research, Bremerhaven, Germany, pp. 33–39.

Buck, B.H. 2004. Farming in a high energy environment: Potentials and constraints of sustainable offshore aquaculture in the German bight (North Sea), PhD Thesis, Alfred Wegener Institute for Polar and Marine Research Bremerhaven, 214 p., Bremen, Germany: University of Bremen.

Buck, B.H. 2007. Experimental trials on the feasibility of offshore seed production of the mussel *Mytilus edulis* in the German bight: Installation, technical requirements and environmental conditions. *Helgol. Mar. Res.* 61: 87–101.

Buck, B.H. and Buchholz, C.M. 2004a. The offshore-ring: A new system design for the open ocean aquaculture of macroalgae. *J. Appl. Phycol.* 16: 355–368.

Buck, B.H. and Buchholz, C.M. 2004b. Support device for the cultivation of macro organisms in marine waters. Patent: PCT/DE2005/000234.

Buck, B.H. and Buchholz, C.M. 2005. Response of offshore cultivated *Laminaria saccharina* to hydrodynamic forcing in the North Sea. *Aquaculture* 250: 674–691.

Buck, B.H. and Krause, G. 2012. Integration of aquaculture and renewable energy systems. In: R.A. Meyers (Ed.), *Encyclopaedia of Sustainability Science and Technology* Vol. 1. Berlin, Germany: Springer Science+Business Media LLC. Chapter No. 180, pp. 511–533.

Buck, B.H. and Langan, R. (Eds.) 2017a. *Aquaculture Perspective of Multi-use Sites in the Open Ocean: The Untapped Potential for Marine Resources in the Anthropocene*. Springer, 404 p.

Buck, B.H. and Langan, R. 2017b. Epilogue—Pathways towards sustainable ocean food production. In: B.H. Buck and R. Langan (Eds.), *Aquaculture Perspective of Multi-Use Sites in the Open Ocean: The Untapped Potential for Marine Resources in the Anthropocene*. Cham, Switzerland: Springer International Publishing AG, pp. 394–404.

Buck, B.H., Ebeling, M., and Griffin, R. 2011. Seaweed aquaculture as a co-use in an offshore wind farm—Feasibilities and constraints of economic viability for cape wind (Massachusetts). AWI-Report – Project No. USA 11/A01—12 p.

Buck, B.H., Krause, G., and Rosenthal, H. 2004. Extensive open ocean aquaculture development within wind farms in Germany: The prospect of offshore co-management and legal constraints. *Ocean Coast. Manag.* 47(3–4): 95–122.

Buck, B.H., Krause, G., Michler-Cieluch, T., Brenner, M., Buchholz, C.M., Busch, J.A., Fisch, R., Geisen, M., and Zielinski, O. 2008. Meeting the quest for spatial efficiency: Progress and prospects of extensive aquaculture within offshore wind farms. *Helgol. Mar. Res.* 62: 269–281.

Buck, B.H., Krause, G., Pogoda, P., Grote, B., Wever, L., Goseberg, N., Schupp, M.F., Mochtak, A., and Czybulka, D. 2017b. Case study German bight: Pioneer projects of aquaculture-wind farm multi-uses. In: B.H. Buck and R. Langan (Eds.), *Aquaculture Perspective of Multi-use Sites in the Open Ocean: The Untapped Potential for Marine Resources in the Anthropocene*. Cham, Switzerland: Springer International Publishing AG, pp. 253–354.

Buck, B.H., Nevejan, N., Wille, M., Chambers, M., and Chopin, T. 2017a. Offshore and multi-use aquaculture with extractive species: Seaweeds and bivalves. In: B.H. Buck and R. Langan (Eds.), *Aquaculture Perspective of Multi-Use Sites in the Open Ocean: The Untapped Potential for Marine Resources in the Anthropocene*. Cham, Switzerland: Springer International Publishing AG, pp. 23–69.

Chopin, T. 2014. Seaweeds: Top mariculture crop, ecosystem service provider. *GAA* 17: 54–56.

Clements, J. and Chopin, T. 2016. Ocean acidification and marine aquaculture in North America: Potential impacts and mitigation strategies. *Rev. Aquacult.* 1–16. doi: 10.1111/raq.12140.

Corey, P., Kim, J.K., Garbary, D.J., Prithiviraj, B., and Duston, J. 2012. Bioremediation potential of *Chondrus crispus* (Basin Head) and *Palmaria palmata*: Effect of temperature and high nitrate on nutrient removal. *J. Appl. Phycol.* 24: 441–448.

Costanza, R. and Folke, C. 1997. Valuing ecosystem services with efficiency, fairness and sustainability as goals. In: G. Daily (Ed.), *Nature's Services: Societal Dependence on Natural Ecosystems*. Washington, DC: Island Press, pp. 49–70.

CRM. 2001. *Wirtschaftliches und ökologisches Potential einer Laminarien-Farm in Deutschland* (Economic and ecologic potentials of algae farming in Germany). Kiel, Germany: Coastal Research, Management.

Edwards, M.D. and Dring, M.J. 2011. Open sea cultivation trails of the red alga, *Palmaria palmata* from seeded tetraspores in Strangford Laugh, Northern Ireland. *Aquaculture* 317: 203–209.

FAO. 2014. *The State of World Fisheries and Aquaculture*. Rome, Italy: Food and Agriculture Organization, 223 p.

FAO. 2016. *The State of World Fisheries and Aquaculture*. Rome, Italy: Food and Agriculture Organization, 190 p.

Forbord, S., Steinhovden, K.B., Rød, K.K., Handå, A., and Skjermo, J. 2017. Cultivation protocol for *Saccharina latissima*. In: B. Charrier, T. Wichard, and C.R.K. Reddy (Eds.), *Protocols for Macroalgae Research*, Chapter 2. Boca Raton, FL: CRC Press, Taylor & Francis Group.

Gimpel, A., Stelzenmüller, V., Grote, B., Buck, B.H., Floeter J., and Núnez-Riboni, I. 2015. A GIS modelling framework to evaluate marine spatial planning scenarios: Co-location of offshore wind farms and aquaculture in the German EEZ. *Mar. Policy* 55: 102–115.

Grote, B. 2016. Bioremediation of aquaculture wastewater: Evaluating the prospects of the red alga *Palmaria palmata* (Rhodophyta) for nitrogen uptake. *J. Appl. Phycol.* 28: 3075–3082.

Holm, P., Buck, B.H., and Langan, R. 2017. Introduction: New approaches to sustainable offshore food production and the development of offshore platforms. In: B.H. Buck and R. Langan (Eds.), *Aquaculture Perspective of Multi-Use Sites in the Open Ocean: The Untapped Potential for Marine Resources in the Anthropocene*. ISBN(1): 978-3-319-51157-3 – ISBN(2) 978-3-319-51159-7 (eBook), pp. 1–20.

Hortimare. 2016. Propagating Seaweed for a Sustainable Future. Accessed online October 22, 2016, at http://www.hortimare.com/.

Jongbloed, R.H., van der Wal, J.T., and Lindeboom, H.J. 2014. Identifying space for offshore wind energy in the North Sea. Consequences of scenario calculations for interactions with other marine uses. *Ener. Policy* 68: 320–333.

Kain, (Jones) J.M. 1991. Cultivation of attached seaweeds. In: M.D. Guiry and G. Blunden (Eds.), *Seaweed Resources in Europe: Uses and Potential*. Chichester, UK: Wiley, pp. 309–377.

Kaiser, J.B. and Chambers, M.D. 2017. Offshore platforms and mariculture in the US. In: B.H. Buck and R. Langan (Eds.), *Aquaculture Perspective of Multi-Use Sites in the Open Ocean: The Untapped Potential for Marine Resources in the Anthropocene*. ISBN(1): 978-3-319-51157-3—ISBN(2) 978-3-319-51159-7 (eBook). pp. 375–391.

Kim, J.K., Kraemer, G.P., and Yarish, C. 2015. Sugar kelp aquaculture in long Island sound and the bronx river estuary for nutrient bioextraction associated with biomass production. *Mar. Ecol. Prog. Ser.* 531: 155–166.

Kraemer, G.P., Kim, J.K., and Yarish, C. 2014. *Seaweed Aquaculture: Bioextraction of Nutrients to Reduce Eutrophication*. Association of Massachusetts Wetland Scientists Newsletter. April 2014 No. 89, pp. 16–17.

Krause, G., Griffin, R.M., and Buck, B.H. 2011. Perceived concerns and advocated organisational structures of ownership supporting offshore wind farm—Mariculture integration. In: G. Krause (Ed.), *From Turbine to Wind Farms: Technical Requirements and Spin Off Products*. Rijeka, Croatia: InTech, Open Access Publisher, pp. 203–218.

Lagerveld, S., Röckmann, C., and Scholl, M. (Eds.) 2014. Combining offshore wind energy and large-scale mussel farming: Background and technical, ecological and economic considerations. IMARES report number C056/14 http://www.maritimecampus.nl/sites/default/files/C056%2014%20Report-Blauwdruk-_SLMS-lcs.pdf

Langan, R. 2009. Opportunities and challenges for offshore farming. Chapter 29 In: G. Burnell and G. Allen (Eds.), *New Technologies in Aquaculture: Improving Production Efficiency, Quality and Environmental Management*. Cambridge, UK: Woodhead Publishing Limited.

Lüning, K. and Buchholz, C. 1996. Massenkultur mariner makroalgen bei helgoland zur gewinnung von phycokolloiden und zur verwendung als biosorptionsmittel. Helgoland, Germany: BAH Report, 75 p.

Mann, K.H. 1972. Ecological energetics of the sea-weed zone in a marine bay on the Atlantic coast of Canada. II. Productivity of the seaweeds. *Mar. Biol.* 14: 199–209.

Marinho, G.S., Holdt, S.L., Birkeland, M.J., and Angelidaki, I. 2015. Commercial cultivation and bioremediation potential of sugar kelp, *Saccharina latissima*, in Danish waters. *J. Appl. Phycol.* doi: 10.1007/s10811-014-0519-8.

McHugh, D.J. 2003. A guide to the seaweed industry. FAO Fisheries Technical Papers T441.

McElwee, J. 1997. Practical experience in off-shore cage rearing, the good, the bad and the ugly. In: C. Hesley (Ed.), *Open Ocean Aquaculture: Chartering the Future of Ocean Farming. Proceedings of an International Conference*, April 23–25, 1997, Maui, Hawaii: UNIHI-Seagrant-CP-98-08, Maui, University of Hawaii Sea Grant College Program, pp. 89–105.

Michler-Cieluch, T., Krause, G., and Buck, B.H. 2009. Reflections on integrating operation and maintenance activities of offshore wind farms and mariculture. *Ocean Coast. Manag.* 52(1): 57–68.

Naylor, R.L., Goldburg, R.J., Primavera, J.H., Kautsky, N., Beveridge, M.C., Clay, J., Folke, C., Lubchenco, J., Mooney, H., and Troell, M. 2000. Effect of aquaculture on world fish supplies. *Nature* 405(6790): 1017–1024.

Neori, A., Chopin, T., Troell, M., Buschmann, A.H., Kraemer, G., Halling, C., Shpigel, M., and Yarish, C. 2004. Integrated aquaculture: Rationale, evolution and state of the art emphasizing seaweed biofiltration in modern aquaculture. *Aquaculture* 231: 361–391.

Neushul, M. and Harger, B.W.W. 1985. Studies of biomass yield from a near-shore macroalgal test farm. *J. Solar Energy Eng.* 107: 93–96.

Neushul, M., Benson, J., Harger, B.W.W., and Charters, A.C. 1992. Macroalgal farming in the sea: Water motion and nitrate uptake. *J. Appl. Phycol.* 4: 255–265.

Perez, P., Kaas, R., Campello, F., Arbault, S., and Barbaroux, O. 1992. *La Culture des Algues Marines Dans le Monde*. Nantes, France: IFREMER.

Poelman, M. and Bolman, B. 2010. Task 1.3 COEXIST: Inventory of possible combinations of different forms of aquaculture and/or fisheries and other activities. Accessed March 24, 2017, at http://www.coexistproject.eu/images/COEXIST/deliverables/WP1/COEXIST_245178_D1.3_FINAL.pdf

Rosenthal, H., Costa-Pierce, B.A., Krause, G., and Buck, B.H. 2012a. *Bremerhaven Declaration on the Future of Global Open Ocean Aquaculture, Part I: Preamble and Recommendations*. Bremerhaven, Germany: The Bremerhaven Economic Development Company.

Rosenthal, H., Costa-Pierce, B.A., Krause, G., and Buck, B.H. 2012b. *Bremerhaven Declaration on the Future of Global Open Ocean Aquaculture—Part II: Recommendations on Subject Areas and Justifications. Aquaculture Forum on Open Ocean Aquaculture Development—From Visions to Reality: The Future of Offshore Farming*. Bremerhaven, Germany: The Bremerhaven Economic Development Company.

Ryan, J. 2005. Offshore aquaculture—Do we need it, and why is it taking so long? International Salmon Farmers Association (Ireland). Expert workshop on sustainable aquaculture, DG JRC European Commission, pp. 17–18. January 2005, Seville, Spain: Institute for Prospective Technological Studies.

SES. 2015a. Patented Seaweed Carrier system by SES, Patent No. EP 09 836 439.1. SES—Seaweed Energy Solutions, Trondheim, Oslo, Norway.

SES. 2015b. Seaweed Carrier system for the cultivation of macroalgae in offshore environments patented by SES—Seaweed Energy Solutions, Trondheim, Norway.

Skjermo, J., Aasen, I.M., Arff, J., Broch, O.J., Carvajal, A., Christie, H., Forbord, S. et al. 2014. A new Norwegian bioeconomy based on cultivation and processing of seaweed: Opportunities and R&D needs. SINTEF Fisheries and Aquaculture, Norway, March 6, 2014, 46 p.

Starchild, A. 1980. Seaward Ho!—The ocean frontier. In: *Office of Technology Assessment* (Ed.), Energy from open ocean kelp farms (pp. 83–98). Honolulu, Hawaii: University Press of the Pacific.

Stelzenmüller, V., Schulze, T., Gimpel, A., Bartelings, H., Bello, E., Bergh, O., Bolman, B. et al. 2013. Guidance on a better integration of aquaculture, fisheries, and other activities in the coastal zone: From tools to practical examples. Coexist project—Interaction in coastal waters.

Stickney, R.R. and McVey, J. 2002. *Responsible Aquaculture*. Wallingford, Oxfordshire: CABI Press.

Stuiver, M., Gerritsen, A.L., Fontein, R.J., and Agricola, H.J. 2012. Multifunctionele platforms: Perspectief voor de toekomst? *Aquacultuur* 27(5): 6–12

Troell, M., Joyce, A., Chopin, T., Neori, A., Buschmann, A.H., and Fang, J.G. 2009. Ecological engineering in aquaculture: Potential for integrated multi-trophic aquaculture (IMTA) in marine offshore systems. *Aquaculture* 297: 1–9.

Van Hoof, L.J.W., Steenbergen, J., Bolman, B.C., Rockmann, C., Kraan, M.L., Piet, G.J., Jak, R.G. et al. 2014. Zee op Zicht: Inzicht; een zoektocht naar een integraal afwegingskader voor het gebruik van de zee. IMARES Report. Accessed March 24, 2017, at http://www.wageningenur.nl/nl/Publicatie-details.htm?publicationId=publication-way-343835313839

chapter two

Cultivation protocol for Saccharina latissima

Silje Forbord, Kristine Braaten Steinhovden, Kaia Kjølbo Rød, Aleksander Handå, and Jorunn Skjermo

Contents

2.1 Introduction

The sugar kelp *Saccharina latissima* (Linnaeus) as discussed by Lane, Mayes, Druehl, and Saunders (synonym: *Laminaria saccharina* [Linnaeus] Lamouroux) is a cold-water species distributed circumpolar in the northern hemisphere (Bolton et al. 1983) where it occurs from the intertidal down to the bottom of the photic zone. It is present on both sides of the Atlantic from Nova Scotia to Europe and in the Pacific along the North American coast and some areas outside Japan and in the Arctic Russia (Wilce 1965; van den Hoek and Donze 1967; Druehl 1970; Druehl and Kaneko 1973; Sjøtun 1985). Temperature, exposure, and competition are the limiting factors of the propagation of the species, and the southern limit is highly determined by temperature (Lüning 1990). Their natural growth sites are clear and turbid coastal waters (Borum et al. 2002), and water temperatures between 10°C and 15°C have been found to give optimum growth (Fortes and Lüning 1980; Bolton and Lüning 1982). *S. latissima* is one of the fastest growing species of kelp in European waters with annual production capacities estimated at 75–170 tons wet weight per hectare at sea (Broch et al. 2013). Cultivation of *S. latissima* is characterized by fast growth, but with low predictability of production volumes and biomass quality due to seasonal-, regional-, and site-dependent growth and composition and different cultivation strategies. In cool-temperate North Atlantic waters, *S. latissima* shows a seasonal- and depth-dependent growth pattern with faster growth close to the surface in autumn and winter and fast growth also at deeper depths in the photic zone in spring and early summer. From late spring to late summer, heavy fouling and necrosis of the distal end of the blade occurs. Harvesting should therefore take place in early summer to avoid destruction of the crop and loss of biomass (Handå et al. 2013; Førde et al. 2016). Europe's contribution to global macroalgae cultivation is neglectable (FAO 2016). However, new consumer trends, market demands, and opportunities for multiple use of macroalgae such as food, bioactive components for functional food and feed ingredients, fertilizers, and biofuels (Shahidi 2009; Holdt and

Kraan 2011; Rajapakse and Kim 2011; Fleurence et al. 2012; Hafting et al. 2012; Tabarsa et al. 2012; Wargacki et al. 2012; Fernand et al. 2016) have strengthened the motivation for industrial macroalgae cultivation in Europe. Industrial, cost-effective cultivation requires novel technology of the whole production line targeting mechanization and automation of the seedling processes, deployment at sea and the harvesting operation, but the necessary equipment for industrial-scale production does not exist. Cultivation equipment must be developed for different environmental conditions, and logistic challenges during deployment and harvesting must be solved. An in-depth understanding of the life cycles and the environmental and biological requirements of macroalgae is a prerequisite to establish protocols for all phases of the cultivation. In the current chapter, we present a protocol for year-round production of *S. latissima* seedlings that is applicable for small-scale laboratory studies and that can be upscaled for mass production.

2.1.1 *The life cycle of laminariales*

The sporophyte of the *S. latissima* consists of a lamina (blade), stipes (stem), and hapter (holdfast). The lamina grows from an intercalary meristem located between the lamina and the stipes, which is short, round, and attached to the substrate with a branched hapter. The sporophyte can reach up to 3 m in length (Hollenberg 1939). *S. latissima* has a diplohaplontic, heteromorphic life cycle, with alternation between a microscopic haploid (n) gametophyte generation and a macroscopic diploid (2n) sporophyte generation, as shown in Figure 2.1 (Kain 1979). The life cycle is common for all the different Laminariales with minor differences. *S. latissima* are fertile during winter (October–January in mid Norway) when the days are short and temperatures are low. Sorus with sporangia develops on the lamina and meiosis produces 32 motile meiospores (zoospores) per sporangium (Schreiber 1930 cited in Kain 1979). When the spores become mature, they are released into the surrounding water and spread with water currents until they settle at a suitable substrate. The meiospores are 4–8 µm long and have two flagella. After settlement, the flagella are lost, and the initial germling growth phase starts with the formation of a germ tube in the distal end of the spore. The germ tubes are filled with the cell content of the spore, making a primary cell, which further develops into male and female gametophytes (Williams 1921 cited in Kain 1979; Sauvageau 1916 cited in Kain 1979). Gametophytes increase in length by cell division while progressively developing branches. Male gametophytes are thin and heavily branched, whereas the female have thicker cells and are less branched (Hollenberg 1939). Gametogenesis in *Laminaria* is induced by blue light (400–512 nm), whereas the optimum temperature varies with geographical distribution (Lüning and Dring 1975; Lüning

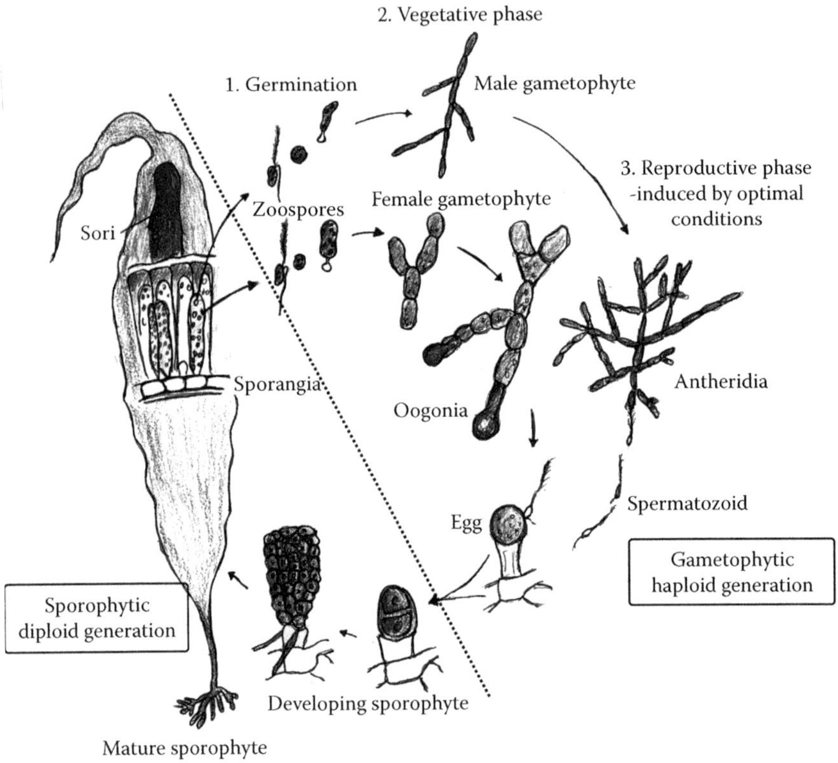

Figure 2.1 Schematic presentation of the life cycle of *S. latissima*. The sporophyte produces unilocular sporangia with meiospores that are released into surrounding waters when mature. These develop into male and female gametophytes. When exposed to blue light, the male gametophytes produce antheridia with spermatozoids, and the female gametophytes produce oogonia with one egg. Fertilization leads to development of small sporophytes that grow into the full-sized sporophytes. (Courtesy of Sanna Matsson.)

and Neushul 1978; Lüning 1980). Once the light and temperature conditions are suitable, the development of reproductive structures is triggered (Lüning 1980). Female gametophytes produce an oogonium, which when fertile produces one nonmotile spherical gamete (oosphere/egg). The male gametophyte develops spermatocysts that produce spermatozoids with two flagella. A signal chemical attracts the spermatozoids, which swim to the oogonium and fertilize the egg (Müller et al. 1979; Müller 1989). The fertilized gametes develop into a microscopic sporophyte that grows into the full-sized sporophyte. The haploid gametophyte stage can be cultivated under controlled environmental conditions in a laboratory. Gametophytes are dark tolerant and can survive in the darkness for up to 18 months (Druehl and Boal 1981; tom Dieck 1993). In weak red light, gametophytes have vegetative growth and will increase in size, without producing gametes or spermatocysts. Fertilization can be induced at any moment by modifying the light regime and expose the gametophytes to blue light. Studies have shown that the ability to produce gametes is still good after 30 years of cultivation under red-light conditions in many kelp species (Druehl et al. 2005).

2.2 *State of the art*

To cultivate kelp at sea requires production of seedlings on substrate to be deployed at sea farms. Cultivation protocols have been established for the most attractive kelp species, such as *S. latissima*, *Alaria esculenta*, and *Laminaria digitata* (Arbona and Molla 2006; Edwards and Watson 2011; Redmond et al. 2014). There are three main strategies for seedling production of kelp, aiming for seeding of the growth substrate with spores, gametophytes, or small sporophytes, whereof seeding with spores and gametophytes thus far is the most incorporated technique. Seeding with spores requires fertile sporophytes and is seasonal dependent if these are collected in natural habitats, but fertility can also be induced by artificial day rhythm and thus enable access to spores independent of natural seasons (Pang and Lüning 2004). Gametophytes can be kept in continuous cultures and can be available for year-through seeding or production of microscopic sporophytes for direct seeding. This is advantageous as the use of incubation facilities can be shortened by several weeks or, in the case of direct seeding, omitted completely.

The presented protocol focus on *S. latissima* seedlings production from spores and gametophytes and is adapted to the infrastructure and equipment that is available at typical aquaculture research laboratories, although some adjustments may be necessary to meet the needs of the different life stages of the species.

2.3 *Materials*

2.3.1 *Collection of sporophytes and induction of sori*

- Boxes or plastic bags
- Coolers (if outdoor or transport temperature is above 15°C)
- Scissors for trimming the sporophytes
- Temperature-regulated room holding 10°C
- Tanks with nutrient-rich seawater and fluorescents lamps for storage and conditioning of sporophytes
- Time switch
- Pressurized air or electrical aquarium pump for aeration

2.3.2 *Sori disinfection, dehydration, spore release, and spore counting*

- Refrigerator holding 4°C
- Fume cupboard or workplace outside
- Light microscope
- Containers (three for disinfection and one for spore release)
- Plastic bag or plastic box with lid for storage
- Scissors
- Paper towel
- Autoclaved or sterilized seawater
- Gloves
- Protecting glasses
- Spatula
- Filter cloth (20–40 μm)
- Counting chamber with glass covers
- Pipette
- Sodium hypochlorite (NaOCl; 200 ppm) or other disinfectant

2.3.3 *Starting gametophyte cultures*

- Climate cabinet or a light-tight climate room holding 10°C
- Pressurized air or electrical aquarium pump for aeration
- Silicon tube for aeration
- Pipette for aeration
- Culture flask
- Rubber top or Parafilm
- Membrane filter (0.2 μm)
- Provasolis-Enriched Seawater (PES) or other culture media
- Germanium dioxide (GeO_2)

2.3.4 *Maintenance of gametophyte cultures*

- Blender
- Blender container
- Culture flask
- Culture flask top or Parafilm
- Pipette for aeration
- Silicon tube for aeration
- Membrane filter (0.2 μm; change if clogged)
- Squeeze bottle
- Funnel

- Filter cloth (20–40 µm)
- Autoclaved seawater
- PES or other culture media
- Germanium dioxide (GeO_2)

2.3.5 *Fertility test*

- Cultivation room with white-light holding 10°C
- Light microscope
- Petri dish
- PES or other culture media
- Germanium dioxide (GeO_2)

2.3.6 *Seeding*

- Temperature-regulated room with white-light holding 10°C
- Filtered and UV-treated nutrient-rich seawater
- Seeding substrate (string, rope, and sheet)
- Container
- Spray bottle

2.4 *Experimental procedures*

2.4.1 *Seed supply and sporulation*

The spores used in the seedling production can be obtained from sporophytes either by releasing them from mature tissue (sorus) or by inducing the spores artificially in the laboratory.

2.4.1.1 *Collection of sporophytes and induction of sorus*

Adult sporophytes are collected from natural stocks or from beach cast. It is preferable to get healthy, big sporophytes with low biofouling, and diving may be necessary to find good specimens. Pack the collected sporophytes in large plastic boxes/bags or in coolers when outdoor temperature is above 15°C. It is important to keep the sporophytes cooled and moist during transport. If the sporophytes are collected during a time of the year when the sporophytes are not fertile, sori can be induced. Sporophytes are trimmed after arrival at the lab. Distal ends and 5–15 cm of the meristem are cutoff the lamina to ensure fresh tissue and prevent putative sporulation inhibitors (Pang and Lüning 2004). After trimming, the sporophytes are kept in flow-through tanks with short-day conditions until sori appear, as shown in Figure 2.2 (*Note 1*). Depending on the

Figure 2.2 The dark area on the thallus of *S. latissima* is called sorus and contains millions of zoospores.

Table 2.1 Key parameters for conditioning of sporophytes

	Key parameters
Light intensity at water surface (μmol m^{-2} s^{-1})	70–100
Water treatment	Sand filter or other coarse filter
Water intake (m depth)	>50
Water flow (times exchange per day)	15
Aeration	Unfiltered
Water temperature (°C)	10
Light regime (L:D) (h)	8:16
Light source	Fluorescent tubes (warm white, cool white, day light)

season, this can take from 4 to 12 weeks but normally from 6 to 8 weeks (Forbord et al. 2012; see Table 2.1 for key parameters).

Tanks of 200–300 L are sufficient for 20–30 sporophytes (*Note 2*). Tanks are illuminated 8 h per day with fluorescent lamps providing a light intensity of around 70–100 μmol photons m^{-2} s^{-1} at the water surface.

The seawater source should ensure water with a high and stable level of nutrients and temperatures not exceeding 10°C. Water intake below 50 m depth may be necessary. The seawater is filtered before use. Aeration is used to keep the sporophytes suspended and circulating in the water column so that all individuals are equally illuminated.

2.4.1.2 *Sorus disinfection*

Sori are always disinfected before spores release to mitigate contaminating organisms in gametophyte cultures and seedling cultivation. NaOCl is a suitable chemical for disinfection, but there are several other chemicals that can be used as well, for instance iodide-based Betadine or Lugol's solution. Optimum concentration, temperature, and reaction time should be evaluated in the case of changing disinfectant.

The areas of mature sori are cut from the sporophytes, and the pieces are immersed in a disinfecting NaOCl bath (200 ppm, 10°C) (*Note 3*). In the case of fouling or sediments on the lamina, the sporophytes are mechanically cleaned with paper towels before disinfection. The sori are left for 2 min with some gentle stirring with a clean spatula. After disinfection, the sori are rinsed two times for 30 s in sterile seawater (10°C). Stir the sporophytes during rinsing so that all parts of the sori are thoroughly rinsed (Rød 2012).

2.4.1.3 *Dehydration*

Disinfected sori are dried with paper towel to remove the surface water without leaving the surface withered. Dried sori are stored in a plastic bag/box with lid for 18–48 h at 4°C to dehydrate.

2.4.1.4 *Spore release*

Dehydrated sori are placed in a beaker or another convenient container filled with sterile seawater holding a temperature of 10°C. Water should just cover the sori, to obtain a dense spore solution. Spores will be released as a response to osmotic shock combined with a temperature shock (Kain 1979). As the spores are released, the seawater becomes brown and cloudy, and the solution can be gently stirred with a clean spatula to enhance the spore release (*Note 4*). The spore solution is filtered through a 20–40 μm filter cloth placed in a funnel to remove bigger plankton and organic debris. If the spore solution is poor, the rehydrated sori can be reused by repeating the above-mentioned dehydrating procedures. Sori are discarded after the spores are released if the color of the sori patches is pale or the lamina/sori have started to decompose. Healthy lamina with dark-brown patches of sori can be placed back in the conditioning tank for a second spore release at a later stage.

2.4.1.5 Spore density

The amount of spores released (spore density) shows great variation and is believed to vary between sporophytes (Kain 1975), within the sorus, with season, and with age of the mature sorus (Joska and Bolton 1987). Spore density is estimated by the use of a Neubauer improved counting chamber in a light microscope with at least 10× magnification (*Note 5*). Density is estimated from the average of the counted squares in the chamber using Equation 2.1:

$$\text{Concentration (cell/mL)} = \frac{\text{Total number of cells} \times 10{,}000}{\text{Number of squares}} \tag{2.1}$$

Other counting chambers can be used, for example, Sedgewick Rafter and Burker's chamber.

2.4.2 Gametophyte cultivation

S. latissima development can be held at the gametophyte life stage by keeping the gametophyte culture in red light under controlled environmental conditions. Fertilization can be induced by modifying the light.

2.4.2.1 Starting-up gametophyte cultures

Spore release is done using the procedure as described in Section 2.4.1, followed by inoculation of the spores in a growth medium. Several growth media can be used in gametophyte cultivation, with the most common being the PES (see Harrison and Berges 2005 for recipe). Divide the spore solution in cultivation flasks dependent on the wanted spore density, and refill with culture medium until the flask is full. Inoculation of ≥250,000 spores mL^{-1} will ensure dense cultures. Add 0.1 mL germanium dioxide (GeO_2) L^{-1} medium for contamination control (for more details see Section 2.4.2.3). Label the flasks with date and origin of the stock. After inoculation, the culture flasks are left in constant red light with light intensity of 20–30 μmol photons m^{-2} s^{-1} at the surface of the culture vessel, as shown in Figure 2.3 (*Note 6*). A light regime of 16–24 h per day can be used. Exposure to white light must be strictly limited as this induces fertility in the cultures. Cultures are kept at 10°C either by use of a climate cabinet or a light-tight climate room. Filtered air (0.2 μm) is provided through silicon tubes (3 mm) attached to a glass pipette for aeration (see Table 2.2 for key parameters). The culture flask is either covered with a rubber top with a central hole for the pipette or with

Figure 2.3 Gametophyte cultures held under red-light conditions.

Table 2.2 Key parameters for gametophyte cultures

	Key parameters
Growth medium	Provasolis-Enriched Seawater (Harrison and Berges 2005)
Light intensity (μmol m^{-2} s^{-1})	20–30
Light regimes (L:D) (h)	24:0 or 16:8
Temperature (°C)	10

Parafilm. Culture flasks of 250–500 mL are used in the initial stage, and as the biomass increases, the culture is either split into two or the total volume of the culture is scaled up, as shown in Figure 2.4a. After minimum four weeks, the growth medium can be renewed the first time, and to ensure a minimal evaporation rate, the cultures are gently aerated. The water level is marked on the culture vessel at inoculation so that evaporation can be monitored and distilled water added in case of heavy evaporation.

(a) (b)

Figure 2.4 (a) Gametophyte cultures after two months of inoculation and (b) gametophytes grown under red-light condition for 32 days (20× magnification).

2.4.2.2 Maintenance of gametophyte cultures

After the first renewing of the growth medium, the cultures are maintained every 7–14 days to keep them in a healthy and steady growing condition following these procedures (*Note 7*):

1. Remove rubber top/Parafilm and silicon tube.
2. Transfer the culture solution from the culture flask to a blender container. Make sure that you scrape the flask with a clean, long spatula to get out the gametophytes settled in the bottom or on the walls.
3. Blend the solution until the gametophyte culture is homogenized (about 30–60 s).
4. Filter the old culture medium back into the culture flask by placing the filter cloth in the funnel. The gametophytes will accumulate in the filter (20–40 μm).
5. Rinse the filter with a squeeze bottle holding sterile seawater or culture medium so that all gametophytes are transferred into a new culture flask.
6. Remove a small sample with a pipette to check the culture for contamination and traces of juvenile sporophytes in the microscope see Figure 2.4b.
7. Fill the new culture flask with fresh growth medium. GeO_2 can be used (0.1 mL GeO_2 L^{-1} medium) if diatoms are present in the culture.
8. Seal the culture flask with a rubber top or Parafilm.

9. Label the flasks with origin and date and mark the water level.
10. Place the flasks back in the red-light culture room/climate cabinet.
11. Allow gentle aeration so that the gametophytes stay suspended.

2.4.2.3 Contamination control

Contaminating organisms such as diatoms and other algae, ciliates, and nematodes should be kept to a minimum in the gametophyte cultures. There are several ways to minimize contamination:

- *Glassware and equipment*: All equipment is autoclaved or sterilized in other ways before used in gametophyte cultivation.
- *Mitigation of contamination from sori*: Sori are always disinfected as described in Section 2.4.1.2. The spore solution is filtered through a plankton filter (20–40 μm) before inoculation to remove plankton and organic debris.
- *Germanium dioxide* (GeO_2): To mitigate the growth of diatoms that may be introduced to the culture from sori, seawater, equipment, and so on, GeO_2 is always added to the growth medium at inoculation. A solution of GeO_2 is made by using 0.894 g GeO_2 dissolved in 200 mL distilled fresh water. Use 0.1 mL GeO_2 solution L^{-1} of culture (Shea and Chopin 2007).
- *Water*: Autoclaved/sterilized seawater is used in all stages of gametophyte cultivation.
- *Air*: Pressurized air or an electrical air pump with a membrane filter (0.2 μm) is used for aeration.

2.4.2.4 Fertility test

A fertility test can be carried out to check if the gametophytes are able to become fertile and grow into viable sporophytes. This is relevant in case the cultures have been maintained for up to one year or more. Transfer gametophytes to Petri dishes containing PES after blending the culture. Use for instance three replicas from each culture. Place the Petri dishes under white light at around 30–60 μmol light intensity, with a 16:8 h light:dark regime and at a surrounding temperature of 10°C. Change the culture medium weekly and monitor the tests with respect to juvenile sporophytes after two weeks and then weekly until sporophytes are present. In a healthy culture, sporophytes should be visible after 2–4 weeks.

2.4.3 Seeding

Different kind of substrates can be used for seeding (strings, ropes, and sheets) with different materials (nylon, polypropylene, polyester, and polyester silk), and the ropes and strings can be both twisted and braided. This protocol is adapted to cultivation on strings (1–2 mm) and ropes (5–10 mm) (*Note 8*).

2.4.3.1 *Seeding with spores*

Spores are acquired as described in Section 2.4.1.4. A spray bottle can be used to distribute the spores evenly onto the substrate, the spore solution can be poured over the substrate, or alternatively the substrate can be bathed in the spore solution. It is recommended to adjust the spore density to 300,000–600,000 spores per mL before seeding. The substrate is left in air for 10–15 min to allow the spores to settle after spraying (*Note 9*).

2.4.3.2 *Seeding with gametophytes*

Fertility can be induced for vegetative gametophytes cultured in red light (see Section 2.4.2.1) by changing the culture conditions. The growth medium is renewed (see Section 2.4.2.2) and the cultures are placed in white light that is provided by fluorescent tubes holding 30–60 μmol light intensity, light regime of 16:8 h light:dark, and surrounding temperature of 10°C for 8–10 days. When a large number of eggs and small sporophytes are visible, the culture is gently blended for 30–60 s until the gametophytes are evenly distributed in the solution. The gametophyte solution is seeded by pouring it over the substrate or by placing the substrate in a gametophyte bath. Alternatively, a spray bottle can be used, but the spray jet can be clogged by the gametophyte tufts. After spray and bath seeding, the substrate is left in air for 15–20 min to allow the gametophytes to settle (*Note 10*).

2.4.3.3 *Hatchery systems*

Different hatchery systems can be used to produce seedlings in a laboratory. Given below is a description of two different setups. The design of the hatchery can be adapted to the space and equipment available.

a. *Cylinder system*: A number of 300 L plastic cylinders filled with seawater are illuminated by four fluorescent tubes evenly distributed on the cylinder wall, as shown in Figure 2.5a. A thin string (1–2 mm) is densely coiled around cylindrical Polyvinyl chloride (PVC) spools (40 cm in length, 6 cm in width) with the use of a lathe or a drilling machine. The spools with the string are washed in hot water without any detergents to get rid of any chemical residues on the string.
b. *Tank system*: Flat, rectangular tanks of 60 (l) × 35 (w) × 17 (d) cm as shown in Figure 2.5b are used. Ropes of 5–10 mm are twined between screws on one side of PVC plates of 57 × 33.5 cm and placed on the bottom of the tanks. Luminous tubes are placed 60 cm above the incubator surface. The ropes are washed in hot water in a dishwasher without any detergents to get rid of any chemical residues on the string.

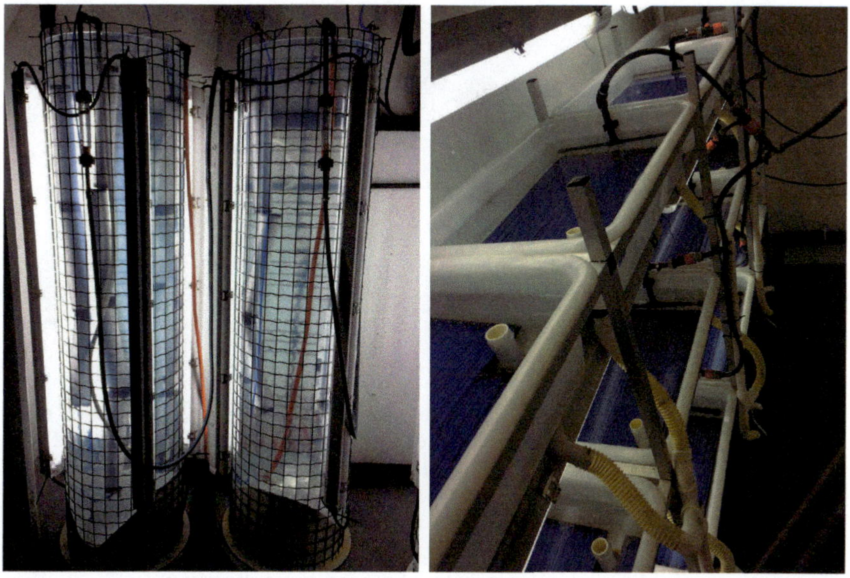

(a) (b)

Figure 2.5 (a) Seedling cultivation systems with cylinders and (b) seedling cultivation systems with tanks.

2.4.3.4 Cultivation under controlled laboratory conditions

For the first three days, the incubators are left stagnant without aeration. Water flow is set to 1.5 L min^{-1} and increased gradually to 2–2.5 L min^{-1} after 3–4 weeks to strengthen the haptera of the juvenile sporophytes before deployment. The light intensity is set to around 70 µmol outside the incubation cylinders. Gentle aeration and water flow is allowed in the incubators after the initial three-day period. Water flow is adjusted with flow meters, or with a stopwatch and a liter measure. The water is preferably taken from >50 m depth and is sand-filtered, particle-filtered (1 µm), and UV-treated (*Note 11*). If the weather conditions are not suited at the time that the sporophytes are ready for deployment in the sea, the light intensity can be reduced to around 20 µmol to keep the sporophytes in a steady state in the hatchery. The seeded substrate is incubated in the hatchery for 4–6 weeks until seedlings have reached a length of about 1–5 mm before deployment, as shown in Figure 2.6a and b. Gametophytes need a shorter time than spores to develop into juvenile sporophytes. See Table 2.3 for key parameters for seedling cultivation.

Figure 2.6 (a) Four-week-old juvenile seedlings grown on a thin string (1, 2 mm twisted nylon) and (b) spools with seedlings ready for deployment at sea after six weeks in the hatchery.

Table 2.3 Key parameters for seedling cultivation

	Key parameters
Light intensity (μmol m^{-2} s^{-1})	20–70
Light regimes (L:D)(h)	16:8
Light source	Fluorescent lamps (warm white, cool white, and day light)
Temperature (°C)	10
Water treatment	Sand-filtered, particle-filtered, UV
Aeration	Unfiltered, cooled
Water flow (L min^{-1})	1.5–2.5

Spools and microscope slides are monitored closely to follow seedlings' development and contamination in the hatchery. If diatom contamination becomes visible (as brown, circular patches), the water flow can be stopped and the stagnant water is treated with GeO_2 (0.1 mL L^{-1}) for three days (*Note 12*). When the water flow is turned on, the chemical is diluted and eventually removed from the system.

2.4.3.5 Development of seedlings under optimal conditions

During the first three days after seeding, the spore will develop into a primary cell. The primary cell is recognized as a brown, round cell with the remains of the spore seen as a tiny black dot. The two are connected through a transparent, elongated germination tube, which might be difficult to see. Over the following days, the primary cell increases in size, and if the conditions are optimal, an egg can be seen as early as on day 8 after seeding. The egg looks similar to the primary cell but is distinguished by its larger size and the empty, transparent membrane of the primary cell left behind it. The empty membrane has the same size and shape as the egg cell and is the best indication of the cell being an egg and not a primary cell. Fertilized eggs will develop rapidly into juvenile sporophytes. The first cell divisions are usually horizontal, and at the size of approximately five cells, the cells start to divide vertically. Figure 2.7 shows the development from spores to seedlings under optimal lab conditions, but one can clearly see the development of contaminating diatoms as the sori have not been disinfected prior to spore release. In commercial production, the development might take slightly more time as larger cultivation systems normally give less degree of control and evenness in the system.

2.4.4 *Deployment*

2.4.4.1 *Transport*

Seedlings must be protected from dehydration and extreme temperatures during transport. Spools and ropes are therefore wrapped in a thin layer of plastic to avoid evaporation of surface water and stored in coolers if the ambient temperature is below 0°C or above 15°C. Transportation of seedlings over a period up to 12 h have shown good results.

2.4.4.2 *Deployment at sea*

S. latissima seeded on ropes can be deployed directly at sea after incubation in the lab. The seeded ropes can be mounted as vertical drop lines or in V- or U-shaped patterns on preinstalled horizontal carrier lines (e.g., 14–22 mm). The horizontal carrier lines and the seeded ropes are placed at depths selected on the basis of seasonal variation in light, temperature, nutrients, and local hydrodynamic conditions to obtain optimum growth while minimizing tearing of the farm equipment.

S. latissima seeded on string coiled on spools must be twisted onto a thicker rope before deployed on preinstalled horizontal carrier lines. This can be done by hand, with handheld tools or with a twisting machine. The twisting machine shown in Figure 2.8a is an example of a mechanical tool for twisting of string meant for R&D and small-scale cultivation. It takes about 20 min to spin a magazine with four spools with 60 m string on each spool onto a thicker rope, as shown in Figure 2.8b, ready for deployment on preinstalled horizontal carrier lines. With a helix length of, for example, 4 cm between each turn of string, it takes 1.8 m string per 1 m of 14 mm thick rope (Figure 2.8b). Accordingly, 12 spools with a total of 720 m string can be twisted onto 400 m thick rope ready for deployment per hour. New machinery with higher capacity is required for industrialization. Sporophytes of

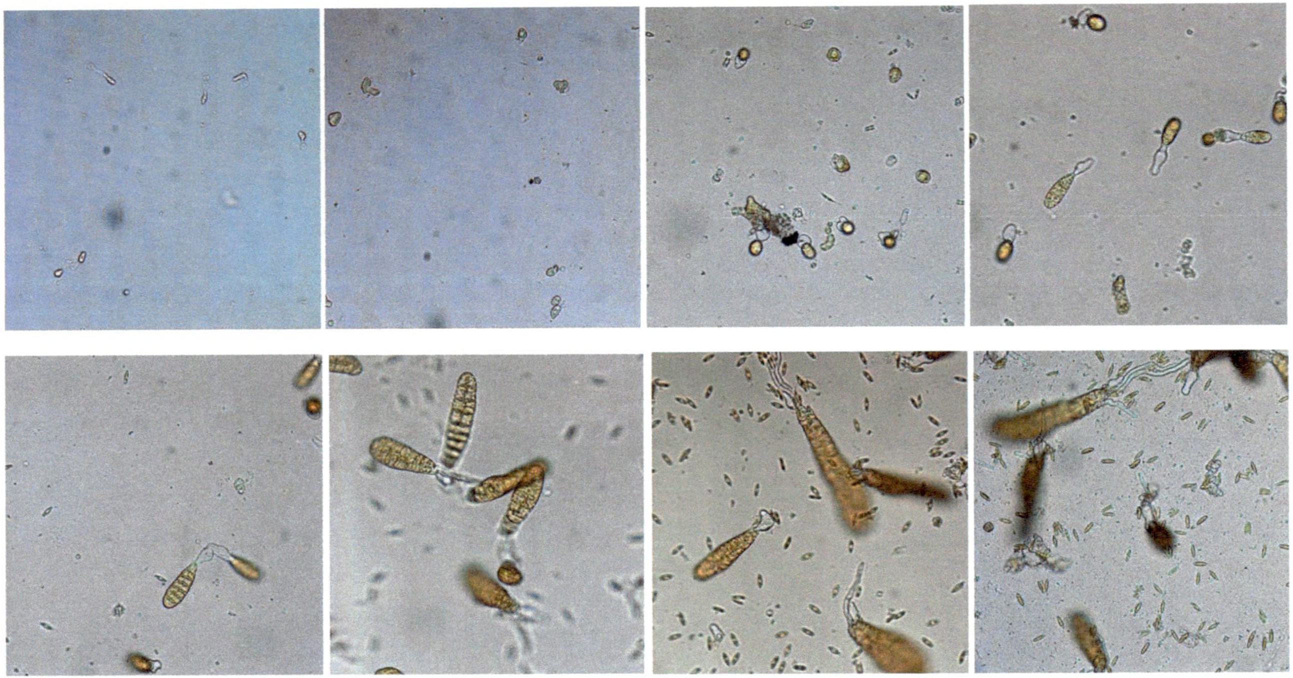

Figure 2.7 Development of spores to sporophytes under optimal lab conditions, but without any sorus disinfection before spore release. Pictures are taken regularly over 21 days (day 3, 6, 8, 11 [picture 1–4, upper row], 14, 16, 18, and 21 [picture 4–8 lower row] post seeding). Pictures 1–2 show primary cells; in picture three eggs are clearly visible and pictures 4–8 show the development of juvenile sporophytes and contaminating diatoms.

(a) (b)

(c)

Figure 2.8 (a) Twisting machine with thick rope in the center and seeded thin strings on spools, (b) thick rope with the thin seedling string twisted onto it, and (c) *S. latissima* cultivated three months in the sea, from mid-February to mid-May.

S. latissima can be deployed from early autumn to late winter before the main growth season in spring. Figure 2.8c shows *S. latissima* after three months of cultivation in the sea. High and predictable productivity and reduced biofouling during sea cultivation claim for novel production technology combined with seasonal and regional cultivation strategies.

2.5 Notes

Note 1: Conditioning tanks are preferably kept separate from the hatchery to avoid transfer of contaminating organisms from sporophytes to gametophyte cultures and seedlings.

Note 2: The size and shape of the conditioning tanks are of minor importance for the sori induction.

Note 3: Always use gloves and protecting glasses when working with NaOCl or other hazardous chemicals and work with good ventilation (fume cupboard or outside).

Note 4: The spore release peaks during the first 35–40 min after initiation (Arbona and Molla 2006) but can be continued for at least 90 min if spore density is scarce after the first 40 min.

Note 5: Spores are counted in the squares, and a minimum of 4 squares should be counted if the density is very high, and all 10 squares should be counted if the density is lower.

Note 6: Red light is acquired either by LED lights (620–750 nm) or by fluorescent tubes covered in red cellophane or red plexiglass.

Note 7: All equipment is cleaned with hot water and detergent and autoclaved/sterilized between each culture maintenance.

Note 8: Make sure that the substrate for seeding is hydrophilic (*water loving*) so that the spore solution/gametophytes is/are absorbed and not repelled by the substrate.

Note 9: If you have a low amount of spore solution available after the spore release or in the case of a very low-spore density, you can collect the spore solution that runs off from the spraying and reuse it or consider bathing the substrate instead of spraying.

Note 10: Microscope slides are seeded and incubated together with the spools/ropes to check the development using a light microscope during the incubation phase.

Note 11: The hatchery normally has activity between August and February, so in the months between production, the water and cooling systems should be turned off to let the hatchery dry out. Disinfect or wash all equipment such as ropes, water hoses, flow meters, and so on, and chlorinate the pipe system one week in advance before you start the production.

Note 12: Treatment with GeO_2 may have a negative effect on the juvenile sporophytes, causing deformation and reduced growth.

References

Arbona J.F and Molla M. 2006. Aquaculture explained: Cultivation of the brown seaweed *Alaria esculenta*. In: Watson L (Ed.), Dublin: Bord Iascaigh Mhara, 50 p.

Bolton, J.J., Germann, I., Lüning, K. 1983. Hybridization between Atlantic and Pacific representatives of the Simplices section of *Laminaria* (Phaeophyta). *Phycologia* 22:133–140.

Bolton, J.J. and Lüning, K. 1982. Optimal growth and maximal survival temperatures of Atlantic *Laminaria* species (Phaeophyta) in culture. *Mar Biol* 66:89–94.

Borum, J., Pedersen, M., Krause-Jensen, D., Christensen, P., Nielsen, K. 2002. Biomass, photosynthesis and growth of *Laminaria saccharina* in a high-arctic fjord, NE Greenland. *Mar Biol* 141:11–19.

Broch, O.J., Ellingsen, I., Forbord, S., Wang, X., Volent, Z., Alver, M.O., Handå, A. et al. 2013. Modelling the cultivation and bioremediation potential of the kelp *Saccharina latissima* in close proximity to an exposed salmon farm in Norway. *Aquac Environ Interact* 4:187–206.

Druehl, L.D. 1970. The pattern of Laminariales distribution in the northwest Pacific. *Phycologia* 9(3/4):237–247.

Druehl, L.D. and Boal, R. 1981. Manipulation of the laminarialean life-cycle and its consequences for kombu mariculture. *Proc Internat Seaweed Symp* 6:161–170.

Druehl, L.D., Collins, J.D., Lane, C.E., Saunders, G.W. 2005. An evaluation of methods used to assess intergeneric hybdridization in kelp using Pacific Laminariales (Phaeophyceae). *J Appl Phycol* 41:250–262.

Druehl, L.D. and Kaneko, T. 1973. On the *Laminaria saccharina* from Hokkaido. *Bot Mag Tokyo* 86:323–326.

Edwards, M. and Watson, L. 2011. *Aquaculture Explained Cultivating Laminaria Digitata*. Irish Sea Fisheries Board, Dunlaoghaire, Ireland.

FAO. 2016. The state of world fisheries and aquaculture 2016. *Contributing to Food Security and Nutrition for All*. Food and Agricultural Organization of the United Nations, Rome, Italy, 200 pp.

Fernand, F., Israel, A., Skjermo, J., Wichard, T., Timmermann, K.R., Golberg, A. 2016. Offshore macroalgae biomass for bioenergy production: Environmental aspects, technological achievements and challenges. *Renew Sus Energ Rev* doi: 10.1016/j.rser.2016.10.046.

Fleurence, J., Morancais, M., Dumay, J., Decottignies, P., Turpin, V., Munier, M., Garcia-Bueno, N., Jaouen, P. 2012. What are the prospects for using seaweed in human nutrition and for marine animals raised through aquaculture? *Trends Food Sci Technol* 27:57–61.

Forbord, S., Skjermo, J., Arff, J., Handå, A., Reitan, K.I., Bjerregaard, R., Lüning, K. 2012. Development of *Saccharina latissima* (Phaeophyceae) kelp hatcheries with year-round production of zoospores and juvenile sporophytes on culture ropes for kelp aquaculture. *J Appl Phycol* 24:393–399.

Førde, H., Forbord, S., Handå, A., Fossberg, J., Arff, J., Johnsen, G., Reitan, K.I. 2016. Development of bryozoan fouling on cultivated kelp (*Saccharina latissima*) in Norway. *J Appl Phycol* 28:1225–1234.

Fortes, M.D. and Lüning, K. 1980. Growth rates of North Sea macroalgae in relation to temperature, PAR and photoperiod. *Helgoländer Meeresuntersuchungen* 34:15–29.

Hafting, J.T., Critchley, A.T., Cornish, M.L., Hubley, S.A., Archibald, A.F. 2012. On-land cultivation of functional seaweed products for human usage. *J Appl Phycol* 24:385–392.
Handå, A., Forbord, S., Wang, X., Broch, O.J., Dahle, S.W., Størseth, T.S., Reitan, K.I., Olsen, Y., Skjermo, J. 2013. Seasonal- and depth-dependent growth of cultivated kelp (*Saccharina latissima*) in close proximity to salmon (*Salmo salar*) aquaculture in Norway. *Aquaculture* 414–415:191–201.
Harrison, P.J. and Berges, J.A. 2005. Marine culture media. In: *Algal Culturing Techniques* R. Andersen (Ed.), pp. 21–33. Academic Press, New York.
Holdt, S. and Kraan, S. 2011. Bioactive compounds in seaweed: Functional food applications and legislation. *J Appl Phycol* 23:543–597.
Hollenberg, G.J. 1939. Culture studies of marine algae. I. *Eisenia arborea*. *Amer J Bot* 26:34–41.
Joska, M. and Bolton, J. 1987. *In situ* measurement of zoospore release and seasonality of reproduction in *Ecklonia maxima* (Alariaceae, Laminariales). *Brit Phycol J* 22:209–214.
Kain, J.M. 1975. The biology of *Laminaria Hyperborea* VII. Reproduction of the sporophyte. *J Mar Biol Assoc U.K.* 55:567–582.
Kain, J.M. 1979. A view of the genus Laminaria. *Oceanogr Mar Biol Ann Rev* 17:101–161.
Lüning, K. 1980. Critical levels of light and temperature regulating gametogenesis of three Laminaria species. *J Phycol* 16:1–15.
Lüning, K. 1990. *Seaweeds. Their Environment, Biogeography, and Ecophysiology*. Wiley, New York.
Lüning, K. and Dring, M.J. 1975. Reproduction, growth and photosynthesis of gametophytes of *Laminaria saccharina* grown in blue and red light. *Mar Biol* 29:195–200.
Lüning, K. and Neushul, M. 1978. Light and temperature demands for growth and reproduction of laminarian gametophytes in southern and central California. *Mar Biol* 45:297–309.
Müller, D.G. 1989. Sexuality and sexual attraction. In: *Algae as Experimental Systems*, A.W. Coleman, L.J. Goff, and J.R. Stein-Taylor (Eds.), pp. 201–213. A.R. Liss, New York.
Müller, D.G., Gassmann, G., Lüning, K. 1979. Isolation of a spermatozoid—Releasing and attracting substance from female gametophytes of *Laminaria digitata*. *Nature* 279:430–431.
Pang, S. and Lüning, K. 2004. Breaking seasonal limitation: Year-round sporogenesis in the brown alga *Laminaria saccharina* by blocking the transport of putative sporulation inhibitors. *Aquaculture* 240:531–541.
Rajapakse, N. and Kim, S.K. 2011. Nutritional and digestive health benefits of seaweed. *Adv Food Nutr Res* 64:17–28.
Redmond, S., Green, L., Yarish, C., Kim, J., Neefus, C. 2014. *New England Seaweed Culture Handbook-Nursery Systems*. Connecticut Sea Grant CTSG-14-01. 92 p. http://nsgl.gso.uri.edu/nhu/nhuh14001.pdf.
Rød, K.K. 2012. Sori disinfection in cultivation of *Saccharina latissima*—Evaluation of chemical treatments against diatom contamination. Master thesis, Norwegian University of Science and Technology, Trondheim, Norway.
Sauvageau, C. 1916. Sur les gametophytes de deux Laminaires. *C R Acad Sci* 56:1–240.

Schreiber, E. 1930. Untersuchungen über Parthenogenetische Geschlechtsbestimmung und Bastardierungsvermögen bei Laminarien. *Planta* 12:331–353.
Shahidi, F. 2009. Nutraceuticals and functional foods: Whole versus processed foods. *Trends Food Sci Technol* 20:376–387.
Shea, R. and Chopin, T. 2007. Effects of germanium dioxide, an inhibitor of diatom growth, on the microscopic laboratory cultivation stage of the kelp, *Laminaria saccharina*. *J Appl Phycol* 19:27–32.
Sjøtun, K. 1985. Ei autoøkologisk undersøkjing av *Laminaria saccharina* (L.) Lamour i Espegrend-området. Master thesis, University of Bergen. In Norwegian, Bergen, Norway.
Tabarsa, M., Rezaei, M., Ramezanpour, Z., Waaland, J.R., Rabiei, R. 2012. Fatty acids, amino acids, mineral contents, and proximate composition of some brown seaweeds. *J Phycol* 48:285–292.
tom Dieck, I. 1993. Temperature tolerance and survival in darkness of kelp gametophytes (Laminariales, Phaeophyta): Ecological and biogeographical implications. *Mar Ecol Prog Ser* 100:253–264.
van den Hoek, C. and Donze, M. 1967. Algal phytogeography of the European coasts. *Blumea* 15 (1):63–89.
Wargacki, A.J., Leonard, E., Win, M.N., Regitsky, D.D., Santos, C.N.S., Kim, P.B., Cooper, S.R. et al. 2012. An engineered microbial platform for direct biofuel production from brown macroalgae. *Science* 335:308–313.
Wilce, R.T. 1965. Studies in the genus Laminaria III. A revision of the north Atlantic species of the Simplices section of the Laminaria. *Bot Gothenburg* 3:247–256.
Williams, J.L. 1921. The gametophytes and fertilization in Laminaria and Chorda. *Ann Bot Lond* 35:603–607.

chapter three

Derivation of clonal stock cultures and hybridization of kelps

A tool for strain preservation and breeding programs

Inka Bartsch

Contents

3.1 Introduction

The term *kelp* is the common name of all species of the marine brown algal order Laminariales, but as common names often do not have a precise definition, in the literature, it is partially also used for other big brown algae of the order Fucales or Desmarestiales, all of which form coastal underwater forests. Here, we only deal with kelps in the restricted definition, that is, the Laminariales. Kelps thrive along rocky shores of polar to cold-temperate waters worldwide, or more seldom in cooler deep-waters of warm-temperate to tropical sites, such as the Mediterranean Sea or the Galapagos islands (Lüning 1990; Bolton 2010). Besides their considerable ecological role as ecosystem engineering species, a few taxa, especially of the genera *Laminaria, Saccharina, Undaria,* and *Macrocystis*, are of major economic importance, either being harvested in the field or cultivated, the latter particularly in Asia and in Chile (FAO 2016). Interest in kelps and kelp cultivation as a resource for food products, alginates, and biofuel production among others is growing steadily and thus strain preservation, in combination with genetic engineering of strains or classical breeding programs, may become more and more important in the future. This is especially significant under the threat of global warming where wide stretches of coasts currently harboring seaweed aquacultures will become tropical to subtropical (Bartsch et al. 2012) and may require warm-tolerant cultivars.

One prerequisite for the easy selection and maintenance of strains is the life cycle of Laminariales. Kelps are diploid macroscopic algae, which consist of a lamina (blade), stipe (stem), and haptera (holdfast), and are mostly perennial. When they become fertile, groups of sporangia (sorus) develop either on the main blade or in some species on smaller special blades, the so-called sporophylls. During spore formation, meiosis takes place resulting in mobile isomorph haploid male and female spores of approximately 4–8 µm length. After spore germination, single-celled gametophytes may either directly enter reproduction (gametogenesis) or initiate vegetative growth depending on environmental conditions. The latter process is used to derive clonal stock cultures for multiuse purposes. For a scheme of the life cycle and further details, see Chapter 2 by Forbord et al. (2017) and Bartsch et al. (2008). The capacity of microscopic clonal gametophytes to grow and preserve their ability for sexual reproduction even after >30 years of maintenance (Druehl et al. 2005; Bartsch, pers. comm.) is a prerequisite for the long-term preservation of clonal male and female gametophytes.

The current chapter elucidates kelp cultivation aspects that have not been addressed in Chapter 2. The skilled researcher or application manager will be able to combine both protocols for specific needs. General seed supply and sporulation protocols described in Chapter 2 are a good starting point to obtain freshly released spores. In addition to this, this chapter will provide details that specifically focus on isolation and vegetative propagation of unialgal (but not axenic) clonal gametophyte strains. As described in Chapter 2, spores can either be obtained by collection of seasonally fertile field sporophytes or by artificial induction of fertile tissue. This enables spore production independent of the season (Buchholz and Lüning 1999; Pang and Lüning 2004).

3.2 State of the art

During the last century, before the advent of viable molecular methods, inter- and intraspecific hybridization experiments of kelps were used to reveal potential relationships within and between kelp genera and families. This interest was mostly driven by basic science. One example of these times highlights the applied potential of hybrid kelp sporophytes: It became evident that digitate species of the genus *Laminaria* from the North and South Atlantic were not interfertile if co-occurring but still retained the ability to hybridize when they were geographically separated. As all these species have a slightly different temperature tolerance, F1-hybrid sporophytes showed different temperature growth characteristics than their parents (tom Dieck and Oliveira 1993). This aspect can be used for aquaculture purposes to develop cultivars better adapted to the required conditions than the wild type. Ethical considerations may, however, prohibit a fruitful application in cases in which respective hybrids may be considered as foreign species introductions.

Full intraspecific hybridization was obtained between geographical isolates of several species or species complexes (e.g., in N-Atlantic *Laminaria digitata,* Atlantic and Pacific *Saccharina* species complex, *Alaria esculenta,* and Pacific *Macrocystis*) (Funano 1980; Lewis and Neushul 1994; overview of hybridization in *Laminaria* and *Saccharina*: Bartsch et al. 2008; Kraan et al. 2000). Successful hybridization is possible even between genera, as has recently been shown by Hwang et al. (2012) for *Undariopsis petersiana* and *Undaria pinnatifida*. On the other hand, even within species, reciprocal crossbreeds are not always successful, especially when involving strains/clones from widely separated localities (Bolton et al. 1983; Müller et al. 2008). Many issues involved in the verification of intergeneric hybrid sporophytes and kelp hybridization, in general, have been discussed in detail by Druehl et al. (2005). Sexual compatibility and the ability to hybridize seems to be a gradual process that probably involves several genes and prezygotic or postzygotic reproductive isolation mechanisms (e.g., Lewis and Neushul 1995; Druehl et al. 2005). As the need for genetic modification of marine macroalgae for industrial purposes has increased (Robinson et al. 2013), the focus on kelp hybridization changed to application issues during the last decades.

The search for species, populations, or cultivars with a high commercial value for aquaculture including respective breeding programs has a long tradition, especially in China. Cultivars propagated through spores from a limited number of sporophytes are however prone to inbreeding depression, leading to the loss of originally well-performing strains (Loureiro et al. 2015). Thus, intra- or interspecific crossbreeds originating from clonal gametophytes are a potential tool to prevent long-term degeneration (*Note 1*). It may help generating first filial hybrid sporophytes with an increased vigor, higher productivity, or growth than the respective parental generations—which is considered to represent a *heterosis effect*.

In Asia and Chile, new superior cultivars with a stable genotype were developed with the help of clonal gametophyte cultures to overcome the restraints from inbreeding. In 2007, selected gametophyte clones from *Saccharina japonica* and *S. longissima* (=*Laminaria japonica* and *L. longissima*) were crossed, and it was an effective tool of utilizing kelp heterozygous vigor (Li et al. 2007). This line of research was constantly pursued in China resulting in several new cultivars with superior commercial value. In *S. japonica*, intraspecific crosses of gametophyte clones from distant sites led to a new cultivar with superior blade quality, high alginate and iodine contents, higher stress tolerance, and reduced biomass loss during the hot summer season. These results are on the basis of data from five seasons. The seeding of the clonal gametophytes successfully prevented inbreeding depression compared with seeding with spores (Li et al. 2016; Zhao et al. 2016). In *U. pinnatifida*, a highly efficient cultivar was also developed through crosses of intraspecific gametophyte clones and further selection (Shan et al. 2016). Similar strategies have been proposed for *Macrocystis* which is a valuable resource of high commercial interest in Chile, to overcome its severe over-exploitation. On a lab-based scale, this was already achieved, but the large-scale application has yet to be accomplished (Westermeier et al. 2010).

3.3 Materials

In general, all listed materials should be clean or where possible sterile.

3.3.1 Facilities and equipment for collection of sporogenous material and release of spores

3.3.1.1 In the field

- Zip-lock bags.
- Labels (for discrimination of specific plants, e.g., for genotyping and phenotyping).

- Coolers or ice packs to keep the temperature at the same or below the environmental seawater temperature.
- Scissors, knife, or cork borer for collecting sorus from sporophytes.
- *Optional*: Cotton or tissue paper for cleaning.
- *For expedition*: Sterile Falcon tubes or cryovessels filled with glass slides.

3.3.1.2 *In the laboratory*

- Tissue paper.
- Distilled water.
- Glass or plastic dishes for preparation of wet chambers of sufficient size with lid, alternatively covering with aluminum foil.
- Temperature-regulated room keeping 5°C–15°C and/or refrigerator (4°C); adapt to your local environmental conditions.
- Filtered seawater (≤2 µm) enriched with 0.5 mL of a saturated GeO_2 solution per liter seawater (Shea and Chopin 2007) to suppress diatom growth.
- Pipettes of any type (e.g., Eppendorf and Pasteur).
- Small plastic Petri dishes.
- Microscope slides, cover glasses, and glass cutter.
- Microscope/ideally inverted microscope.
- Extendable closing foil (such as Parafilm).
- Permanent marker.

3.3.2 *Facilities and equipment list for isolation and propagation of clonal cultures*

- Dissecting microscope.
- Inverted microscope (optional).
- Plastic Petri dishes (do not use microwell plates as they are difficult to monitor and do not contain sufficient medium).
- Pasteur glass pipettes.
- Pipette suction ball (silicon) (only silicon balls have the right pressure, avoid Peleus balls or other bigger pipette aids as they do not allow for the fine movement skills you need for isolation). Alternatively, you can use a silicon tube, which is connected with the pipette and your mouth.
- Gas flame.
- Forceps.
- Pestle and mortar.
- Seawater enriched with ½ PES (Provasoli nutrient solution) (Provasoli 1968).
- Extendable closing foil (such as Parafilm).
- Steady hand.

3.3.3 *Facilities and equipment list for performing laboratory-scale hybridization programs*

- Vegetative clonal male and female gametophyte cultures of different parental origin representing different phenotypes or genotypes.
- Pasteur pipettes.
- Mortar and pestle.
- Glassware, tubing, and air filters.
- Filtered nutrient-enriched seawater (½ PES).

3.4 *Experimental procedures*

This protocol focuses on kelps in general and applies to any species as long as adjustments are made according to the temperature demands of the respective species. With this protocol as a baseline, modifications can easily be applied depending on the local situation and the preferred personal procedures. Similar, but less detailed protocols are available elsewhere (e.g., Druehl et al. 2005).

3.4.1 *Collection and preparation of fertile sporophytes*

Sorus development is gradual, and a ripe sorus ready to release spores is discernable by a uniform dark-brown color (Figure 3.1), it elevates above the surface of the blade and thereby is palpable and sometimes a transparent membrane is detaching from the surface. See *Note 2* for the fertility of sporophytes.

- Collect fertile kelp tissue (sorus) in the field.
- Take care that the sorus does not suffer from high irradiance, temperatures >10°C–15°C, or frost and remains wet during transportation. Adapt size of transport container to size or quantity of material and field situation.
- Cooling can be achieved by use of ambient seawater; otherwise, add ice packs, but take care to protect kelp tissue from freezing.
- Sporogenous tissue keeps fresh and healthy for two days under these conditions.

3.4.2 *Induced spore release*

As the aim is to obtain uncontaminated healthy unialgal spores, spread out well separated on the substrate, spores have to be sown in a rather low density, and only a small piece of healthy and ripe sorus is needed.

Figure 3.1 Sporophylls of *Alaria esculenta* showing different stages of sorus ripening. Scale bar = 1 cm.

The smaller the piece of sorus is, the higher the probability that it is not contaminated by diatoms or other organisms.

1. Cut out sorus; small pieces of ripe sori are sufficient for the purpose.
2. Clean sorus with tissue paper wetted with distilled water and wipe dry. Chemical disinfection is not needed.
3. Avoid material that is infected by endophytes or epiphytes (Figure 3.2) or shows signs of biofouling.
4. Store the sorus in a humid chamber (e.g., by addition of wet paper towels) overnight at low temperatures (e.g., 4°C in the refrigerator). Adapt size of the chamber to the size of the sorus. In our case, a Petri dish is sufficient. Take care that the sorus does not come into contact with freshwater (*Note 3*).
5. Rehydrate sorus piece in a clean Petri dish with sterile seawater (10°C–15°C) the next day.
6. Subject to light and relatively warm but nonlethal temperatures (~15°C). Spores are usually released within 10–30 min.
7. After spore release, remove the sorus and discard.

Figure 3.2 Fertile sorus of *Saccharina latissima* infected by endobiota.

8. Check for mobility and density of spores (if possible).
9. Dilute a few drops of the spore suspension in a new Petri dish filled with sterile seawater and GeO_2.
10. Ideally, a low spore density is generated (<10,000 spores cm^{-2}) so that spores will settle separately from each other in distances of >200 μm (Figure 3.3). It can easily be achieved if you initiate a dilution series from which you can choose later. This is particularly useful if you cannot check for spore density after release.
11. Label Petri dish and seal with Parafilm.

The shorter the release period, the less chance there is for any unwanted contamination. In contrast to the protocol by Forbord et al. 2017 (Chapter 2), a mass release of spores is not essential.

During expeditions/field—work without direct laboratory access, the following protocol may be useful to collect spores:

- Fill Falcon tubes, cryovessels, or similar equipment with two clean glass slides firmly pressed together or cut appropriate pieces of slides to fit into the sterile tubes.

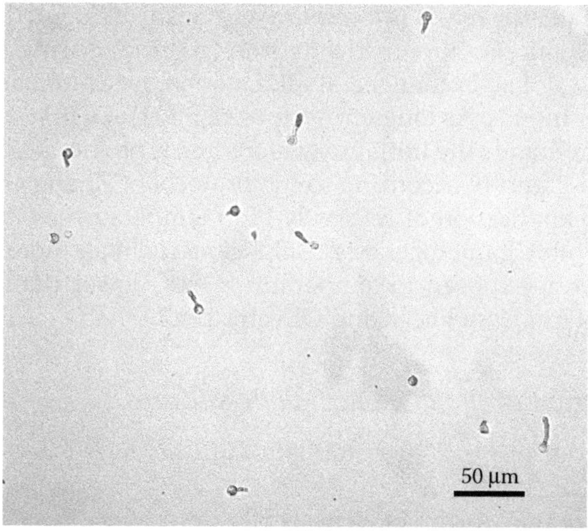

Figure 3.3 Germinated kelp spores in a density that should ideally be less than shown for isolation purposes.

- Fill up with cool, sterile seawater and GeO_2.
- Place a small piece of freshly cleaned sorus (as above-mentioned) into prelabeled tubes.
- Protect tubes from heat and direct sun light.
- Discard the sorus after approximately 30 min–2 h.
- Back in the laboratory, transfer the slides with the outer side up into sterile plastic Petri dishes freshly filled with sterile seawater and GeO_2 after spores have settled (approx. 24 h).
- Label and seal the dishes by using Parafilm.

After sowing the spores, these germinate autonomously into unicellular gametophytes within a few days.

3.4.3 Conditions to keep gametophytes vegetative

Unicellular gametophytes have to be prevented from becoming fertile. Otherwise, a rapid transition into sporophytes takes place, and no gametophytes are left for isolation of clones. Blue light initiates gametogenesis and even low intensities, especially if combined with low temperatures, are extremely inducing (Lüning and Dring 1975; Bartsch pers. obs.). Red light or very low-light intensities should be applied. Modern white LED lamps often have a much higher irradiance in the blue spectrum than fluorescent lamps that were predominantly used in laboratories until recently. Thus, care should be taken that no blue light source is close to your gametophyte cultures. Nutrients, especially those including iron, also induce gametogenesis (Motomura and Sakai 1981; Lewis et al. 2013). Thus, the nutrient medium is ideally iron-free and, during the first four weeks, should not be exchanged at all. Later on, medium change is sufficient once per month. As long-term exposition to GeO_2 may also be harmful to kelps, exchange the initial seawater after approximately four weeks, and only add P and N according to the protocol of Zhang et al. (2008) or ≤½ PES. The application of relatively high temperatures is another possibility to inhibit gametogenesis. Kelp gametophytes normally do not become fertile if exposed to temperatures ~4°C lower than their upper survival limit (e.g., tom Dieck and Oliveira 1993).

3.4.4 Isolation of clonal gametophyte cultures

- Check the original gametophyte cultures every 2–4 weeks after sowing.
- Take care that no gametogenesis takes place.

One to two months after inoculation, gametophytes normally have grown to sufficient size to be isolated under the dissecting microscope or the inverted microscope. By then, the gametophytes have differentiated into multicellular bigger female and smaller male gametophytes. Ideally, you begin isolating when gametophytes are big enough to be seen and treated under the dissecting microscope.

- Draw out glass Pasteur pipettes to get a narrower mouth or just melt the tip of the glass pipette to smooth the outer margin of the mouth or take a microcapillary tube. Avoid forceps for isolation purposes. Even fine forceps are not suitable as the gametophytes easily stick to them.
- Identify well-separated male and female gametophytes.
- Select one single gametophyte with your pipette.
- Transfer the droplet of water with the single gametophyte into a new sterile Petri dish.
- Scrutinize this droplet under the (inverted) microscope to verify that you selected a single gametophyte and verify the sex.
- Fill up with sterile ½ PES (ideally iron free) or sterile seawater just enriched with N and P (Zhang et al. 2008).
- Seal with Parafilm, label, and replace into conditions allowing for vegetative growth.
- Repeat this process with the opposite sex so that you generate at least one clonal pair of gametophytes per parental sporophyte.

It should be noted that single isolated gametophytes may not be visible in the Petri dish until they have grown to a larger size.

3.4.5 *Propagation of clonal cultures*

Approximately 3–6 months later, the separated gametophyte has formed a small vegetative cushion of approximately 1–3 mm in diameter. You can now use this clone to propagate it further (Figure 3.4):

- Take the gametophyte cushion with a sterile one-way pipette and place it into a sterile mortar.
- Very carefully crush the cushion, without applying a lot of pressure, add a bit of sterile seawater, and pour the fragments into a bigger Petri dish.

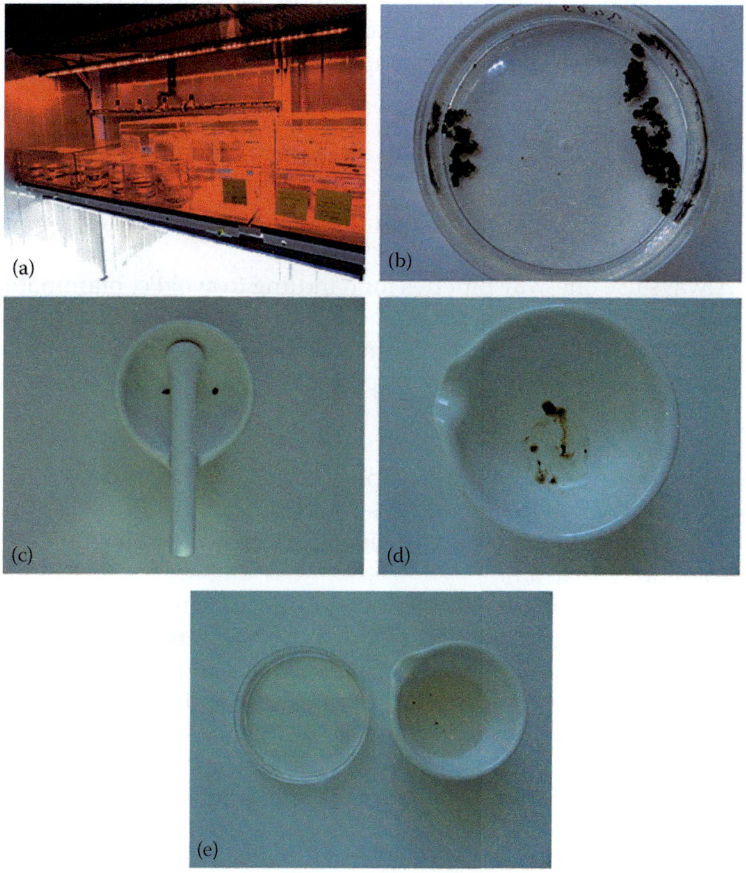

Figure 3.4 Vegetative clonal gametophytes and their propagation. (a) Red light grown gametophytes, (b) vegetative gametophyte cushions, (c) sterile mortar with small vegetative gametophyte cushions, (d) carefully fragmented gametophytes, and (e) fragmented gametophytes filled up with seawater for sowing.

- Fill up with sterile iron-free nutrient-enriched seawater or ≤½ PES.
- Label the dishes and seal with Parafilm.
- Transfer back into conditions allowing for vegetative growth.

Another several months later, there is enough material to initiate either sporophyte production for hybridization experiments or to further propagate the clonal gametophytes for aquaculture purposes (e.g., Zhang et al. 2008).

3.4.6 Hybridization of clonal gametophyte cultures

Once established, clonal gametophytes allow for controlled experiments or induction of cultivars with specific characteristics. This is useful for various applied purposes or investigations of trait inheritance through classical or molecular methods.

- Choose the male and female clonal gametophyte strains to be crossed.
- Take a similar amount of vegetative gametophyte material from all strains and sexes.
- Always use one-way pipettes for handling to avoid contamination of clones with opposite sex.
- Carefully crush each gametophyte clone separately in a sterile mortar (as above-mentioned) and produce separate stock solutions for each clone.

To observe the phenology and/or physiology of sporophytes resulting from crosses, the following cultures have to be initiated by combining similar amounts of the respective clonal male and female gametophyte stock solutions (Table 3.1).

A parthenogenetic control is needed as all female gametophytes that become fertile, release eggs and those that are not fertilized by male spermatia may also develop haploid or polyploid parthenosporophytes. An apogamic control only produces sporophytes in seldom cases but should be monitored for if the response of the species or strain is unknown. Parthenosporophytes and apogamic sporophytes are often misshaped without a clear basipetal orientation and often lack the development of basal rhizoidal filaments (tom Dieck and Oliveira 1993; Figure 3.5). In some species, parthenosporophytes with normal

Table 3.1 Breeding scheme of gametophyte crossings and controls for detection of performance effects in the filial F1-sporophyte generation

Breeding treatment	Crossing scheme of parental genotypes	
Selfings[1]	Strain 1 ♂ × strain 1 ♀	Strain 2 ♂ × strain 2 ♀
Hybrids[2]	Strain 1 ♂ × strain 2 ♀	Strain 2 ♂ × strain 1 ♀
Parthenogenesis control[3]	Strain 1 ♀	Strain 2 ♀
Apogamic control[4]	Strain 1 ♂	Strain 2 ♂

[1] Combination of single parental genotypes; control of original phenotype.
[2] Combination of two parental genotypes; resulting phenotypes of sporophytes will be controlled against selfings, parthenogenetic, and apogamic control sporophytes.
[3] Control whether female gametophytes may convert unfertilized eggs into sporophytes.
[4] Control whether male gametophytes may develop sporophytes asexually through transformation of gametophyte cells.

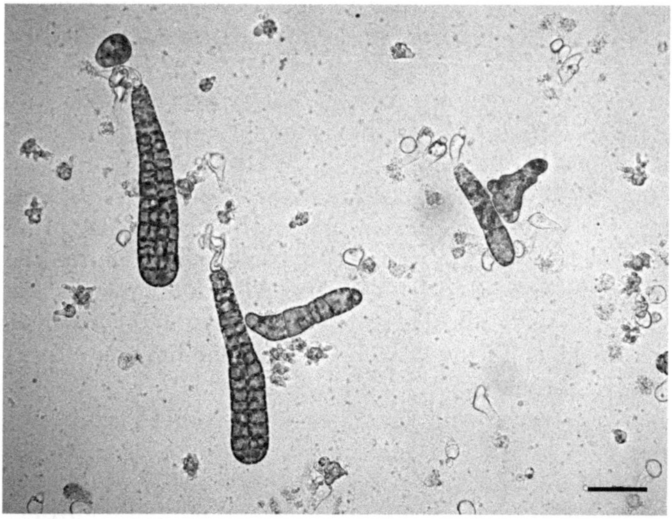

Figure 3.5 *Saccharina latissima:* Microscopic sporophytes (with rhizoids) and presumable parthenosporophytes (deformed sporophytes without rhizoids). Scale bar = 50 µm.

morphology have been reported growing into fertile adult sporophytes and producing only haploid female gametophytes, whereas misshaped parthenosporophytes were polyploid (Fang et al. 1978; Lewis et al. 1993). The comparison of the morphological development of sporophytes allows for the visual interpretation of successful hybridization without molecular genotyping techniques, although they cannot replace the latter.

- After initiation of crosses, subject cultures to conditions inducing gametogenesis and growth of juvenile sporophytes (see Forbord et al. 2017 in Chapter 2).
- Microscopic sporophytes develop within 14 days under optimum conditions, which differs for different kelp species. In general, temperatures between 5°C and 15°C and an irradiance of 20–40 μmol photons m^{-2} s^{-1} in a 16:8 light:dark cycle will produce good results.
- Avoid continuous daylengths for the induction of gametogenesis and sporophyte production as egg release will no longer be synchronized (Lüning 1981). As a consequence, the formation of sporophytes is greatly reduced, and a considerable number of presumable parthenosporophytes with abnormal morphology appear (Figure 3.5).
- When sporophytes are discernible by the naked eye (approximately 1–2 mm in length) they should be transferred into aerated sterile glass beakers.
- Change the nutrient-enriched seawater at least once per week.
- Adapt the size of the glassware according to the size of the sporophytes (rule of thumb example: approximately hundred 2 cm sporophytes grow well in 5 L beakers). Also take care that the density of sporophytes is adjusted during their growth. Sporophytes need a lot of nutrients during this phase. Thus, the number of beakers will increase continuously.
- Replace glassware and tubing every 2–3 weeks with clean, sterile items.
- Within 3–4 months, you obtain sporophytes of 2–5 cm length (Figure 3.6), which can be used for subsequent controlled physiological or transcriptomic experiments, enabling the comparison of strain characteristics with the evaluation of possible heterosis effects.

Figure 3.6 Sporophyte cultures in aerated 5 L glass beakers.

3.5 Notes

Note 1: Advantages and disadvantages of clonal gametophytes: Clonal gametophytes are rather easy to obtain and to maintain for requirements of basic research and commercial applications. The advantages of clonal gametophytes are that (1) they can be kept for decades, (2) probably remain genetically relatively stable (final evidence is still missing), and (3) sporophytes can be initiated at any time of the year and can thus uncouple seeding from the seasonal availability of spores. The disadvantages of clonal gametophytes are that (1) much time is needed for their establishment, that (2) they have to be protected from fatal contamination or disease, and that (3) care has to be taken that clones stay vegetative but grow sufficiently.

Note 2: Fertility of sporophytes: The season of sporophyte fertility is different from species-to-species (see, e.g., Bartsch et al. 2008). Most species are fertile during autumn to winter, some over summer, and in some species/regions, fertile material may be found year-round but with different intensities. In instances where environmental conditions have been stressful, it may occur that sori looking perfect do not release viable spores, and germination capacity is low. This has been shown for *L. digitata* in summer (Bartsch et al. 2013).

Note 3: Spore release methods: Each laboratory has its own methods to release spores. Some protocols keep cleaned sori for 2 h in the air, whereas others keep them for two days in the refrigerator without the wet chamber. The disadvantage of the forced spore release procedure is that not only ripe but also unripe sporangia burst and release healthy motile and unripe spores with no swimming activity. The motility status of spores should thus be checked under the microscope after release. In some cases, the tissue may release considerable amounts of mucus (alginate) that can be harmful to the spores and may require immediate attention (e.g., filtration through net-gauze or dilution with sterile seawater or the transfer of the spore solution into new tubes/Petri dishes).

Acknowledgment

I thank Klaus Lüning who initiated my interest in kelps many years ago, provided the initial training in the described methods, and fostered my early career. I thank Bertrand Jacquemin for introducing me to the spore capture method in the Falcon tubes without induced spore release. I thank A. Wagner and D. Liesner for checking of inconsistencies and J. Bartsch for polishing the English of the manuscript.

References

Bartsch, I., J. Vogt, C. Pehlke, and D. Hanelt. 2013. Prevailing sea surface temperatures inhibit summer reproduction of the kelp *Laminaria digitata* at Helgoland (North sea). *J. Phycol.* 49:1061–1073.

Bartsch, I., C. Wiencke, K. Bischof et al. 2008. The genus Laminaria *sensu lato*: Recent insight and development. *Eur. J. Phycol.* 43:1–86.

Bartsch, I., C. Wiencke, and T. Laepple. 2012. Global seaweed biogeography under a changing climate: The prospected effects of temperature. In *Seaweed Biology. Novel Insights into Ecophysiology, Ecology and Utilization*, Wiencke, C. and Bischof, K. (Eds.), Springer, Heidelberg, Germany, pp. 383–406.

Bolton, J.J. 2010. The biogeography of kelps (Laminariales, Phaeophyceae): A global analysis with new insights from recent advances in molecular phylogenetics. *Helgol. Mar. Res.* 64:263–279.

Bolton, J.J., I. Germann, and K. Lüning. 1983. Hybridization between Atlantic and Pacific representatives of the Simplices section of *Laminaria* (Phaeophyta). *Phycologia* 22:133–140.

Buchholz, C. and K. Lüning. 1999. Isolated, distal blade discs of the brown alga *Laminaria digitata* form sorus, but not discs near to the meristematic transition zone. *J. Appl. Phycol.* 11:579–584.

Druehl, L.D., J.D. Collins, C.E. Lane, and G.W. Saunders. 2005. An evaluation of methods used to assess intergeneric hybridization in kelp using Pacific Laminariales (Phaeophyceae). *J. Phycol.* 41:250–262.
Fang, T.C., J.H. Tai, Y.L. Ou, C.C. Tui, and T.C. Chen. 1978. Some genetic observations on the monoploid breeding of *Laminaria japonica*. *Sci. Sin.* 21:401–408.
Food and Agriculture Organization of the United Nations. 2016. *Fisheries and Aquaculture Information and Statistics Services*. Retrieved from http://www.fao.org/figis/ (accessed May 3, 2016).
Forbord, S., K.B. Steinhovden, K.K. Rød, A. Handå, and J. Skjermo. 2017. Cultivation protocol for *Saccharina latissima*. In *Protocols for Macroalgae Research*, B. Charrier, T. Wichard, and C.R.K. Reddy (Eds.), Chapter 2. Boca Raton, FL: CRC Press, Taylor & Francis Group.
Funano, T. 1980. Crossing experiments between several species of *Laminaria* in Hokkaido. *Hokusuishi-Geppo* 37:181–207 (in Japanese).
Hwang, E.K., Y.G. Gong, and C.S. Park. 2012. Cultivation of a hybrid of free-living gametophytes between *Undariopsis peterseniana* and *Undaria pinnatifida*: Morphological aspects and cultivation period. *J. Appl. Phycol.* 24:401–408.
Kraan, S., A.V. Tramullas, and M.D. Guiry. 2000. The edible brown seaweed *Alaria esculenta* (Phaeophyceae, Laminariales): Hybridization, growth and genetic comparisons of six Irish populations. *J. Appl. Phycol.* 12:577–583.
Lewis, R.J. and M. Neushul. 1994. Northern and southern hemisphere hybrids of *Macrocystis* (Phaeophyceae). *J. Phycol.* 30:346–353.
Lewis, R.J. and M. Neushul. 1995. Intergeneric hybridization among five genera of the family Lessoniaceae (Phaeophyceae) and evidence for polyploidy in a fertile *Pelagophycus* × *Macrocystis* hybrid. *J. Phycol.* 31:1012–1017.
Lewis, R.J., M. Green, and M.E. Azfal. 2013. Effects of chelated iron on oogenesis and vegetative growth of kelp gametophytes (Phaeophyceae). *Phycol. Res.* 61:46–51.
Lewis, R.J., B. Jiang, M. Neushul, and X.G. Fei. 1993. Haploid parthenogenetic sporophytes of *Laminaria japonica* (Phaeophyceae). *J. Phycol.* 29:363–369.
Li, X., Y. Cong, G. Yang et al. 2007. Trait evaluation and trial cultivation of Dongfang No. 2, the hybrid of a male gametophyte clone of *Laminaria longissima* (Laminariales, Phaeophyta) and a female one of *L. japonica*. *J. Appl. Phycol.* 19:139–151.
Li, X., Z. Zhang, S. Qu et al. 2016. Breeding of an intraspecific kelp hybrid Dongfang no. 6 (*Saccharina japonica*, Phaeophyceae, Laminariales) for suitable processing products and evaluation of its culture performance. *J. Appl. Phycol.* 28:439–447.
Loureiro, R., C.M.M. Gacon, and C. Rébours. 2015. Seaweed cultivation: Potential and challenges of crop domestication at an unprecedented pace. *New Phytologist* 206:489–492.
Lüning, K. 1981. Egg release in gametophytes of *Laminaria saccharina*: Induction by darkness and inhibition by blue light and UV. *Br. Phycol. J.* 16:579–593.
Lüning, K. 1990. *Seaweeds: Their Environment, Biogeography, and Ecophysiology*. New York: Wiley.
Lüning, K. and M.J. Dring. 1975. Reproduction, growth and photosynthesis of gametophytes of *Laminaria saccharina* grown in blue and red light. *Mar. Biol.* 29:195–200.
Motomura, T. and Y. Sakai. 1981. Effect of chelated iron in culture media on oogenesis in *Laminaria angustata*. *Bull. Japan Soc. Sci. Fish.* 47:1535–1540.

Müller, R., C. Wiencke, and K. Bischof. 2008. Interactive effects of UV radiation and temperature on microstages of Laminariales (Phaeophyceae) from the Arctic and North Sea. *Clim. Res.* 37:203–213.

Pang, S.J. and K. Lüning. 2004. Breaking seasonal limitation: Year-round sporogenesis in the brown alga *Laminaria saccharina* by blocking the transport of putative sporulation inhibitors. *Aquaculture* 240:531–541.

Provasoli, L. 1968. Media and prospects for the cultivation of marine algae. In *Cultures and Collections of Algae*, A. Watanabe and A. Hattori (Eds.), *Proceedings of the US Japan Conference*, Hakone 1966. Tokio, Japan: Japanese Society of Plant Physiology, pp. 63–75.

Robinson, N., P. Winberg, and L. Kirkendale. 2013. Genetic improvement of macroalgae: Status to date and needs for the future. *J. Appl. Phycol.* 25:703–716.

Shan, T.F., S.F. Pang, J. Li et al. 2016. Breeding of an elite cultivar Haibao No. 1 of *Undaria pinnatifida* (Phaeophyceae) through gametophyte clone crossing and consecutive selection. *J. Appl. Phycol.* 28:2419–2426.

Shea, R. and T. Chopin. 2007. Effects of germanium dioxide, an inhibitor of diatom growth, on the microscopic laboratory cultivation stage of the kelp, *Laminaria saccharina*. *J. Appl. Phycol.* 19:27–32.

tom Dieck, I. and E.C. de Oliveira. 1993. The section Digitatae of the genus *Laminaria* (Phaeophyta) in the northern and southern Atlantic: Crossing experiments and temperature responses. *Mar. Biol.* 115:151–160.

Westermeier, R., D.J. Patiño, H. Müller, and D.G. Müller. 2010. Towards domestication of giant kelp (*Macrocystis pyrifera*) in Chile: Selection of haploid parent genotypes, outbreeding, and heterosis. *J. Appl. Phycol.* 22:357–361.

Zhang, Q.S., S.C. Qu, Y.Z. Cong, S.J. Luo, and X.X. Tang. 2008. High throughput culture and gametogenesis induction of *Laminaria japonica* gametophyte clones. *J. Appl. Phycol.* 20:205–211.

Zhao, X.B., S.J. Pang, F. Liu et al. 2016. Intraspecific crossing of *Saccharina japonica* using distantly related unialgal gametophytes benefits kelp farming by improving blade quality and productivity at Sanggou Bay, China. *J. Appl. Phycol.* 28:449–455.

chapter four

Cryopreservation of macroalgae

John G. Day

Contents

4.1 Introduction

Cryopreservation is most usually considered to be the storage of living cells at ultra-low cryogenic temperatures, most commonly in liquid, or vapor phase, liquid nitrogen (–196°C), and it has achieved a status of routine and confident application for many organisms (Fuller et al. 2004). However, in the vast majority of cases, purely plunging samples straight into liquid nitrogen will result in total disruption of cellular and intracellular architecture and function. Therefore, methodologies have been developed, which reduce or prevent cryoinjury; these can be categorized under two main approaches: (1) traditional controlled cooling-rate freezing and (2) vitrification (Figure 4.1).

In the first approach, cells are cryoprotected using colligative, penetrating cryoprotectants, often described as cryoprotective additives, which lower the freezing point and ameliorate the damaging effects of excessive solute concentration (Mazur, 2004). These are commonly used in association with osmotic additives, which are sometimes applied in a growth phase, before cryopreservation, in which the freezable water content is reduced (Withers, 1975). Carefully regulated slow cooling is

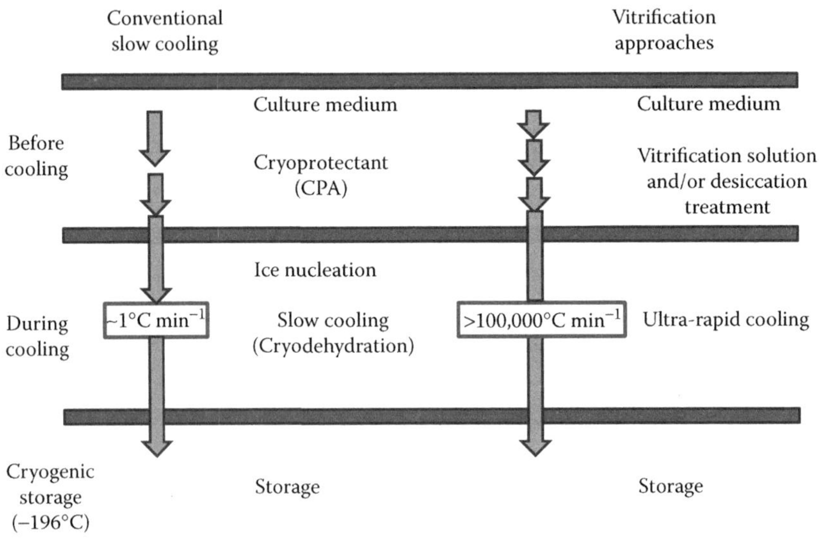

Figure 4.1 Schematic outlining the principal stages in conventional colligative cryopreservation and vitrification.

achieved using a computer-programmed freezer, or alternatively a solvent containing chiller/passive freezer, designed to have an average cooling rate of approximately –1°C min^{-1}, may be employed. In traditional cryogenic protocols, the cells are dehydrated as a result of a nonequilibrium vapor pressure gradient being formed when extracellular ice nucleation has occurred. This gradient results in the movement of unfrozen water from the inside of the cell to the extracellular space, thus reducing the amount of freezable water within individual cells. The cells are successfully cryopreserved when a critical balance between ice nucleation and dehydration is achieved, such that when plunged into liquid nitrogen the ice crystals that are spontaneously formed are too small to cause damage, or alternatively they form an amorphous glass. Where this is not achieved, intracellular ice may be nucleated, and ice crystal growth will almost inevitably occur. The resultant chemical and physical stresses within the alga will result in severe damage to cellular architecture and ultrastructure (Figure 4.2). Successful traditional controlled rate freezing requires the careful application/choice of colligative cryoprotectants that include methanol, dimethyl sulfoxide (DMSO), and glycerol (Day and Brand, 2005) and commonly involves an empirical experimental design approach, sequentially optimizing the various factors involved (Withers, 1975; Morris, 1976; Mazur, 2004). Success or failure to retain viability is affected by the rate of cooling, the temperature of ice nucleation, the terminal transfer temperature before exposure to liquid nitrogen, and the holding times at these respective steps.

(a) (b)

Figure 4.2 Cryophotomicrograph of the coenocytic alga *Vaucheria sessilis* pretreated with 5% (w/v) DMSO cooled at –1°C min^{-1} to –40°C. (a) Image taken using phase contrast microscopy. From the top: sample at room temperature; sample cooled to –20°C dehydrated and embedded in extracellular ice; sample during rewarming at –6°C with mechanical damage to the tip of the bud; thawed sample with ruptured membrane and obvious mechanical damage to the tip of the bud. (b) Image taken using differential interference contrast (DIC) microscopy/Nomarski microscopy. Dendritic ice crystals touching the algal thallus and inducing ice nucleation within the alga, as evidenced by the presence of *flashing*, that is, the formation of refractive ice crystals, which appear black under DIC.

The second approach, that is vitrification, involves the formation of a stable *glass* (i.e., the solidification of a liquid in the absence of crystallization). Complete vitrification (i.e., that of the entire system including both the extra- and intracellular components of the sample) may be achieved through increasing the samples' viscosity by evaporative desiccation (e.g., in sterile air streams or over silica gel) and/or employing osmotic dehydration in some instances in conjunction with the use of penetrating cryoprotectants (Harding et al. 2004). For both higher plants and some algae, vitrification may involve the application of multiple-component chemical protective additives (Sakai, 2004), such as plant vitrification solution 2 (PVS2) with osmotic and/or evaporative dehydration and desiccation (Fabre and Dereuddre, 1990). These procedures may include alginate encapsulation and combinations of encapsulation–vitrification (Sakai, 2004). Vitrification has facilitated the cryopreservation of previously cryogenic-storage recalcitrant germplasm (Engelmann and Takagi 2000; Benson 2004) and has been successfully applied to both eukaryotic algae and cyanobacteria (Harding et al. 2004). The process of vitrification is critically dependent on the formation of a stable glass on both cooling and rewarming, and it is important to optimize the cryoprotective dehydration step within the natural desiccation tolerance range of the germplasm to be cryopreserved, which may be a significant issue for aquatic organisms such as algae (Harding et al. 2010). A fundamental understanding of the physical aspects of cryobiology may assist in developing improved cryoprotection strategies, and the use of analytical approaches such as differential scanning calorimetry may enable the precise profiling of water-phase behavior in tissues (Dumet et al. 2000; Harding et al. 2004).

4.2 State of the art

Over the past 40 years, cryopreservation has been employed by algal Biological Resource Centers (BRCs) to conserve their holdings of microalgae (Morris 1976; Watanabe et al. 1992; Bodas et al. 1995; Wood et al. 2008), although today well in excess of the >3000 strains conserved in the COBRA project (Day et al. 2005) are held by collections worldwide in a cryopreserved state, many taxa remain recalcitrant to conventional cryopreservation methodologies. However, to date, relatively very few macroalgae have been held by collections in this format (Heesch et al. 2012), although there is considerable potential to do so. This section provides an overview of the current status of macroalgal cryopreservation, and Table 4.1 summarizes exemplar methodologies that have been successfully applied. It is clear from the relatively limited literature that there is, in some cases, a lack of understanding of the basic principles of cryopreservation, and to some extent, this may explain some of the problems that have been observed. For example, where materials are held at high subzero temperatures, samples will not remain in a vitreous state, so over time, ice crystal formation and growth will undoubtedly result in damage at a cellular level. This will inevitably reduce the capacity of any sample to revive and regenerate a normal thallus.

Table 4.1 Reports of successful cryopreservation of macroalgae with storage at −196°C

Alga	Procedure	Storage duration	Reported viability[a]	Reference
Undaria pinnatifida (gametophytes)	Conventional two-step cooling + cryoprotective additives (CPA)	9 days	Survival	Arbault et al. 1990
Porphyra yezoensis	Conventional two-step cooling + CPA	Up to 300 days	60%	Kuwano et al. 1993
Porphyra linearis	Conventional two-step cooling + CPA	20 min at −196°C	up to 70%	Arbault and Delanoue, 1994
Porphyra miniata	Conventional two-step cooling + CPA	Long-term storage (>10 years)	Viable	Day, 1998
Vegetative thalli (apical tips) of: *Gracilaria corticata* *Ulva lobata Hypnea musiformis*	Conventional two-step cooling + CPA		Viable, regeneration of thalli	Lalrinsanga et al. 2009
Gametophytic thalli of *Ulva prolifera*	Conventional two-step cooling + CPA	120 days	>90%	Lee and Nam, 2016
Laminaria japonica spores	Conventional two-step cooling + CPA	24 h	50%	Zhang et al. 2007
Ectocarpus thalli and spores	Conventional two-step cooling + CPA	1 month	5%–50%	Heesch et al. 2012
Female gametophytic cells of: *Laminaria japonica*	Conventional two-step cooling + wide range of CPA	>12 h	5%–53%	Kuwano et al. 2004

(*Continued*)

Table 4.1 (Continued) Reports of successful cryopreservation of macroalgae with storage at −196°C

Alga	Procedure	Storage duration	Reported viability[a]	Reference
L. longissima			37%–60%	
Kjellmaniella crassifolia			8%–67%	
Ecklonia stolonifera			28%–60%	
E. kurome			1%–42%	
Undaria pinnatifida			1%–60%	
Gametophytic cells of *Eisenia bicyclis*	Conventional two-step cooling + CPA	200 days	27% Male 31% female	Kono et al. 1998
Gametophytes of *Undaria pinnatifida*	Conventional two-step cooling + CPA			Renard et al. 1992
Sporelings and apical segments of mature thalli of the marine red alga *Gracilaria tikvahiae*	Conventional two-step cooling + CPA	>1 h	0%–93%	van der Meer and Simpson, 1984
Gracilaria foliifera			Thallus 36% Sporelings 43%	
Devaleraea ramentaceae			Thallus 100% Sporelings 100%	
Palmaria palmata			Thallus 100% Sporelings 100%	
Chondrus crispus			Thallus 70% Sporelings 86%	
Ulva lactuca			Thallus 100%	

(*Continued*)

Table 4.1 (Continued) Reports of successful cryopreservation of macroalgae with storage at –196°C

Alga	Procedure	Storage duration	Reported viability[a]	Reference
Enteromorpha intestinalis thallus	Conventional two-step cooling + CPA	>1 week	100%	Fleck, 1998
Gametophytic thalli of *Porphyra yezoensis*	One-step plunge in cryostraw + CPA		>80%	Choi et al. 2013
Protoplasts of *Porphyra yezoensis*	Conventional vitrification with a range of vitrification solutions	2 days	67%	Liu et al. 2004
Undaria pinnatifida gametophytes	Encapsulation dehydration with PVS2, vitrification		31% Male 26% female	Wang et al. 2011
Gametophytes of *Laminaria digitata*	Encapsulation dehydration in sucrose followed by slow cooling then plunge		25%–75%	Vigneron et al. 1997

[a] In many cases, these data are on the basis of positive assessment using a vital and/or a mortal stain to differentiate between live and dead cells rather than the capacity of individual spores, cells, or thalli to regrow. The available data are from samples assessed at least 24 h after thawing/rewarming, as materials immediately post-thaw will inevitably provide an overestimate of viability levels.

The author has observed this phenomenon for a range of eukaryotic algal taxa stored at temperatures of –80°C and higher (unpublished data). This has also been reported by Taylor and Fletcher (1999) who stored *Enteromorpha intestinalis* spores for up to five weeks at –20°C. There are a number of similar reports in the literature summarized by Taylor and Fletcher (1998), but the majority of these involved short-term storage, usually in the order of days, and it is, from the authors experience, unlikely that samples stored at high subzero temperatures would remain viable for longer than a few months. These reports and more recent papers, including Zhuang et al. (2015), who successfully stored filaments of the brown alga, *Scytosiphon lomentaria*, for a month at –20°C, have been omitted from Table 4.1.

4.3 Materials

4.3.1 Materials required for both methods

- Incubator/growth cabinet, a class I biological safety cabinet/laminar flow cabinet
- Appropriate cryosafety personal protective equipment, including goggles/face shield/visor, gloves (rated for liquid nitrogen emersion), and long cryo apron
- Disposable pipettes/disposable plastic Pasteur pipettes (Pastettes™)
- 2 mL cryogenic tubes
- A small (~1 L) dewar
- A heated water bath
- Storage cryostat/refrigerator with an appropriate inventory system
- Long forceps
- Culture medium: Modified Provasoli, f/2, or another appropriate medium
- 70% (v/v) ethanol
- Liquid nitrogen

4.3.2 Materials for conventional colligative cryopreservation

- A programmable, controlled rate cooler, or alternatively a passive cooling unit and a –80°C freezer (*Note 1*)
- *Cryoprotectant solutions*: 10% (v/v) DMSO and 9% (v/v) D-sorbitol in an appropriate culture medium
- Plasticware: membrane filters (0.5 μm pore size)
- Petri dishes (5 cm diameter)
- Universal tubes (20 mL)

4.3.3 Materials required for a vitrification-based approach

- Alginate (5% w/v) encapsulation solution is made from Sigma low-viscosity (3%) sodium salt. The alginate solution is prepared in preheated (60°C) culture medium in a 250 mL Schott bottle; 5 g of sodium alginate is directly added to the bottle containing deionized water (which should contain a magnetic stirring rod). The bottle contents are vigorously shaken (not stirred) until the alginate is wet (it will not completely dissolve) and thoroughly dispersed in the solution. Autoclave the alginate 121°C for 15 min. Remove the bottle between 80°C and 100°C and place it on a magnetic stirrer until the alginate dissolves (this can be overnight).
- Calcium chloride (100 mM) solution is prepared in deionized water.
- Sucrose (0.5 and 0.75 M) dehydration medium is made in the appropriate culture medium.
- Sterile 9 cm filter papers.
- Sterile 9 cm Petri dishes.
- Multiwell recovery plates (Biddy Sterilin Ltd., 100 mm square Petri dish, or equivalent).
- Sterile forceps.
- Heated magnetic stirrer.
- Oven for determining dry weights (set at 105°C).
- Thermometer and humidity meter (optional).

4.4 Experimental procedures

4.4.1 Method for conventional colligative cryopreservation

The method outlined below is based on Heesch et al. (2012) and has been successfully employed at the Culture Collection of Algae and Protozoa (CCAP) for a number of different macroalgal taxa. As for all cryopreservation procedures, it is normally optimal to employ a healthy, vigorous, relatively dense, late log phase or early stationary phase, and cultures. Furthermore, if available axenic strains, or those with low levels of bacterial contaminants should be used and, those contaminated with other eukaryotes should be avoided.

1. One to two weeks before cryopreservation separate thalli into smaller pieces (approximately 1–2 mm in length). Then transfer to Erlenmeyer flasks containing 100 mL of culture medium under the same culture conditions (*Note* 2).
2. Prepare 10 mL aliquots of cryoprotectant solution (10% [v/v] DMSO and 9% [v/v] D-sorbitol in the appropriate culture medium) and filter-sterilize into a sterile universal tube.

3. Aseptically transfer 1 mL aliquots of sterilized cryoprotectant solution into 2 mL cryovials.
4. Aseptically transfer intact sections of thalli to each cryovial, seal the cryovials, and incubate for 15–30 min at room temperature (*Note 3*).
5. Program the controlled cooler: Start temperature 20°C; ramp, cool at −1°C min^{-1} to −40°C; dwell, hold at −40°C for 10 min (*Notes 1 and 4*).
6. Start the program to purge the system with nitrogen vapor and to allow the system to stabilize at the start temperature.
7. On reaching the start temperature (most systems have an audible alarm), transfer the cryovials to the cooling chamber of the programmable cooler and initiate the cooling ramp.
8. After the end of the program, an alarm will sound; rapidly transfer the cryovials to a small dewar containing liquid nitrogen using long forceps (*Note 5*).
9. Samples for storage should be transferred to the cryostat (ultra-cold freezer) in the liquid nitrogen containing dewar. Transfer of cryovials to the storage system should be performed rapidly using long forceps (*Note 6*).
10. To recover cultures, vials are thawed by placing in a preheated water bath (40°C) and agitated until the last ice crystal has just melted (*Note 7*).
11. On thawing, rapidly transfer to a laminar flow/biological safety cabinet and wipe the outside of the vial with 70% (v/v) ethanol (*Note 8*).
12. By using a disposable plastic pipette (Pastette), the cryoprotectant solution is aseptically removed, discarded, and replaced by 1 mL of fresh sterile medium. The thallus is then aseptically transferred to a labeled Petri dish containing 10 mL of an appropriate medium. Then cover in aluminum foil and relabel with the strain designation and date (*Note 9*).
13. Incubate at the standard culturing temperature for the cryopreserved organism; after 24 h, partially remove the foil and after a further 24–96 h, remove all the foil covering (*Note 10*).
14. After one week, by using standard aseptic techniques, replace the medium (see *Note 11*).
15. After an appropriate period (2–8 weeks, depending of the strain), a normal culture should be obtained. This may be maintained by routine serial transfer, or employed for experimental use.

4.4.2 Method for a vitrification-based approach

The below-mentioned method outlined is based on Harding et al. (2008) and has been successfully employed for a large number of different microalgal taxa (Elster et al. 2008; Harding et al. 2008; Lukešová et al. 2008), as well as for *Laminaria* gametophytes at the CCAP. In addition, the principles involved and the methodology are very similar to the method of Vigneron et al. (1997) who successfully cryopreserved gametophytes of *L. digitata* using this approach.

1. One to two weeks before cryopreservation, separate thalli into smaller pieces (approximately 1 mm in length). Then transfer to Erlenmeyer flasks containing 100 mL of culture medium, under the same culture conditions (see *Note* 2).
2. Remove the supernatant and add ~10 mL of 5% (w/v) alginate (see *Note* 12).
3. Gently swirl the mixture of algae and alginate to evenly distribute the sections of thalli, but ensure that air bubbles are not formed.
4. By using a disposable sterile plastic Pastette and holding it in a vertical position, slowly dispense the alginate/algal solution dropwise into a 150 mL of 100 mM $CaCl_2$ and allow the beads to equilibrate for 60 min.
5. Carefully, decant off the calcium solution and transfer the beads to 0.5 M sucrose in culture medium to osmotically dehydrate the cells for 24 h (under standard light and temperature conditions).
6. Decant this medium and replace with 0.75 M sucrose in culture medium for another 24 h (under standard light and temperature conditions).
7. Remove the beads from the preculture sucrose solution and remove excess liquid sucrose medium by blotting the surface of the beads on sterile filter (7–8 cm) papers placed on Petri dishes.
8. Transfer the beads to sterile Petri dishes (9 cm) ensuring that they are not touching each other and that they are evenly distributed throughout the dish.
9. Place the open Petri dishes with the beads (closest to the air source) in a horizontal flow laminar airflow cabinet. Note and monitor the temperature (~25°C optimum) and laboratory relative humidity.
10. Air-dry the beads in the airflow (~1 m s^{-1}) for 3 to 4 h. This normally results in beads with a residual moisture content of ~25%–30% (w/w) (*Note 13*).
11. Transfer the desiccated beads into cryovials (10 beads per cryovials) and plunge directly into liquid nitrogen in a 1 L dewar.
12. Samples for storage should be transferred to the cryostat (ultra-cold freezer) in the liquid nitrogen containing dewar. Transfer of cryovials to the storage system should be performed rapidly using long forceps (*Note* 6).
13. Remove samples from the storage dewar; if necessary, transport them in a small dewar containing liquid nitrogen to the lab. Then rapidly remove the cryovials and place them in a heated water bath

(+40°C) to rewarm for 2–3 min (do not completely cover the cryovials in water). Carefully, wipe the cap and tube with a tissue containing 70% ethanol to ensure sterile conditions.

14. Transfer the (now very sticky) beads from the cryovial (with a small spoon spatula) to 5 mL of culture medium for 1 h (for bead pre-swelling and rehydration), and then transfer each bead (10 beads per cryovial) into one square of the (5 × 5 squares) Petri dish (Biddy Sterilin, or equivalent), with each square containing 1 mL of fresh medium. Alternatively, place beads in a small Petri dish containing fresh sterile medium. Then cover in aluminum foil and relabel with strain designation and date (*Note 9*).
15. Incubate under the standard culturing temperature for the cryo-preserved organism; after 24 h, partially remove the foil, and after another 24–96 h, remove all the foil coverings (*Notes 10 and 14*).
16. After one week, using standard aseptic techniques, replace the medium (*Note 11*).
17. After an appropriate period (2–8 weeks, depending of the strain), a normal culture should be obtained. This may be maintained by routine serial transfer, or employed for experimental use.

4.5 Notes

Note 1: Alternatively, for many materials, a passive freezer unit such as Mr Frosty™ may be used. These units have usually been designed to have a nonlinear cooling rate of ~1°C min^{-1} when placed in a –80°C freezer. Follow the manufacturer's instructions provided as these may vary between units.

Note 2: Exchange culture medium once a week prior to cryopreservation, to maximize culture quality and minimize the bacterial numbers in nonaxenic strains. This is to prevent overgrowth of the alga by bacteria stimulated by nutrient released on cutting the algal thalli. All algae interact both positively and negatively with their associated bacterial flora. For some taxa, if the associated flora are removed, altered, or diminished, growth and phenotypic abnormalities may be observed (Wichard and Oertel, 2010; Spoerner et al. 2012).

Note 3: At the CCAP, 15 vials are normally filled and processed (see *Note 6*).

Note 4: In general, mechanical seeding of ice has not been used for algal cryopreservation, but if it is an option in the system employed, empirical experimentation may be used to ascertain if it results in higher post-thaw viability.

Note 5: It is essential that the cryovials are not allowed to warm up before plunging into liquid nitrogen.

Note 6: Storage temperature is critical and should be <–130°C. It is advisable that replicate samples are stored in at least two separate *refrigerators*. At CCAP, 10 cryovials are normally stored in a working bank, 2 vials are stored in a master/backup bank, and 3 are used to check viability/efficacy of the protocol.

Note 7: For most marine taxa, it is important not to prolong their incubation at 40°C. Alternatively, slower warming, for example, in a 25°C water bath may be beneficial for some strains, but in general rapid warming is optimal as it avoids/minimizes ice crystal regrowth.

Note 8: There may be high levels of viable bacterial and fungal spores in liquid nitrogen that may contaminate recovered cultures.

Note 9: It is optimal to prevent further biochemical-based injuries; to do this as standard practice is to incubate revived samples for a period in the dark.

Note 10: It is important to ensure that cells are not subjected to light levels likely to induce photo-oxidative stress during the recovery phase.

Note 11: Because of the lysis of cells killed by the cryopreservation procedure, it is necessary to exchange culture medium after a week to reduce the bacterial numbers in nonaxenic strains. For more heavily bacterized cultures, this should be done more frequently to maximize the possibility of recovering a healthy culture.

Note 12: For freshwater taxa, one makes up the alginate in the appropriate medium; however, for marine strains, the high levels of Na^+ in the medium prevent gelation, and as a result, the bead formation and strength is poor, so alginate is prepared in deionized water.

Note 13: The level of dehydration is critical to the formation of a stable *glass*; if residual moisture levels significantly exceed 30%, the material will not successfully vitrify.

Note 14: On incubation, the high levels of Na^+ in marine media will result in depolymerization of the calcium alginate, and the beads will soften and eventually disintegrate releasing the encapsulated alga.

Acknowledgment

The author acknowledges National Capability funding from the UK Natural Environment Research Council for the CCAP. In addition, he would like to thank Roland Fleck (King's College London) for providing the images used in Figure 4.2 and Rachel Saxon (SAMS) for technical assistance in creating the figure.

References

Arbault, S. and Delanoue, G. 1994. Cryopreservation of *Porphyra linearis* Conchocelis phase. *Cryptogamie Algologie* 15:65–72.

Arbault, S., Renard, P., Perez, R., and Kaas, R. 1990. Cryopreservation trials on gametophytes of the food alga *Undaria pinnatifida* (Laminariales). *Aquatic Living Resources* 3:207–215.

Benson, E. E. 2004. Cryoconserving algal and plant diversity: Historical perspectives and future challenges. In *Life in the Frozen State,* B. Fuller, N. Lane, and E. E. Benson (Eds.), pp. 299–328. Boca Raton, FL: CRC Press.

Bodas, K., Brennig, C., Diller, K. R., and Brand, J. J. 1995. Cryopreservation of blue-green and eukaryotic algae in the culture collection at the University of Texas at Austin. *CryoLetters* 16:267–274.

Choi, Y. H., Nam, T. J., and Kuwano, K. J. 2013. Cryopreservation of gametophytic thalli of *Porphyra yezoensis* (Rhodophyceae) by one-step fast cooling. *Journal of Applied Phycology* 25:531–546.

Day, J. G. 1998. Cryo-conservation of microalgae and cyanobacteria. *CryoLetters* Supplement 1:7–14.

Day, J. G., Benson, E. E., Harding, K., Knowles, B., Idowu, M., Bremner, D., Santos, F. et al. 2005. Cryopreservation and conservation of microalgae: The development of a pan-european scientific and biotechnological resource (the COBRA project). *CryoLetters* 26:231–238.

Day, J. G. and Brand, J. J. 2005. Cryopreservation methods for maintaining cultures. In *Algal Culturing Techniques,* R. A. Andersen (Ed.), pp. 165–187. New York: Academic Press.

Dumet, D. J., Block, W., Worland, R., Reed, B. M., and Benson, E. E. 2000. Profiling cryopreservation protocols for *Ribes ciliatum* using differential scanning calorimetry. *CryoLetters* 21:367–378.

Elster, J., Lukavský, J., Harding, K., Benson, E. E., and Day, J. G. 2008. Deployment of the encapsulation/dehydration protocol to cryopreserve polar microalgae held at the Czech Republic academy of sciences institute of botany. *CryoLetters* 29:27–28.

Engelmann, F. and Takagi, T. 2000. *Cryopreservation of Tropical Plant Germplasm; Current Research Progress and Application.* JIRCAS International Agriculture Series No. 8. Rome, Italy: IPGRI.

Fabre, J. and Dereuddre, J. 1990. Encapsulation-dehydration: A new approach to cryopreservation of potato shoot-tips. *CryoLetters* 11:413–426.

Fleck, R. A. 1998. *The Assessment of Cell Damage and Recovery in Cryopreserved Freshwater* Protists. PhD thesis. Dundee, Scotland: University of the Abertay Dundee.

Fuller, B., Lane, N., and Benson, E. E. 2004. *Life in the Frozen State.* Boca Raton, FL: CRC Press.

Harding, K., Day, J. G., Lorenz, M., Timmerman, H., Friedl, T., Bremner, D. H., and Benson, E. E. 2004. Introducing the concept and application of vitrification for the cryo-conservation of algae- a mini-review. *Nova Hedwigia* 79:207–226.

Harding, K., Müller, J., Lorenz, M., Timmerman, H., Friedl, T., Day, J. G., and Benson, E. E. 2008. Deployment of the encapsulation/dehydration protocol to cryopreserve microalgae held at SammLung von Algenkulturen, Universität Göttingen Germany. *CryoLetters* 29:15–20.

Harding, K., Müller, J., Timmermann, H., Lorenz, M., Day, J. G., and Friedl, T. 2010. The use of amplified fragment length polymorphism (AFLP) markers to verify the identity and genetic stability of euglenoids cryopreserved using encapsulation dehydration cryoprotective strategies. *CryoLetters* 31:460–472.

Heesch, S., Day, J. G., Yamagishi, T., Kawai, H., and Küpper, F. C. 2012. Cryopreservation of the model alga *Ectocarpus* (Phaeophyceae). *CryoLetters* 33:327–336.

Kono, S., Kuwano, K., and Saga, N. 1998. Cryopreservation of *Eisenia bicyclis* (Laminariales, Phaeophyta) in liquid nitrogen. *Journal of Marine Biotechnology* 6:220–223.

Kuwano, K., Argua, Y., and Saga, N. 1993. Cryopreservation of the conchocelis of the marine alga *Porphyra yezoensis* Ueda (Rhodophyta) in liquid nitrogen. *Plant Science* 94:215–225.

Kuwano, K., Kono, S., Jo, Y. H., Shin, J. A., and Saga, N. 2004. Cryopreservation of the gametophytic cells of Laminariales (Phaeophyta) in liquid nitrogen. *Journal of Phycology* 40:606–610.

Lalrinsanga, P. L., Deshmukhe, G., Chakraborty, S. K., Dwivedi, A., Barman, N., and Kumar, N. 2009. Preliminary studies on cryopreservation of vegetative thalli of some economically important seaweeds of India. *Journal of Applied Aquaculture* 21:250–262.

Lee, Y. N. and Nam, K. W. 2016. Cryopreservation of gametophytic thalli of *Ulva prolifera* (Ulvales, Chlorophyta) from Korea. *Journal of Applied Phycology* 28:1207–1213.

Liu, H. Q., Yu, W. G., Dai, J. X., Gong, Q. H., Yang, K. F., and Lu, X. Z. 2004. Cryopreservation of protoplasts of the alga *Porphyra yezoensis* by vitrification. *Plant Science* 166:97–102.

Lukešová, A., Hrouzek, P., Harding, K., Benson, E. E., and Day, J. G. 2008. Deployment of the encapsulation/dehydration protocol to cryopreserve diverse microalgae held at the Institute of Soil Biology, Academy of Sciences of the Czech Republic. *CryoLetters* 29:21–26.

Mazur, P. 2004. Principles of cryobiology. In *Life in the Frozen State*, B. Fuller, N. Lane, and E. E. Benson (Eds.), pp. 3–66. Boca Raton, FL: CRC Press.

Morris, G. J. 1976. The Cryopreservation of *Chlorella* 1. Interactions of cooling, protective additive and warming rate. *Archives of Microbiology* 107:57–62.

Renard, P., Arbault, S., Kaas., R., and Perez, R. 1992. A method for the cryopreservation of the gametophytes of the food alga *Undaria pinnatifida* (Laminariales). *Comptes Rendus De I Academie des Science Serie III Life Science* 315:445–451.

Sakai, A. 2004. Plant cryopreservation. In *Life in the Frozen State*, B. Fuller, N. Lane, and E. E. Benson (Eds.), pp. 329–346. Boca Raton, FL: CRC Press.

Spoerner, M., Wichard, T., Bachhuber, T., Stratmann, J., and Oertel, W. 2012. Growth and thallus morphogenesis of *Ulva mutabilis* (Chlorophyta) depends on a combination of two bacterial species excreting regulatory factors. *Journal of Phycology* 48:1433–1447.

Taylor, R. and Fletcher, R. L. 1998. Cryopreservation of eukaryotic algae—A review of methodologies. *Journal of Applied Phycology* 10:481–501.

Taylor, R. and Fletcher, R. L. 1999. A simple method for the freeze-preservation of zoospores of the green macroalga *Enteromorpha intestinalis*. *Journal of Applied Phycology* 11:257–262.

van der Meer, J. P. and Simpson F. J. 1984. Cryopreservation of *Gracilaria tikvahiae* (Rhodophyta) and other macrophytic marine algae. *Phycologia* 23:195–202.
Vigneron, T., Arbault, S., and Kaas, R. 1997. Cryopreservation of gametophytes of *Laminaria digitata* (L) lamouroux by encapsulation dehydration. *CryoLetters* 18:93–98.
Wang, B., Zhang, E., Gu, Y., Ning, S., Wang, Q., and Zhou, J. 2011. Cryopreservation of brown algae gametophytes of *Undaria pinnatifida* by encapsulation–vitrification. *Aquaculture* 317:89–93.
Watanabe, M. M., Shimizu, A., and Satake, K. 1992. NIES-microbial culture collection at the national institute of environmental studies: Cryopreservation and database of culture strains of microalgae. In *Proceedings of Symposium on Culture Collection of Algae,* M. M. Watanabe (Ed.), pp. 33–42. Tsukuba, Japan: NIES.
Wichard, T. and Oertel, W. 2010. Gametogenesis and gamete release of *Ulva mutabilis* and *Ulva lactuca* (Chlorophyta): Regulatory effects and chemical characterization of the "swarming inhibitor." *Journal of Phycology* 46:248–259.
Withers, L. A. 1975. Freeze-preservation of cultured cells and tissues. In *Frontiers of Plant Tissue Culture,* T. A. Thorpe (Ed.), pp. 297–306. Calgary, Alberta: International Association of Plant Tissue Cultures, University of Calgary.
Wood, S. A., Rhodes, L. L., Adams, S. L., Smith, J. F., Tervit, H. R., and Cary S. C. 2008. Maintenance of cyanotoxin production by cryopreserved cyanobacteria in the New Zealand culture collection. *New Zealand Journal of Marine and Freshwater Research* 42:277–283.
Zhang, Q. S., Cong, Y. Z., Qu, S. C., Luo, S. J., Li, X. J., and Tang, X. X. 2007. A simple and highly efficient method for the cryopreservation of *Laminaria japonica* (Phaeophyceae) germplasm. *European Journal of Phycology* 42:209–213.
Zhuang, Y., Gong, X., Zhang, W., and Gao, W. 2015. Cryopreservation of filaments of *Scytosiphon lomentaria* by vitrification. *Journal of Applied Phycology* 27:1337–1342.

chapter five

Unraveling seaweeds bacteriomes

From field site to computer screen

Tânia Aires, Gerard Muyzer, Ester A. Serrão, and Aschwin H. Engelen

Contents

5.1 Introduction

Evidence shows that marine macroalgae (*seaweeds*) rely on their associated microbiome for basic functions, such as growth, development, nutrient supply, protection, and, in some cases, adaptation and acclimation to environmental stresses (Singh and Reddy 2014; Wichard 2015; Dittami et al. 2016). The term *holobiont* was first coined for corals because of their dependency on associated microbes for survival (Rosenberg et al. 2007), but empirical evidence broadened this term to other organisms, such as seaweeds (Barott et al. 2011). It is now widely recognized that it is essential to consider the entire holobiont (i.e., the host and its associated microbiota) to understand how seaweeds adapt and evolve (Singh and Reddy 2014; Dittami et al. 2016).

The use of state-of-the-art molecular techniques as high throughput 16S rRNA amplicon sequencing and *omics* allow not only to reveal the importance of seaweed–bacteria interactions but also to explore the role of the different community players in seaweeds ecology and evolution.

Regardless of their location (surface biofilm, endophytically, or both), seaweed-associated bacteria have been shown to be primarily determined by host species (Lachnit et al. 2009; Eigemann et al. 2013; Aires et al. 2015; Aires et al. 2016; Vieira et al. 2016). A continental-scale study of the bacterial community associated with the dominant kelp *Ecklonia radiata* (Marzinelli et al. 2015) displayed variations along the latitudinal distribution. Host health conditions were shown to be determining in shaping this kelp's bacterial community, suggesting that stress can disrupt this stable bacteria/host association. Other biotic (e.g., disease and herbivory) and abiotic factors influence seaweed-associated bacterial communities. Local environment factors, such as pollution and/or contamination, can impact the associated bacterial guild: A shift to bacterial taxa related to hydrocarbon degradation and heavy metals detoxification was associated with the invasive species *Asparagopsis taxiformis* growing in polluted waters (Aires et al. 2016).

Seaweed-associated bacterial assemblages are also not temporally fixed. Studies on brown seaweeds (*Laminaria hyperborea* by Bengtsson et al. 2010 and *Fucus vesiculosus* by Stratil et al. 2013) have shown changes across seasons. Those changes can be related to, for example, seawater temperature, which may influence the *antifouling* capacity of the seaweed and result in some unwanted pathogenic associations (Stratil et al. 2013), or to the growth cycle of the seaweed in which the aging process leads to a mature and more diverse biofilm (Bengtsson et al. 2010, Mancuso et al. 2016). Hence, the seaweed-associated bacterial community is the result of several interacting factors whereby host phylogeny takes the lead, but environment and geography may change the way (Hollants et al. 2013).

The vital importance of seaweed-associated bacterial communities has also been demonstrated in experimental studies in which manipulation revealed their determining role in seaweed development and adaptation/acclimation (Matsuo et al. 2005; Dittami et al. 2016). The normal development and morphogenesis of most Ulvales is, for example, dependent on bacteria from the *Cytophaga–Flavobacterium–Bacteroides* group (Matsuo et al. 2005; Spoerner et al. 2012; Weiss et al. 2017), and the capacity of *Ectocarpus* to cope with transitions from salt to freshwater is lost in the absence of its associated bacterial community (Dittami et al. 2016). These (and several others) studies provide empirical evidence on how microbes can be essential to algae, impacting their development, physiological response, metabolism regulation, and adaptation to different environments. The importance of seaweed-associated bacteria probably goes beyond the immediate effects; their role on driving/facilitating speciation is now being considered in evolutionary studies, and this has also raised interest from the seaweed industry (Charrier et al. 2017). The way seaweed–bacterial associations are established and how microbial recruitment is initiated is still an ongoing discussion. The most accepted theory relies on a combination of stochastic processes and niche selection consistent with the competitive lottery model as proposed by Burke et al. (2011). In Burke's study, bacterial community taxonomic structure was different between samples and over time but functionally equivalent. Thus, it is proposed that a set of functional characteristics is required to allow colonization of a certain niche (in this case, the seaweed itself and its own niche) and that determines which bacteria are recruited from the available *guild*. But the specific phylogenetic shifts directly related to environmental stresses remain to be explained; experimental studies are needed to fully understand the recruitment process.

Besides their invaluable ecological role as habitat-structuring species, seaweeds are rich in biologically active compounds, which led to their exploitation in several economical areas, including their use as food resource (Gupta and Abu-Ghannam 2011), source of medicinal and pharmaceutical components (Smit 2004), and antifouling properties (Qian et al. 2010). Seaweed-associated microbiome studies can improve seaweed aquaculture by increasing productivity and fitness (Singh and Reddy 2016) and by preventing diseases, just as manipulation of bacteria associated with human gut and crops can improve human health and agricultural yield, respectively.

The growing need to understand these vital interactions and the role played by bacterial communities on seaweed evolution, adaptation, and development has promoted development of laboratory and informatics techniques. In the past decades, next-generation sequencing (NGS) and powerful bioinformatic tools allowed one to unveil the previously hidden world of seaweeds.

5.2 State of the art

The study of seaweed–bacteria associations and their main drivers goes back to the early 1970s when culturing and microscopy techniques already detected species, tissues, and seasonal-specific associations (Bolinches et al. 1988 and Cundell et al. 1977 in Egan et al. 2013). The obvious advantage of culturing is the possibility of sequencing the complete genome of bacteria of interest and to perform manipulative studies to provide information about their ecological roles and the understanding of their interactions (Allen-Vercoe 2013). However, standard culturing techniques applied to environmental samples generally only provide about 1% of the real bacterial diversity, and those numbers just go slightly up when specific conditions are applied to specific environments (Riesenfeld et al. 2004).

Yet, when allied to microscopy, these early techniques managed to reveal important seaweed/bacteria associations as the famous endosymbiotic bacteria found in *Caulerpa taxifolia*, allowing seaweed survival in oligotrophic environments. These endosymbionts collaborate with biofilm bacteria from the rhizoids, which take up carbon, nitrogen, and phosphorus from the substrata and translocate their products to the photoassimilatory organs (Chisholm et al. 1996). The development of culture-independent methods revealed an amazing diversity of the microbial world, and more than half of the recognized microbial phyla have no cultured representatives (Riesenfeld et al. 2004). The most popular culture-independent technique is PCR amplification, cloning, and sequencing of the 16S rRNA gene, a phylogenetic powerful marker that allowed recognition of the bacterial diversity as known today (Schäfer and Muyzer 2001). However, when studying seaweed-associated bacterial communities, as mentioned earlier, different factors should be considered. The inclusion of seasonal and environmental dynamics and spatial variation in studies of seaweed-associated microbial communities will result in an enormous number of samples to be analyzed. So, cloning and sequencing of 16S rRNA genes, a laborious and time-consuming method, are not the best choice to analyze such a large number of samples (Schäfer and Muyzer 2001). With these limitations, the genetic fingerprinting techniques emerged, and several seaweed-associated bacteria were characterized mainly through *denaturing gradient gel electrophoresis (DGGE)* and *terminal restriction fragment length polymorphism (T-RFLP)* (See e.g., Meusnier et al. 2001; Delbridge et al. 2004; Burke et al. 2009; Bengtsson et al. 2010; Nylund et al. 2010; Tujula et al. 2010; Hollants et al. 2011; Lachnit et al. 2011). Still, there are some drawbacks associated with fingerprinting techniques. Besides the higher number of replicates required for accuracy and reproducibility, a bias towards the most abundant taxa, which likely hides less abundant ones, is observed. Results/banding pattern interpretation can also be dubious and difficult.

With the emergent necessity of getting a more complete picture of the seaweed holobiont, state-of-the-art high-throughput techniques known as *omics* have been widely used in the latest seaweed holobiont studies. Along with powerful bioinformatics tools and data storage, massive 16S rRNA amplicon pyrosequencing, metagenomics, metatranscriptomics, metaproteomics, and metabolomics (see Figure 5.1 for a complete description) are widely applied. The isolated use and integration of these new techniques allow one to answer the questions "Who is there? What are they doing? How do they do it?" in a more reliable way, increasing our knowledge on the identity and functionality of seaweed-associated microbiomes.

In the current chapter, we describe the methodologies from field sampling to the comprehensive *in silico* analysis of seaweed-associated

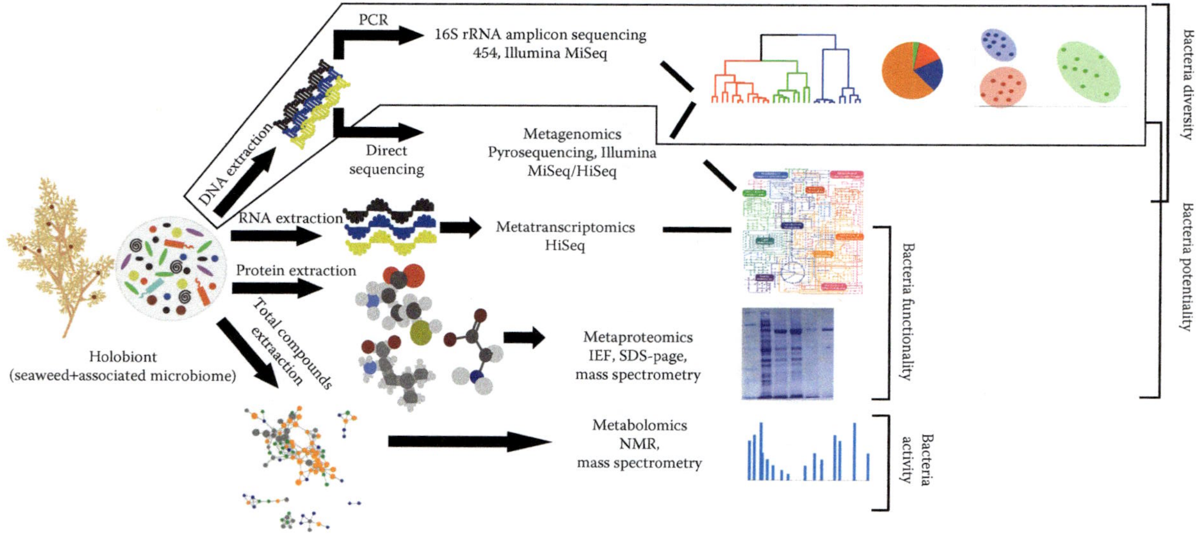

Figure 5.1 Schematic diagram representing the different metaomics techniques and respective outputs, used to uncover seaweed microbiome identity/diversity, potentiality, functionality, and activity. 16S amplicon sequencing, metagenomics, and metatranscriptomics analyses can be achieved using 454 Roche GS FLX+ and Illumina (Mi-Seq or Hi-Seq) sequencing methods. Subsequently, software programs such as QIIME and Mothur can be used for bacterial community diversity studies, whereas MG-RAST, IMG/M, METAREP, CAMERA, and MEGAN4 can be used for metagenomes' functional annotation. Metaproteomic data can be analyzed by isoelectric focusing (IEF), sodium dodecyl sulfate polyacrylamide gel electrophoresis (SDS–PAGE), matrix-assisted laser desorption/ionization time-of-flight mass spectrometry (MALDI–TOF–MS), and liquid chromatography–mass spectrometry (LC–MS). Bacterial activity (metabolome) can be assessed by nuclear magnetic resonance spectroscopy and diverse mass spectrometry methods. Here we describe the sequencing of the 16S amplicon used to determine the diversity of the microbial communities and the identity of their community members. Image and legend are adapted from Singh and Reddy (2016).

bacterial communities. Our focus is on bacterial identity, diversity, and predicted functional profiles based on 16S rRNA amplicon sequencing (although the same methodology can be used for metagenomics up to PCR amplification). This can be considered as a starting point for further functional studies using more powerful approaches, such as metagenomics and metatranscriptomics.

5.3 Materials

- Latex or nitrile gloves
- Scalpel and blades or scissors
- 96% (v/v) ethanol
- Paper tissue
- Permanent marker
- 2 mL tubes (or cryotubes in case you use liquid nitrogen)
- Xpedition™ Lysis/Stabilization Solution (Zymo Research) or liquid nitrogen
- Storage box for 2 mL tubes
- DNA decontaminating solution (e.g., 10% [v/v] bleach)
- Nitrile gloves
- Micropipettes (different volumes) and their respective pipette tips with filters
- 96% (v/v) ethanol
- Beaker
- Tweezers
- Scalpel and blades
- Bunsen burner
- Bead-beater
- Ice bath
- Microcentrifuge
- Vortex
- UV light (optional, but recommended)
- Water bath or heating block (*Note 1*)

5.4 Experimental procedures

5.4.1 Sampling

Sampling seaweeds for associated bacterial community studies implies some more aseptic measures than regular sampling. Sometimes, field sampling is carried out in rough conditions, although simple steps should be followed to avoid undesirable further contamination of any source.

1. Always use gloves to handle samples.
2. Samples should be thoroughly cleaned from attached sediment particles and epiphytes. Use the surrounding seawater to help in this process.
3. As a rule of thumb, use a minimum of five replicates (within the same location/treatment) that should represent the variation present (so, in the field, no neighboring individuals should be sampled).
4. Associated bacterial communities (*Note 2*) tend to be tissue specific. When sampling seaweeds featuring differentiated structures (e.g., holdfast, stipe and blades, and reproductive structures when present), these should be separated without representing different replicates of the same individual (for each sample/replicate use distinct or ethanol-disinfected scissors or scalpel). A small piece of the tissue fitting in a 2 mL tube is enough.
5. Sampling preservation should be performed by separating the above-mentioned structures into different tubes. Make sure that individual replicates are clearly labeled (using a permanent marker): for example, after tissue separation in three different tubes, replicate 1 from species X should be labeled *X1_tissue_structure* (*Notes 3 and 4*).
6. Samples should be preserved fresh in the field. There are two options: (1) emerge the tissue completely in a preservation buffer as the Xpedition™ Lysis/Stabilization Solution (Zymo Research) in individual tubes (regular 2 mL tubes with no more than 500 μL of buffer), or (2) flash freeze the individualized tissues in liquid nitrogen (use adequate cryotubes).
7. Moreover, environmental reference samples as seawater and sediment should be sampled (replicates of 0.5 L of seawater [depending on the bacterial abundance] and about 1 mL of sediment). Seawater should preferably be filtered directly in the field using 0.2 μm filters. Seawater filters and sediment should be stored using the same method as for the seaweed tissues. The current chapter does not include details on dealing with these samples.

5.4.2 DNA extraction

Several commercial kits can be used to perform the extraction of seaweed-associated bacterial DNA. In this protocol, we have chosen to describe the MO BIO's PowerSoil™ DNA Isolation Kit protocol, which is probably the most widely used for several other host organisms (e.g., the Seagrass Microbiome Project—https://seagrassmicrobiome.org). Other extraction kits, as the FastDNA™ SPIN Kit for soil from MP biomedicals, and the Quick-gDNA™ MiniPrep from Zymo Research also performed well to our experience (*Notes 5 and 6*). Changes to the users' manual are referred in *Notes 2, 5 and 6*.

Before starting the extraction, make sure that the workspace is clean by disinfecting all surfaces with bleach or other DNA decontaminating solution (consider placing pipettes, scalpel, and tweezers under UV light before the procedure). Prepare a beaker with ethanol in which your handling material (tweezers and scalpel) will be placed. Use gloves during the entire process.

Sample preparation should be carried out right before extraction, and transfer of samples into the MO BIO bead-beating tubes should be performed close to a flame. If samples stored in Xpedition™ Lysis/Stabilization solution are used, vortex the sampling tubes and transfer a small piece of tissue into the MO BIO bead-beating tube, followed by 100 μL of the stabilization solution. Samples stored in liquid nitrogen ideally should not be thawed before the tissue transference. Disinfect the scalpel and tweezer using ethanol and flame in between transfers to avoid contamination from one sample to another.

After sample preparation, start with step 3 of the MO BIO PowerSoil® DNA Isolation Kit's Experienced User Protocol (*Note 7*).

5.4.3 16S rRNA amplification

In contrast to other hosts, the biggest drawback when studying bacterial communities associated with photosynthetic organisms is the cyanobacterial origin of the chloroplast lineage (Whatley et al. 1979) (*Note 8*). The use of the conserved, and most widely used, 16S rRNA primers (Wang and Qian 2009; Apprill et al. 2015) will lead to the cross-amplification of plastid DNA interfering with the 16S rRNA characterization of bacteria and masking the real bacterial contribution. If one does not aim to study the seaweed bacterial biofilm (in which the swab sampling technique itself avoids chloroplast contamination), selective methods need to be used to eliminate (or turn to residual) the chloroplast interference. Most recent studies have been mainly using two different techniques to avoid PCR amplification of 16S rRNA genes from plastids: (1) the use of blocking primers based on peptide nucleic acid (PNA) PCR clamps, developed by Lundberg et al. (2013), which bind to chloroplast amplicons inhibiting their further amplification and (2) the use of primers that were specifically designed and modified to not bind to the chloroplast DNA. The first method is promising, although still seldom used, but can be time-consuming (each host species might require specific PNAs) and expensive. The use of modified primers is, for some, considered an added bias to the PCR. However, Hanshew et al. (2013) reviewed and improved a collection of *chloroplast avoiding* primer testing and attesting their efficiency and undetectable bias. Here, we suggest two sets of primers, which have been working efficiently in our lab, depending on the sequencing technology used

(454 pyrosequencing or Illumina MiSeq). For 454 technology, which allows fragment sequencing up to 600 bp, we have selected the less biased and best on chloroplast amplicons avoidance from Hanshew study (2013)—the primer pair 799Fmod3 (5′-CMGGATTAGATACCCKGG-3′) and 1392R (5′-ACGGGCGGTGTGTRC-3′) amplifying the V5–V8 region of the 16S. This primer pair has shown no bias toward any taxa and the chloroplast contamination was a low as 0%–0.88% and has been proven to work well for seaweed associated bacteria (Aires et al. 2016).

For Illumina MiSeq technology (paired end), in which fragments sequencing can nowadays go up to 2 × 250 bp (~500 bp) and the choice of primers is limited, our lab has tested the primer pair used by Bodenhausen et al. (2013)—799F (5′-AACMGGATTAGATACCCKG-3′) and 1193r (5′-ACGTCATCCCCACCTTCC-3′) that amplify a ~400 bp amplicon from the V5–V7 region of the 16S rRNA gene. These primers were also successfully tested in our lab for the amplification of bacteria from brown seaweeds (Vieira et al. 2016), and from marine angiosperms (unpublished data).

Despite the suggestions given, we strongly encourage a literature search related to the objectives and the seaweed species addressed in the study before deciding on the 16S rRNA amplification strategy (*Note 9*).

5.4.4 16S rRNA amplicon sequencing data analysis

As molecular techniques advance to deliver more data, the software programs also have to keep up with ways to analyze that ever increasingly large amount of information (*Note 10*). When it comes to 16S rRNA amplicon sequencing, there are two popular analysis programs: (1) Mothur (Schloss et al. 2009) and (2) QIIME (Caporaso et al. 2010). The later has been widely used in several studies, from human to thermal vents-associated microbiomes. QIIME's popularity is due to its user-friendly native Python coding, supplemented with very detailed and easy-to-understand tutorials and an extremely helpful and responsive discussion forum. Here, we are going to guide readers through the initial 16S rRNA data QIIME (version 1.8.0) processing and suggesting possible statistical programs to use afterwards.

First, computer must be suitable for QIIME installation along with all the dependencies (http://QIIME.org/install/install.html). Depending on the sequencing technology used (454 or Illumina) and the format of the sequences, the *quality filtering/demultiplexing* step will require different arguments. Before starting, the respective tutorials should be read, and data should be prepared accordingly—454 tutorial (http://QIIME.org/tutorials/tutorial.html) and Illumina tutorial (http://nbviewer.jupyter.org/github/biocore/QIIME/blob/1.9.1/examples/ipynb/illumina_overview_tutorial.ipynb).

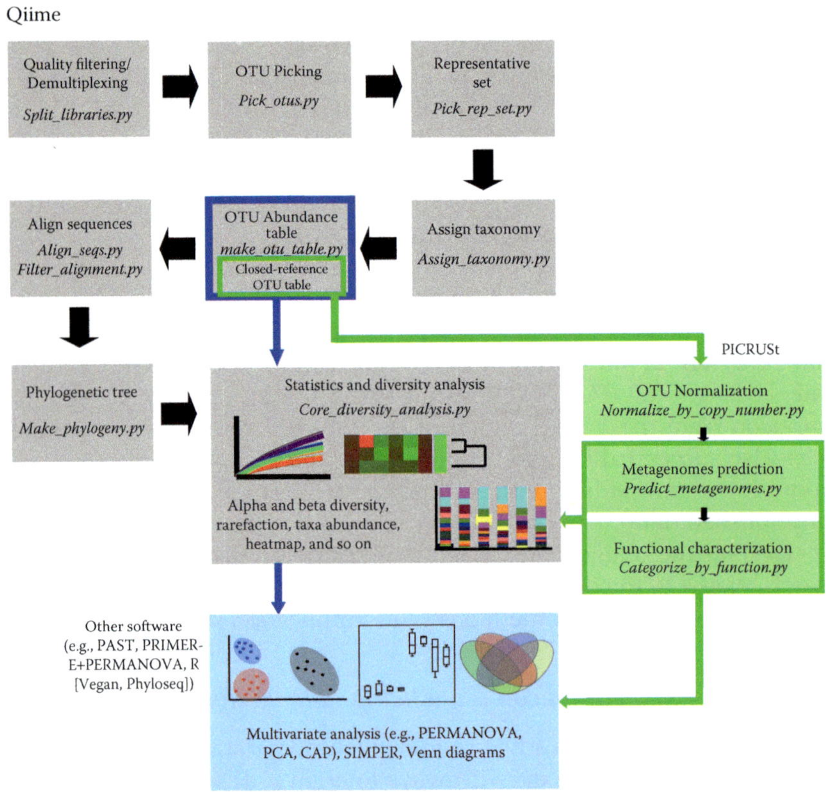

Figure 5.2 Schematic diagram representing the initial and main steps (along with the respective scripts) to analyze 16S amplicon sequencing data in QIIME (gray shaded area) and subsequent functional profiles prediction using PICRUSt (dark-gray shaded area). Downstream statistical and diversity analysis, and respective visualization using QIIME, are represented by gray-shaded areas. Other suggested software (with QIIME/PICRUSt generated OTU/KEGG pathways table) are represented by the white-shaded area.

Once data and metadata are prepared, the following steps can be followed (also illustrated in Figure 5.2):

1. *split_libraries.py*: This script assigns the reads according to samples and on the basis of each differentiated barcode (given on your metadata file). Quality filtering is also carried out in this step by removing any low quality or ambiguous reads.
2. *pick_otus.py*: This script should be adapted to the operational taxonomic unit (OTU) picking strategy of your choice (see http://QIIME.org/tutorials/otu_picking.html). This script clusters all the

sequences into OTUs on the basis of their similarity. In QIIME, OTUs are clusters of sequences representing taxonomic similarity.
3. *pick_rep_set.py*: The file created previously will be used in this script to extract a representative sequence from each OTU that will be used in downstream analysis.
4. *assign_taxonomy.py*: This script will assign each representative sequence, picked in step 3, to a certain taxon.
5. *make_otu_table.py*: In this step, the OTU map created in step 2 and the taxonomic assignments from step 4 will be assembled as a table of OTU abundances for each sample with taxonomic predictions for each OTU (*Note 11*).
6. *align_seqs.py*: The representative set of sequences created in step 3 can be aligned using this script through a method of your choice.
7. *filter_alignment.py*: This script will remove gaps created during alignment to further generate a useful phylogenetic tree relating the sequences.
8. *make_phylogeny.py*: This script uses the filtered alignment created in the previous step and builds a phylogenetic tree in which the relatedness of the OTUs representative sequences can be visualized (*Note 12*).
9. *core_diversity_analysis.py*: This script represents a workflow gathering several QIIME diversity analyses beginning with an OTU table, a metadata file, and an optional phylogenetic tree. We strongly advise the individual use of each script comprising this workflow script.

In every step, arguments and parameters can be changed to fit the data and methodologies of choice (for details, see individual QIIME scripts here: http://QIIME.org/scripts/index.html).

The analysis process described in this point is illustrated and summarized in Figure 5.2. As an example of a possible output from the process explained above, Figure 5.3 depicts part of the results' analysis of a study on bacterial community associated with two different red seaweed species (based on 16S rRNA amplicon sequencing).

5.4.5 *Functional prediction based on 16S rRNA gene sequencing*

Metagenomics and metatranscriptomics are still expensive and time-consuming techniques. Some tools have been developed to help giving the first insight into the bacterial community potential function. Phylogenetic Investigation of Communities by Reconstruction of Unobserved States (PICRUSt) is a software program that allows one to predict the functional composition of a metagenome using 16S rRNA gene data and a known database of reference genomes (Langille et al. 2013). Similar to QIIME, PICRUSt uses Python coding, and its installation process can be consulted in the following website: https://picrust.github.io/picrust/install.html#install.

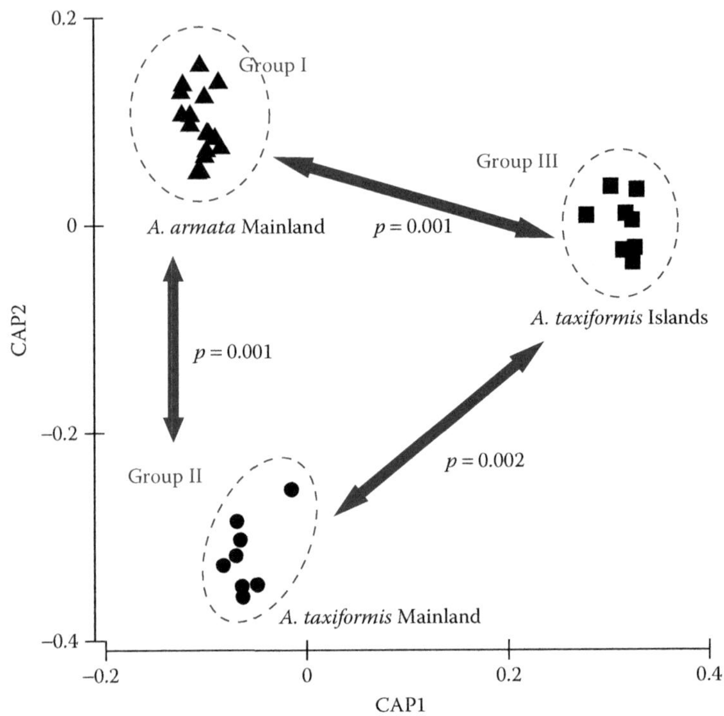

Figure 5.3 Example of a possible output of seaweed-associated bacterial community analysis using QIIME and PRIMER-E+PERMANOVA. In this example, the table with OTU abundances, created in QIIME (step 5 of point 2.4), was used for Canonical Analysis of Principle coordinates (CAP) (based on Bray–Curtis distances) in PRIMER-E+PERMANOVA. The figure shows the clustering of the different seaweed samples (two Asparagopsis species in two contrasting locations), on the basis of bacterial communities' composition. Associated bacteria cluster by both species and habitat, showing the influence of these two factors on shaping the communities. Statistical analyses (PERMANOVA) p values are depicted between the different groups (groups considered significantly different if $p < 0.05$). For more detail, see the complete study by Aires et al. (2016).

Once the necessary files are in the right format and the adequate database is used (*Note 13*), PICRUSt predicts the functional profile of the seaweed bacterial community. The main steps (scripts) to obtain a predicted metagenome table are the following:

1. *normalize_by_copy_number.py*: This script uses the OTU table created in QIIME and normalizes it by dividing each OTU by the known/predicted 16S copy number abundance.
2. *predict_metagenomes.py*: This script uses the normalized table created in the previous step to create the final metagenome functional

predictions. The output is a table with the Kyoto Encyclopedia of Genes and Genomes (KEGG) Orthologous and respective proteins predictions for each sample.

3. *categorize_by_function.py*: The metagenome predictions calculated in the step before are collapsed, by this script, to a specified level within the pathway hierarchy (e.g., KEGG Orthologouss will be collapsed into KEGG pathways).

Similar to QIIME, arguments and parameters can be changed in every step according to the user's preferences (see https://picrust.github.io/picrust/tutorials/metagenome_prediction.html). The output table from the metagenome predictions and function characterization are in BIOM format that makes possible to use it for core analysis in QIIME (see step 9 of point 2.4) and have access to the same diversity analysis as for the OTU data (please be aware of the BIOM formats incompatibilities mentioned above). Using the *-f* argument in steps 2 and 3 delivers a text file table that can be used in several other statistic and diversity analysis programs (see point 2.4). For data visualization and basic statistics, STAMP (Parks et al. 2014) is popular within PICRUSt users. The use of PICRUSt is schematized in Figure 5.2. An example of a possible output, after PICRUSt analysis and using the table generated in step 3, is shown in Figure 5.4.

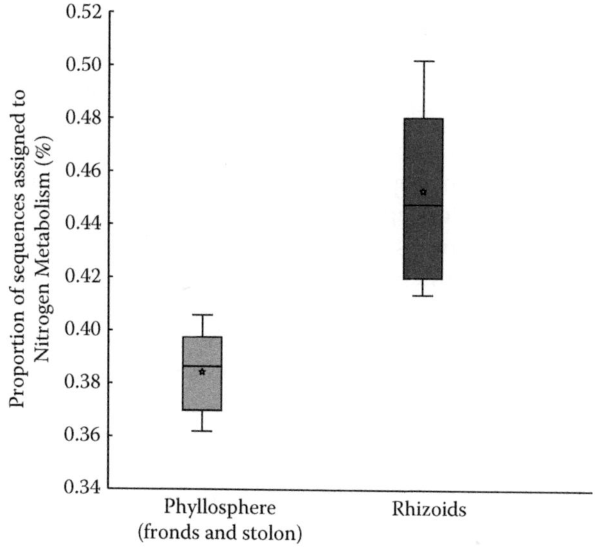

Figure 5.4 Example of a possible output of seaweed-associated bacterial community analysis using PICRUSt level 3 KEGG pathways' predictions. Box plot showing the distribution in the proportion of sequences assigned to nitrogen metabolism. The top of the box indicates the third quartile, the bottom the first quartile, and the line in the middle is the median. (Unpublished results).

5.5 Notes

Note 1: For the DNA extraction step, the listed materials only refer to those not provided in the MO BIO's PowerSoil™ DNA Isolation Kit.

Note 2: This protocol comprehends total bacterial community (both endo- and epiphytic bacteria). If one is interested in just sampling the bacterial biofilm, cotton swabs should be rubbed on the tissue surface and stored as described earlier. For sampling endophytic bacteria, please consider the protocol provided by, for example, Delbridge et al. (2004), Hollants et al. (2010), Aires et al. (2012).

Note 3: Sampling strategy and labeling will depend on the researcher's question/hypothesis. This protocol is designed to sample the entire individual by considering tissue specificity.

Note 4: Field sampling tends to be messy and makes the required aseptic conditions challenging. Organizing all the storage material in advance facilitates the process and avoids undesirable contaminations. Example: Sampling tubes can be labeled in advance according to sampling expectations, samples to be preserved in the stabilization solution should be placed in a plastic box and kept vertical to avoid spills and, when possible, tubes should be isolated using Parafilm tape.

Note 5: In step 4, if Zymo solution is being used as a preservation method, it might interact with solution C1 and form a white precipitate. The precipitate should be removed by placing the tubes at 65°C for 5 min.

Note 6: Immediately place the tubes in the bead-beater for, at least, two min (the time might depend on the tissue type and facility of disruption) and resume to step 6 of the experienced user protocol.

Note 7: Extraction process should be performed as quickly as possible in a sterile environment. To do so, prepare and label all the required tubes beforehand, keep all the materials you need close by, and keep the Bunsen burner flame on during the entire extraction. To avoid undesirable spills and subsequent contamination, in step 15 of the experienced user protocol for PowerSoil® DNA Isolation Kit, load 600 µL of the supernatant mixed with the C4 buffer onto the spin filter (instead of the recommended 675 µL). More than three loads might be needed. Always make a blank extraction to evaluate possible contamination.

Note 8: PCR specifications should be optimized in each lab according to the reagents, primers, and the thermocycler machine used. However, the entire PCR process should always be carried out under sterile conditions (PCR chamber or laminar flow hood), and a proofreading Taq polymerase should be used.

Note 9: From our experience, bacterial communities associated with some seaweed species might be difficult to amplify, either because of the small size of the sample, or because of the high content of polyphenols in seaweeds, that could hamper the PCR and result in low-yield amplifications. Performing a nested PCR using the 16S universal primers 27F and 1492r (Lane, 1991) before the modified primers above-mentioned can solve this problem. To avoid a large bias, the second PCR should have a reduced number of cycles (Aires et al. 2016).

Note 10: *In silico* analysis' methodologies depend on the addressed question. Therefore, only advice on the software usage and initial guidance on taxonomic identification of bacterial communities is given in this chapter.

Note 11: After performing step 5, the OTU table (in BIOM format) can be converted into, for example, a text or excel file and used in several other programs for statistic and diversity analysis, as well as for visualization of the results (e.g., Excel, STAMP, PAST, PRIMER-E+PERMANOVA and the R packages Vegan and Phyloseq). The conversion can be done using a command line in QIIME, see: http://biom-format.org/documentation/biom_conversion.html.

Note 12: The alignment and phylogeny inference steps are only necessary if phylogenetic metrics (e.g., UniFrac) are used in downstream analysis. The use of the next suggested script might require the use of a phylogenetic tree.

Note 13: Some important tips must be noted before using PICRUSt: (a) PICRUSt uses the bacterial community taxonomic profile and abundances (OTU table generated in QIIME) to predict functional potential, so all the reads have to be assigned and get hit to a particular reference sequence collection. To obtain a fully taxonomy assigned OTU table, the OTU picking strategy (step 2 of the previous point) is the *closed-reference* OTU picking. Using the default 97% identity assignment is highly recommended; (b) to date, PICRUSt works best with the 16S rRNA sequence database Greengenes v13.5 as a reference sequence collection (download can be done here: http://greengenes.secondgenome.com/downloads), and that is the database to be used for *closed-reference* OTU picking (using the files 97_otus.fasta and 97_otu_taxonomy.txt as *–r* and *–t* arguments, respectively, in the QIIME script *pick_otus.py*); (c) QIIME latest versions use BIOM 2.1 version (hdf5 format) to create the OTU table, whereas PICRUSt version 1.0.0 uses an older

BIOM 1.3.1 version (json format) so, some incompatibilities may arise and conversion of different tables versions may be needed, for that please consult http://biom-format.org/documentation/biom_conversion.html. In the latest PICRUSt version 1.1.0, the BIOM version should not be an issue.

Acknowledgments

We thank the Phycomorph COST action (FA1406) and acknowledge the financial support of FCT (SFRH/BPD/107878/2015 (AE), SFRH/BPD/116774/2016 (TA) and GENEKELP).

References

Aires, T., Marbà, N., Serrao, E. A., Duarte, C. M., Arnaud-Haond, S. 2012. Selective elimination of chloroplastidial DNA for metagenomics of bacteria associated with the green alga *Caulerpa taxifolia* (Bryopsidophyceae). *J Phycol* 48 (2): 483–490.

Aires, T., Moalic, Y., Serrao, E. A., Arnaud-Haond, S. 2015. Hologenome theory supported by co-occurrence networks of species-specific bacterial communities in siphonous algae (*Caulerpa*). *FEMS Microbiol Ecol* 91: 1–14.

Aires, T., Serrão, E. A., Engelen, A. H. 2016. Host and environmental specificity in bacterial communities associated to two highly invasive marine species (Genus *Asparagopsis*). *Front Microbiol* 7: 559.

Allen-Vercoe, E. 2013. Bringing the gut microbiota into focus through microbial culture: Recent progress and future perspective. *Curr Opin Microbiol* 16 (5): 625–629.

Apprill, A., Mcnally, S., Parsons, R., Weber, L. 2015. Minor revision to V4 region SSU rRNA 806R gene primer greatly increases detection of SAR11 bacterioplankton. *Aquat Microb Ecol* 75 (2): 129–137.

Barott, K. L., Rodriguez-Brito, B., Janouškovec, J. et al. 2011. Microbial diversity associated with four functional groups of benthic reef algae and the reef-building coral *Montastrea annularis*. *Environ Microbiol* 13 (5): 1192–1204.

Bengtsson, M. M., Sjøtun, K., Øvreås, L. 2010. Seasonal dynamics of bacterial biofilms on the Kelp *Laminaria hyperborea*. *Aquat Microb Ecol* 60 (1): 71–83.

Bodenhausen, N., Horton, M. W., Bergelson, J. 2013. Bacterial communities associated with the leaves and the roots of *Arabidopsis thaliana*. *PLoS One* 8 (2): e56329.

Bolinches, J., Lemos, M. L., Barja, J. L. 1988. Population dynamics of heterotrophic bacterial communities associated with *Fucus vesiculosus* and *Ulva rigida* in an estuary. *Microb Ecol* 15 (3): 345–357.

Burke, C., Kjelleberg, S., Thomas T. 2009. Selective extraction of bacterial DNA from the surfaces of macroalgae. *Appl Environ Microbiol* 75 (1): 252–256.

Burke, C., Steinberg, P., Rusch, D., Kjelleberg, S., Thomas, T. 2011. Bacterial community assembly based on functional genes rather than species. *PNAS* 108 (34): 14288–14293.

Caporaso, J. G., Kuczynski, J., Stombaugh, J. et al. 2010. Correspondence QIIME allows analysis of high-throughput community sequencing data intensity normalization improves color calling in SOLiD sequencing. *Nature* 7 (5): 335–336.

Charrier, B., Coates, J. C., Stavridou, I. 2017. Surfing amongst oil-tankers: Connecting emerging research fields to the current international landscape. *Trends Plant Sci* 22 (1): 1–3.

Chisholm, J. R. M., Dauga, C., Ageron, E., Grimont, P. A. D., Jaubert, J. M. 1996. Roots' in mixotrophic algae. *Nature* 381: 382.

Cundell, A. M., Sleeter, T. D., Mitchell, R. 1977. Microbial populations associated with the surface of the brown alga *Ascophyllum nodosum*. *Microb Ecol* 4 (1): 81–91.

Delbridge, L., Coulburn, J., Fagerberg, W., Tisa, L. S. 2004. Community profiles of bacterial endosymbionts in four species of *Caulerpa*. *Symbiosis* 37: 335–344.

Dittami, S. M., Duboscq-Bidot, L., Perennou, M. et al. 2016. Host–microbe interactions as a driver of acclimation to salinity gradients in brown algal cultures. *ISME J* 10: 51–63.

Egan, S., Harder, T., Burke, C., Steinberg, P., Kjelleberg, S., Thomas, T. 2013. The seaweed holobiont: Understanding seaweed-bacteria interactions. *FEMS Microbiol Rev* 37 (3): 462–476.

Eigemann, F., Hilt, S., Salka, I., Grossart, H. P. 2013. Bacterial community composition associated with freshwater algae: Species specificity versus dependency on environmental conditions and source community. *FEMS Microbiol Ecol* 83 (3): 650–663.

Gupta, S., Abu-Ghannam, N. 2011. Bioactive potential and possible health effects of edible brown seaweeds. *Trends Food Sci Technol* 22 (6): 315–326.

Hanshew, A. S., Mason, C. J., Raffa, K. F., Currie, C. R. 2013. Minimization of chloroplast contamination in 16S rRNA gene pyrosequencing of insect herbivore bacterial communities. *J Microbiol Methods* 95 (2): 149–155.

Hollants, J., Decleyre, H., Leliaert, F., De Clerck, O., Willems, A. 2011. Life without a cell membrane: Challenging the specificity of bacterial endophytes within *Bryopsis* (Bryopsidales, Chlorophyta). *BMC Microbiol* 11: 255.

Hollants, J., Leliaert, F., De Clerck, O., Willems, A. 2010. How endo- is endo-? Surface sterilization of delicate samples: A *Bryopsis* (Bryopsidales, Chlorophyta) case study. *Symbiosis* 51: 131–138.

Hollants, J., Leliaert, F., Verbruggen, H., Willems, A., De Clerck, O. 2013. Permanent residents or temporary lodgers: Characterizing intracellular bacterial communities in the siphonous green alga *Bryopsis*. *Proc Biol Sci* 280: 20122659.

Lachnit, T., Blümel, M., Imhoff, J. F., Wahl, M. 2009. Specific epibacterial communities on macroalgae: Phylogeny matters more than habitat. *Aquat Biol* 5: 181–186.

Lachnit, T., Meske, D., Wahl, M., Harder, T., Schmitz, R. 2011. Epibacterial community patterns on marine macroalgae are host-specific but temporally variable. *Environ Microbiol* 13: 655–665.

Lane, D. J. 1991. 16S/23S rRNA sequencing. In *Nucleic Acid Techniques in Bacterial Systematic*, E. Stackebrandt and M. Goodfellow (Eds.), pp. 115–175. New York: John Wiley & Sons.

Langille, M. G. I., Zaneveld, J., Caporaso, J. G. et al. 2013. Predictive functional profiling of microbial communities using 16S rRNA marker gene sequences. *Nature Biotechnol* 31 (9): 814–821.

Lundberg, D. S., Yourstone, S., Mieczkowski, P., Jones, C. D., Dangl, J. L. 2013. Practical innovations for high-throughput amplicon sequencing. *Nat Meth* 10 (10): 999–1002.
Mancuso, F. P., D'Hondt, S., Willems, A., Airoldi, L., De Clerck, O. 2016. Diversity and temporal dynamics of the epiphytic bacterial communities associated with the canopy-forming seaweed *Cystoseira compressa* (Esper) Gerloff and Nizamuddin. *Front Microbiol* 7: 476.
Marzinelli, E. M., Campbell, A. H., Valdes, E. Z. et al. 2015. Continental-scale variation in seaweed host-associated bacterial communities is a function of host condition, not geography. *Environ Microbiol* 17(10): 4078–4088.
Matsuo, Y., Imagawa, H., Nishizawa, M., Shizuri, Y. 2005. Isolation of an algal morphogenesis inducer from a marine bacterium. *Science* 307: 1598.
Meusnier, I., Olsen, J. L., Stam, W. T., Destombe, C., Valero, M. 2001. Phylogenetic analyses of *Caulerpa taxifolia* (Chlorophyta) and of its associated bacterial microflora provide clues to the origin of the Mediterranean introduction. *Mol Ecol* 10: 931–946.
Nylund, G. M., Persson, F., Lindegarth, M., Cervin, G., Hermansson, M., Pavia, H. 2010. The red alga *Bonnemaisonia asparagoides* regulates epiphytic bacterial abundance and community composition by chemical defence. *FEMS Microbiol Ecol* 71 (1): 84–93.
Parks, D. H., Tyson, G. W., Hugenholtz, P., Beiko, R. G. 2014. STAMP: Statistical analysis of taxonomic and functional profiles. *Bioinformatics* 30 (21): 3123–3124.
Qian, P., Xu, Y., Fusetani, N. 2010. Natural products as antifouling compounds: Recent progress and future perspectives. *Biofouling* 26 (2): 223–234.
Riesenfeld, C. S., Schloss, P. D., Handelsman, J. 2004. Metagenomics: Genomic analysis of microbial communities. *Annu Rev Genet* 38: 525–552.
Rosenberg, E., Koren, O., Reshef, L., Efrony, R., Zilber-Rosenberg, I. 2007. The role of microorganisms in coral health, disease and evolution. *Nature Rev Microbiol* 5: 355–362.
Schäfer, H., Muyzer, G. 2001. Denaturing gradient gel electrophoresis in marine microbial ecology. *Method Microbiol* 30: 425–468.
Schloss, P. D., Westcott, S. L., Ryabin, T. et al. 2009. Introducing mothur: Open-source, platform-independent, community-supported software for describing and comparing microbial communities. *Appl Environ Microbiol* 75 (23): 7537–7541.
Singh, R. P., Reddy, C. R. K. 2014. Seaweed-microbial interactions: Key functions of seaweed-associated bacteria. *FEMS Microbiol Ecol* 88: 213–230.
Singh, R. P., Reddy, C. R. K. 2016. Unraveling the functions of the macroalgal microbiome. *Front Microbiol* 6: 1–8.
Smit, A. J. 2004. Medicinal and pharmaceutical uses of seaweed natural products: A Review. *J Appl Phycol* 16 (4): 245–262.
Spoerner, M., Wichard, T., Bachhuber, T., Stratmann, J., Oertel, W. 2012. Growth and thallus morphogenesis of *Ulva mutabilis* (Chlorophyta) depends on a combination of two bacterial species excreting regulatory factors. *J Phycol* 48 (6): 1433–1447.

Stratil, S. B., Neulinger, S. C., Knecht, H., Friedrichs, A. K., Wahl, M. 2013. Temperature-driven shifts in the epibiotic bacterial community composition of the brown macroalga *Fucus vesiculosus*. *Microbiologyopen* 2: 338–349.
Tujula, N. A., Crocetti, G. R., Burke, C., Thomas, T., Holmström, C., Kjelleberg, S. 2010. Variability and abundance of the epiphytic bacterial community associated with a green marine Ulvacean alga. *ISME* 4 (2): 301–311.
Vieira, C., Engelen, A. H., Guentas, L. et al. 2016. Species specificity of bacteria associated to the brown seaweeds *Lobophora* (Dictyotales, Phaeophyceae) and their potential for induction of rapid coral bleaching in *Acropora muricata*. *Front Microbiol* 7: 316.
Wang, Y., Qian, P. Y. 2009. Conservative fragments in bacterial 16S rRNA genes and primer design for 16S ribosomal DNA amplicons in metagenomic studies. *PLoS One* 4 (10): e7401.
Weiss, A., Costa, R., Wichard, T. 2017. Morphogenesis of *Ulva mutabilis* (Chlorophyta) induced by *Maribacter* species (*Bacteroidetes, Flavobacteriaceae*). *Botanica Marina* 60 (2): 197–206.
Whatley, J. M., John, P., Whatley F. R. 1979. From extracellular to intracellular: The establishment of mitochondria and chloroplasts. *Proc Biol Sci* 204: 165–187.
Wichard, T. 2015. Exploring bacteria-induced growth and morphogenesis in the green macroalga order Ulvales (Chlorophyta). *Front Plant Sci* 6: 86.

chapter six

Heavy metal ecotoxicity on the early life history stages of macroalgae

Pablo P. Leal and Michael Y. Roleda

Contents

6.1 Introduction

Marine ecotoxicology is a branch of toxicology dealing with the study of the toxic effects, caused by natural or synthetic pollutants, with fundamental focus on the members of the marine ecosystems, for example, flora and fauna. The objective of ecotoxicological studies is to determine the impacts of anthropogenic toxic waste discharged to marine environments (Chapman 1995). The main tool used in ecotoxicological research is a short-term assay to measure the toxicity of a pollutant (e.g., heavy metals) on different physiological parameters of a key species (Burridge and Bidwell 2002). Empirical data can be used to estimate the long-term negative effects of the pollutants on the population of the investigated species (Anderson et al. 1990, Contreras et al. 2007).

Macroalgae are the foundation species in coastal environments that create three-dimensional habitats. They are at the base of the food web providing energy to higher trophic levels and provide nursery grounds and habitats for diverse assemblages of invertebrates and fish. All macroalgae have microscopic phases within their life cycle, and these phases may be the most susceptible life stage to environmental stressors (Roleda et al. 2007, 2009). Macroalgal spores and gametes are liberated into the water column, settle on the rocky coastline, undergo sexual reproduction, and give rise to the next macroscopic generation. Macroalgae produce enormous numbers of propagules, for example, different types of spores and gametes, but only a small fraction germinates and survive to maturity.

Performing ecotoxicity tests on the early life-history stages of ecologically relevant marine benthic macroalgae will provide fundamental knowledge on the relative susceptibility of their microscopic stages that give rise to macroscopic, multicellular, and usually morphologically complex adult stages. Negative effects on the early life-history stages and the consequent failure to establish the next macroscopic generation of key macroalgal phyla will have cascading effects on coastal ecosystems as they provide habitat and food to invertebrates and fish. Thus, knowing the species' survival thresholds is relevant for policy-makers and managers to help in preserving marine biodiversity and preventing the local extinction of key marine species.

Ecotoxicity tests are performed in controlled laboratory conditions. The test organism is exposed to a range of concentrations of the toxicant, for example, heavy metal, to estimate the concentration at which it can negatively affect the physiology and ontogenetic development of the test organism (Burridge and Bidwell 2002, Eklund and Kautsky 2003, van Dam et al. 2008). To facilitate comparison between studies, we recommended reporting the half maximal effective concentration EC_{50}, that is, the effective concentration at which 50% of the physiological and/or metabolic processes are affected by the toxicant (Leal et al. 2016a, b).

6.2 State of the art

The field of heavy metal ecotoxicology has become increasingly important over the past few decades to assess the impacts of industrial pollution on marine coastal ecosystems. For example, bioaccumulation of heavy metals in algae can cause health risks to other organisms in the food chain, including humans. Hence, generation of robust data from rigorous ecotoxicological studies is paramount not only for risk assessment but also for rehabilitation measures. Moreover, ecotoxicity tests will provide basic information on the species' responses and survival thresholds to heavy metals.

In the absence of trace metal clean procedures, most published studies on trace metal ecotoxicology prior to the mid-1970s, especially in the field of marine biology, are considered inaccurate and unreliable. In this regard, Maienthal and Becker (1976) recommended the routine use of trace metal clean techniques to produce accurate and reliable data on ecotoxicology of marine algae. However, the early recommendations have largely been ignored in subsequent trace metal studies. For example, recent review appraising copper ecotoxicological studies on marine algae covering studies from 1977 to 2015 showed alarming methodological deficiencies (Leal et al. 2016a). Despite the availability of trace metal clean procedures for more than 40 years, ~70% of the literatures reviewed did not follow the recommended protocols. Half of the studies did not specify the labware, whereas the remaining 50% reported the use of glassware (~25%) and plasticware (~25%) and only ~30% of the studies specified cleaning protocols to remove trace metal impurities of the labware. The copper form used to prepare the stock solutions was specified in most studies (~80%) but acidification to stabilize the dissolved copper was performed in only ~20%. These studies generally reported the nominal concentrations, but only ~40% measured the actual dissolved copper concentrations (Leal et al. 2016a).

A rigorous exotoxicity test procedure must include trace metal clean techniques and stabilization of the heavy metal stock solution: This will substantially reduce the subsequent procedural contamination and the depletion of the toxicant's actual dissolved concentration, respectively (Leal et al. 2016a). The protocol includes proper labware cleaning and storage, stock solution preparation and stabilization with acidification, and trace metal clean experimental and sampling procedures (Leal et al. 2016a). The labware composition is also important, because glass can adsorb some metals (e.g., copper) and leach others, altering metal concentrations in the solution (Richter 2003). Therefore, the use of plastic bottles and culture dishes (e.g., Teflon or polyethylene) during stock solution preparation and storage, and performance of toxicity test, respectively, are recommended (Garman et al. 1994, Burridge and Bidwell 2002, Leal et al. 2016a, b). Failure to follow these steps will compromise the validity of any ecotoxicity test result as the negative effect of a toxicant, for example, a heavy metal, may be underestimated because of metal contamination or overestimated because of metal loss by adsorption (Leal et al. 2016a).

6.3 Materials

6.3.1 Trace metal cleaning

- Distilled and deionized water (~18 MΩ cm): Several liters are required for trace metal cleaning.
- Nonabrasive soap, detergent, or cleaning powder.
- Acid solution: Prepare 10% 2 M HCl in deionized water.
- 5–10 L basin for soaking labwares.
- 500 mL polyethylene plastic bottles.
- Polystyrene tissue culture vessels: Six-well plate (Costar 3516, Corning Inc.).
- All labwares for handling algal materials, spore release, and ecotoxicity test: For example, beakers, graduated cylinder, screw-capped tubes, trays, and containers, among others.
- Laminar flow hood.
- Laboratory gown.
- Hand gloves.

6.3.2 Algal material for spore release and toxicity test

- *Reproductive algal material*: for example, sorus (Figure 6.1).
- *Natural seawater*: Collect into polystyrene containers preferably within the vicinity and simultaneous with algal material collection (Synthetic seawater, e.g., Sea salts S9883 Sigma, may also be used).
- Seawater filtration system with 0.2 µm pore size filters: Filtered natural seawater siphoned into new acid-washed polystyrene containers should be stored in the dark and under low temperature (e.g., 5°C).
- High-precision balance.
- Beakers.
- Graduated cylinder.
- Tissue paper.
- Soft-bristled brush.
- *Hemocytometer*: (Neubauer-improved, Marienfeld Superior).
- Compound microscope for counting spores.
- *Environmental chamber*: Set temperature, irradiance, and light:dark cycle according to the experimental design, for example, 12°C, 50 µmol photons m^{-2} s^{-1}, and 12:12 light:dark cycle, respectively.
- *Metal-stock solution*: Dissolve the toxicant chemical in deionized water (e.g., 2 g Cu^{2+} L^{-1}). To stabilize the solution and avoid adsorption into the container walls, reduce the solution to pH 2 with 1 M HCl.
- 10 mL screw-capped polyethylene plastic tubes: To store media for the analysis of total dissolved heavy metal concentration.

(a) (b)

Figure 6.1 Reproductive tissue aka sorus from (a) sporophylls of *Alaria esculenta* and (b) lamina of *Saccharina latissima*.

- *HNO_3*: To acidify media for the analysis of total dissolved heavy-metal concentration.
- Micropipettes (2–200 µL; 1–1000 µL) and corresponding tips.
- Inverted microscope with digital camera for documenting meiospores ontogenetic development.
- Personal computer and digital imaging software.

6.4 Experimental procedures

6.4.1 Trace metal clean techniques

Labware cleaning must be performed in a trace metal clean area (e.g., laminar flow hood). Wear gloves and proper laboratory garments during the cleaning process.

1. Wash labware with nonabrasive soap and rinse thoroughly with distilled water and then with deionized water.
2. Soak labware in an acidic solution (10% 2 M HCl) for at least one week. Then rinse labware with deionized water three times.
3. Dry labware inside the trace metal clean area (e.g., laminar flow hood).

6.4.2 *Algal material*

To test macroalgal reproductive cells, for example, spores and gametes, fertile individuals are collected to induce spore or gamete release. For kelps, for example, reproductive tissue (sorus) occurs on different parts of the thallus (Figure 6.1; Leal et al. 2014). Spore release can be induced from dehydrated mature sorus kept overnight at 4°C and subsequent temperature shock on submersion into seawater at 12°C (Leal et al. 2016b). Use at least five replicates (*Note 1*).

1. After collection of reproductive plants, cutoff blades with mature sori and clean of epibionts by soft brushing under filtered seawater (*Note 2*).
2. Blot dry blades and wrap in moist tissue paper overnight at 4°C. The next day, partial spore release may be observed as brown staining on the tissue paper.
3. From each reproductive blade excise soral disks (e.g., 2 cm Ø) using a cork borer or weigh soral tissue.
4. Immerse a defined number of soral disks or tissue weight in a known volume of filtered (12°C) seawater (e.g., 2 × of 2 cm^2 disks in 50 mL or 50 g tissue in 500 mL filtered seawater) (Figure 6.2a).
5. After a defined period, for example, 15–30 min, remove the soral disks or blades (Figure 6.2b). Slightly mix the suspension, draw a subsample (*Note 3*), and measure spore density using a hemocytometer (Guillard and Sieracki 2005).
6. Dispense × μL of known density and adjust to 25,000 spores mL^{-1} (final volume = 12 mL) onto each well of the six-well culture vessels (*Note 4*). Allow meiospores to settle for 3 h. Then, renew the culture medium to eliminate unsettled meiospores and rest detritus.

6.4.3 *Toxicity test*

After settlement, meiospores can now be exposed to the predefined nominal metal concentrations (*Note 5*). For comparison between studies, it is essential to report the experimental exposure frequency (chronic or one-time exposure) and period. For one-time exposure, the duration, that is, the number of hours or days in some cases, must be clearly specified in subsequent report or publication.

1. Prepare the nominal metal concentrations in different polyethylene plastic bottles. After 3 h settlement and elimination of unsettled meiospores and detritus, renew the culture medium in each well of the six-well culture vessels with corresponding predefined nominal metal concentrations.

Figure 6.2 Induction of meiospore release by temperature shock: (a) Rehydration of soral blades in 12°C seawater after overnight storage of the blotted-dry blades in cold chamber (4°C) and (b) spore released after 15 min submersion of the blades.

2. Renew the culture medium with the appropriate metal concentration every 2–3 days to simulate chronic metal exposure and to avoid inorganic carbon and nutrient depletion. Analytical blanks (i.e., treatment without meiospores) corresponding to each copper treatment must also be prepared.
3. To measure the total dissolved metal concentrations, collect fresh and old media (2–3 days old) corresponding to each nominal concentration into 10 mL polyethylene screw-capped tubes using trace metal clean procedure. From each replicate of each heavy metal treatment (with meiospores) and the corresponding analytical blanks, collect 0.15 mL of culture medium, dilute in 4.25 mL of ultrapure water, and acidify with 0.09 mL of HNO_3 (acidified sample, 4.5 mL 2% HNO_3) and store until analysis (*Note 6*). This will allow one to calculate and distinguish metal loss because of adsorption in stock solution bottle and in culture vessels and absorption by the cell, and possible metal contamination because of unsterile environment. The metal absorbed by the cell can then be calculated as metal loss in *treatment A*–metal loss in *blank A*.

6.4.4 End points

End points such as meiospore germination, germling growth rate, gametophyte sex differentiation, and sex ratio can be measured from photographs of the meiospores cultivated under different metal concentrations.

1. Take digital photographs of at least five visual fields of each experimental unit under an inverted microscope. The visual field may be haphazardly chosen or determined by random sampling. A 10× objective is recommended. Photographs in time series, for example, every two days, are necessary to quantify different response variables. Photographs can be analyzed by using different software such as ToupView 3.5 (Abràmoff et al. 2004).
2. The following response variables can be qualified and quantified in the photographs:
   a. Meiospore germination is visible after two days. Meiospores are considered germinated when the germ tube is visible. Germination percentage (%) is calculated from 350 individuals in each experimental unit.
   b. Germling growth rate can be measured from day three and thereafter. Measure the size of at least 30 germlings (i.e., sexually ambiguous growing meiospores) in each experimental unit. Germling growth rate in each treatment can be calculated using the equation described by Yong et al. (2013): Growth rate ($\%\cdot\text{day}^{-1}$) = $([(W_t/W_0)^{1/t} - 1]) \times 100$, where W_0 is the initial size, W_t is the final size, and t is days of culture.
   c. Sexually differentiated gametophyte development and sex ratio can be measured after 15 days and thereafter. Measure the size of at least 30 male and 30 female gametophytes in each experimental unit. Gametophyte growth rate can be calculated using the above-mentioned formula. From haphazardly chosen visual fields, count at least 200 gametophytes and classify them as male or female (Figure 6.3). A sex ratio can be expressed as the frequency of males per progeny, that is, number of males/(number of males + number of females) (Roleda et al. 2012). The sex ratio of 0.5: equal number of males and females, >0.5: sex ratio is biased toward males, and <0.5: sex ratio is biased toward females.
3. The EC_{50}, that is, the concentration of the toxicant (e.g., metal) that caused the inhibition of 50% of an end point (e.g., meiospore germination or germling growth) can be calculated using statistical software (Figure 6.4).

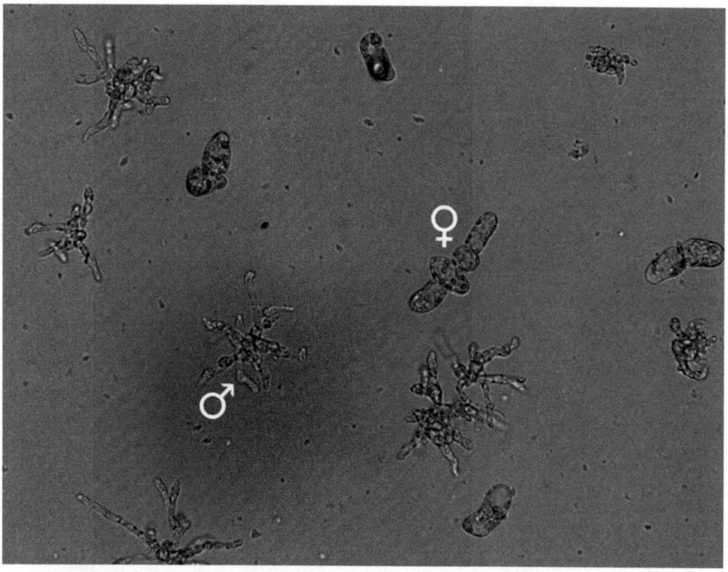

Figure 6.3 Sexually differentiated kelp gametophytes after 15 day cultivation.

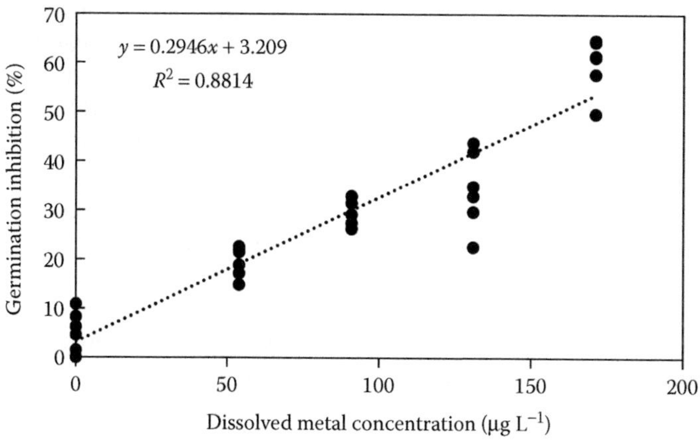

Figure 6.4 Relationship between dissolved heavy metal concentration and germination rate expressed as percentage. Linear regression was used to obtain dose–response relationship. The calculated half maximal effective concentration EC_{50} is 159 µg L^{-1} dissolved metal.

6.4.5 Adaptation of the protocol to microalgae and adult macroalgal specimens

The earlier protocol may also be used for toxicity tests on microalgae and adult macroalgal specimens with corresponding modifications.

1. During exponential growth phase, microalgal cells are transferred to different media with known concentrations of the toxicant (Levy et al. 2007). Most of the microalgal toxicity tests measure the decrease in growth rates (aka cell division rate) after 48–72 h of exposure to the toxicant (Stauber 1995, Levy et al. 2007). The inhibitory effects of the toxicant on the photosynthetic rates can also be measured (Abalde et al. 1995, Magnusson et al. 2008) using standard protocols for dissolved oxygen measurement using oxygen electrode or optode.
2. For adult macroalgal specimens, whole juvenile individual or meristematic pieces of the adult thalli (e.g., excised discs or tips) are exposed to a range of known nominal concentrations of the toxicant. When using excised tissue, the marginal wounds must be allowed to heal first before experimental exposure to the toxicant. This can be performed by cultivating the macroalgal pieces under optimal condition for at least 1–2 days. Then the toxicity assay can be performed under standard culture conditions with constant aeration and/or water movement. The starting biomass per unit volume of seawater (biomass to volume ratio) must be optimized (e.g., 1 g fresh weight L^{-1}). The starting biomass, for example, tissue size or weight, of each experimental unit is recommended to be comparable among different treatments, that is, concentrations of the toxicant (Haglund et al. 1996). The surface area or weight of the sample can be measured every 2–3 days to calculate growth inhibition and biomass loss/reduction (Haglund et al. 1996). Inhibition of photosynthetic efficiency can be measured using a pulse amplitude modulated (PAM) chlorophyll fluorometer (Baumann et al. 2009). For chronic exposure to toxicants (e.g., heavy metal), the ambient or nutrient replete seawater with the respective nominal metal concentration must be renewed every two days; this will avoid nutrient (e.g., nitrogen, phosphorous, and carbon) depletion (Stauber 1995, Leal et al. 2016b), which may confound the effects of the toxicant.

6.5 Notes

Note 1: Ideally, it is recommended to use different individuals as replicates to measure individual variability to a given stress factor. However, in the case of extensive experimental variables and levels of treatment to contend with, working with several spore suspensions released from different individuals (e.g., $n = 5$) may be logistically unworkable. Therefore, spores released from sori of at least 10 individuals may be pooled into one stock spore suspension, and experimental replication of $n = 5$ can be dispensed from the pooled spore suspension (Leal et al. 2017).

Note 2: After removing visible epibionts and thorough cleaning, blades may be treated with germanium oxide solution to eradicate nonvisible microalgae, for example, diatoms (Andersen and Kawashi 2005). The concentration must be carefully adjusted such that chemical residues on the blade during subsequent spore release should not have any negative effects on the vitality and viability of spores.

Note 3: We recommend sampling the uppermost layer of the spore suspension in which swimming meiospores are likely to be concentrated, whereas exudates, detritus, and unviable meiospores precipitate to the bottom of the container.

Note 4: Recommended spore suspension density to ensure well-dispersed and even spore settlement density (Figure 6.3).

Note 5: Determination of the concentration range (low and high range) of the heavy metal must consider the ecological significance of the treatment or the mechanistic effect of the toxicant. The stock solution (with nominal concentration) must be prepared during the scheduled day for the renewal of the medium and must be used within 1 h of preparation.

Note 6: Total dissolved metal concentrations can be quantified using inductively coupled plasma mass spectroscopy (Leal et al. 2016b), among other standard protocols.

References

Abalde, J., Cid, A., Reiriz, S., Torres, E., and Herrero, C. 1995. Response of the marine microalga *Dunaliella tertiolecta* (Chlorophyceae) to copper toxicity in short time experiments. *Bull. Environ. Contam. Toxicol.* 54:317–324.

Abràmoff, M.D., Magalhães, P.J., and Ram, S.J. 2004. Image processing with image. *J. Biophotonics Int.* 11:36–42.

Andersen, R.A. and Kawashi, M. 2005. Traditional microalgae isolation techniques. *In* Andersen, R. A. (Ed.) *Algal Culturing Techniques*. 1st ed. Elsevier Academic Press, Burlington, NJ, pp. 83–100.

Anderson, B.S., Hunt, J.W., Turpen, S.L., Coulon, A.R., and Martin, M. 1990. Copper toxicity to microscopic stages of giant kelp *Macrocystis pyrifera*: Interpopulation comparisons and temporal variability. *Mar. Ecol. Prog. Ser.* 68:147–156.

Baumann, H.A., Morrison, L., and Stengel, D.B. 2009. Metal accumulation and toxicity measured by PAM-Chlorophyll fluorescence in seven species of marine macroalgae. *Ecotoxicol. Environ. Saf.* 72:1063–1075.

Burridge, T.R. and Bidwell, J.R. 2002. Review of the potential use of brown algal ecotoxicological assays in monitoring effluent discharge and pollution in southern Australia. *Mar. Pollut. Bull.* 45:140–147.

Chapman, P.M. 1995. Ecotoxicology and pollution—Key issues. *Mar. Pollut. Bull.* 31:167–177.

Contreras, L., Medina, M.H., Andrade, S., Oppliger, L.V., and Correa, J.A. 2007. Effects of copper on early developmental stages of *Lessonia nigrescens* Bory (Phaeophyceae). *Environ. Pollut.* 145:75–83.

Eklund, B.T. and Kautsky, L. 2003. Review on toxicity testing with marine macroalgae and the need for method standardization—Exemplified with copper and phenol. *Mar. Pollut. Bull.* 46:171–181.

Garman, G.D., Pillai, M.C., and Cherr, G.N. 1994. Inhibition of cellular events during early algal gametophyte development: Effects of select metals and an aqueous petroleum waste. *Aquat. Toxicol.* 28:127–144.

Guillard, R.R.L. and Sieracki, M.S. 2005. Counting cells in cultures with the light microscope. *In* Andersen, R. A. (Ed.) *Algal Culturing Techniques*. 1st ed. Elsevier Academic Press, Burlington, NJ, pp. 239–252.

Haglund, K., Björklund, M., Gunnare, S., Sandberg, A., Olander, U., and Pedersén, M. 1996. New method for toxicity assessment in marine and brackish environments using the macroalga *Gracilaria tenuistipitata* (Gracilariales, Rhodophyta). *Hydrobiologia*. 326/327:317–325.

Leal, P.P., Hurd, C.L., Fernández, P.A., and Roleda, M.Y. 2017. Ocean acidification and kelp development: reduced pH has no negative effects on meiospore germination and gametophyte development of *Macrocystis pyrifera* and *Undaria pinnatifida*. *J. Phycol.* 53:557–566.

Leal, P.P., Hurd, C.L., and Roleda, M.Y. 2014. Meiospores produced in sori of non-sporophyllous laminae of *Macrocystis pyrifera* (Laminariales, Phaeophyceae) may enhance reproductive output. *J. Phycol.* 50:400–405.

Leal, P.P., Hurd, C.L., Sander, S.G., Armstrong, E.A., and Roleda, M.Y. 2016a. Copper ecotoxicology of marine algae: A methodological appraisal. *Chem. Ecol.* 32:786–800.

Leal, P.P., Hurd, C.L., Sander, S.G., Kortner, B., and Roleda, M.Y. 2016b. Exposure to chronic and high dissolved copper concentrations impede meiospore development of the kelps *Macrocystis pyrifera* and *Undaria pinnatifida* (Ochrophyta). *Phycologia*. 55:12–20.

Levy, J.L., Stauber, J.L., and Jolley, D.F. 2007. Sensitivity of marine microalgae to copper: The effect of biotic factors on copper adsorption and toxicity. *Sci. Total Environ.* 387:141–154.

Magnusson, M., Heimann, K., and Negri, A.P. 2008. Comparative effects of herbicides on photosynthesis and growth of tropical estuarine microalgae. *Mar. Pollut. Bull.* 56:1545–1552.

Maienthal, E. and Becker, D. 1976. A survey of current literature on sampling, sample handling, and long term storage for environmental materials. *Natl. Bur. Stand. (U.S.), Tech. Note* 929:40 p.

Richter, R. 2003. *Clean Chemistry Techniques for the Modern Laboratory*. First. Milestone Press, Shelton, CT. 96 p.

Roleda, M.Y., Campana, G.L., Wiencke, C., Hanelt, D., Quartino, M.L., and Wulff, A. 2009. Sensitivity of Antarctic *Urospora penicilliformis* (Ulotrichales, Chlorophyta) to ultraviolet radiation is life-stage dependent. *J. Phycol.* 45:600–609.

Roleda, M.Y., Morris, J.N., McGraw, C.M., and Hurd, C.L. 2012. Ocean acidification and seaweed reproduction: Increased CO_2 ameliorates the negative effect of lowered pH on meiospore germination in the giant kelp *Macrocystis pyrifera* (*Laminariales, Phaeophyceae*). *Glob. Chang. Biol.* 18:854–864.

Roleda, M.Y., Wiencke, C., Hanelt, D., and Bischof, K. 2007. Sensitivity of the early life stages of macroalgae from the northern hemisphere to ultraviolet radiation. *Photochem. Photobiol.* 83:851–862.

Stauber, J.L. 1995. Toxicity testing using marine and freshwater unicellular algae. *Australas. J. Ecotoxicol.* 1:15–24.

van Dam, R.A., Harford, A.J., Houston, M.A., Hogan, A.C., and Negri, A.P. 2008. Tropical marine toxicity testing in Australia: A review and recommendations. *Australas. J. Ecotoxicol.* 14:55–88.

Yong, Y.S., Yong, W.T.L., and Anton, A. 2013. Analysis of formulae for determination of seaweed growth rate. *J. Appl. Phycol.* 25:1831–1834.

chapter seven

A simple protocol for a rapid and consistent production of a large number of viable protoplasts from the Ulvophycean *species*

Vishal Gupta and C.R.K. Reddy

Contents

7.1 Introduction

Protoplasts, the plant somatic cells devoid of cell wall, offer a unique experimental material for cellular biology and biotechnology (Davey et al. 2005). The most remarkable achievement of protoplast research has been the development of various agricultural crop varieties with improved agronomic traits by somatic hybridization (Bajaj 2004). Protoplasts have also been extensively employed in genetic transformation studies because of their amenability for genetic manipulation techniques (Eeckhaut et al. 2013). The recent developments in modern biotechnology techniques have further broadened the scope of application of protoplast research in further understanding the genetic regulations involved in dedifferentiation (Xiao et al. 2012), organelle distribution and epigenetic changes (Jiang et al.

2013), signal transduction (Costa et al. 2012), transient gene expression system (Miao and Jiang 2007; Yoo et al. 2007; Zhang et al. 2011; Pitzschke and Persak 2012; Thévenin et al. 2012), interaction between transcription factors and genes (Li et al. 2011; Wehner et al. 2011; Huha et al. 2013; Ge et al. 2013), cell-wall biosynthesis (Zhao et al. 2008; Bhargava et al. 2013), gene deletion, and mutation library creation (Ge et al. 2013). Recent interest on cell-specific expression profiling has further provided impetus to the protoplast research (Robert et al. 2007; Tohge et al. 2011). In addition, plant protoplasts have also been employed as simple, fast, reliable, and versatile one-hybrid like screening system to investigate the interaction studies between transcription factors and their associated genes offering advantages over the yeast two-hybrid system (Thévenin et al. 2012).

The protoplast isolation from terrestrial plants was first initiated by Von Hannstein in 1880 by using the mechanical method and was subsequently refined by Plowe in 1931 (Cocking 1961). Cocking in 1961 showed the first successful isolation of protoplasts from tomato seedling root tips using cellulase extracted from fungus *Myrothecium verrucariae* (Cocking 1961). The successful regeneration of plant protoplasts with the enzymatic treatment of tissue was first published by Takebe et al. (1971) in tobacco, almost a decade later from the first report of protoplasts isolation. As compared with higher plants, marine macroalgal protoplast research is in a nascent stage and dates back to 1970.

7.2 State of the art

Similar to higher plants, the initial studies have followed mechanical methods for isolation of protoplasts from siphonaceous coenocytic algae (Tatewaki and Nagata 1970; Enomoto and Hirose 1972; Kobayashi 1975). The success in producing a large number of viable protoplasts has become possible only after the development of an enzymatic method by Millner et al. (1979) for *Enteromorpha intestinalis* (Linnaeus) Nees. Since then, considerable efforts have been made to isolate and culture protoplasts from diverse marine macroalgal species (Reddy et al. 2008, 2010). The protoplasts' yield, viability, and regeneration rate have been attributed to several factors, including the enzyme constituents, their concentrations, pH, osmotic conditions and ionic strength of the isolation medium, incubation temperature, physiological state and age of donor plant, and its culture conditions (Reddy and Fujita 1989, 1991; Krishna Kumar et al. 1999; Reddy et al. 2006). The complexity and associated variations in cell-wall composition of seaweeds further pose difficulties in protoplast preparation using the enzymes commonly used for terrestrial plants. The enzyme compositions thus far employed for protoplast preparation from seaweeds have been complex and include either one or a mixture of crude enzymes that were mostly sourced from microbes or marine animals (abalone, aplysia, and top shell) guts (Reddy et al. 2008). In addition to these complexities, most of the enzyme formulations previously employed for isolation of protoplasts from *Ulvophycean* species were made in seawater (Saga 1984; Saga and Kudo 1989; Uppalapati and Fujita 2002). The high-ionic potential of the medium was reported to interfere with studies on electrofusion, and other investigation on ion channels. Consequently, enzyme formulations made in distilled water supplemented with a higher concentration of osmolyte, for example, mannitol (1 M), have been preferred (Zhang 1983; Fujita and Migita 1985). However, the viability (30%) and regeneration abilities (50% of viable protoplasts) of such isolated protoplasts have been found to be markedly low as compared with those obtained with seawater-based enzyme mixture. The high-osmotic medium has been reported to affect the regeneration rate and subsequent morphogenesis of cultured protoplasts (Saga et al. 1986). The use of higher concentrations of osmolytes (sugar alcohols) is prone to cause microbial contamination imposing a gradual dilution of it in protoplast culture medium to improve and induce morphogenesis. Therefore, the development of a facile, reliable, and unified protocol for routine protoplast preparation with reproducible yield is of paramount importance to circumvent the inherent problems associated with the use of previously reported enzyme mixtures regarding variability in yield, regeneration rate, and developmental morphogenesis.

7.3 *Materials*

7.3.1 *Reagents*

- Sodium chloride (NaCl).
- 2-(*N*-Morpholino) ethanesulfonic acid.
- Dextran sulfate sodium salt.
- Mannitol.
- Cellulase Onozuka R-10 (Yakult Co. Ltd., Tokyo, Japan).
- Macerozyme R-10 (Yakult Co. Ltd., Tokyo, Japan).
- Deionized water (Millipore water).

7.3.2 *Equipment*

- Weighing balance (analytical type).
- Weighing boats.
- Magnetic stirrer with stirring rod and bar.
- pH meter.
- Refrigerated centrifuge.
- Centrifuge tubes (polypropylene).
- Tube racks.
- Cryovials.

- Cryovial storage box.
- Tissue chopping plates.
- Single-edge surgical blades.
- Orbital platform shaker.
- Glass funnel.
- Glass beakers.
- Nylon mesh (30 μm pore diameter).
- Centrifuge glass tubes (Pyrex).
- Petri dish (tissue culture grade).
- Glass pipettes.
- Microscope upright.
- Refrigerator (4°C and –20°C).

7.3.3 Reagent setup

Enzyme mixture is prepared by dissolving 2% of cellulase Onozuka R-10 in precooled (4°C) deionized water with 1% of NaCl, 25 mM of 2-(*N*-Morpholino) ethanesulfonic acid (Sigma-Aldrich, USA), and 0.5% dextran sulfate (Sigma-Aldrich, USA) adjusted to pH 6.0. Later, the mixture was centrifuged at 10,000 × g for 20 min at 4°C, and the clear supernatant was used for isolating protoplasts. Mannitol was added as osmoticum to a final concentration of 0.8 M in the enzyme solution just before commencing the protoplast isolation experiment (*Notes 1 and 2*).

7.4 Experimental procedures

1. Take clean and healthy thallus of 100 mg of fresh weight (*Note 3*).
2. Chop the tissue with a sharp razor blade to a size of <2 mm.
3. Transfer the chopped material to a glass tube.
4. Add autoclave seawater to it to remove debris.
5. Repeat step 4 for 3–4 times till clear supernatant was obtained.
6. Decant sweater completely from the glass tube.
7. Add enzyme solution to chopped tissue in a glass tube and transfer the material to a Petri dish.
8. Incubate the Petri dish at a temperature of 22°C–25°C on an orbital shaker at a speed of 60 rpm under dark.
9. After 2 h of incubation, take first microscopic observation for the release of protoplasts.
10. After the release of protoplasts, filter the solution using a nylon mesh of pore size of 30 μm and collect the solution in a centrifuge tube.
11. Centrifuge the solution at 800 rpm for 3–5 min (*Note 4*).
12. Remove the supernatant in another tube without disturbing the pellet (*Note 5*).
13. Wash the pellet with sterile seawater.

14. Repeat steps 11 and 12.
15. Gently dispense the pellet containing protoplasts in 1 mL of sterile seawater.
16. Estimate the protoplasts yield by traditional hemocytometer method.

The schematic description of the protocol is shown in Figure 7.1. The isolated protoplasts following this protocol from different species of genus *Ulva* were spherical in shape, ranging from 20 to 32 μm in size and contained a cup-shaped parietal chloroplast. The resultant protoplast yields, and regeneration rate are summarized in Table 7.1. Regeneration rate of protoplasts was calculated on the basis of the number of cells dividing and regenerating into germlings. Most of the protoplasts under optimal culture conditions (temperature 25°C and photon flux intensity of 15 μM m^{-2} s^{-1}) regenerated cell wall after 72 h and subsequently initiated cell division. The dedifferentiation of protoplast to develop germlings is shown in Figure 7.2. Cell-wall development was confirmed as the emission of blue fluorescence after staining with calcofluor white MR2 (Fluorescent brightener 28, Sigma Aldrich, USA) (Figure 7.2).

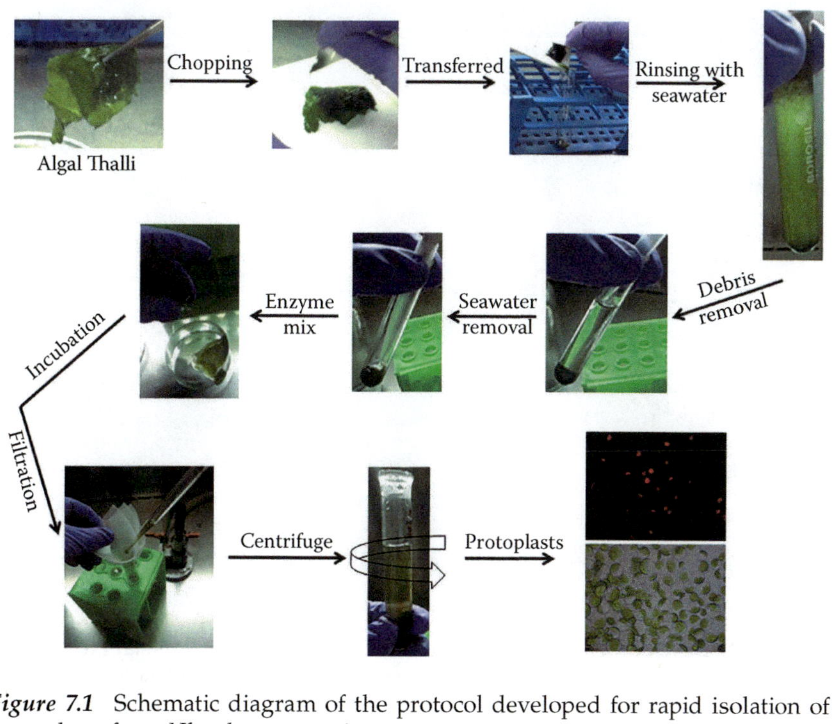

Figure 7.1 Schematic diagram of the protocol developed for rapid isolation of protoplasts from *Ulvophycean* species.

Table 7.1 Protoplasts yields and regeneration rates obtained from different *Ulvophycean* species following the described protocol

Species	Protoplasts yield (×10^6 cells/g fresh wt.)	Regeneration rate (%)
Ulva conglobata	85.2 ± 2.9	92.3 ± 6.8
U. fasciata	81.4 ± 2.1	90.7 ± 7.3
U. lactuca (wild)	74.5 ± 1.6	89.4 ± 5.6
U. lactuca (sterile mutant)	76.6 ± 1.2	91.2 ± 4.5
U. pertusa (sterile mutant)	74.4 ± 1.9	92.3 ± 6.2
U. reticulata	88.6 ± 2.3	90.3 ± 6.7
U. rigida	79.6 ± 1.1	89.5 ± 2.4
Enteromorpha tubulosa	89.7 ± 1.9	91.4 ± 4.7
E. flexuosa	95.4 ± 2.2	91.4 ± 4.8
E. compressa	82.5 ± 1.5	90.5 ± 5.9
Monostroma oxyspermum	102.8 ± 1.8	90.5 ± 6.5

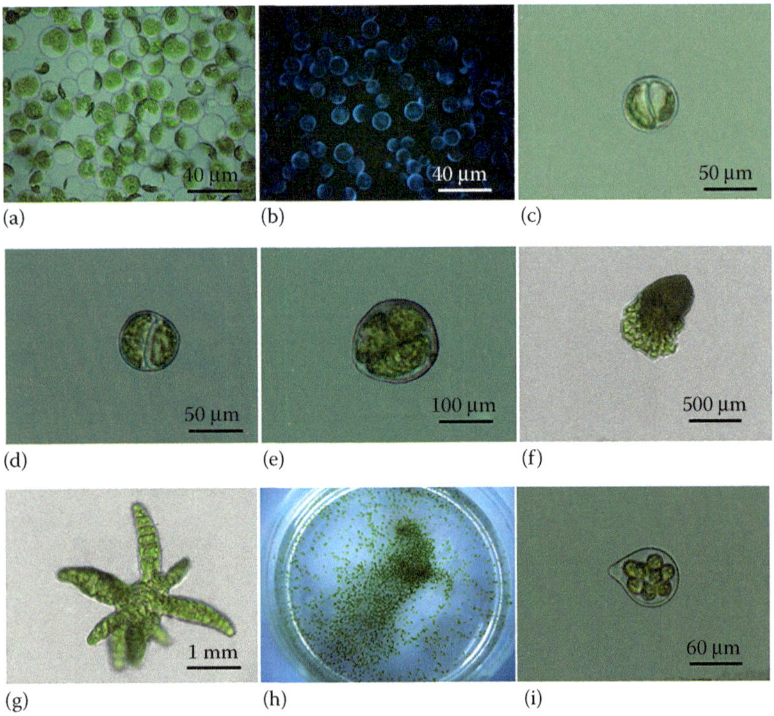

Figure 7.2 Developmental morphogenesis of cultured protoplasts. (a) Freshly isolated protoplasts, (b) Cell wall regeneration in 3 day old protoplast, (c) Protoplast regeneration after 4 days, (d) 6 days, (e) 8 days, (f) 12 days, and (g) 18 days, (h) small germlings in the Petri dish after 25 days, (i) 4 day old protoplast with spore formation.

7.5 Notes

Note 1: While preparing and handling the enzyme mix, it is advised to maintain the temperature at 4°C. Exposing the enzyme mix at room temperature for long durations is detrimental to its activity and results in poor or no yields of protoplasts.

Note 2: Store enzyme mix at –20°C.

Note 3: The exponentially growing algal tissue is required to obtain good yields of protoplasts (Table 7.2).

Note 4: Protoplasts are highly fragile because of the lack of rigid cell wall; therefore, they advised to handle very carefully. Shake or accidental jerks may lead to the rupture of released protoplast.

Note 5: The supernatant after step 12 can be collected and stored at –20°C for further reuse as enzyme mix for protoplast preparation. This can be continued for five times.

Table 7.2 Troubleshooting table

Problem	**Possible reason**	**Solution**
Poor yields	Health of the algal tissue	Assure the exponential growth of the alga before processing for protoplasts isolation
	Chopping	Assure fine chopping of the tissue to the level of size of <2 mm
	pH of the enzyme mix	Check and adjust the pH of the enzyme mix to 6.0
	Enzyme handling	Enzyme mix should be stored at low temperature
	Incubation temperature	Check and incubate the algal tissue with enzyme mix at temperature of 25°C
Bursting or shrinking of released protoplasts	Osmotic balance	Check and adjust the osmotic balance of protoplasts isolation medium with 0.8 M mannitol along with 2% NaCl (w/v)
Low-regeneration rate	Handling error	Assure gentle handling of protoplasts as they rupture with shake and jerks
	Centrifugation	Check and centrifuge at low speed in swing-bucket rotor
	Use of pipettes	Check and use wide mouth glass pipettes instead of micropipette

Acknowledgments

The financial support received from the FP7 program of European Commission under grant agreement no. 241383 is gratefully acknowledged. VG thanks, CSIR-NIO, Goa for infrastructural support in writing this protocol.

References

Bajaj, Y.P.S. 2004. Biotechnology in agriculture and forestry. In *Somatic Hybridization in Crop Improvement I*, vol. 27. Springer-Verlag, Berlin, Germany.

Bhargava, A., Ahad, A., Wang, S., Mansfield, S.D., Haughn, G.W., Douglas, C.J., and Ellis, B.E. 2013. The interacting MYB75 and KNAT7 transcription factors modulate secondary cell wall deposition both in stems and seed coat in *Arabidopsis*. *Planta* 237: 1199–1211.

Cocking, E.C. 1961. Properties of isolated plant protoplasts. *Nature* 191: 780–782.

Costa, A. et al. 2012. The *Arabidopsis* central vacuole as an expression system for intracellular transporters: Functional characterization of the Cl^-/H^+ exchanger CLC-7. *J. Physiol.* 590: 3421–3430.

Davey, M.R., Anthony, P., Power, J.B., and Lowe, K.C. 2005. Plant protoplasts: Status and biotechnological perspectives. *Biotechnol. Adv.* 23: 131–171.

Eeckhaut, T., Lakshmanan, P.S., Deryckere, D., van Bockstaele, E., and van Huylenbroeck, J. 2013. Progress in plant protoplast research. *Planta* 238: 991–1003.

Enomoto, K. and Hirose, H. 1972. Culture studies on artificially induced aplanospores in the marine alga *Boergesenia forbesii* (Harvey) Feldman (Chlorophyceae, Siphonocladales). *Phycologia* 11: 119–122.

Fujita, Y. and Migita, S. 1985. Isolation and culture of protoplasts from seaweeds. *Bull. Fac. Fish. Nagasaki Univ.* 57: 39–45.

Ge, C.Y., Duan, Y.B., Zhou, M.G., and Chen, C.J. 2013. A protoplast transformation system for gene deletion and complementation in *Sclerotinia sclerotiorum*. *J. Phytopathol.* 161: 800–806.

Huha, S.U., Kimb, M.J., and Paeka, K.H. 2013. *Arabidopsis* Pumilio protein APUM5 suppresses Cucumber mosaic virus infection via direct binding of viral RNAs. *Proc. Natl. Acad. Sci. USA.* 110: 779–784.

Jiang, F., Zhu, J., and Liu, H.L. 2013. Protoplasts: A useful research system for plant cell biology, especially dedifferentiation. *Protoplasma* 250: 1231–1238.

Kobayashi, K. 1975. Growth of extra cellular protoplast of *Bryopsis maxima* in an agar medium. *Bull. Tokyo Gakugei Univ. Ser.* 27: 1–5.

Krishna Kumar, G.R., Addepalli, M.K., and Reddy, C.R.K. 1999. Regeneration of the thallus of *Monostroma oxyspermum* (Chlorophyta) from protoplasts in axenic culture. *Phycologia* 38: 503–507.

Li, E., Wang, S., Liu, Y., Chen, J.G., and Douglas, C.J. 2011. Ovate family protein4 (OFP4) interaction with KNAT7 regulates secondary cell wall formation in *Arabidopsis thaliana*. *Plant J.* 67: 328–341.

Miao, Y. and Jiang, L. 2007. Transient expression of fluorescent fusion proteins in protoplasts of suspension cultured cells. *Nat. Protoc.* 2: 2348–2353.

Millner, P.A., Callow, M.E., and Evans, L.V. 1979. Preparation of protoplasts from the green alga *Enteromorpha* intestinalis (L.). *Planta* 147: 174–177.

Pitzschke, A. and Persak, H. 2012. *Poinsettia* protoplasts—A simple, robust and efficient system for transient gene expression studies. *Plant Methods* 8: 14.

Reddy, C.R.K., Dipakkore, S., Krishnakumar, G.R., Jha, B., Cheney, D.P., and Fujita, Y. 2006. An improved enzyme preparation for rapid mass production of protoplasts as seed stock for aquaculture of macrophytic marine green algae. Aquaculture 260: 290–297.

Reddy, C.R.K. and Fujita, Y. 1989. Protoplasts isolation and fusion of *Ulvapertusa* and *U. conglobata*. In *Current Topics in Marine Biotechnology,* Miyachi, S., Karube, I., and Ishida, Y. (Eds.), pp. 235–238. The Japanese Society of Marine Biotechnology, Tokyo, Japan.

Reddy, C.R.K. and Fujita, Y. 1991. Regeneration of plantlets from *Enteromorpha* (Ulvales, chlorophyta) protoplasts in axenic culture. *J. Appl. Phycol.* 3: 265–275.

Reddy, C.R.K., Gupta, M.K., Mantri, V.A., and Jha, B. 2008. Seaweed protoplasts: Status, biotechnological perspectives and needs. *J. Appl. Phycol.* 20: 619–632.

Reddy, C.R.K., Gupta, V., and Jha, B. 2010. Developments in biotechnology of red algae. In *Red Algae in the Genomic Age, Cellular Origin, Life in Extreme Habitats and Astrobiology,* Seckbach., J. and Chapman, D.J. (Eds.), pp. 1309–1341. Springer Publication, Dordrecht, the Netherlands.

Robert, S., Zouhar, J., Carter, C., and Raikhel, N. 2007. Isolation of intact vacuoles from *Arabidopsis* rosette leaf–derived protoplasts. *Nat. Protoc.* 2: 259.

Saga, N. 1984. Isolation of protoplasts from edible seaweeds. *Bot. Mag. Tokyo* 97: 423–427.

Saga, N. and Kudo, T. 1989. Isolation and culture of protoplasts from the marine green alga *Monostroma angicava*. *J. Appl. Phycol.* 1: 25–30.

Saga, N., Polne-Fuller, M., and Gibor, A. 1986. Protoplasts from seaweeds: Production and fusion. *Beih Nova Hedwegia* 83: 37–43.

Takebe, I., Labib, G., and Melchers, G. 1971. Regeneration of whole plants from isolated mesophyll protoplasts of tobacco. *Naturwissenschaften* 58: 318–320.

Tatewaki, M. and Nagata, K. 1970. Surviving protoplasts in vitro and their development in *Bryopsis*. *J. Phycol.* 6: 401–403.

Thévenin, J. et al. 2012. A new system for fast and quantitative analysis of heterologous gene expression in plants. *New Phytol.* 193: 504–512.

Tohge, T. et al. 2011. Toward the storage metabolome: Profiling the barley vacuole. *Plant Physiol.* 157: 1469–1482.

Uppalapati, S.R. and Fujita, Y. 2002. A simple method for mass isolation of protoplasts from species of *Monostroma, Enteromorpha* and *Ulva* (chlorophyta, Ulvales). *J. Appl. Phycol.* 14: 165–168.

Wehner, N., Hartmann, L., Ehlert, A., Bottner, S., Onate-Sanchez, L., and Droge-Laser, W. 2011. High-throughput protoplast transactivation (PTA) system for the analysis of *Arabidopsis* transcription factor function. *Plant J.* 68: 560–569.

Xiao, L., Zhang, L., Yang, G., Zhu, H., and He, Y. 2012. Transcriptome of protoplasts reprogrammed into stem cells in *Physcomitrella patens*. *PLoS One* 7(4): e35961.

Yoo, S., Cho, Y., and Sheen, J. 2007. *Arabidopsis* mesophyll protoplasts: A versatile cell system for transient gene expression analysis. *Nat. Protoc.* 2: 1565–1572.

Zhang, D. 1983. Study on the protoplasts preparation, culture and fusion of somatic cells from two species of green algae *Ulva linza* and *Monostroma angicava* Kjellm. *J. Shandong Coll. Oceanol.* 13: 57–65.

Zhang, Y. et al. 2011. A highly efficient rice green tissue protoplast system for transient gene expression and studying light/chloroplast-related processes. *Plant Methods* 7: 30.

Zhao, M., Morohashi, K., Hatlestad, G., Grotewold, E., and Lloyd, A. 2008. The TTG1-bHLH-MYB complex controls trichome cell fate and patterning through direct targeting of regulatory loci. *Development* 135: 1991–1999.

chapter eight

Purification of sporulation and swarming inhibitors from Ulva

Application in algal life-cycle controlling

Ralf W. Kessler, Taghreed Alsufyani, and Thomas Wichard

Contents

8.1 Introduction

Ulva species (Chlorophyta), also known as sea lettuce, are the most abundant representatives of green macroalgae and can be found in coastal benthic areas worldwide. They cause seasonal algal blooms in eutrophic waters and even in areas contaminated with heavy metals (Gao et al. 2010). The model organism, *Ulva mutabilis*, is a multicellular haplodiplontic alga, which differentiates into a thallus and a holdfast. The thallus is made of blade cells and the holdfast of stem and rhizoid cells (Løvlie 1964). The blade cells develope into gametangia (sporangia) and release 16 biflagellated gametes per gametangium (16 tetraflagellated zoospores) (Løvlie 1964, 1968; Wichard and Oertel 2010). After merging of the two mating types of the gametes mt+ and mt–, zygotes are formed which develop into the sporophytes. However, if the gametes do not find a mating partner, they develop into new gametophytes, which allow the maintenance of haploid strains (Wichard 2015, see *Ulva*'s life cycle in Chapter 9 by Califano and Wichard 2018).

The sporulation processes, gametogenesis and zoosporogenesis, are key parts of *Ulva* reproduction and still an understudied field. Both the gametophytes and sporophytes of the haplodiplontic *Ulva* have the same morphological characteristics but form either biflagellated gametes or quadriflagellated zoids. Sporulation can artificially be induced by cutting the mature algae into smaller pieces and washing them. This treatment removes sporulation inhibitors (SI-1 and SI-2), which regulate the differentiation of blade cells into gametangia and sporangia in *Ulva*. (Nilsen and Nordby 1975; Løvlie 1978; Stratmannet al. 1996). In addition, further studies revealed that switching temperatures, as experienced by certain tropical *Ulva* species, can induce the cell differentiation (Carl et al. 2014). Moreover, Lüning et al. (2008) found that the reproduction rhythmicity of *U. pseudocurvata* is also controlled by the moon phases (Lüning et al. 2008).

The gametogenesis is subdivided into a determination phase (36 h), within vegetative growth could be resumed by addition of SIs, and a differentiation phase afterwards, insensitive to SIs when gametogenesis occurred. (Wichard and Oertel 2010; Katsaros et al. 2017). After the point of no return (36 h after induction), 16 progametes are formed and the pore caps are completed during the third night after induction. The molecular changes during cell differentiation processes can also be monitored through specific biomarkers in *Ulva* by mass spectrometric imaging (Kessler et al. 2017). The gamete discharge is controlled by the swarming inhibitor (SWI), a low-molecular weight factor (Wichard and Oertel 2010). On removal of the SWI, the gametes are discharged in the light. As the orchestrated action of the inhibitors synchronizes both the sporulation processes and the release of gametes, the mating probability of the gametes can be highly increased. Sporogenesis and zoid release in sporophytes are induced principally in the same way as in gametophytes, but the latter are mainly used in this protocol for the purification of regulative factors.

8.2 State of the art

Originally, sporulation inhibiting substances were extracted from a suspension of living fragments of *U. mutabilis* (Nilsen and Nordby 1975). The factor can be washed out and purified for further characterization by cutting *U. mutabilis* into small pieces (Stratmann et al. 1996). Stratmann et al. (1996) showed, by using bioassay-guided size-exclusion chromatography (SEC), that the SI-2 activity belongs to a bouquet of at least three different factors, equally biologically active, with molecular weights around 1000 Da. Moreover, the same research group extracted another SI, a glycoprotein (SI-1), from both the *Ulva* cell wall and *Ulva* culture medium (UCM). This protein is unusually stable and showed an extremely high-apparent molecular mass of $1–4 \times 10^7$ Da extracted from axenic callus-like cultures of *Ulva*. However, when *Ulva* developed normally in the presence of their symbiotic bacteria, smaller forms of SI-1 accumulated in the growth medium because of the bacterial decomposition processes. In this context, gametogenesis-inhibiting glycoproteins ($2–4 \times 10^4$ Da) were also found in the water-soluble extracts of the thallus of *U. prolifera* (Jonsson et al. 1985). Further attempts to characterize the glycoprotein revealed that the substance binds to an anionic ion exchanger and to concanavalin A-Sepharose, which captures glycosylated biomolecules.

Significantly, cross-testing studies between *Ulva* species showed that the SI-2 factor was interchangeable between *U. mutabilis* and *U. rigida*, but not the SI-1 factor, whereas both factors were interchangeable between *U. mutabilis* and *U. linza* (Stratmann et al. 1996; Vesty et al. 2015). Gametes are finally discharged by the gametangium when an SWI, which is released into the growth

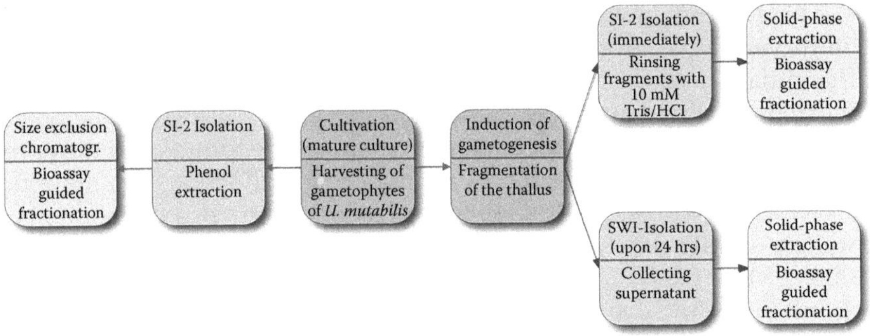

Figure 8.1 *All-in-one workflow for the partial purification of known inhibitors* (sporulation inhibitor 1 [SI-1], sporulation inhibitor 2 [SI-2], swarming inhibitor [SWI]) regulating the gametogenesis, and release of gametes in *Ulva*. On removal of SI-1 and SI-2, the SWI production can be induced by putting the fragmented algae into fresh UCM. The best SI-1 extraction results can be achieved with axenic or with young *Ulva* thalli.

medium during gametogenesis, declines below a threshold concentration. After C_{18} solid-phase extraction, the mass spectrometric analysis revealed the molecular ion of SWI at *m/z* 293 [M+H$^+$] (Wichard and Oertel 2010).

The presented extraction protocol aims to purify all inhibitors from one upscaled *Ulva* culture (Figure 8.1) and is based on earlier studies (Stratmann et al. 1996; Vesty et al. 2015). Overall, the approach provides partially purified regulative factors, for example, for studies in algal aquaculture for life-cycle controlling.

8.3 Materials

8.3.1 Ulva *culture*

The material necessary for the standard cultivation of *Ulva* was described previously (Stratmann et al. 1996; Alsufyani et al. 2017). Gametophytes of the same species, *U. mutabilis* (Føyn), were also used in the protocols of the Chapters 9, 18, and 31.

8.3.2 *Culturing of* U. mutabilis *on a large scale*

Standardized cultures grown preferentially under axenic conditions or with defined associated bacteria are recommended for reproducible extraction yields of the inhibitors (Stratmann et al. 1996, Alsufyani et al. 2017).

- 10 L UCM (see Chapter 9).
- 25 L polycarbonate bottle and sterile filters for air bubbling system (Alsufyani et al. 2017).

- About 1000–5000 two-week-old *U. mutabilis* germlings (ca. 5 mm length) as seed stock (see Chapter 9 for detailed information about the preparation of the seed stock).
- Bacterial cultures of *Maribacter* sp. MS6 and *Roseobacter* sp. MS2 cultured in half-strength Marine Broth (Carl, Roth, 40 g L^{-1} solved in water) in sterile culture flasks (50 mL). A quantity of 1 mL of each bacterial culture (optical density at 620 nm = 1.0) is added to 10 L of UCM.
- An ultraviolet–visible (UV–vis) spectrometer for monitoring the optical density of bacterial growth in Marine Broth.
- Autoclavable plastic sieve or mesh nets for harvesting *Ulva.*

8.3.3 *Harvesting of* Ulva *biomass for the extraction and the bioassay-guided fractionation of inhibitors*

On account of the inducibility of gametogenesis and the determination of SWI activity, three phases (Table 8.1) of growth can be defined in the tripartite community of *U. mutabilis* (gametophytes of the natural developmental mutant, slender) (Alsufyani et al. 2017). The time point of harvesting of the *Ulva* samples has to be carefully selected depending on the purpose of the experiment (Table 8.1).

8.3.4 *Induction of gametogenesis/sporogenesis and bioassay-guided fractionation of the inhibitors*

- Growth stage of *Ulva*: Artificially inducible gametogenesis (A.I.G.) (Table 8.1).
- *Ulva* culture medium (Stratmann et al. 1996).
- Reverse osmosis water.
- Herb chopper (e.g., Zick-Zick Food Chopper, Zyliss, Switzerland).

Table 8.1 Growth stage-dependent inducibility of sporulation in *Ulva mutabilis*

Growth stage	Age of culture on inoculation	Recommended for
Noninducible gametogenesis (N.I.G.)	<14 days	- SI-1 and SI-2 extractions.
Artificially inducible gametogenesis (A.I.G.)	14–21 days	- Combined extraction of SI-2 and SWI. - Preparation of any bioassay described.
Spontaneously inducible gametogenesis (S.I.G.)	>21 days	- The culture already contains a mixture of different growth stages. Gametogenesis of thalli still not differentiated can be induced to produce gametes for other purposes.

- Pipettes (1–100 μL) for preparing a dilution series.
- A total of 24- or 96-well plate (Sarstedt, Nümbrecht, Germany) for performing multiple bioassays in parallel.
- Inverted microscope for monitoring the differentiation of blade cells into gametangia (or sporangia) and the discharge of the gametangia.

8.3.5 Sporulation inhibitor 1: Extraction

- *Growth stage of* Ulva: *Noninducible gametogenesis (N.I.G.)* or axenic cultures (Table 8.1).
- Liquid nitrogen.
- Pestle and mortar.
- Dounce tissue grinder set (Sigma Aldrich, München, Germany).
- *Buffer*: 10 and 50 mM Tris–HCl (pH 8.0) at room temperature.
- *Extraction solution*: Phenol saturated with 100 mM Tris–HCl and 1 mM EDTA adjusted to pH 7.5 in a brown glass bottle at room temperature.
- Three-necked flask.
- Acetone.
- Absolute ethanol at –20°C.
- Bench top refrigerated centrifuge (at 0°C).
- Camaprene® (Honeywell, USA) or equivalent security gloves.

8.3.5.1 Sporulation inhibitor 1: Size-exclusion chromatography for cleanup

- Sephadex G-50-Superfine (GE Healthcare, Germany); size: 18.8 × 1.1 cm column; solvent: 30 mM NaCl, 10 mM Tris–HCl (pH 8) at 5.3 mL h^{-1}.
- Peristaltic pump.
- Bradford protein quantification assay Rotiquant® (Carl Roth, Karlsruhe, Germany).
- UV–vis spectrometer.

8.3.6 Sporulation inhibitor 2: Extraction and bioassay-guided fractionation

- *Growth stage of Ulva*: We recommend using *Ulva* material of the growth stage A.I.G. to combine the extraction of SI-2 and SWI (Table 8.1).
- 10 mM Tris–HCl (pH 8.0).
- *Different filters and membranes*: Stainless steel filter in a funnel, Whatman GF/C glass microfiber filter (GE Healthcare, Germany) and cellulose acetate membranes (CA, 0.45 μm) (Carl Roth, Karlsruhe, Germany).

- Biotech dialysis membrane tubing (cutoff >500 Da <1000 Da) (SpectrumLabs, Rancho Dominguez, USA).
- Sephadex LH-20 (GE Healthcare, Freiburg, Germany), column size: 27.5 × 1.7 cm, solvent: 1 mM ammonium acetate buffer at 5.3 mL h^{-1}.
- Bench top refrigerated centrifuge (at 0°C).

8.3.7 *Swarming Inhibitor: Extraction and bioassay-guided fractionation*

- *Growth stage of* Ulva: A.I.G. (Table 8.1).
- *Solid-phase extraction (SPE)*: Sep-Pak C18 Plus Long Cartridge, 900 mg sorbent per cartridge, 37–55 μm particle size (Waters, UK).
- *Solvent for elution from SPE*: Aqueous methanol.
- *Fast protein liquid chromatography (FPLC)*: Column Superdex™ Peptide 10/300 GL (Sigma Aldrich, München, Germany), 30 × 1 cm.
- *Ultrahigh pressure liquid chromatography*: The column was a Polar-Premium solvent (10 cm × 3 mm, particle size 2.6 μm; ThermoFisher Scientific, Massachusetts, USA). The linear gradient was performed with water (A) and acetonitrile (B) at a flow rate of 0.4 mL min^{-1} over 11.5 min. The gradient was from 0% to 40% B in 1.0 min, from 40% to 52% B in 8 min, and from 52% to 100% (B) in 0.5 min, followed by 1 min at initial conditions (0% B). The LC was coupled to a QExactive–Quadrupole–Orbitrap LC–MS (Thermo Scientific, Massachusetts, USA). The temperature of the source was maintained at 120°C and the desolvation temperature at 400°C. The cone voltage was 10 V.
- *High-pressure liquid chromatography (HPLC)*: The column was a Purospher® Star RP-18e (25 × 1 cm, particle size 5 μm; MerckMillipore, Massachusetts, USA). The linear gradient was performed with water (A) and acetonitrile (B) at a flow rate of 2 mL min^{-1} over 36 min. The gradient was from 0% to 40% B in 8 min; from 40% to 52% B in 20 min, and from 52% to 100% in 2 min, followed by a hold for 5 min at initial conditions (0% B).
- SpeedVac concentrator (SpeedVac RVC 2-25, Christ, Osterode, Germany) to evaporate solvent.

8.4 *Experimental procedures*

8.4.1 *Cultivation of* U. mutabilis *under standardized conditions*

1. Haploid gametophytes from the fast-growing developmental mutant *slender* (sl) of *U. mutabilis* Føyn (mt+) are cultured. A seed stock of axenic gametes should be prepared for the cultivation of *Ulva* in bioreactors (see Chapter 9).

2. Gametes are separated from the accompanying bacteria according to the well-established procedure by Spoerner et al. (2012) and are finally used for the inoculation of an axenic (bacteria free) culture of *Ulva* in UCM.
3. Axenic gametes (about 10^3 gametes) are incubated overnight in 250 mL sterile UCM in polycarbonate tissue culture flasks (BD Falcon, Franklin Lake, NJ, USA) in the dark, allowing settlement of the germ cells.
4. The bacterial strains, *Roseovarius* sp. MS2 (Genbank EU359909) and *Maribacter* sp. MS6 (Genbank EU359911), essential for *Ulva*'s growth and morphogenesis, are cultivated at 20°C in Marine Broth medium on an orbital shaker. These bacteria were originally isolated from *U. mutabilis*. The bacterial pellets were washed three times by resuspending them with sterile UCM before axenic germlings were inoculated with the two bacteria to establish the tripartite community.
5. Axenic germlings (feedstock) are used for the inoculation of the 25 L bioreactors filled with 15 L UCM (Figure 8.2).
6. *Standard culture conditions*: Cultures are illuminated in a 17:7 h light/dark regime at 20°C with 60–120 µmol photons ($m^{-2} s^2$) (50% GroLux, 50% daylight fluorescent tubes; OSRAM, München, Germany).

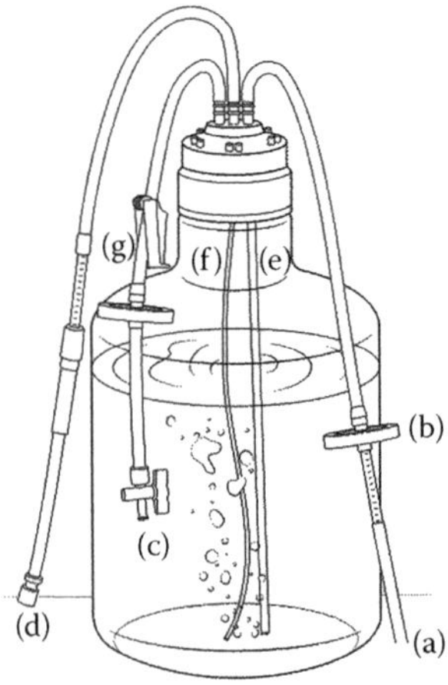

Figure 8.2 *Drawing of the 25 L bioreactor used for cultivation of Ulva mutabilis*: (a) air inlet; (b) sterile HEPA-Vent (Ø = 50 mm, Whatman) filter; (c) air outlet; (d) sampling outlet; (e) bubbling tube (Duran glass, Ø = 4 mm); (f) sampling tube (Teflon, Ø = 1 mm); and (g) hose clamp to control the sampling flow. (Adapted from Alsufyani, T. et al. 2017, With permission, CC-BY.)

8.4.2 Induction of the gametogenesis/ zoo sporogenesis and preparation of the bioassays

With the described conditions (8.4.1), spontaneous gametogenesis does not occur until algae are older than four weeks, but gametogenesis can be artificially induced by the removal of SIs (Alsufyani et al. 2017) (Table 8.1). Culture conditions certainly have a strong impact on the age-dependent growth stages influencing the inducibility of the gametogenesis.

1. Wash five *U. mutabilis* fertile thalli (3–5 cm length) with UCM and incubate them in 0.2 L UCM for 15 min at room temperature.
2. Mince the alga into 3–5 mm^2 by using the herb chopper and incubate the fragments in 50% UCM for 15 min at room temperature; repeat the washing step twice (for further details see Chapter 9). Recover the algal fragments through a mesh of 100 μm.
3. Prepare the dilution series of the potentially active test sample with UCM in the 96-well plate. The total volume per well is 200 μL (Figure 8.3a). The ratio of a standard dilution series might be, for example, 1:5, 1:10, 1:50, and 1:100, but the dilution factor must be adjusted if the SWI concentration is higher.
4. *Do not forget a negative control*: Only UCM (i.e., no inhibitor) is added and, thus, all blade cells should differentiate into gametangia. In the case of the positive control, a defined amount of SI (100 U mL^{-1}, see definition in step 9) of the laboratory quality control sample is added. The blade cells should, thus, not differentiate into gametangia or sporangia, respectively (Figure 8.3a). The quality control samples should be treated in the same manner as the test samples and are used to validate the test run.
5. Transfer the sieved algal pieces with forceps into the multiwell plates (40–80 pieces per well). When transferring the fragments (from step 2) into the test well, remove the leftover liquid with a paper towel to avoid the dilution of the test solution.
6. The algal fragments should be cultured under standard conditions for the next three days.
7. On the morning of the third day after induction, the fragments must be inspected in each test well under an inverted microscope. Those fragments that show on blade cells are classified as noninduced; all others with gametangia are classified as induced (Figure 8.3b). *If the positive control does not show any inhibition, the experiment needs to be repeated with a new batch of younger algae, which can still perceive the sporulation inhibitor.*

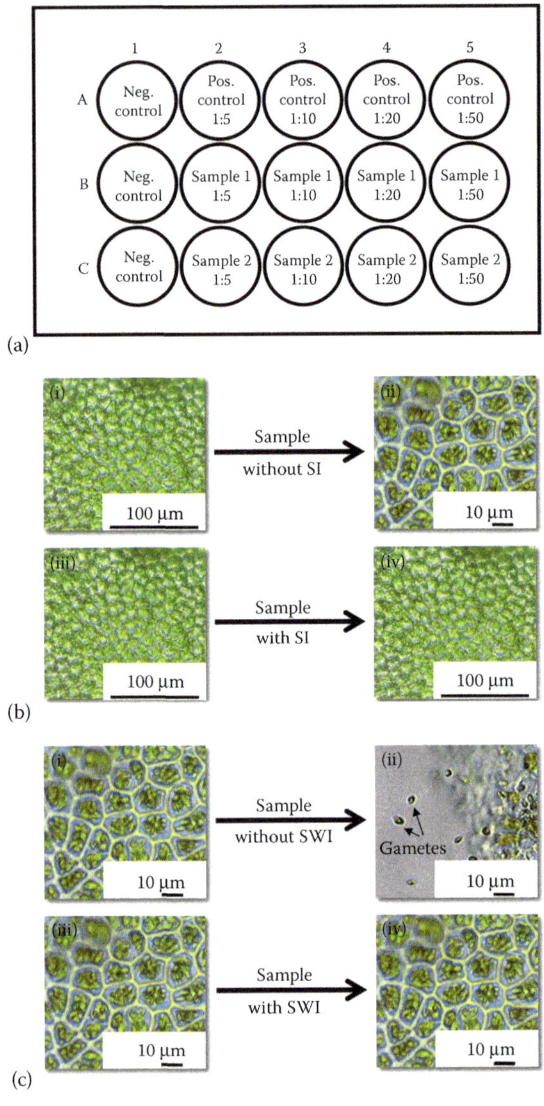

Figure 8.3 *Set up for both sporulation and swarming inhibitor bioassays.* (a) Multiwell-based bioassay-guided fractionation and semiquantification through dilution series of the biologically active samples. (a) The general principle of the bioassay activity test for the sporulation inhibitors (SI-1 or SI-2) or the swarming inhibitor (SWI) is shown: The negative control contains no inhibitor, all blade cells develop into gametangia. The positive control contains a defined amount of SI or SWI. (b) Light microscopy reveals the blade cells before the experiment (i, iii). Without SI-1 or SI-2 in the test sample, (ii) cells develop into gametangia within three days. (c) Gametangia are shown (i, iii). Without SWI in the test sample, (ii) gametangia are completely discharged within 15 min. The effect of inhibitory samples for both bioassays (b, c) is shown in (iv).

8. As a proof of principle, induced fragments discharge the gametes after an additional medium exchange. The whitish fragments can be counted easily and subtracted from the total number of fragments investigated (Equation 8.1):

$$\text{Inhibitory rate}\,(\%) = \frac{\left(\text{Total amount of fragments} - \text{Amount of fragments with gametangia}\right)}{\text{Total amount of fragments}} * 100 \quad (8.1)$$

9. 1 U of the SI-1 and SI-2 is hereby defined as the minimal amount of the factor that inhibits differentiation of a blade cell into a gametangium (=gametogenesis) with an inhibitory rate >50% in 1 mL UCM within three days under standard conditions. If the test sample, for example, diluted in a ratio of 1:5 reveals an inhibitory rate of 85%, but only 40% in a ratio of 1:10, the sample contains 5 U mL^{-1} of the inhibitor. Fragments (size: 1–5 mm^2) sometimes show a mixture of the blade and differentiated cells. Those fragments counted as induced, which harbor more than 50% of differentiated cells (i.e., gametangia).

8.4.3 Sporulation inhibitor 1

Blade cells of *U. mutabilis* and *U. linza* excrete regulatory high-molecular weight factors, which are probably unknown glycoproteins of various molecular weights, into their cell wall and into the environment. As SI-1 accumulates in the growth medium, it can be purified from media of (preferentially) axenically growing algae, whose life cycle and morphogenesis were abnormal (Spoerner et al. 2012). Mostly smaller SI-1 forms were detected in the growth medium and in the cell walls under nonaxenic conditions (Stratmann et al. 1996), because of potentially enzymatic digestion by the bacteria.

8.4.3.1 Extraction of the sporulation inhibitor 1 from thallus

1. The extraction of SI-1 must be carried out with a phenol solution. Therefore, gloves (e.g., Campren®) are recommended for safety reasons.
2. Wash 1 g of thalli with distilled water and grind in liquid nitrogen using a pestle and mortar.
3. Add 5 mL 50 mM Tris–HCl (pH 8.0) and treat the suspension with a Dounce homogenizer (*Note 1*).
4. Insoluble material can be recovered by centrifugation (3800 rpm, 15 min, 0°C).
5. Wash the insoluble material with the same buffer once more and then suspend it in 2 mL of 10 mM Tris–HCl (pH 8.0) buffer.
6. Mix the suspension at 60°C with the same volume of phenol (saturated with 100 mM Tris–HCl, 1 mM EDTA, pH 7.5) and mix vigorously for 20 min on an orbital shaker.

7. After cooling down the sample, the phases are separated by centrifugation.
8. The phenol phase must be washed with 10 mM Tris–HCl (pH 8.0) by again mixing, separating, and collecting the phenol phase.
9. Three volumes of acetone are added to the phenol phase and incubated at –20°C for ≥30 min (*Note 2*).
10. The precipitate can be recovered by centrifugation (3800 rpm, 20 min, 0°C).
11. Wash the precipitate twice with absolute ethanol (*Note 3*).
12. After drying in the vacuum, resuspend the precipitate with 100 mM Tris–HCl (pH 8.0) and store the sample at –20°C for further purification steps.

8.4.3.2 Extraction of the sporulation inhibitor 1 from (axenic) growth medium

1. SI-1 should be prepared from growth medium of an *Ulva* specimen, whose gametogenesis is not inducible (N.I.G.).
2. Growth medium must be filtered under sterile conditions through a cellulose-acetate filter (0.22 μm).
3. Mix vigorously 1 L conditioned medium with 100 mL phenol (saturated with 100 mM Tris–HCl, 1 mM EDTA, pH 7.5) on a shaker for 15 min at room temperature.
4. After centrifugation, transfer the phenol phase to a (siliconized) plastic centrifuge tube.
5. This extraction should be repeated once and the phenol phases can be combined afterward (*Note 4*).
6. After reextraction with the same volume of 10 mM Tris–HCl (pH 8.0), the phenol phase was mixed with three volumes of acetone and incubated at –20°C for >30 min (*Note 2*).
7. The precipitate can be recovered by centrifugation (3800 rpm, 20 min, 0°C) and washed twice with absolute ethanol (*Note 3*).
8. After drying in the vacuum, suspend the precipitate with 100 mM Tris–HCl (pH 8.0) and store the sample at –20°C for further purification steps.

8.4.3.3 Bioassay-guided cleanup of sporulation inhibitor 1 with size-exclusion chromatography

1. Prepare the Sephadex G-50 column and equilibrate the gel bed with the mobile phase (10 mM Tris–HCl).
2. Add the SI-1 extract carefully into the column and follow the standard procedure for SEC.
3. Perform the bioassay as described in 8.4.2 to identify the fraction with SI-1 activity (Figure 8.4a).

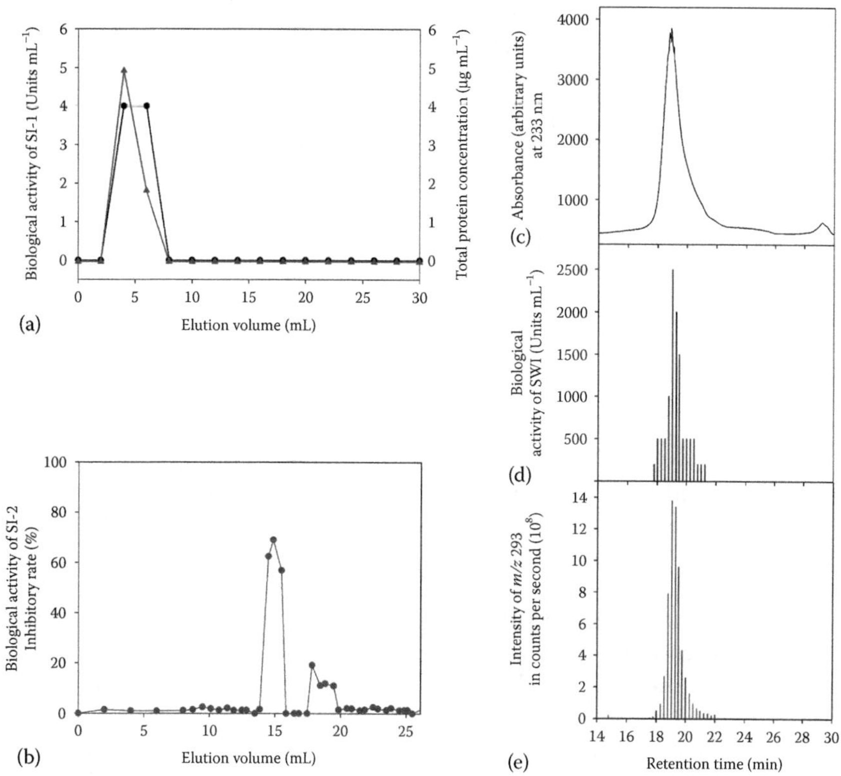

Figure 8.4 *Bioassay-guided fractionations and cleanup of the various inhibitors by (a, b) size-exclusion chromatography (SEC) and (c–e) fast performance liquid chromatography (FPLC)*: (a) SEC (Sephadex G-50) of the sporulation inhibitor 1 (SI-1) extracted from UCM of axenic *Ulva* cultures. Biological activity (dark-black line) was determined by the sporulation inhibitor bioassay, and the total protein concentration was determined by a Bradford assay (light-gray line). (b) SEC of the sporulation inhibitor 2 (SI-2) extract (50 U mL^{-1}). The bioassay was performed in dilution steps of 1:5. The SI-2 was biologically determined, if the inhibitory rate was higher than 50%. (c–e) FPLC UV–vis of the SWI extracted 24 h on induction of the gametogenesis. The elution was monitored (c) by a UV detector at 233 nm, (d) by the SWI-activity through the bioassay, and (e) by a mass spectrometric analysis (UHPLC-QExactive-Orbitrap-MS) at *m/z* 293 [M+H⁺]. The elution rate was 0.25 mL min^{-1} and fractions were collected every 0.25 min.

4. Prepare the Bradford assay:
   - Add 10 µL Rotiquant® to 180 µL of each SI-1 fraction.
   - Dilute with 200 µL distilled water and homogenize the solution.
   - Measure the absorbance at 595 nm.
   - Perform the calibration of the analysis with albumin by using five equidistant concentrations and a minimum of three replicates for each calibration standard.
5. The partially purified inhibitor can now be applied, for example, for experiments of life-cycle controlling in *Ulva*.

8.4.4 *Sporulation inhibitor 2*

8.4.4.1 *Extraction of the sporulation inhibitor 2*

The SI-2 can be found in the lumen of the bilayered thallus of *U. mutabilis*, *U. linza*, and *U. rigida* (Stratmann et al. 1996; Vesty et al. 2015). As the SI-2 concentration is constant during the life time of the alga (Stratmann et al. 1996), the time point of harvesting is not as critical as in the case of purification of the SI-1 and SWI. However, the extraction should be carried out with a dense solution of *Ulva* fragments (2–5 mm^2). After the exhaustive extraction of both SI-1 and SI-2, the gametogenesis is induced and the SWI production starts. The following procedure for SI-2 extraction is based on 1 g fresh weight of *U. mutabilis* fragments. *Ulva* harvested directly from natural seawater samples must be cleaned more carefully before processing. Alga should be washed and spin-dried with a salad spinner in the first place for the processing of larger amounts of biomass.

1. Harvest *Ulva*, dry it carefully with paper towels, and weigh the biomass.
2. Wash the alga with UCM twice and incubate the biomass for 15 min in 0.2 L UCM at room temperature to remove any leftover traces of the SI-1.
3. An amount of 1 g of dried *Ulva* is suspended in 1 mL of 10 mM Tris–HCl (pH 8) and minced into pieces with the herb chopper.
4. Add 3 mL of 10 mM Tris–HCl (pH 8) to the suspension and incubate the alga for 15 min at room temperature.
5. Filter the algae with a stainless steel filter in a funnel.
6. Transfer the suspension to a 5 mL plastic cup (Eppendorf, Germany) for centrifugation at 3800 rpm for 10 min at 4°C.
7. The supernatant is passed initially through a GF/C filter and then through a cellulose-acetate membrane (0.45 µm).
8. All samples should be stored at –20°C.

8.4.4.2 Bioassay-guided cleanup of the sporulation inhibitor 2

1. The extracted SI-2 from *U. mutabilis* must be freeze-dried carefully for concentration, but not to dryness, and filtered through a cellulose-acetate membrane (*Note 5*).
2. An amount of 40 mL of SI-2 is used with dialysis tubing with a cutoff >500 Da <1000 Da for 24 h, which retains the SI-2, but removes the high amounts of Tris–HCl (pH 8.0) from the sample (*Note 6*).
3. Due to the dilution effects of the membrane, cleaned up SI-2 has to be freeze-dried again, but not to dryness (*Note 5*).
4. Prepare the Sephadex LH-20 column and the mobile phase (1 mM ammonium acetate). (As the SI-2 is smaller than 1000 Da, the cleanup should be carried out with Sephadex G-25-superfine or Sephadex LH-20.)
5. Add the SI-2 sample carefully onto the SEC column (diameter of the column: 1.7 cm, height: 27.5 cm). The elution at a flow rate of 0.2 mL min^{-1} should be carried out with 1 mM ammonium acetate under neutral conditions.
6. Perform the bioassay (as described in *Section 8.4.2*) to identify those fractions with SI-2 activity (Figure 8.4b).
7. After pooling the active fractions, the partially purified inhibitor can now be applied for experiments of life-cycle controlling in *Ulva.*

8.4.5 Swarming inhibitor

8.4.5.1 Extraction and bioassay-guided cleanup of the swarming inhibitor

Washing and mincing remove SI-1 and SI-2, the gametogenesis will be induced. The SWI can be already measured about 12 h after induction and is accumulating in the growth medium. To avoid potential decomposition, the SWI should be isolated from the growth medium at an early phase of gametogenesis (Wichard and Oertel 2010).

1. Wash the algal fragments, already used for the SI-2 extraction, three times with half-strength UCM.
2. Transfer the fragments into UCM and incubate them for 24 h under standard conditions (*Note 7*).
3. Remove the algal fragments, for example, with a stainless steel filter in a funnel.
4. Pass the growth medium through a GF/C filter and subsequently through a cellulose-acetate membrane (0.45 μm).
5. Load the filtrate onto a SepPaktC_{18} Plus Long Cartridge at 2 mL min^{-1}. Up to 2 L can be loaded onto one cartridge.

6. A stepwise elution with 15 mL should be adopted as follows: (1) water, (2) 25% methanol, (3) 50% methanol, (4) 75%, and (5) 100% methanol has to be performed (v:v).
7. All samples should be stored at –20°C.

8.4.5.2 *Bioassay-guided cleanup of the swarming inhibitor*

1. A dilution series that has at least six steps ranging from 10 to 1000 U mL^{-1} of the SWI is recommended. One unit of the SWI is defined as the concentration that inhibits the swarming of gametes out of the gametangia (i.e., fragmented thallus) completely in 1 mL of UCM for 15 min at 20°C.
2. Induction of gametogenesis: Wash the *Ulva* with 50% UCM for 15 min.
3. Mince the alga into pieces with an herb chopper, transfer the fragments (3–5 mm^2) into fresh UCM for 15 min, and repeat the washing step twice (see Chapter 9).
4. Transfer the minced algae into a Petri dish filled with 20 mL UCM and incubate the algal fragments for three days (*Note 8*).
5. On the morning of the third day after induction, pipet the test samples extracted in 8.4.5.1 in a dilution series into the 96-well plate and add 20–30 algal fragments (Figure 8.3c). Before transferring the gametangia into the 96-well plates, dip the fragments quickly into fresh UCM and blot them dry softly with paper.
6. Always keep a negative control (no SWI, gametes should swarm out) and positive control (SWI inside, gametes should not swarm out).
7. After 15 min, check the well plate for released gametes and the discharged gametangia under the microscope.
8. Determine the amount of SWI in the test sample performing a dilution series of the fractions with UCM. Here, we define 1 U of SWI as the minimal amount of the factor-inhibiting gamete released in 1 mL UCM under standard conditions (Figure 8.4c–e).
9. Dry the extracted SWI carefully using a SpeedVac and dissolve it in 500 μL water. The partially purified SWI can already be used for further physiological tests.
10. Samples must be cleaned up with an FPLC Superdex™ Peptide 10/300 column for further purification and structure elucidation. Elution at a rate of 0.25 mL min^{-1} was conducted by using methanol, monitored by a UV detector at 233 nm, and fractionated in 0.25 mL steps (Figure 8.4a).
11. The fractions collected should be analyzed, for example, by a high-resolution Q-Exactive-Orbitrap mass spectrometer and by the SWI bioassay (Figure 8.4d, e).

12. Combine the FPLC fractions with high SWI activity.
13. Clean up the biologically active samples through HPLC (Polar-Premium C_{18} column) and monitor the absorbance of the SWI at 233 nm (Wichard and Oertel 2010).
14. Analyze the fractions again by UPLC-Q-Exactive-Orbitrap-MS measurements and bioassays.
15. Combine the HPLC fractions with high SWI activity and determine the final SWI concentrations in U mL^{-1}.

8.5 Notes

In several *Ulva* species, gametogenesis and subsequent gamete release can be artificially induced by the removal of sporulation and swarming inhibitors, but thus far, neither these inhibitors nor the signaling pathways inducing gametogenesis have been characterized. Ongoing attempts to elucidate the structure of the SIs will result in a deeper understanding of the molecular mechanism in a midterm run. However, the compounds already partially purified can be used as a tool to manipulate *Ulva*'s life cycle to extend the vegetative growth of algal aquaculture.

Note 1: Extract thoroughly with the Dounce homogenizer to get higher yields.

Note 2: It is recommended to precipitate the protein overnight for the best yields.

Note 3: Additional centrifugation might be necessary to get rid of the ethanol. It is important to use pure ethanol because the protein is soluble in water.

Note 4: The second SI-1 extraction step increases the yields.

Note 5: Attention: Freeze-drying of the SI-2 might result in loss of yields.

Note 6: It is important to exchange the rinsing water several time to remove most of the buffer.

Note 7: After 24 h, the highest SWI concentration is reached during gametogenesis, which was shown by Wichard and Oertel (2010).

Note 8: The density of thallus fragments applied to the SWI bioassay should be 500 ± 100 fragments mL^{-1}. Too many fragments of the thallus might result in self-inhibition because of the *de novo* production of inhibitors during the determination phase of the gametogenesis. Few fragments would also produce only minor amounts of the SWI. Therefore, gametes would be released immediately on the morning of the third day.

Acknowledgments

This work was funded by the Deutsche Forschungsgemeinschaft (DFG) through the Collaborative Research Centre 1127 "Chemical Mediators in complex Biosystems" (T.W., R.K.) and the Deutsche Bundesstiftung Umwelt for the PhD fellowship (R.K.).

References

Alsufyani, T., A. Weiss, and T. Wichard. 2017. Time course exo-metabolomic profiling in the green marine macroalga *Ulva* (Chlorophyta) for identification of growth phase-dependent biomarkers. *Mar. Drugs* 15 (1):14. doi: 10.3390/md15010014.

Califano G. and T. Wichard. 2018. Preparation of axenic cultures in *Ulva* (Chlorophyta). In *Biotechnology Protocols for Macroalgae Research*, B. Charrier, T. Wichard, and C.R.K. Reddy (Eds.), Chapter 9. Boca Raton, FL: CRC Press, Taylor & Francis Group.

Carl, C., R. de Nys, R. J. Lawton, and N. A. Paul. 2014. Methods for the induction reproduction in a species of filamentous *Ulva*. *PLoS One* 9 (5). doi: 10.1371/journal.pone.0097396.

Gao, S., X. Y. Chen, Q. Q. Yi, G. C. Wang, G. H. Pan, A. P. Lin, and G. Peng. 2010. A Strategy for the proliferation of *Ulva prolifera*, main causative species of green tides, with formation of sporangia by fragmentation. *PLoS One* 5 (1): e8571. doi: 10.1371/journal.pone.0008571.

Jonsson, S., M. H. Laur, and L. Pham-Quang. 1985. Mise en évidence de differentes types de glycoprotéines dans un extrait inhibiteur de la gamétogenése chez *Enteromorpha prolifera* Chlorophycée marine. *Cryptogamie/Algologie* 6:253–264.

Løvlie, A. 1964. Genetic control of division rate and morphogenesis in *Ulva mutabilis* Føyn. *CR Trav. Lab. Carlsb. Comptes.* 34:77–168.

Løvlie, A. 1968. On the use of a multicellular alga (*Ulva mutabilis* Føyn) in the study of general aspects of growth and differentiation. *Nytt. Magasin. Zoologi.* 16:39–49.

Løvlie, A. 1978. Genetic control of cell cycles during morphogenesis in *Ulva mutabilis*. *Develop. Biol.* 64 (1):164–177.

Katsaros, C., A. Weiss, I. Llangos, I. Theodorou, and T. Wichard. 2017. Cell structure and microtubule organisation during gametogenesis of *Ulva mutabilis* Føyn (Chlorophyta). *Bot. Mar.* 60 (2):123–135. doi: 10.1515/bot-2016-0073.

Kessler, R. W., A. C. Crecelius, U. S. Schubert, and T. Wichard. 2017. *In situ* monitoring of molecular changes during cell differentiation processes in marine macroalgae through mass spectrometric imaging. *Anal. Bioanal. Chem.* 409 (20): 4893–4903. doi: 10.1007/s00216-017-0430-7.

Lüning, K., P. Kadel, and S. J. Pang. 2008. Control of reproduction rhythmicity by environmental and endogenous signals in *Ulva pseudocurvata* (Chlorophyta). *J. Phycol.* 44 (4):866–873. doi: 10.1111/j.1529-8817.2008.00535.x.

Nilsen, G. and O. Nordby. 1975. Sporulation inhibiting substance from vegetative thalli of green alga *Ulva mutabilis* Føyn. *Planta* 125 (2):127–139.

Spoerner, M., T. Wichard, T. Bachhuber, J. Stratmann, and W. Oertel. 2012. Growth and thallus morphogenesis of *Ulva mutabilis* (Chlorophyta) depends on a combination of two bacterial species excreting regulatory factors. *J. Phycol.* 48:1433–1447. doi: 10.1111/j.1529–8817.2012.01231.x.

Stratmann, J., G. Paputsoglu, and W. Oertel. 1996. Differentiation of *Ulva mutabilis* (Chlorophyta) gametangia and gamete release are controlled by extracellular inhibitors. *J. Phycol.* 32 (6):1009–1021.

Vesty, E. F., R. W. Kessler, T. Wichard, and J. C. Coates. 2015. Regulation of gametogenesis and zoosporogenesis in *Ulva linza* (Chlorophyta): Comparison with *Ulva mutabilis* and potential for laboratory culture. *Front. Plant Sci.* 6:15. doi: 10.3389/fpls.2015.00015.

Wichard, T. 2015. Exploring bacteria-induced growth and morphogenesis in the green macroalga order Ulvales (Chlorophyta). *Front. Plant Sci.* 6:86. doi: 10.3389/fpls.2015.00085.

Wichard, T. and W. Oertel. 2010. Gametogenesis and gamete release of *Ulva mutabilis* and *Ulva lactuca* (Chlorophyta): Regulatory effects and chemical characterization of the swarming inhibitor. *J. Phycol.* 46 (2):248–259. doi: 10.1111/j.1529-8817.2010.00816.x.

chapter nine

Preparation of axenic cultures in Ulva *(Chlorophyta)*

Gianmaria Califano and Thomas Wichard

Contents

9.1 Introduction

The green marine alga *Ulva*, also commonly known as the *sea lettuce*, is a cosmopolitan seaweed. It alternates between the isomorphic life stages of gametophyte (haploid) and sporophyte (diploid). Gametophytes produce biflagellate phototactic germ cells (gametes) through mitosis. Meanwhile, diploid sporophytes generate photophobic tetraflagellate germ cells (zoospores or zoids) (Figure 9.1) (Løvlie 1964). Hereby, sporulation inhibitors control the gametogenesis/zoosporogenesis and production of germ cells (Chapter 8 by Kessler et al. 2018).

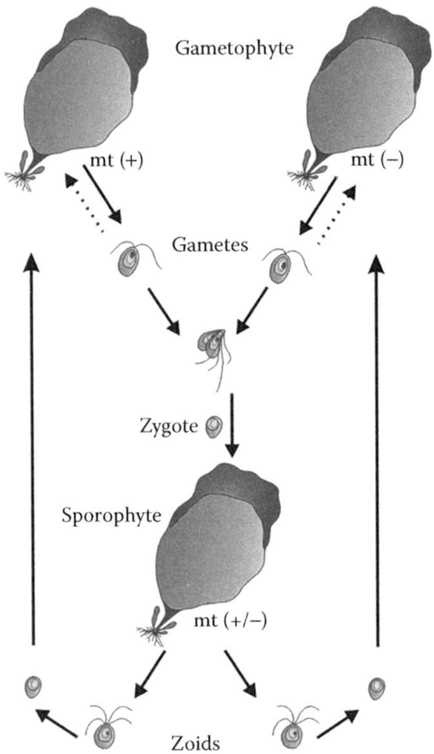

Figure 9.1 The isomorphic life cycle of *Ulva*. All *Ulva* species are isomorphic and alternate between gametophytic and sporophytic life stages with similar morphologies. The gametophytes are haploid (n) and the sporophytes are diploid (2n). The mating types (mt) are indicated by (+) and (–). Dashed arrows show the parthenogenetic development of gametophytes derived from unfused gametes. Blade cells differentiate into gametangium or sporangium (dark-gray margins). (From Wichard, T., *Front. Plant Sci.*, 6, 86, 2015. With permission CC-BY.)

Axenic gametes of *Ulva*, exemplified by *Ulva mutabilis* (Chlorophyta), develop into callus-like colonies consisting of undifferentiated cells. However, the complete morphogenesis can be recovered either by the combination of the two bacterial strains *Roseovarius* sp. MS2 and *Maribacter* sp. MS6 or by morphogenetic compounds (MS2 and MS6 factor) extracted from the bacterial supernatant (Spoerner et al. 2012; Wichard 2015; Weiss et al. 2017). In addition, the *Roseovarius* species exhibits a specific chemotactic affinity to the rhizoid cells of *U. mutabilis* (Kessler et al. 2018) and seems to cooperate with both the *Maribacter* strain and the alga by chemical communication, forming a symbiotic tripartite community. Rhizoid cells attach to the substratum by an adhesive poly-glycoprotein for the formation and maintenance of this cross-kingdom assembly. Various effects on the algal development by changing the microbiome (through the application of antibiotics) have been observed in the past by various research groups (see review by Wichard 2015); however, the definite growth- and morphogenesis-promoting effects of the associated bacteria can only be studied if truly axenic cultures are available.

9.2 *State of the art*

The methods used mostly in phycology facilitate a mix of antibiotics and chemical treatments to inhibit bacterial growth. However, this does not result in the elimination of all marine bacteria that are associated to macroalgae. Antibiotics might also cause negative side effects on the macroalgal development (Andersen 2005). Here, we present a detailed protocol to separate germ cells from their accompanying bacteria (as reviewed in Wichard 2015). In principle, the axenicity of gametes can be achieved by their phototactic movement through glass capillary pipettes (Tatewaky and Provasoli 1964; Stratmann et al. 1996; Spoerner et al. 2012, Vesty et al. 2015). The process of obtaining gametes from algal thalli is defined as gametogenesis. It can be artificially started on the removal of the sporulation inhibitors by mincing the mature thallus into small single-layered fragments and washing them intensively. During the second day after induction, the cells differentiate into gametangia containing 16 progametes, which mature during the following night into fully developed gametes ready for swarming. However, these gametes cannot be released the next morning until a further swarming inhibitor (SWI), which accumulates in the medium during gametogenesis, had been removed (Wichard and Oertel 2010) (see also Chapter 8). The gametes assemble at the brightest spot of the sporulation dish where they can be isolated. A dense suspension of gametes can then be harvested and loaded into a horizontally placed glass capillary filled with sterile *Ulva* culture medium (UCM). Gametes move horizontally to the tip (conical opening) of the pipette, where they can be collected in a few microliters of sterile UCM before the purification has to be repeated (Figure 9.2). The gametes out-compete the accompanying microflora on a total length of about 0.5 m after three purification rounds (Wichard 2015).

The confirmation of axenicity is practically achieved, in concise time, by the application of the 16S rRNA gene amplification testing the supernatant of the purified gametes. When the purification of gametes from bacteria does not succeed, denaturing gradient-gel electrophoresis (Alsufyani et al. 2017) or next-generation sequencing techniques are compelling

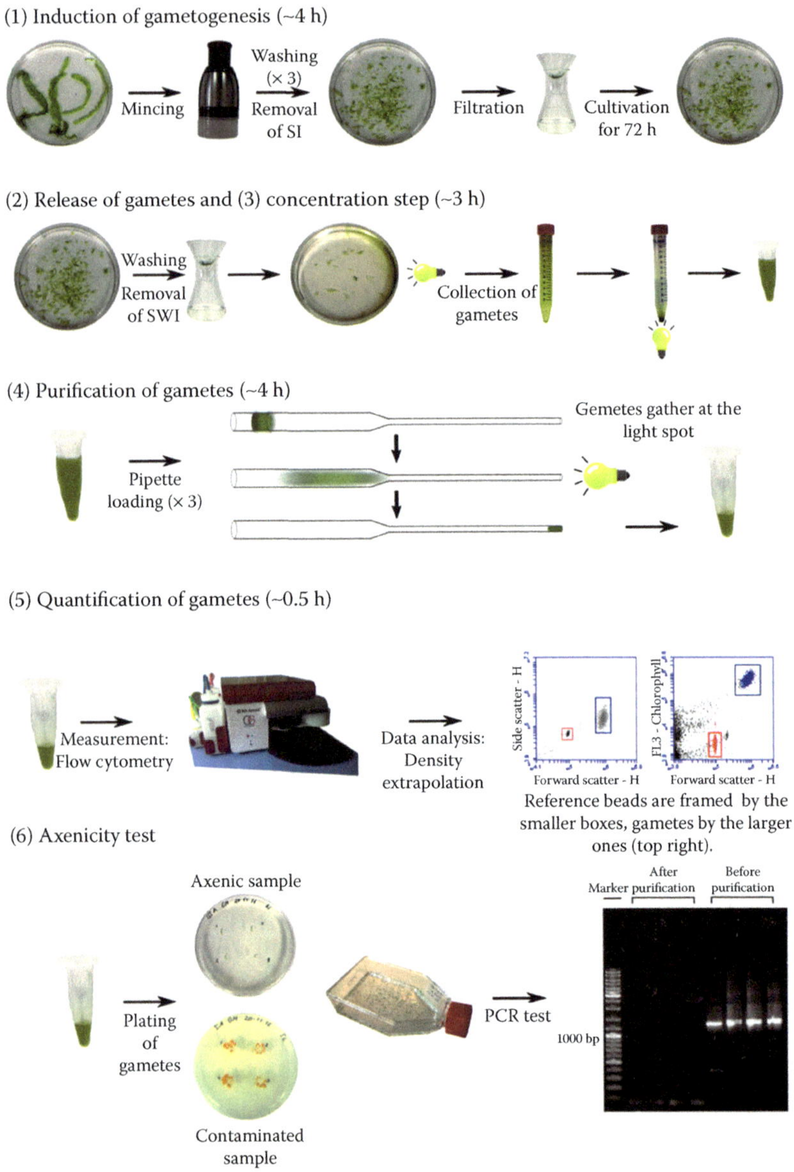

Figure 9.2 The workflow of obtaining axenic gametes from *Ulva* includes six operational steps. The working time is given in parenthesis. Gametogenesis is induced by removal of sporulation inhibitors (SI). On the morning of the third day after induction and removal of a swarming inhibitor (SWI), phototactic gametes can be collected and purified from the accompanying bacteria. Axenicity test: Contaminated samples can be identified by plating gametes on agar (cultivable bacteria) or by PCR (non-cultivable bacteria).

techniques to characterize the diversity and to identify the accompanying bacterial community.

Axenic cultures are an essential tool, for example, (1) to study the chemical ecology of macroalgal–bacterial interactions, (2) to identify (allelopathic) compounds released by the macroalgae, (3) to study the effects of bacterial morphogenetic compounds on the algal growth, and (4) to have a controlled feedstock for algal aquacultures. In all cases, the precise counting of gametes ensures the experimental accuracy and reproducibility of the suggested approach, based on the work of Spoerner et al. (2012).

9.3 Materials

9.3.1 Equipment

- Sterile laminar flow cabinet, providing a minimum working area of 0.55 m^2.
- Benchtop centrifuge with rotor adaptation for plastic microtubes (Eppendorf-Centrifuge 5415D, maximum speed of 16,100 g).
- Thermocycler device for polymerase chain reaction (PCR).
- Flow cytometer (BD Accuri C6, USA) equipped with 488 nm argon laser and emission long pass filters for the wavelength of 670 nm.
- Portable light source (e.g., fluorescent tube).
- Automatic pipettes.

9.3.2 Consumables and reagents

9.3.2.1 Consumables

- Nitrile gloves.
- Autoclaved and dried borosilicate glass Pasteur pipette long tips (23 cm).
- Membrane filter tips for automatic pipettes.
- Polystyrene filter-capped tissue culture flask 50 mL (Sarstedt, Germany).
- Sterile support to hold glass Pasteur pipettes in a horizontal position.
- Plastic microtubes with a volume of 1.5 mL (VWR, Germany).
- Plastic tubes 15 mL (VWR, Germany).
- Fluorescent reference beads for a more accurate quantification to apply to the samples for the flow cytometer (*Note 1*).
- Autoclaved and sterile-filtered *Ulva* culture medium (UCM).
- Aqueous glutaraldehyde, 25% (v:v) (Sigma-Aldrich, Germany).
- Marine broth (Carl Roth, Germany).

- Agar for solid microbiological medium (Agar–Agar Kobe I, Carl Roth, Germany).
- Bleaching solution for chemical inactivation of the algae.

9.3.2.2 PCR Reagents

- 10% of buffer Taq polymerase (10 × DreamTaq buffer, which includes 20 mM of $MgCl_2$, [Thermo Scientific, Germany]).
- 5 U μL^{-1} DNA polymerase (Dream Taq, Thermo Scientific, Germany).
- Primers (27f [GGG TTT GAT CCT GGC TCA G] and 1390r [ACG GGC GGT GTG TRC AA]), 0.2 mM of dNTPs-mix (Thermo Scientific, Germany).
- Dimethyl sulfoxide (DMSO; Sigma-Aldrich, Germany).
- Bovine serum albumin (BSA; Thermo Scientific, Germany).

9.3.3 Ulva *culture medium*

A specific UCM was developed that allows *Ulva* and its associated bacteria to develop sufficiently (Table 9.1) (Stratmann et al. 1996).

Table 9.1 Ingredients of *Ulva* culture medium (UCM)

Solution	Ingredients
I	19.14 g L^{-1} NaCl, 7.28 g L^{-1} $Na_2SO_4 \cdot 10\ H_2O$, 8.68 g L^{-1} $MgCl_2 \cdot 6\ H_2O$, 1.24 g L^{-1} $CaCl_2 \cdot 6\ H_2O$, 85 mg L^{-1} $NaNO_3$, 6.6 mg L^{-1} $(NH_4)_2SO_4$.
II	0.7 g L^{-1} $NaH_2PO_4 \cdot H_2O$, 8.8 g L^{-1} $NaHCO_3$, 10.0 g L^{-1} Tris–OH.
III	7.84 g L^{-1} KBr, 54.2 g L^{-1} KCl, 1.95 g L^{-1} $SrCl_2 \cdot 6\ H_2O$.
IV	668.4 mg L^{-1} EDTA, 1140 mg L^{-1} H_3BO_3, 199 mg L^{-1} $FeSO_4 \cdot 7\ H_2O$, 3.9 mg L^{-1} $CuSO_4 \cdot 5\ H_2O$, 12.6 mg L^{-1} $Na_2MoO_4 \cdot 2\ H_2O$, 36 mg L^{-1} $MnCl_2 \cdot 4\ H_2O$, 44 mg L^{-1} $ZnSO_4 \cdot 7\ H_2O$, 3.3 mg L^{-1} $Co(NO_3)_2 \cdot 6\ H_2O$, 2.3 mg L^{-1} NH_4VO_3, 3.9 mg L^{-1} KJ, 263 μg L^{-1} $Na_2SeO_3 \cdot 5\ H_2O$, 9.3 μg L^{-1} As_2O_3, 6.6 μg L^{-1} $Na_2WO_4 \cdot 2\ H_2O$, 3.4 μg L^{-1} TeO_2, pH is adjusted to 8.0 with NaOH.
V	0.05 mg L^{-1} B_{12}, 0.2 g L^{-1} thiamine · HCl, 0.1 g L^{-1} niacin, 0.1 g L^{-1} Ca–panthothenate, 0.04 g L^{-1} pyridoxine · HCl, 0.01 g L^{-1} p–aminobenzoic acid, 5 mg L^{-1} biotin, 0.8 g L^{-1} thymine, 1.0 g L^{-1} inositol, 0.26 g L^{-1} orotic acid, 0.2 mg L^{-1} folinic acid (citrovorum), 2.5 mg L^{-1} folic acid, 0.04 g L^{-1} putrescine · 2HCl, 5 mg L^{-1} riboflavin, 0.02 g L^{-1} pyridoxamine · 2HCl, 0.36 g L^{-1} choline chloride.

Note: One L of solution I has to be supplemented with 10 mL of solutions II–IV and 2 mL of solution V. The pH is adjusted to 8.1 with HCl. Solution I is autoclaved; solution II–V are sterile filtered.

9.4 Experimental procedures

9.4.1 Induction of gametogenesis and release of gametes

1. Collect the algae from the growth medium (up to 2 g fresh weight) in the morning and dry them softly with paper towels (a bit of moisture is necessary for the following process).
2. Mince the algae into small fragments (3–5 mm^2) using a chopper. The two cell layers of the blade tissue must be separated for the efficient induction of the gametogenesis (see also Chapter 8).
3. Wash the algal fragments with a solution of 50% UCM for 15 min.
4. Recover the algal fragments through a mesh of 100 μm.
5. Repeat *steps 3* and *4* twice to remove all the sporulation inhibitors.
6. Transfer the fragments of the thallus into a filter-capped tissue culture flask with UCM (e.g., add 50 mL of UCM for a tissue culture flask with an area of 355 cm^2).
7. Incubate the flask with fragmented algae at *Ulva* standard growing conditions (i.e., at 18°C with a light–dark rhythm of 17/7 h; at an illumination of 60–120 μmol photons m^2 s^{-1} using 50% GroLux and 50% daylight fluorescent tubes [OSRAM, München, Germany] without additional aeration) for three days.
8. On the morning of the third day after the induction of the gametogenesis, replace the growth medium with UCM to remove the SWI (*Note 2*).
9. Expose the flask containing the *Ulva* fragments to the daylight (e.g., in front of a window) or bulb light. Gametes are released and assembled at the brightest spot close to the light source (Figure 9.2).

9.4.2 Enrichment of gamete density

Before the gametes can be loaded into a Pasteur pipette for purification from the accompanying bacteria, the density of the gametes should be enriched.

1. Gametes gather at the spot light, forming a visible green line on the flask wall directly exposed to the light (*Note 3*).
2. Collect the gametes with a pipette into a 15 mL plastic tube (the plastic tube is covered with an aluminum foil except for the bottom part) and again expose the gametes to light.
3. After approximately 10 min, remove the clear supernatant (without gametes) carefully and transfer the green fraction into a 1.5 mL tube (Figure 9.2).

Repeat this procedure until an adequate number of gametes have been collected. The number of axenic gametes that are sufficient depends on each experiment. 1–2 × 10^7 gametes, for example, are required to obtain

1 μg of total DNA. Therefore, under optimal conditions, 1 cm^2 of algal thallus releases about 1×10^7 gametes. Considering that up to 90% of the gametes are lost during purification, 10 cm^2 of the thallus has to be used at the beginning of the purification process (*Note 4*).

9.4.3 *Purification of gametes from bacteria*

1. Place a light source directly behind the transparent wall (if available) of the sterile laminar flow cabinet. Alternatively, the light source can be sterilized and placed inside the hood.
2. Use a plastic tube holder or a homemade support to keep the Pasteur pipettes horizontal (Figure 9.3).
3. Fill the Pasteur pipettes carefully with the sterile UCM from the wide side of the pipette, leaving about 1 cm of space for the gamete suspension (*Note 5*). Avoid any air bubbles in the Pasteur pipette.

Figure 9.3 Photographs of the two pipette supports: (a) Supports are used during the purification process. The Pasteur glass pipette is placed on the supports inside the laminar flow cabinet, and the light source is placed outside the enclosure. (b) A plastic tube rack can be also used as safe support for the glass pipettes.

4. Load between 100 and 150 μL (depending on the density of the gamete suspension) into each Pasteur pipette. Transfer the gamete suspension very carefully into the pipette, which is already in place horizontally, without any disturbance (*Note 6*).
5. Wait until the gametes gather at the opposite opening, forming a green spot (*Note 7*). It should not take longer than 30 min!
6. Incline the pipettes slightly and rotate them to collect up to 5 drops of the bright green suspension with gametes into a sterile 1.5 mL plastic tube (*Note 8*).
7. Repeat steps 2 to 5 twice (*Note 9*) to purify the same batch of gametes 3 times sequentially. Ten glass pipettes can be handled at the same time. Considering that gametes have to swim 3 times through the Pasteur pipette, a minimum number of 30 sterile pipettes filled with UCM are necessary (*Note 10*).
8. Finally, a bright green suspension is expected. Split the sample into several aliquots.

9.4.4 Measuring gamete density: Flow cytometry

The period in which germ cells conserve their motile state is relatively short (10 to 12 h) before they settle onto the surface of the vessel. During this time, the cell density of gametes must be quantified and distributed in the final vessels, where experiments are carried out. At this time, the determination of the cell density is an essential step to guarantee the same conditions between replicates and the experimental reproducibility.

1. Homogenize the gamete suspension by turning the plastic tube several times (9.4.3 step 8) (*Note 11*).
2. Collect 10 μL of the gamete suspension and mix it with 200 μL UCM and glutaraldehyde (final concentration of 2% [v:v]).
3. Prepare four replicates.
4. Analyze at a flow rate of 14 μL min^{-1}.
5. Choose the far-red wavelength region (670-LP) filter to quantify the gamete (Figure 9.2).

9.4.5 Axenicity proof of gametes

The missing axenicity test is a pitfall of many studies (Wichard 2015). Therefore, Spoerner et al. (2012) recommended amplifying the highly conserved 16S rRNA of bacteria collected from the supernatant *via* PCR using specific primer pairs optimized for marine bacteria. In line with this recommendation, we suggest two tests that should be performed to prove the axenicity of the gamete suspension.

9.4.5.1 *Test of axenicity on agar medium*

1. Plate 50 μL of the gamete suspension in triplicate on marine broth (agar 1%) (*Note 12*).
2. Incubate the agar plates at room temperature for three days.
3. Examine the plate to assure the absence of colony-forming units.

9.4.5.2 *Test of axenicity by PCR*

1. Ensure gametes settle on the surface of the experimental vessel (wait for more than 48 h after the inoculation of the gametes).
2. Collect 2 mL of the supernatant (*Note 13*).
3. Spin down the potential contaminant bacteria at a maximum speed 16,100 × g for 5 min (Centrifuge 5415 D, Eppendorf, Germany).
4. Remove the supernatant and add 50 μL of ultrapure sterile water to the (non-visible) pellet.
5. Boil the sample for 15 min.
6. Keep the sample on ice for 5 min and then centrifuge down the debris and proteins (maximum speed for 5 min).
7. Take 3.5 μL of the supernatant as a DNA template for a PCR reaction (total volume = 25 μL) (Table 9.2).
8. PCR cycle: Preheating lid at 99°C, initial denaturation at 95°C for 5 min, 30 cycles of 0.5 min at 95°C, 0.5 min at 58°C, 1.5 min at 72°C, and a final elongation for 7 min at 72°C.
9. Perform the PCR also with positive controls, which are samples from non-axenic cultures of *Ulva* (Figure 9.2 and Table 9.2).
10. A typical contaminated culture with bacteria releasing the *Roseovarius*-factor is shown in Figure 9.4.

Table 9.2 PCR list of reagents and volumes

Reagents	Volume (μL)	Final concentration
Ultrapure water	14.6	N/A
10 × Dream taq buffer	2.5	10%
Primer 27f	1	0.4 μM
Primer 1392r	1	0.4 μM
dNTPs-mix	0.5	N/A
BSA	1.6	0.64 mg mL^{-1}
DMSO	0.5	2% (v:v)
TAQ polymerase	0.15	0.75 units
DNA template	3.5	N/A
Total volume	25	N/A

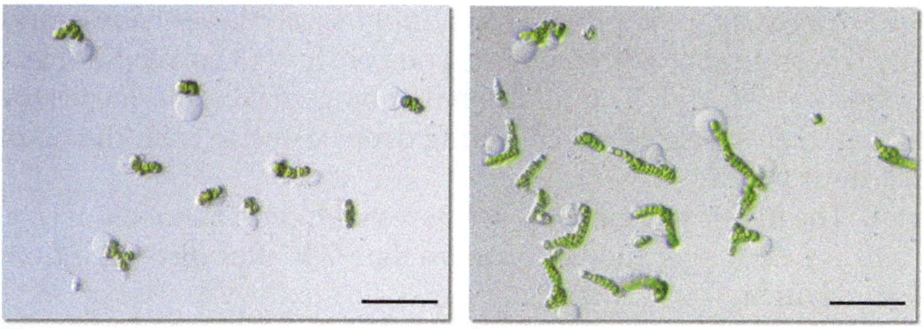

(a) (b)

Figure 9.4 Ten–day–old cultures of *Ulva mutabilis* are shown: (a) axenic culture, (b) typically contaminated culture (after purification) with bacteria possessing the same activity as the *Roseovarius* sp. MS2. Although the typical cell wall protrusion of axenic cultures are visible in (b) as well, the longitudinal cell divisions indicate the presence of morphogenetic compounds released by bacteria. Magnification bar = 100 μm.

9.5 Notes

Note 1: An exact amount of reference beads is necessary for an accurate quantification, depending on the flow cytometer available. Choose the beads on the basis of the specific excitation and emission wavelength of the laser and filters installed in the equipment.

Note 2: The SWI accumulates in the growth medium during gametogenesis and prevents the discharge of the gametes on the morning of the third day after induction. Induced fragments will change their color from the dark green of the blade cell to the brownish of the gametangium.

Note 3: Most gametes are discharged from the gametangia on medium exchange within the first minutes. However, if the cell division of the cultures is not completely synchronic, the formation and release of gametes will be affected, and the gametes will be continuously released from the gametangium over a period of 2–4 h.

Note 4: There is a significant loss of gametes during the purification steps (up to 90%). Not all gametes will reach the tip of the pipette. If the density of gametes is too high, self-shading of the gametes will prevent them from going through the pipette.

Note 5: Avoid the formation of gas bubbles, which will hinder the movement of the gametes' during the filling of UCM and loading of gametes in the Pasteur pipettes.

Note 6: In the case of gamete suspension with low density, apply up to 800 μL of the suspension into the wide opening of the pipette.

Note 7: If the movement of the gametes takes longer than 30 min for 15 cm because of slow-swimming, the separation from the accompanying bacteria will be insufficient.

Note 8: Alternative sampling: The ends of the capillaries were connected by silicon tubing with the open ends of short (3 cm) capillaries with closed other end. The additional tubing makes the harvesting of the gametes easier than collecting drops directly from the tip of the Pasteur pipette.

Note 9: The motility of gametes decreases with time. Gametes of *U. mutabilis* are no longer motile for more than 7 h after discharge from the gametangia.

Note 10: Alternative purification step: Spoerner et al. (2012) suggested, as a prepurification step, "that the gametes assembling at the brightest spot of the sporulation dish should be aspirated with a pipette in a small volume and diluted with twofold concentrated UCM to a 1.6-fold concentration, to increase the density of the medium. A volume of 1–2 mL of this suspension was placed at the bottom of a sterile 160/15 mm standard glass test tube using a sterile pipette, thoroughly avoiding any contact with the upper sterile inner tube surface. The gamete suspension of higher density was then cautiously over-layered with less dense sterile standard UCM and filled up to the top. After placement for at least 15 min in a dark room, 25 cm below a fluorescent lamp, the bulk of the positive phototactic gametes had moved vertically to the meniscus. Residual algal debris, bacteria, and other undesired contaminants, and zoids and zygotes, remained within the dense liquid pad at the bottom of the test tube."

Note 11: As gametes move toward the light source, homogenization of the gamete suspension is hard to achieve. Four technical replicates are recommended to estimate the density of the gametes in suspension.

Note 12: The culture-dependent technique used here is an inexpensive way to check for common heterotrophic bacteria. By contrast, a culture-independent 16S rRNA gene-based method can detect uncultivable bacteria by example, denaturing gradient gel electrophoresis (Wichard 2015, 2016).

Note 13: Alternatively, a direct PCR of the intact bacteria in the supernatant can be performed, according to Woodman (2008).

Acknowledgments

This work was funded by the Deutsche Forschungsgemeinschaft (DFG) through the Collaborative Research Centre 1127 *Chemical Mediators in complex Biosystems* and by the European Union's Horizon 2020 research and innovation program under the Marie Sklodowska–Curie grant agreement No.642575—The ALgal Microbiome: Friends and Foes (ALFF) (G.C. and T.W.).

References

Alsufyani, T., A. Weiss, and T. Wichard. 2017. Time course exo-metabolomic profiling in the green marine macroalga *Ulva* (Chlorophyta) for identification of growth phase-dependent biomarkers. *Mar. Drugs* 15: 14. doi: 10.3390/md15010014.

Andersen, A.R. 2005. *Algal Culturing Techniques: A Book for All Phycologists*. London, UK: Elsevier Academic Press.

Kessler, R.W., T. Alsufyani, and T. Wichard. 2018. Purification of sporulation and swarming inhibitors from *Ulva*: Application in algal life cycle controlling. In *Protocols for Macroalgae Research*, Charrier, B., Wichard, T., and Reddy, C.R.K. (Eds.), Chapter 8. Boca Raton, FL: CRC Press.

Løvlie, A. 1964. Genetic control of division rate and morphogenesis in *Ulva mutabilis* Føyn. *C. R. Trav. Lab. Carlsberg* 34: 77–168.

Spoerner, M., T. Wichard, T. Bachhuber, J. Stratmann, and W. Oertel. 2012. Growth and thallus morphogenesis of *Ulva mutabilis* (Chlorophyta) depends on a combination of two bacterial species excreting regulatory factors. *J. Phycol.* 48: 1433–1447. doi: 10.1111/j.1529-8817.2012.01231.x.

Stratmann, J., G. Paputsoglu, and W. Oertel. 1996. Differentiation of *Ulva mutabilis* (Chlorophyta) gametangia and gamete release are controlled by extracellular inhibitors. *J. Phycol.* 32(6): 1009–1021. doi: 10.1111/j.0022-3646.1996.01009.x.

Tatewaky, M. and L. Provasoli. 1964. Vitamin requirements of three species of antithamnion. *Bot. Mar.* 6: 193–203. doi: 10.1515/botm.1964.6.3-4.193.

Vesty, E.F., R. Kessler, T. Wichard, and J. Coates. 2015. Regulation of gametogenesis and zoosporogenesis in *Ulva linza* (Chlorophyta): Comparison with *Ulva mutabilis* and potential for laboratory culture. *Front. Plant Sci.* 6: 15. doi: 10.3389/fpls.2015.00015.

Wichard, T. 2015. Exploring bacteria-induced growth and morphogenesis in the green macroalga order Ulvales (Chlorophyta). *Front. Plant Sci.* 6: 86. doi: 10.3389/fpls.2015.00086.

Wichard, T. 2016. Identification of metallophores and organic ligands in the chemosphere of the marine macroalga *Ulva* (Chlorophyta) and at land-sea interfaces. *Front. Mar. Sci.* 3: 131. doi: 10.3389/fmars.2016.00131.

Wichard, T. and W. Oertel. 2010. Gametogenesis and gamete release of *Ulva mutabilis* and *Ulva lactuca* (Chlorophyta): Regulatory effects and chemical characterization of the "Swarming Inhibitor". *J. Phycol.* 46(2): 248–259. doi: 10.1111/j.1529-8817.2010.00816.x.

Weiss, A., R. Costa, and T. Wichard. 2017. Morphogenesis of *Ulva mutabilis* (Chlorophyta) induced by *Maribacter* species (Bacteroidetes, Flavobacteriaceae). *Bot. Mar.* 60(2): 197–206. doi: 10.1515/bot-2016-0083.

Woodman, M.E. 2008. Direct PCR of intact bacteria (colony PCR). *Curr. Prot. Microbiol.* 9: A.3D.1–A.3D.6.

section two

Chemical composition

chapter ten

Biochar production from seaweeds

Loretto Contreras-Porcia, Matías Araya,*
Elizabeth Garrido-Ramírez, Cristian Bulboa,
Jean Pierre Remonsellez, Javier Zapata,
Camila Espinoza, and Jorge Rivas

Contents

10.1 Introduction

Biochar is a carbon-rich solid derived from thermally stabilized organic material, meaning that this product can be stored and applied to soils for improving its quality. Biochar differs from other solid products derived from thermochemical conversion, such as charcoal or active carbon, in that long-term carbon storage is an objective, than the generation of raw material for processing industries or fuels (Mašek et al. 2013). In technical terms, biochar is produced through the thermal decomposition of organic

* Loretto Contreras-Porcia and Matías Araya: Both the authors have contributed equally.

material under oxygen (O_2)-limited conditions and at relatively low temperatures (<900°C) (Lehmann and Joseph 2009; Cha et al. 2016). Through a process termed pyrolysis, biochar production results in subproducts such as gases (e.g., synthesis gas) and liquids (e.g., tars and oils) (Bridgwater and Peacocke 2000; Cha et al. 2016). At lower temperatures (<550°C), pyrolysis tends to produce more biochar, whereas at higher temperatures, more synthesis gas is generated.

In recent years, biochar has been a subject of important scientific and commercial interest because of its role in carbon sequestration, ability to improve nutrient uptake efficiency in soils, and capacity to adsorb pollutants from diverse matrices. In relation to improving soils, several studies demonstrate that biochar incorporation can improve (1) nutrient retention; (2) cationic interchange capacity; (3) chemical, physical, and biological soil properties; and (4) plant health and growth, in addition to reducing nitrogen loss in fertilized soils through volatilization and lixiviation (Jassal et al. 2015; Mandal et al. 2016). Adding biochar to soils also reduces the emission of greenhouse gases such as CH_4 and N_2O (Qian et al. 2015). For example, Case et al. (2015) reported that biochar suppresses the accumulated production of N_2O by 91% in near-saturated fertilized soils. Furthermore, denitrification is reduced by 37%, which represents between 85% and 95% of N_2O emissions in soil.

The use of low-cost raw materials is of vital importance for future biochar applications. Consequently, biochar production using seaweeds has garnered widespread attention as a promising alternative. The low cost of this raw material is because of macroalgae abundance and easy collection from natural fields or detachments, and from seaweed farms. Research associated with the use of biochar originating from macroalgae primarily focuses on the use of this raw material as a low-cost adsorbent of pollutants, mainly in the removal of metals (Cho et al. 2013; Park et al. 2016) and phosphate (Jung et al. 2015a, b: 2016a, b) from aquatic systems.

Scientific studies assessing the use of seaweed-derived biochar in soils are fewer and more recent. For example, Bird et al. (2011) studied the potential of different types of seaweeds on biochar generation and, subsequently, applications in carbon sequestration, as a soil amender, and as a fertilizer. Roberts et al. (2015a, b) used biochar generated from freshwater macroalgae (*Oedogonium*) cultivated in a wastewater of a coal-fired power station to enhance the rehabilitation of soils close to a coal mine. Cole et al. (2017) used the same species to recover nitrogen (N) and phosphorus (P) from municipal wastewater and then used the cultivated algae to produce biochar as a soil amendment. As such, algal biomass use is now considered as a new biotechnological alternative (Lee and Woo 2014) from an environmental point of view.

10.2 State of the art

Biochar can be obtained from different biomass feedstock, including wood residue, crop residue, grass, manure, sludge and municipal organic waste, and seaweeds, among others. Successful biochar use depends on the properties of the originating biomass and operational conditions, including residency time, temperature, heating rate, reactor type, and so on, among others (Rajapaksha et al. 2016). As such, biochar produced from seaweeds has a low carbon content compared with biochar from lingocellulosic materials and, conversely, higher concentrations of macronutrients (i.e., N and P) and interchangeable trace elements (i.e., Ca, Mg, K, and Mo), which are essential for plant growth (Bird et al. 2011; Roberts et al. 2015a, b). Nevertheless, the final composition of biochar, and the physicochemical properties varies depending on the seaweed used for production (Table 10.1). Bird et al. (2012) compared biochar derived from freshwater (FW) and saltwater (SW) macroalgae with lignocellulosic biochar (commercial charcoal). Lignocellulosic biochar was similar to the algal biochar in terms of pH, CEC, and surface area characteristic, but has considerable higher carbon content (72%) compared with FW biochar (11.6%) and SW biochar (17.4%). In turn, algal biochar presents higher ash content and considerable highest nutrient (N, P, and K) contents than lignocellulosic biochar. This final trait permits using algal biochar as a soil amender (Bird et al. 2011, 2012).

10.3 Materials

10.3.1 Extraction and pre-drying of seaweed biomass

- Plastic bags.
- Raschel mesh.
- Tap water.
- Distilled water.
- Hemp or cotton rope, 10–20 mm in diameter.
- Ice pack.
- Knife.
- Bag sealer.

10.3.2 Sample preparation by pyrolysis

- Drying oven.
- Plastic or stainless-steel baskets.
- Sealable bags.
- Stainless-steel mill.
- No. 18 (1.0 mm) and N° 3.5 (5.7 mm) sieve.
- Bag sealer.
- Semi-analytical balance.

Table 10.1 Main physicochemical characteristics of biochar generated from different biomass sources and different pyrolysis conditions

Feedstock	PT[a] (°C)	C (%)	H (%)	O (%)	N (%)	S (%)	pH[b]	CEC[c] (cmol kg^{-1})	Ash (%)	SA[d] (m^2 g^{-1})	References
Algae											
Cladophora vagabunda	450	21.80	1.20	–	2.00	–	9.87	23	54.20	–	Bird et al. (2011)
Caulerpa taxifolia	450	24.80	1.20	–	2.40	–	9.65	17	20.90	–	–
Cladophoropsis sp.	450	23.60	1.50	–	2.80	–	10.07	27	46.50	–	–
Chaetomorpha linum	450	23.60	1.30	–	2.40	–	9.61	36	16.00	–	–
Chaetomorpha indica	450	10.20	0.80	–	1.10	–	7.83	16	73.50	–	–
Cladophora patentiramea	450	20.30	1.20	–	1.70	–	9.12	21	47.00	–	–
Cladophora coelothrix	450	34.60	1.50	–	3.3	–	8.72	19	32.10	–	–
Ulva flexuosa	450	22.60	1.20	–	2.70	–	10.00	41	42.6	–	–
	414	25.70	1.90	–	3.90	–	9.80	39	29.2	–	–
	305	28.90	2.80	–	5.00	–	8.00	29	26.5	–	–
Undaria pinnatifida	500	56.60	2.70	33.00	4.80	3.00	–	–	47.70	–	Cho et al. (2013)
Gracilaria edulis	450	30.90	2.20	16.5	2.80	4.40	7.60	–	–	2.02	Roberts et al. (2015c)
Eucheuma spinosum	450	25.60	1.8	24.90	0.80	9.30	8.20	–	–	30.30	–
Kappaphycus alvarezii	450	31.30	2.10	23.80	0.70	7.00	8.80	–	–	2.24	–
Sargassum sp.	450	28.90	2.10	18.20	1.10	2.80	10.80	–	–	7.46	–
Unadaria pinnatifida	450	34.80	2.80	15.60	2.40	0.80	9.90	–	–	1.33	–
Saccharina japonica	450	28.00	1.90	16.40	2.20	1.00	11.00	–	–	1.29	–

(Continued)

Table 10.1 (Continued) Main physicochemical characteristics of biochar generated from different biomass sources and different pyrolysis conditions

Feedstock	PT[a] (°C)	C (%)	H (%)	O (%)	N (%)	S (%)	pH[b]	CEC[c] (cmol kg^{-1})	Ash (%)	SA[d] (m^2 g^{-1})	References
Undaria pinnatifida	200	30.58	3.820	34.70	3.40	–	7.37	–	32.83	1.909	Jung et al. (2016b)
	400	31.92	1.74	13.56	2.35	–	8.38	–	41.85	70.29	–
	600	36.64	0.99	11.76	2.31	–	10.36	–	48.71	61.80	–
	800	40.51	0.70	10.02	2.00	–	11.09	–	50.41	44.49	–
Ulva compressa	500	57.60	2.60	22.40	5.60	11.80	–	157	43.30	–	Kim et al. (2016)
Porphyra tenera	500	74.70	3.00	9.00	9.10	4.20	–	–	21.30	–	Park et al. (2016)
Oedogonium	450	49.80	–	–	5.50	1.00	10.60	21.30	–	–	Cole et al. (2017)
Others											
Poultry litter	400	–	3.08	15.62	4.48	–	9.49	48	–	–	Jassal et al. (2015)
	500	–	2.04	8.60	3.99	–	9.33	36	–	–	–
	600	–	1.49	5.20	3.27	–	9.91	47	–	–	–
Spruce–pine–fir wood waste	400	–	3.60	17.70	0.28	–	5.47	1	–	–	–
	500	–	2.99	9.42	0.40	–	7.13	0.9	–	–	–
	600	–	2.42	4.75	0.54	–	7.53	1.3	–	–	–
Poultry litter	550	77.50	–	–	2.74	0.39	8.66	–	–	12.80	Mandal et al. (2016)
	460	72.90	–	–	4.95	0.13	10.84	–	–	202.50	–

[a] PT, Final pyrolysis temperature.
[b] May differ due to the method for pH measurement, see references.
[c] Cation exchange capacity.
[d] Surface area.

10.3.3 Biochar generation by pyrolysis

- Pyrolysis reactor with speed/temperature controls, supplemented with thermocouples and gas inlet/outlet with condenser.
- Pure gaseous nitrogen (N_2).

10.3.4 Storage of biochar samples

- No. 35 (0.5 mm) and 18 (1.0 mm) sieve.
- Sealable bags.
- Bag sealer.
- Stainless-steel mill.
- Porcelain mortar.

10.4 Experimental procedures

10.4.1 Extraction and pre-drying of seaweed biomass (the case of Macrocystis pyrifera)

1. Store extracted biomass samples in large or small plastic bags with an ice pack, depending on the amount (*Note 1*).
2. Remove biomass from the plastic bags, manually remove any large debris, and immediately wash three times with tap water. Repeat this process three times with distilled water (*Note 2*).
3. The algae biomass can be open-air dried either by (1) evenly distributing the algae mass on a clean mesh (Figure 10.1a) to allow water drainage during the day with algae covered by the mesh at night to maintain drying, or (2) take the seaweed from the holdfast and hanging these evenly along a rope (Figure 10.1b). The drying process should be monitored (*Note 3*).
4. Remove the holdfast and store the rest of the tissue in sealed plastic bags (*Note 4*).

10.4.2 Sample preparation for pyrolysis

1. Remove the remaining moisture by oven drying at 40°C for 24 h (Jung et al. 2016b). For this, take 1 kg of biomass and place it in plastic or aluminum containers (*Note 5*).
2. The dry product must be milled manually or in a stainless-steel mill and sieved to retain particles 1.0–5.7 mm (Cho et al. 2013; Roberts et al. 2015c; Jung et al. 2016b; Kim et al. 2016) (*Note 6*).
3. Store the ground biomass in sealed plastic bags at room temperature until pyrolysis.

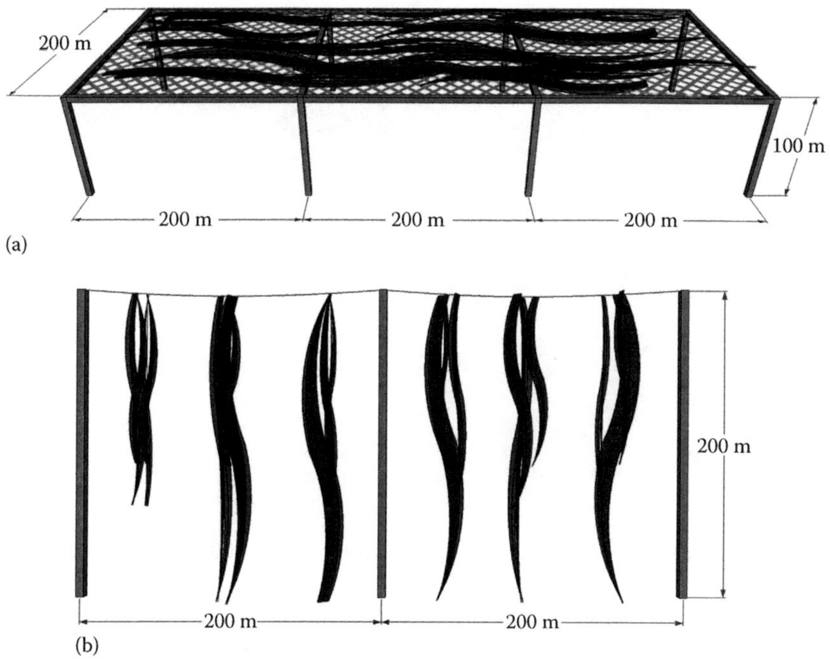

Figure 10.1 Open-air biomass dryer systems on a clean mesh (a) or hanging along a rope (b).

10.4.3 *Biochar generation by pyrolysis*

1. Put the samples in the pyrolyzer and open the N_2 gas passage at a flow of 3–4 L min^{-1} for until achieving an inert atmosphere (Bird et al. 2011; Roberts et al. 2015c).
2. Begin increasing the temperature at a rate of 3°C –5°C min^{-1}, starting at room temperature until reaching 450°C–500°C (Bird et al. 2011, 2012; Cho et al. 2013; Jung et al. 2015a; Kim et al. 2016; Lee et al. 2016). Maintain the N_2 flow and the final temperature and other pyrolysis conditions (see *Note 1*) for 1 h. Finally, cool to room temperature while maintaining the N_2 flow constant (Johansson et al. 2016) (*Note 7*).
3. Remove the biochar sample from the pyrolyzer and store at room temperature in sealed plastic bags.

10.4.4 *Storage of biochar samples*

1. Grind the samples until small enough to standardize the sample size according to the study purposes (Figure 10.2) (*Note 8*).

Figure 10.2 Biochar obtained from *Macrocystis pyrifera* at 300°C, 450°C, and 600°C.

2. Wash the samples three times with deionized water (1:20 sample:water ratio). The samples must then be filtered to obtain the solid product, which must be subsequently dried in an oven at 60–80°C for 24 h (Jung et al. 2015a; 2016b).
3. The obtained biochar must be cooled in a desiccator for 1 h and then stored in sealed plastic bags at room temperature.

10.5 Notes

Note 1: To prevent the decomposition of algae biomass, refrigerated storage and/or transportation should not exceed 48 h, whereas non-refrigerated storage and/or transportation should not exceed 24 h.

Note 2: Initial washing is performed primarily to remove sand and any other organisms. Maintaining the holdfast will facilitate algae handling.

Note 3: Monitor and periodically move the algae biomass to ensure uniform drying (for ground or hanging drying methods). Algae biomass must be covered with mesh for ground drying under windy conditions. If a moisture content > 15% persists after two days, and if environmental conditions will not improve, directly proceed to Step 4, which will remove any moisture excess.

Note 4: If the next step will be carried out in less than 12 h, it is advisable to store the biomass in a dry place so that they do not get wet again.

Note 5: Use perforated boxes or baskets to ensure that heat reaches all algal tissue surfaces. If posterior experiments measure heavy metal contents, plastic boxes are advisable for the drying process. If posterior experiments are related to the organic matter, stainless-steel baskets are advisable for the drying process.

Note 6: If following analyses focus on heavy metals adsorption, grinding the samples manually or with a mortar are recommended.

Note 7: An end pyrolysis temperature of 450°C is more efficient and results in less solid biomass loss in the pyrolysis process (Bird et al. 2011). This temperature may vary depending on the purpose of the study and pyrolysis conditions.

Note 8: To perform adsorption tests, it is advisable to grind up to pass through a 0.5–1.0 mm sieve (Zhang et al. 2013). If the biochar will be used to assess heavy metals adsorption, avoid crushing the samples with metallic instruments. In the case of applying the biochar to the soil, it a particle size of 0.5–2.0 mm is recommended (Bird et al. 2012; Roberts et al. 2015c; Jung et al. 2016b).

Acknowledgments

This work was supported by FIC 30397482-0 Gobierno Regional de Valparaíso and Project CONICYT FB-0002-2014 to LC-P. We acknowledge the language support provided by Ashley VanCott, BioPub Ltd. We also thank Ignacio Perez-Massad for providing Figure 10.1.

References

Bird, M.I., Wurster, C.M., de Paula, S.P.H., Bass, A.M., and de Nys, R. 2011. Algal biochar production and properties. *Bioresource Technol.* 102:1886–1891.

Bird, M.I., Wurster, C.M., de Paula Silva, P.H., Paul, N.A., and de Nys, R. 2012. Algal biochar: Effects and applications. *GCB Bioenergy* 4:61–69.

Bridgwater, A. and Peacocke, G. 2002. Fast pyrolysis processes for biomass. *Renew. Sust. Energy Rev.* 4:1–73.

Case, S., McNamara, N.P., Reay, D., Stott, A., Grant, H., and Whitaker, J. 2015. Biochar suppresses N_2O emissions while maintaining N availability in a sandy loam soil. *Soil Biol. Biochem.* 81:178–185.

Cha, A.B., Sung, H.P.C., Sang-Chul, J.C., Changkook, R.D., Jong-Ki, J.E. Min-Chul, S.B., and Young-Kwon, P.A. 2016. Production and utilization of biochar: A review. *J. Ind. E. Chem.* 40:1–15.

Cho, H.J., Baek, K., Jeon, J.-K., Park, S.H., Suh, D.J., and Park, Y.-K. 2013. Removal characteristics of copper by marine macro-algae-derived chars. *Chem. Eng. J.* 217:205–211.

Cole, A.J., Paul, N.A., de Nys, R., and Roberts, D. 2017. Good for sewage treatment and good for agriculture: Algal based compost and biochar. *J. Environ. Manag.* 200:105–113.

Jassal, R., Johnson, M., Molodovskaya, M., Black, T.A., Jollymore, A., and Sveinson, K. 2015. Nitrogen enrichment potential of biochar in relation to pyrolysis temperature and feedstock quality. *J. Environ. Manag.* 152:140–144.

Johansson, C.H., Paul, N.A., Nys, R., and Roberts, D.A. 2016. Simultaneous biosorption of selenium, arsenic and molybdenum with modified algal-based biochars. *J. Environ. Manag.* 165:117–123.

Jung, K.-W., Hwang, M.-H., Jeong, T.-U., and Ahn, K.-H. 2015a. A novel approach for preparation of modified-biochar derived from marine macroalgae: Dual purpose electro-modification for improvement of surface area and metal impregnation. *Bioresource Technol.* 191:342–345.

Jung, K.-W., Jeong, T.-U., Hwang, M.-H. Kim, K., and Ahn, K.-H. 2015b. Phosphate adsorption ability of biochar/Mg–Al assembled nanocomposites prepared by aluminum-electrode based electro-assisted modification method with $MgCl_2$ as electrolyte. *Bioresource Technol.* 198:603–610.

Jung, K.W., and Kyu-Hong, A.K.-H. 2016a. Fabrication of porosity-enhanced MgO/biochar for removal of phosphate from aqueous solution: Application of a novel combined electrochemical modification method. *Bioresource Technol.* 200:1029–1032.

Jung, K.-W, Kim, K., Jeong, T.-U., and Ahn, K.-H. 2016b. Influence of pyrolysis temperature on characteristics and phosphate adsorption capability of biochar derived from waste-marine macroalgae (*Undaria pinnatifida* roots). *Bioresource Technol.* 200:1024–1028.

Kim, B.-S., Lee, H.W., Park, S.H., Baek, K., Jeon, J.-K., Cho, H.J., Jung, S.-C., Kim, S.C., and Park, Y.-K. 2016. Removal of Cu^{2+} by biochars derived from green macroalgae. *Environ. Sci. Pollut. Res.* 23:985–994.

Lee, H., Park, R.-S., Lee, H.W., Hong, Y., Lee, Y., Park, S.H., Jung, S.C., Yoo, K.-S., Jeon, J.-K., and Park, Y.-K. 2016. Adsorptive removal of atmospheric pollutants over *Pyropia tenera* chars. *Carbon Lett.* 19:79–88.

Lee, M.N. and Woo, S.H. 2014. Application of biochar produced from wood and seaweed for the removal of dye in wastewater. *2nd FORBIOM Workshop Potentials and Biochar to Mitigate Climate Change,* Hanbat National University, Korea.

Lehmann, J. and Joseph, S. 2009. Biochar for environmental management: An Introduction. In *Biochar for Environmental Management: Science and Technology,* Lehmann, J. and Joseph, S. (Eds.), pp. 1–9, Earthscan, London, UK.

Mandal, S., Thangarajan, R., Bolan, N.S., Sarkar, B., Khan, N., Ok, Y.S., and Naidu, R. 2016. Biochar-induced concomitant decrease in ammonia volatilization and increase in nitrogen use efficiency by wheat. *Chemosphere.* 142:120–127.

Mašek, O., Brownsort, P., Cross, A., and Sohi, S. 2013. Influence of production conditions on the yield and environmental stability of biochar. *Fuel.* 103:151–155.

Park, S.H., Cho, H.J., Ryu, C., and Park, Y.-K. 2016. Removal of copper (II) in aqueous solution using pyrolytic biochars derived from red macroalga *Porphyra tenera. J. Ind. Eng. Chem.* 25:314–319.

Qian, K., Kumar, A., Zhang, H., Bellmer, D., and Huhnke, R. 2015. Recent advances in utilization of biochar. *Renew. Sustain. Energy Rev.* 42:1055–1064.

Rajapaksha, A.U., Chen, S.S., Tsang, D.C.W., Ming, Z.M., Vithanage, M., Mandal, S., Gao, B., Bolanm N.S., and Ok, S.Y. 2016. Engineered/designer biochar for contaminant removal/immobilization from soil and water: Potential and implication of biochar modification. *Chemosphere.* 148:276–291.

Roberts, D.A., Paul, N.A., Bird, M.I., and de Nys, R. 2015a. Bioremediation for coal-fired power stations using macroalgae. *J. Environ. Manag.* 153:25–32.

Roberts, D.A., Paul, N.A., Cole, A.J., and de Nys, R. 2015b. From waste water treatment to land management: Conversion of aquatic biomass to biochar for soil amelioration and the fortification of crops with essential trace elements. *J. Environ. Manage.* 157:60–68.

Roberts, D.A., Paul, N.A., Dworjanyn, S.A., Bird, M.I., and de Nys, R. 2015c. Biochar from commercially cultivated seaweed for soil amelioration. *Sci. Rep.* 5:1–6.

Zhang, M., Gao, G., Varnoosfaderani, S., Hebard, A., Yao, Y., and Inyang, M. 2013. Preparation and characterization of a novel magnetic biochar for arsenic removal. *Bioresource Technol.* 130:457–462.

chapter eleven

Identification and quantification of laminarins in brown algae

Angelika Graiff, Wolfgang Ruth, and Ulf Karsten

Contents

11.1 Introduction

Brown algae exhibit a unique central storage carbon metabolism (Michel et al. 2010). The photoassimilated D-fructose 6-phosphate is not used by these algae to produce sucrose as in higher plants but is rather converted into the polyol D-mannitol (Stenhouse 1844, Yamaguchi et al. 1966). Brown algae lack or contain only very low concentrations of free sugars such as glucose, fructose, or sucrose (Yamaguchi et al. 1966). These free sugars most likely undergo immediate biochemical conversion into D-mannitol and polysaccharides.

The long-term carbon storage polysaccharide in brown algae is laminarin, a vacuolar β-1,3-glucan, which chemically differs from storage products of most other living organisms that typically use glycogen or starch (α-1,4-glucans). Laminarin, also known as laminaran or leucosin, was first isolated from Laminariaceae by Schmiedeberg (1885). Unlike other algal polysaccharides (e.g., agar, alginate, and carrageenan), laminarin does not have thickening and gelling properties (Rupérez et al. 2002). Laminarins are a class of low-molecular mass storage β-glucans, which are composed of (1,3)-β-D-glucan (Rioux et al. 2007) and consisting of chains of β-1,3′-linked glucose units with occasional β-1,6′-linkages (Percival and Ross 1951, Beattie et al. 1961). These β-1,6′-linkages are present in a linear chain of β-1,3′-linked glucose residues and mainly as interchain linkages, which lead to a ramification of the molecule (Peat et al. 1958, Annan et al. 1965). The molecular weight of laminarin is approximately 5000 Da and is dependent on the degree of polymerization, which is in the range of 20–25 glucose units (Nelson and Lewis 1974, Alderkamp et al. 2007). Furthermore, this polysaccharide is polydisperse, consisting of a minor G-series with polymers containing only glucose residues, and a more abundant M-series with glucans terminated with a 1-linked D-mannitol residue (Read et al. 1996, Figure 11.1).

Consequently, the chemical variation in this storage molecule is based on the number of β-1,6′-linkages, the degree of branching, and the presence or absence of terminal mannitol residues (Craigie 1974). The structure and ratio of the different types of laminarin vary according to algal species, tissue type, reproductive conditions, and seasonal growth and environmental factors (Read et al. 1996, Chizhov et al. 1998, Rioux et al. 2010).

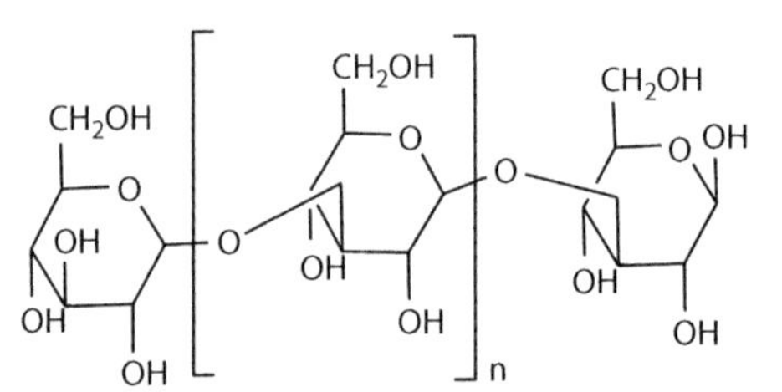

(a) G-chain laminarin

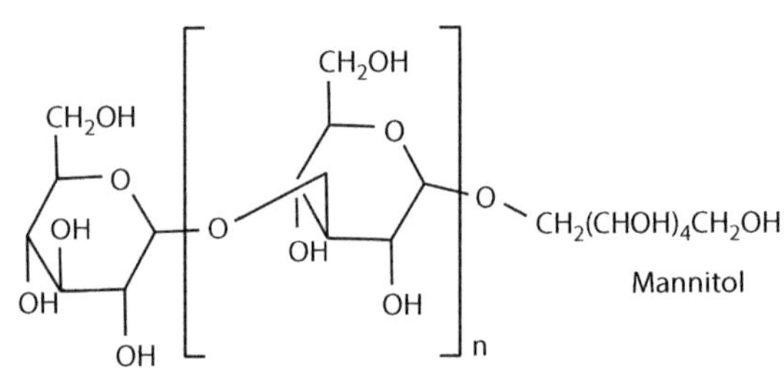

(b) M-chain laminarin

Figure 11.1 Chemical structures of one unit of G- and M-chain of laminarin.

Laminarin shows not only chemical variations but also quantitative changes. Over the course of a year, some brown algae shift their major storage compounds from mannitol in spring to laminarin in autumn (Dethier and Williams 2009). Mannitol is remobilized and translocated in perennial species of brown algae from mature storage tissues to supply the rapidly growing meristematic parts with carbon for biomass formation (e.g., reviewed by Gómez and Huovinen 2012). Stored polysaccharides are of particular ecological relevance for perennial brown algae because they often allow the uncoupling of growth from photosynthesis. This is a crucial adaptation to ecosystems with a strong seasonal variation such as the polar regions, which are characterized by continuous darkness in winter and continuous daylight in summer.

Laminarins are also of increasing economic interest because some of their derivatives are used for high-end products such as in cosmetics and pharmacology (Yvin et al. 1999).

11.2 State of the art

Different studies quantifying laminarin from the algal material used specific kinds of extraction methods, hydrolysis conditions, and tests for glucose as equivalent units for laminarin (see, e.g., Gagné et al. 1982, Gómez and Wiencke 1998, Dethier and Williams 2009). Laminarin was extracted from the dried algal material using ethanol (20%) at 75°C for 2–3 h and subsequent hot acidic hydrolysis (HCl or H_2SO_4) of ethanol extracts (Chapman and Craigie 1977, Gómez and Wiencke 1998) or enzymatic hydrolysis (Johnston et al. 1977, Devillé et al. 2004). Another used method is the direct extraction of laminarin from dried algal biomass with sulfuric or hydrochloric acid and the use of the supernatant after neutralization for quantification of laminarin as total sugar equivalents (e.g., Devillé et al. 2004, Schiener et al. 2015). However, the hydrolysis of the complete algal material will most probably result in a strong overestimation of laminarin, because acidic hydrolysis will cleave all sugars or sugar residues present in the biomass such as, for example, glycolipids, glycoproteins, and celluloses. Furthermore, the acidic hydrolysis of pure commercial laminarin is incomplete, and hence not quantitative, because a larger amount of oligomer carbohydrate units than single glucose units is received under widely applied hydrolysis conditions (for details see Graiff et al. 2016). In addition, the hydrolysis of polysaccharides is leading to uncontrolled cleavage of sugar units (Templeton et al. 2012). Thus, the received concentration of glucose from laminarin varied depending on the hydrolysis duration (Graiff et al. 2016).

Laminarin quantification was performed, either indirectly after hydrolysis by measuring the cleaved glucose units with enzymatic assays (Gómez and Wiencke 1998) or with the anthrone reagent method

(Yemm and Willis 1954), both calibrated against glucose standards. So far, neither intercalibration nor any test for reproducibility or sensitivity of the different protocols have been conducted to ensure that these analytical approaches are comparable.

The commercial laminarin, which is used as a reference substance for identification and quantification is available from a company as a lyophilized powder and was extracted from *Laminaria digitata*. However, the company is not able to provide any information about the extraction procedure and purity of the laminarin product. Hence, it is not certified as a standard. It might be reasonable to assume that this commercially available laminarin is a product being obtained from algal material by chemical or enzymatic extraction procedures according to the U.S. patent for the use of laminarin in cosmetics and pharmacology (Yvin et al. 1999). Thus, the use of a liquid chromatography–mass spectrometry method (LC–MS) is essential for identification and characterization of laminarins, as the high-performance liquid chromatography separates the mixture of laminarin species and the mass spectrometry analyzes their molar masses (Figure 11.2). The LC–MS measurements based on the method of Graiff et al. (2016) demonstrated clearly that the commercial laminarin consists of a mixture of four different chemical species of laminarin showing the typical molar mass distribution of 2000–7000 Da. Quantification depending on the commercially available laminarin from *L. digitata* is problematic, because of the species-specific variability and number of the chemical structures of laminarin, which can vary depending on environmental and individual conditions of the harvested algae.

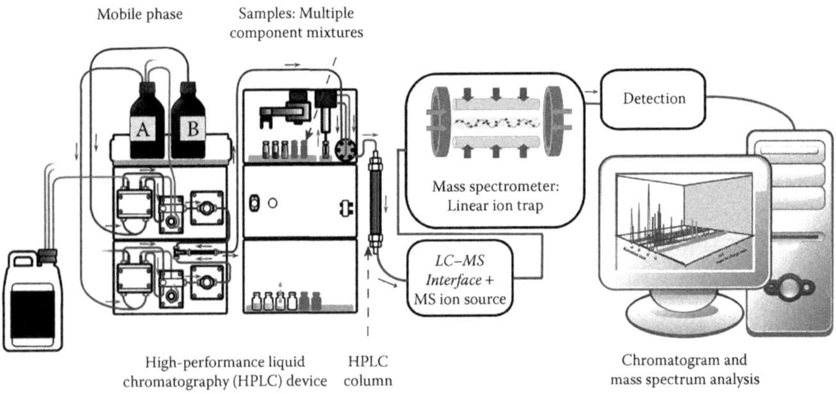

Figure 11.2 Scheme of a liquid chromatography–mass spectrometry (LC–MS) system. (Courtesy of Daniel Norena-Caro, Creative Commons, CC0 1.0 Universal Public Domain Dedication.)

These points also represent a great challenge working with natural substances as a reference compound for calibration because the chemical structure and quantity depend on the biological source and environmental factors. The resulting aspect of laminarin variability has to be considered in all applied analytical approaches. A second commercially available laminarin substance from another company showed no detectable chemical laminarin species by LC–MS; however, the signals indicated that the substance consisted of disaccharides. Thus, according to our experiences, LC–MS measurements provide the most precise and reliable data as they detect (1) the presence, (2) the different chemical structures of laminarin, and (3) their quantities.

The described LC–MS method is an analytical tool to identify and characterize carefully and finally, quantify laminarin in extracts from different brown algal species to overcome the methodological problems described. The cold extraction of this storage polysaccharide from brown algae is a very simple method and is suitable for further exact quantitative determination of these complex compounds. The obtained extract from the dried algal material can be analyzed directly by LC–MS. The direct characterization and quantification of laminarin by LC–MS is a helpful approach to modern biochemical analytics to verify the importance of laminarin for brown algae.

11.3 Materials

11.3.1 Extraction of laminarin from macroalgal material

- Freeze drier
- Laboratory mill
- Spatula
- Laboratory balance
- Pipettes
- 1 ml reaction vessels
- Ultrapure water
- Laboratory shaker
- Centrifuge (14,000 rpm) (*Note 1*)

11.3.2 Precipitation of alginate

- Formic acid (98%, MS-grade)
- Ethanol (>99.7%)
- Vortex mixer
- Refrigerator (*Note 2*)

11.3.3 Liquid chromatography–mass spectrometrical method measurement

- Commercial laminarin from *Laminaria digitata* (Sigma L9634, Sigma-Aldrich Chimie S.a.r.l., France).
- 1.5 ml microvials (MS certified).
- LTQ Velos Pro mass spectrometer with Accela HPLC (Thermo Scientific, USA).
- Kinetex column (C18, 150 × 2.1 mm, 2.6 μm core shell, Phenomenex, USA).
- Eluents:
  - Ultrapure water
  - Acetonitrile (MS-grade)
  - Modifier: Formic acid (MS-grade)
- The linear ion trap mass spectrometer has to work in the ESI negative ion mode in a mass range of 100–4000 atomic mass units (amu). The use of an ion trap equipped mass spectrometer has the advantage to accomplish several experiments simultaneously. For analyzing laminarin, the simultaneous use of the full scan and zoom scan mode are recommended.
- Instrument control and evaluation software (Xcalibur 2.2, Thermo Scientific, USA).

11.4 Experimental procedures

11.4.1 Laminarin extraction

Two extraction steps are used to guarantee a complete extraction of the different laminarin species from the algal samples. The first extraction with ultrapure water supplies mainly unbranched short polysaccharide chains (90%), the second mainly branched long polysaccharide chains (10%), and the third no laminarin at all.

1. For extraction, grind the samples (<1 mm grain size) with an analytical mill (Ika MF 10 Basic, Germany).
2. Extract the algal material (0.2 g DW) with ultrapure water (1.5 ml) on a shaker (250 rpm) at 25°C for 2 h.
3. Centrifuge 1 ml of the water extract at 14,000 rpm for 5 min (HettichMikro 22 R, Germany).
4. Transfer the supernatant into a 4 ml vial and add 1.5 ml of ultrapure water. Then the procedure is repeated, and both supernatants are mixed and converted into a microvial (300 μl) for direct LC–MS analysis.
5. If the algal sample contains a large amount of alginate, see *Note 2*.

11.4.2 LC–MS measurements

Laminarin is analyzed using LC–MS, Figure 11.2 to separate and identify the different laminarin species contained in extracted algal samples (*Note 3*).

1. For separation of the different laminarin species use a Kinetex column (2.6 μm C18, 150 × 3 mm) (*Note 4*).
2. Use ultrapure water and acetonitrile (90% water with 0.1% formic acid (v:v) and 10% acetonitrile with 0.1% formic acid) as mobile phase and run isocratically at a flow rate of 0.2 ml min^{-1}.
3. For determination of molar masses of different laminarin species run the MS in ESI negative ion mode (Cone energy of the electrospray: 3.5 kV) in a mass range of 100–4000 amu.
4. Accomplish simultaneous full scan and special zoom scan measurements on representative laminarin masses to receive qualitative data of the laminarin species and quantitative data of the total laminarin content.
5. Apply the zoom scan measurements to achieve exact information about the charge state of the measured representative molar masses as the x-axis of the MS measurements is the mass to charge signal (m/z). These measurements concerning the charge state are based on the abundance ratio of the carbon isotopes: ^{12}C and ^{13}C. Thus, the actual molecular mass of, for example, 3256 Da is calculated with the detected mass of 1628 amu on the x-axis and a charge state of 2 (Figure 11.3). Then calculate the mean chain length and branched chains of individual laminarin species in the reference substance and in the algal samples.
6. Achieve quantitative information of the total laminarin content in the algal samples on the comparison of full scan measurements of the commercial reference laminarin to full scan measurements of the extracted algal samples.
7. Perform a smoothing of the full scan peaks using the evaluation software. Most likely, the quantitative information will be afflicted with an error because an absolute quantification is only possible with a licensed laminarin standard substance, which accurately corresponds to the different laminarin species in the respective algal samples. However, natural substances are characterized by the presence of various chemical species of laminarin. Nevertheless, comparative measurements because of the trends are possible and offer good evidence for quantification of total laminarin contents from algal samples.

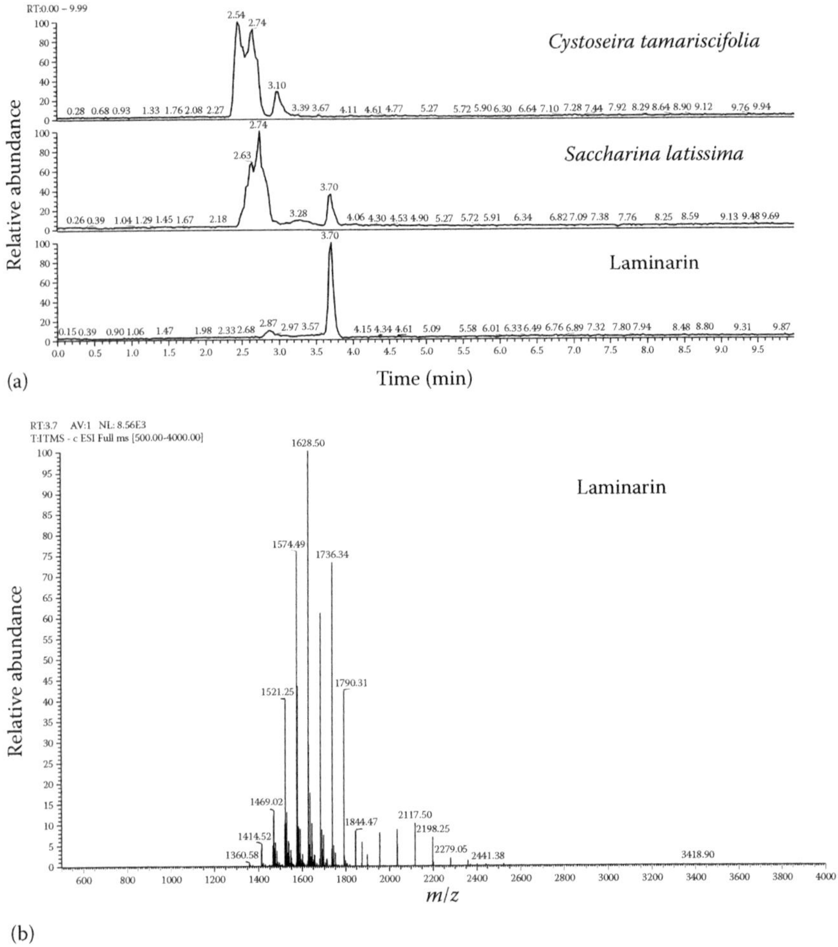

Figure 11.3 Chromatograms of algal extracts and commercial laminarin from *Laminaria digitata* (a). The commercial laminarin shows a distinctive single laminarin peak at a retention time of 3.70 min. The chromatogram of *Cystoseira tamariscifolia* reveals no laminarin in contrast to *Saccharina latissima* showing a clear laminarin peak at 3.70 min. LC–MS mass spectrum of commercial laminarin as a reference and for calculation of its molar masses (b).

11.4.3 Laminarin reference and quantification

For the calibration of the LC–MS and for a reliable determination and analysis of laminarin from natural algal samples, the identification and characterization of laminarin of commercially available substances are necessary (*Note 5*).

1. Dissolve commercial laminarin from *L. digitata* in ultrapure water, which will provide clear LC–MS signals (Figure 11.3). Ultrapure water suits best for dissolving laminarin compared to solvents such as methanol, ethanol, ethanol/water (1:1, v/v), ethyl acetate, and tetrahydrofuran (THF) (for details see Graiff et al. 2016).
2. Prepare standard solutions of the commercial laminarin in different concentrations (0.7, 0.375, 0.14, 0.07, and 0.014 mg ml^{-1}) and analyze them with the LC–MS to receive a calibration curve.
3. In addition to the full-scan measurements of the LC–MS, apply the special simultaneous zoom scan measurement to identify charge states of laminarin and to calculate the molar mass distribution. Molar mass is determined at the maximum of the statistical molar mass distribution.
4. Identify commercial laminarin in different concentrations by its retention time of 3.70 min and its distinctive mass spectra of different chain lengths (Figure 11.3).
5. Apply the special simultaneous zoom scan measurement, to identify laminarin as doubly charged. On the basis of this evidence, the typical molar mass distribution of 2000–7000 Da can be confirmed for the commercial laminarin by LC–MS analysis. Then use the commercial laminarin as a reference for comparison with natural algal samples.
6. This method will allow quantifying laminarin in very low concentrations (<0.014 mg ml^{-1}).
7. Quantify laminarin from the algal samples by comparison of the retention time and peak area with standard solutions (calibration range 0.375–0.07, 0.7–0.07, and 0.14–0.014 mg ml^{-1}) using the evaluation software.

11.5 Notes

Note 1: A hot standard extraction method completely extracts alginate in addition to the laminarins, which results in very viscous extracts. These extracts are completely unsuitable for LC–MS analysis. The hot extraction follows a standardized microwave extraction method for plant components in ultrapure water as a solvent. The conditions for microwave extraction are 20% of 400 W (MarsXpress 5 CEM) at a rate of 5°C min^{-1} to 60°C for 5 min. This microwave-based method is very effective and can be used for fresh and freeze dried raw material with water as the solvent. It takes only 20 min for a complete extraction. However, this standard extraction method is proved to be completely unsuitable for extraction of laminarin from brown macroalgal samples.

Note 2: Certain brown macroalgal species (e.g., Fucales) contain very high amounts of alginates. Thus, even the application of the described cold water method extracts substantial amounts of alginate. The extraction of alginates can be recognized by a higher viscosity and a visible turbidity of the extract. To avoid disturbance of the LC–MS measurements, a method was developed to precipitate alginates. From the aqueous extract, 500 μL is transferred into a reaction vessel, and the alginates are precipitated with 450 μL ethanol (>99.7%), and 50 μL formic acid (98%). This mixture is mixed for 10 s (Vortex mixer), then stored for 12 h in a refrigerator and finally centrifuged for 5 min at 14,000 rpm (Hettich micro 22 R, Germany). The clear and liquid supernatant is transferred into a microvial (300 μl) for later LC–MS analysis. To achieve comparable analysis conditions, the commercial laminarin substance used for the reference measurements is treated in the same way.

Note 3: The received extracts have to be measured either immediately or after 4 h. Longer storage of the extracts, even in the refrigerator, leads to noticeable decomposition of the polysaccharides, and thus lower laminarin content will be detected.

Note 4: LC–MS analytics have some particular characteristics, which have to be considered. If chemically different laminarin species, number of different chain lengths, and presence of branched chains have to be studied and investigated with the HPLC-measuring system, we recommend the use of a column for size-exclusion chromatography. Therefore, we use for the chromatographic separation of individual laminarin chains a Zenix SEC 150 column (Sepax Technologies, Inc., 250 mm length, 4.6 mm ID, 3 μm particle size, 150 An intermediate volume) with ultrapure water and 0.1% formic acid as eluents. However, for the quantitative determination of the total laminarin content, this column is not suitable because an integration of split or shoulder peaks is not possible even after smoothing. For quantification purposes, the use of the column described in Methods is better suitable.

Note 5: During an analysis sequence with the LC–MS system, a laminarin reference sample has to be inserted as a quality control-run (QC-run) after at least six algal sample runs. Despite careful sample preparation, both the spray shield and the transfer capillary of the mass spectrometer get blocked relatively fast, which leads to a smaller ion transfer into the mass spectrometer. A QC-run permits clear measurements because of an unhindered ion transfer.

Acknowledgment

This research was funded by the Project BIOACID Phase II of the German Federal Ministry of Education and Research (BMBF; FKZ 03F0655L).

References

Alderkamp, A.-C., M. van Rijssel, and H. Bolhuis. 2007. Characterization of marine bacteria and the activity of their enzyme systems involved in degradation of the algal storage glucan laminarin. *FEMS Microbiol. Ecol.* 59:108–117.

Annan, W. D., E. Hirst, and D. J. Manners. 1965. 162. The constitution of laminarin. Part V. The location of 1,6-glucosidic linkages. *J. Chem. Soc.* 885–891. doi:10.1039/JR9650000885.

Beattie, A., E. L. Hirst, and E. Percival. 1961. Studies on the metabolism of the Chrysophyceae. Comparative structural investigations on leucosin (Chrysolaminarin) separated from diatoms and laminarin from the brown algae. *Biochem. J.* 79:531–537.

Chapman, A. R. O., and J. S. Craigie. 1977. Seasonal growth in *Laminaria longicruris*: Relations with dissolved inorganic nutrients and internal reserves of nitrogen. *Mar. Biol.* 40:197–205.

Chizhov, A. O., A. Dell, H. R. Morris et al. 1998. Structural analysis of laminarans by MALDI and FAB mass spectrometry. *Carbohydr. Res.* 310:203–210.

Craigie, J. S. 1974. Storage products. In *Algal Physiology and Biochemistry*, W. P. D. Stewart (Ed.), pp. 206–235. Berkley, CA: University of California Press.

Dethier, M. N., and S. L. Williams. 2009. Seasonal stresses shift optimal intertidal algal habitats. *Mar. Biol.* 156:555–567.

Devillé, C., J. Damas, P. Forget, G. Dandrifosse, and O. Peulen. 2004. Laminarin in the dietary fibre concept. *J. Sci. Food Agricult.* 84:1030–1038.

Gagné, J. A., K. H. Mann, and A. R. O. Chapman. 1982. Seasonal patterns of growth and storage in *Laminaria longicruris* in relation to differing patterns of availability of nitrogen in the water. *Mar. Biol.* 69:91–101.

Gómez, I., and P. Huovinen. 2012. Morpho-functionality of carbon metabolism in seaweeds. In *Seaweed Biology: Novel Insights into Ecophysiology, Ecology and Utilization*, C. Wiencke, and K. Bischof (Eds.), pp. 25–46. Berlin, Germany: Springer Verlag.

Gómez, I., and C. Wiencke. 1998. Seasonal changes in C, N and major organic compounds and their significance to morpho-functional processes in the endemic Antarctic brown alga *Ascoseira mirabilis*. *Polar Biol.* 19:115–124.

Graiff, A., W. Ruth, U. Kragl, and U. Karsten. 2016. Chemical characterization and quantification of the brown algal storage compound laminarin—A new methodological approach. *J. Appl. Phycol.* 28:533–543.

Johnston, C. S., R. G. Jones, and R. D. Hunt. 1977. A seasonal carbon budget for a Laminarian population in a scottish sea-loch. *Helgoländer Wissenschaftliche Meeresuntersuchungen* 30:527–545.

Michel, G., T. Tonon, D. Scornet, J. M. Cock, and B. Kloareg. 2010. Central and storage carbon metabolism of the brown alga *Ectocarpus siliculosus*: Insights into the origin and evolution of storage carbohydrates in eukaryotes. *New Phytol.* 188:67–81.

Nelson, T. E., and B. A. Lewis. 1974. Separation and characterization of the soluble and insoluble components of insoluble laminaran. *Carbohydr. Res.* 33:63–74.

Peat, S., W. J. Whelan, and H. G. Lawley. 1958. 141. The structure of laminarin. Part I. The main polymeric linkage. *J. Chem. Soc.* 724–728. doi:10.1039/JR9580000724.

Percival, E. G. V., and A. G. Ross. 1951. 156. The constitution of laminarin. Part II. The soluble laminarin of *Laminaria digitata*. *J. Chem. Soc.* 720–726. doi:10.1039/JR9510000720.

Read, S. M., G. Currie, and A. Bacic. 1996. Analysis of the structural heterogeneity of laminarin by electrospray-ionisation-mass spectrometry. *Carbohydr. Res.* 281:187–201.

Rioux, L.-E., S. L. Turgeon, and M. Beaulieu. 2007. Characterization of polysaccharides extracted from brown seaweeds. *Carbohydr. Polm.* 69:530–537.

Rioux, L.-E., S. L. Turgeon, and M. Beaulieu. 2010. Structural characterization of laminaran and galactofucan extracted from the brown seaweed *Saccharina longicruris*. *Phytochem.* 71:1586–1595.

Rupérez, P., O. Ahrazem, and J. A. Leal. 2002. Potential antioxidant capacity of sulfated polysaccharides from the edible marine brown seaweed *Fucus vesiculosus*. *J. Agr. Food Chem.* 50:840–845.

Schiener, P., K. D. Black, M. S. Stanley, and D. H. Green. 2015. The seasonal variation in the chemical composition of the kelp species *Laminaria digitata*, *Laminaria hyperborea*, *Saccharina latissima* and *Alaria esculenta*. *J. Appl. Phycol.* 27:363–373.

Schmiedeberg, W. 1885. *Über die Bestandteile der Laminaria*. Gesellschaft deutscher Naturforscher und Ärzte, Leipzig, Tageblatt der 58. Versammlung, p. 427.

Stenhouse, J. 1844. Über das Vorkommen von Mannit in *Laminaria saccharina* und einigen anderen Seegräsern. *Justus Liebigs Annalen Der Chemie* 51:349–354.

Templeton, D. W., M. Quinn, S. van Wychen, D. Hyman, and L. M. L. Laurens. 2012. Separation and quantification of microalgal carbohydrates. *J. Chromatogr. A.* 1270:225–234.

Yamaguchi, T., T. Ikawa, and K. Nisizawa. 1966. Incorporation of radioactive carbon from $H^{14}CO_3^-$ into sugar constituents by a brown alga, *Eisenia bicyclis*, during photosynthesis and its fate in the dark. *Plant Cell Physiol.* 7:217–229.

Yemm, E. W., and A. J. Willis. 1954. The estimation of carbohydrates in plant extracts by anthrone. *Biochem. J.* 57:508–514.

Yvin, J.-C., F. LeVasseur, and F. Hud`Homme. 1999. Use of laminarin and oligosaccharides derived there from in cosmetics and for preparing a skin treatment drug. US Patent 5980916A.

chapter twelve

Determination of carbohydrate composition of macroalgae

Wouter J.J. Huijgen, E.M. Cobussen-Pool, B.F. van Egmond, and J.W. van Hal

Contents

12.1 Introduction

Carbohydrates are a major constituent of macroalgae (in addition to minerals and proteins) (Van Hal et al. 2014). In addition, carbohydrates are key target components in the application of macroalgae for production of, for example, biobased chemicals and biofuels (Van Hal et al. 2014). However, a generally accepted protocol for adequate determination of the content of individual carbohydrates is lacking. The EU-H2020 MacroFuels project aims to develop such a protocol.

The carbohydrates that occur in seaweeds vary between the major classes of seaweeds: that is, the brown seaweeds (*Phaeophycaea* including Kelps), the red seaweeds (*Rhodophyta*), and the green seaweeds (*Chlorophyta*). Within the group of red seaweeds, at least three different main structural carbohydrates occur: that is, (1) xylan (e.g., *Palmaria palmata*), (2) agar (e.g., *Gracilaria* sp.), and (3) carrageenan (e.g., *Chondrus crispus*). Table 12.1 gives an overview of carbohydrates reported in the literature for the seaweeds used in our study.

Here, we present a method for determination of the carbohydrate composition of macroalgae based on hydrolysis and subsequent determination of resulting monosaccharides by high-performance anion-exchange chromatography with pulsed amperometric detector (HPAEC–PAD). The presented approach is intended to be uniform for all seaweeds, with optimum hydrolysis conditions that have to be determined for each (class of) seaweed. As an example, the optimum hydrolysis times for three different seaweeds have been identified. This optimization procedure can also be extended to other seaweeds. In addition, a gradient and elution order for the HPAEC–PAD is described that allows determination of the major

Table 12.1 Overview of (structural) carbohydrates present in seaweeds used (Van Hal et al. 2016) (see also *Note 12*)

Seaweed	Laminaria digitata	Palmaria palmata	Ulva lactuca
Seaweed class	Brown	Red	Green
Carbohydrates in seaweed (corresponding monosaccharides on hydrolysis)	Mannitol Alginate (mannuronic and guluronic acid) Laminarin (glucose) Cellulose (glucose) Fucoidan (fucose)	Floridoside (galactose and glycerol) Xylan (xylose) Cellulose (glucose)	Ulvan (rhamnose, xylose, glucuronic acid, and iduronic acid) Cellulose (glucose)

sugar alcohols, reducing sugars, deoxy sugars, and uronic acids occurring in tested seaweeds in a single run.

12.2 State of the art

A generally accepted protocol for determination of carbohydrate composition of seaweeds is lacking. However, for other types of biomass, such protocols exist, such as the method for lignocellulosic biomass developed by the U.S. National Renewable Energy Laboratory (NREL). This method is based on two-stage hydrolysis of structural carbohydrate polymers using H_2SO_4 and subsequent determination of resulting monosaccharides (Sluiter et al. 2010). In a comparison of different chemical and enzymatic hydrolysis methods, the NREL method has been selected in literature for compositional analysis of brown seaweeds (Manns et al. 2014). In addition to the standard NREL protocol, in which second-stage hydrolysis is performed at 121°C, an alternative protocol with hydrolysis at 100°C has been reported in the literature for summative composition analysis of lignocellulosic biomass (TAPPI 2009, Wildschut et al. 2013). This modified approach has been applied in literature to, for example, *Ulva lactuca* (Van der Wal et al. 2013). Anticipating that more labile carbohydrates such as uronic acids would give a higher recovery at a lower temperature, the starting point for this study was based on this approach, that is, as follows:

- *Prehydrolysis*: 12 M H_2SO_4, 30°C, 1 h (to hydrolyze the polymeric structural carbohydrates into soluble [oligomeric] carbohydrates).
- *Hydrolysis*: 1.2 M H_2SO_4, 100°C, 3 h (to convert oligomeric carbohydrates into monosaccharides).
- Analysis of resulting monomeric carbohydrates by HPAEC–PAD.

The reaction times of both steps have been varied to determine optimum conditions. It was found that prehydrolysis is required to have adequate hydrolysis of structural components such as cellulose and ulvan. Optimum prehydrolysis time determined for all the three seaweeds was 60 min (identical to lignocellulosic biomass).

The optimum hydrolysis time was found to depend on both the seaweed type and specific monosaccharides of interest. For each monosaccharide, a balance between formation and degradation (e.g., to furans), which are both acid catalyzed, needs to be found. Recommended optimum hydrolysis times based on the maximum total carbohydrate response are given in Table 12.2.

Table 12.2 Optimum hydrolysis times recommended

Seaweed	*Laminaria digitata*	*Palmaria palmata*	*Ulva lactuca*
Optimum hydrolysis time (h)	3	2	3[a]
Too short, that is, underestimation of:	Mannuronic acid Glucose (Chemically bound) mannitol		Glucuronic acid Rhamnose
Too long, that is, degradation of:	Guluronic acid	Xylose	

[a] Iduronic acid is very sensitive to degradation by hydrolysis. The optimum response of all monosaccharides, including iduronic acid, could not be obtained with a single hydrolysis time.

12.3 *Materials*

12.3.1 *Seaweed*

- In the case of nondried seaweed, immediately dry it to prevent compositional changes.
  - Optionally, remove large contaminants such as stones and snails.
  - Cut seaweed, if needed.
  - Dry at mild conditions, preferably by freeze-drying (alternatively, in a [vacuum] oven at ≤50°C).
  - Check moisture content. Ambient-dry conditions (about 5%–15% moisture content, depending on seaweed species) are sufficient.
- Crush or cut dried seaweed to <1 cm and homogenize.
- Store the dried seaweed in closed containers at ambient temperature until compositional analysis.

12.3.2 *Reagents*

- 72% w/w (12 M) sulfuric acid (stored at 4°C)
- $BaCO_3$
- Bromophenol blue indicator (20 mg L^{-1})
- Standards analysis monomeric carbohydrates:
  - *Sugar alcohols*:
    - Mannitol
    - Glycerol
  - *Reducing sugars*:
    - Glucose
    - Xylose
    - Arabinose (might be left out as not detected in seaweeds)
    - Galactose
    - Fructose (internal standard)

- *Deoxy sugars*:
  - Rhamnose (as L-rhamnose monohydrate)
  - Fucose
- *Uronic acids*:
  - Guluronic acid as the sodium salt (CarboSynth Limited, UK)
  - Mannuronic acid as the sodium salt (CarboSynth Limited, UK)
  - Iduronic acid (CarboSynth Limited, UK)
  - Galacturonic acid
  - Glucuronic acid

12.3.3 *Equipment*

- Halogen moisture analyzer.
- 50 mL centrifuge plastic tubes with a screw cap.
  - Use centrifuge tubes that are heat resistant to maximum operating temperature (i.e. 100°C).
  - Use centrifuge tubes with a cap that are liquid tight at elevated temperature, to prevent leaking while shaking. Centrifuge tubes with a flat screw cap tend to work fine.
- 10 mL plastic centrifuge tubes.
- Water bath set at 30°C.
- Water bath set at 100°C.
- Centrifuge applicable for 50 mL centrifuge tubes.
- Centrifuge applicable for 10 mL centrifuge tubes.
- HPAEC setup (Dionex, Sunnyvale, CA, USA) consisting of
  - ICS3000 Ion Chromatography System
  - CarboPac PA1 column (2 × 250 mm)
  - CarboPac guard column (2 × 50 mm)
  - Pulsed amperometric detector (PAD)
  - Pump 1 (used for gradient over column, 0.25 mL/min)
    - *Eluent A*: Ultrapure water
    - *Eluent B*: 0.1 M NaOH in ultrapure water
    - *Eluent C*: 0.1 M NaOH and 1 M NaAc in ultrapure water
  - Pump 2 (used for postcolumn addition of NaOH, 0.15 mL min^{-1})
    - 0.25 M NaOH in ultrapure water.

12.4 *Experimental procedures*

Determination of the carbohydrate composition of macroalgae consists of the following steps (see also the general *Notes 1–5*):

1. Preparation of seaweed sample.
2. Two-stage hydrolysis (to hydrolyze the polymeric structural carbohydrates into monosaccharides).

3. Analysis of resulting monosaccharides by HPAEC–PAD using a gradient that allows determination of the major sugar alcohols, reducing sugars, deoxy sugars, and uronic acids occurring in tested seaweeds in a single run.

12.4.1 *Preparation seaweed sample*

1. Take a representative sample of ambient-dry seaweed from the closed container (see also 3.1.). Use at least about 2 g, but in the case of inhomogeneous sample use a larger amount.
2. Optionally, let seaweed sample equilibrate with ambient humidity overnight. Normally, equilibration takes place during subsequent milling, and this step can be skipped.
3. Mill seaweed sample to <250 μm:
   a. Use, for example, a laboratory cutting mill, knife mill, or a vibratory disc mill.
   b. Avoid heating of the sample while milling (i.e., mill in steps using short milling times).
   c. After each milling step, sieve sample over <250 μm and mill part that does not pass again.
4. Transfer seaweed sample quantitatively to a plastic container:
   a. Use a brush to get the finer part of the sample off the inside of the mill.
   b. Avoid separation of inorganics particles that might be present.
5. Homogenize seaweed sample well by mixing.
6. Take representative subsample and determine moisture content using a moisture analyzer or by drying till constant weight (x_{moisture}).
7. Weigh empty 50 mL centrifuge plastic tube (m_{empty}).
8. Weigh in 300 ± 5 mg of representative subsample in the centrifuge tube (m_{seaweed}).
9. Make sure all the sample is at the bottom of the tube.

12.4.2 *Hydrolysis*

12.4.2.1 *Prehydrolysis*

1. Place tube with the sample in ice water and let it cool down (proper cooling is crucial!).
2. Open the tube and carefully add 3 mL of prechilled 72% w/w sulfuric acid.
3. Carefully mix the sample with the sulfuric acid using a sealed glass Pasteur pipette. Leave the pipette in the tube.
4. Transfer the open tube to a preheated water bath and incubate for 60 min at 30°C (*Note 6*).

5. Mix sample every 10–15 min using the sealed glass Pasteur pipette.
6. After incubation, transfer the tube to ice water to stop the prehydrolysis reaction.

12.4.2.2 *Hydrolysis*

1. Add demineralized H_2O to the tube to a final volume of 30 mL. While adding the water, clean the sealed glass Pasteur pipette and take it out.
2. Close the tube with the screw cap and mix well by shaking.
3. Transfer tube, placed in a rack, to a preheated water bath, and incubate at 100°C for 180 min (*Laminaria digitata* and *Ulva lactuca*) or 120 min (*Palmaria palmata*) (*Notes 7 and 8*).
4. Mix every 45–60 min by shaking. Be cautious and shake gently as centrifuge tube is hot and tube and/or screw cap might have become softer at elevated temperature (Note!)
5. After incubation, transfer the tube to ice water to cool down and stop the hydrolysis.
6. Take tube out of ice water, dry the outside, and weigh the tube (m_{final}).
7. Centrifuge tube for 5 min at 2500–3000 g at 20°C.
8. Alternatively, the tube can be placed overnight in a refrigerator after initial cooling in ice water to cool down further and allow the residual solids to settle. In general, thus a clear supernatant is obtained, and centrifugation is not needed.
9. Take a sample from the supernatant (*Note 9*).
10. Store sample in the freezer at –20°C until analysis. The sample should preferably be analyzed within a few weeks but might be kept for substantially longer times.

12.4.3 *Analysis monomeric carbohydrates*

Analysis of monomeric carbohydrates resulting from seaweed hydrolysis as described above can be performed with any adequate technique. However, GC and HPLC techniques are generally not able to adequately distinguish all individual seaweed constituents and require the use of multiple columns and/or derivatization steps. Here, we describe HPAEC, given its ability to separate sugar alcohols, reducing sugars, deoxy sugars, and uronic acids in a single run.

12.4.3.1 *Sample preparation*

1. Thaw hydrolysate sample.
2. Add 0.2 mL seaweed hydrolysate to a 10 mL plastic centrifuge tube.
3. Add 4.8 mL 20 mg L^{-1} bromophenol blue indicator to the tube.

4. Neutralize by adding small amounts of $BaCO_3$ (total about 75 mg) until the solution turns blue while vortexing (final pH >4.6, max 8.2).
5. Centrifuge tube for 30 min at 2500–3000 g at 20°C.
6. Take 1.0 mL of the clear supernatant and add to a vial.
7. Add 0.5 mL 150 ppm fructose as an internal standard to the vial (*Note 13*).
8. Homogenize the solution and place vial in the autosampler of the HPAEC–PAD.

12.4.3.2 *Gradient*

After sample injection, while using 15 mM NaOH as mobile phase, pure water is used first as eluent for successively sugar alcohols, deoxy sugars, and reducing sugars and, subsequently, a gradient of NaOH and Na acetate is used as eluent for uronic acids (*Notes 10 and 11*).

1. Condition column with 0.1 M NaOH and 1 M NaAc (100% eluent C) (*Note 10*).
2. Condition column with 15 mM NaOH (85% eluent A [H_2O] and 15% eluent B [100 mM NaOH]) (*Note 10*).
3. Inject sample (injection volume 10 μL).
4. Run gradient as given in Table 12.3.

Table 12.3 HPAEC–PAD multigradient optimized for determination of the major sugar alcohols, deoxy sugars, reducing sugars, and uronic acids occurring in tested seaweeds in a single run (including conditioning for next injection)

t (min)	Eluent A (H_2O) (%)	Eluent B (0.1 M NaOH) (%)	Eluent C (0.1 M NaOH and 1 M NaAc) (%)
0	85	15	0
0	85	15	0
2	85	15	0
2	100	0	0
12.5	100	0	0
17	85	15	0
17.5	73	16	11
29.5	45	40	15
30	0	0	100
35	0	0	100
35.5	85	15	0
48	85	15	0

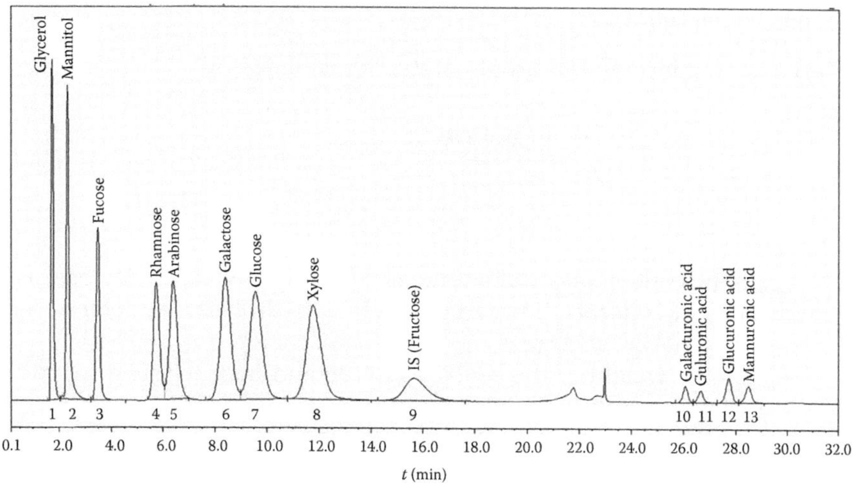

Figure 12.1 HPAEC-PAD chromatograph of a mixture of standards (excluding iduronic acid).

Figure 12.1 shows the HPAEC chromatograph of a mixture of standards and the gradient applied. Table 12.4 gives the elution order of components and the indicative retention times.

For component labeling, see Table 12.4.

Table 12.4 Elution order and indicative retention times of the components (*Note 14*)

#	Component	t (min)
1	Glycerol	1.7
2	Mannitol	2.3
3	Fucose	3.5
4	Rhamnose	5.7
5	Arabinose	6.4
6	Galactose	8.4
7	Glucose	9.6
8	Xylose	11.8
9	Fructose (IS)	15.7
10	Galacturonic acid	26.1
11	Guluronic acid	26.7
12	Glucuronic acid	27.7
13	Mannuronic acid	28.5
14	Iduronic acid	31.0

12.4.4 Calculations

The content of individual carbohydrates is calculated by

$$c_{\text{carb}} = \frac{[\text{carb}] \cdot (m_{\text{final}} - m_{\text{empty}})}{m_{\text{seaweed}} \cdot (1 - x_{\text{moisture}})} \cdot 100\% \tag{12.1}$$

where:

c_{carb} is the content of carbohydrate in seaweed (% w/w on dry basis).
[carb] is the concentration carbohydrate in hydrolysate as determined by HPAEC–PAD (g/g).
m_{final} is the weight of the centrifuge tube and hydrolysate after two-stage hydrolysis (g).
m_{empty} is the weight of the empty centrifuge tube (g).
m_{seaweed} is the weight of the seaweed sample used for hydrolysis (g).
x_{moisture} is the moisture content of the seaweed sample (–).

The carbohydrate content is based on corresponding monosaccharides. Optionally, a molar mass correction to anhydrous monomeric equivalents can be made in case the carbohydrate is known to be part of a polymer.

12.5 Notes

12.5.1 General remarks

Note 1: Perform compositional analysis at least in duplicate.

Note 2: The summative composition of seaweeds might be further completed by determination of, among others, proteins (e.g., based on total nitrogen) (Safi et al. 2017), phenolic components (Hou et al. 2017), and ash (by calcination of seaweed at 550°C).

Note 3: In a compositional analysis of lignocellulosic biomass, recovery factors are sometimes used to correct for acid degradation of formed monosaccharides (Sluiter et al. 2010). A similar approach might be followed for seaweeds, although it should be realized that (1) the use of recovery factors tends to lead to an overestimation of the content of individual carbohydrates (Whitfield et al. 2016) and (2) uronic acids are more sensitive to acid degradation than reducing sugars. (Moreover, the sensitivity of each uronic acid toward acid degradation differs.)

Note 4: No distinction between different glucans in, for example, Kelps, that is, laminarin and cellulose, can be made with this method. If required, see Chapter 11 by Graiff et al. (2017).

Note 5: The presented protocol has not been verified for agarophytes, such as *Gracilaria* sp., carrageenan-containing *Rhodophyta* such as

Chondrus crispus and fucoidan-rich brown seaweeds, such as *Fucus* sp. One can use the optimization procedure described to adapt the protocol for the desired seaweed.

12.5.2 *Seaweed hydrolysis*

Note 6: For *Palmaria palmata*, the prehydrolysis time might be shortened to 30 min.

Note 7: To prevent excessive boiling during the hydrolysis step, which tends to disturb the centrifuge tubes in the rack, the water bath can best be set at 99.5°C–99.8°C.

Note 8: To prevent pressure built-up during the hydrolysis step, the screw cap of the centrifuge tube might be slightly loosened before the tube is placed in the water bath.

Note 9: Optionally, the dry weight of the solid residue remaining after two-stage hydrolysis might be determined, after washing, to verify the completeness of the liquefaction.

12.5.3 *High-performance anion-exchange chromatography with pulsed amperometric detector*

Note 10: Adequate conditioning of the column is crucial for proper analysis and stable retention times.

Note 11: Calibration standards are perishable.

Open solutions of calibration standards as least as possible to reduce the risk of microbiological contamination.

Store calibration standards adequately.

Check calibration curves using an independent standard.

Note 12: In case other sugar alcohols than only mannitol and glycerol are present in the hydrolysate, separation of individual sugar alcohols over the CarboPac PA1 column might be insufficient. In that case, additional analysis of the hydrolysate with another analytical method is recommended such as HPAEC–PAD using a CarboPac MA1 column.

Note 13: In (tested) seaweeds, as far as known, no fructose is present. However, the absence of fructose should always be verified for each type of seaweed by HPAEC–PAD analysis of its hydrolysate without the addition of an internal standard. As an alternative to fructose, for example, ribose might be used as internal standard. The retention time of ribose is slightly longer than that of fructose.

Note 14: Mannose cannot be separated adequately from xylose with the presented HPAEC–PAD protocol. Please be aware when applying this protocol to other types of biomass such as nonmarine aquatic plants such as Azolla and water hyacinth. In seaweeds, as far as known, no mannose is present.

Acknowledgments

The authors thank Mark Smit for his contribution to the experimental work and Arjan Smit for his ideas and input. The current chapter is part of the MacroFuels project. This project has received funding from the European Union's Horizon 2020 research and innovation program under grant agreement No 654010.

References

Graiff, A., W. Ruth, and U. Karsten. 2017. Identification and quantification of laminarin in brown algae. In *Protocols for Macroalgae Research*, B. Charrier, T. Wichard, and C. R. K. Reddy, (Eds.), Chapter 11. Boca Raton, FL: CRC Press, Taylor & Francis Group.

Hou, X., R. Neerup, and A. B. Bjerre. 2017. Total phenol content and antioxidant capacity analysis of seaweed biomass. In *Biotechnology Protocols for Macroalgae Research*, B. Charrier, T. Wichard, and C. R. K. Reddy (Eds.), Chapter 15. Boca Raton, FL: CRC Press, Taylor & Francis Group.

Manns, D., A. L. Deutschle, B. Saake, and A. S. Meyer. 2014. Methodology for quantitative determination of the carbohydrate composition of brown seaweeds (Laminariaceae). *RSC Advances* 4:25736–25746.

Safi, C., J. van Leeuwen, Y. Telleman, N. Engelen-Smit, L. van den Broek, and P. Harmsen. 2017. Quantification of proteins in seaweeds. In *Biotechnology Protocols for Macroalgae Research*, B. Charrier, T. Wichard, and C. R. K. Reddy (Eds.), Chapter 13. Boca Raton, FL: CRC Press, Taylor & Francis Group.

Sluiter, J. B., R. O. Ruiz, C. J. Scarlata, A. D. Sluiter, and D. W. Templeton. 2010. Compositional analysis of lignocellulosic feedstocks. 1. Review and description of methods. *J. Agric. Food Chem.* 58:9043–9053.

TAPPI (Technical Association of the Pulp and Paper Industry) 2009. Standard test methods. http://www.tappi.org. Norcross, GA.

Van der Wal, H., B. L. H. M. Sperber, B. Houweling-Tan, R. R. C. Bakker, W. Brandenburg, and A. M. López-Contreras. 2013. Production of acetone, butanol, and ethanol from biomass of the green seaweed *Ulva lactuca*. *Bioresour. Technol.* 128:431–437.

Van Hal, J. W., A. B. Bjerre, A. M. López-Contreras, M. Stanley, G. O. Hreggvidsson, and W. J. J. Huijgen. 2016. Carbohydrate analysis of seaweed in the biorefinery to chemicals and fuel context. *Paper Presented at the International Seaweed Symposium*, Copenhagen, Denmark.

Van Hal, J. W., W. J. J. Huijgen, and A.M. López-Contreras. 2014. Opportunities and challenges for seaweed in the biobased economy. *Trends Biotechnol.* 32:231–233.

Whitfield, M. B., M. S. Chinn, and M. W. Veal. 2016. Improvement of acid hydrolysis procedures for the composition analysis of herbaceous biomass. *Ener. Fuels* 30:8260–8269.

Wildschut, J., A. T. Smit, J. H. Reith, and W. J. J. Huijgen. 2013. Ethanol-based organosolv fractionation of wheat straw for the production of lignin and enzymatically digestible cellulose. *Bioresour. Technol.* 135:58–66.

chapter thirteen

Quantification of proteins in seaweeds

Carl Safi, Jelle van Leeuwen, Yvette Telleman, Nicole Engelen-Smit, Lambertus van den Broek, and Paulien Harmsen

Contents

13.1 Introduction

Seaweeds have been long exploited as a marine vegetable and food additive owing to their rich composition of unique functional polysaccharides, minerals, and proteins (Fujiwara-Arasaki et al. 1984; Fleurence 1999). Though proteins are an important component of seaweed composition, additional value can be created by extracting proteins together with high-value compounds, such as health-improving components and hydrocolloids (Jung et al. 2013). Aside from the economic benefits of protein extraction, seaweed-based proteins can also alleviate the growing global demand for protein that is expected in the coming years. As seaweeds only need water, CO_2, and the fertilizers present in the sea, they can also contribute to reducing the environmental hazards associated with the current production of animal proteins (Steinfeld et al. 2006; Gerber et al. 2013). The protein content of seaweeds varies through species, cultivation location, and season (Black 1950, Patarra et al. 2011). Studies have shown that the protein content in green species can fluctuate from 5% to 32% (w/dw), brown species from 3% to 16% (w/dw), and red seaweeds can contain from 2% to 29% (w/dw) protein at 90% confidence level (Angell et al. 2016). Combining the interest in seaweed protein with these observed fluctuations stresses, the need for reliable methods to determine the protein content in seaweed samples and fractions.

13.2 State of the art

Proteins are polypeptides formed from amino acids (AAs) through amide linkages. These so-called peptide bonds are synthesized by the carboxyl group of the AA at the end of the growing polypeptide chain, which reacts with the amino group of an incoming free AA and thereby releasing a water molecule. For extraction, seaweed proteins can be found in soluble or insoluble form, with a ratio that is dependent on the species selected and the extraction method applied (Barbarino and Lourenco 2005; Angell et al. 2017; Vilg and Undeland 2017). Many methods are available for the determination of total soluble proteins. These can be based on colorimetric assays such as Lowry and Bradford (Barbarino and Lourenco 2005). Here, the presence of the peptide bonds or the presence of certain AAs plays a role. Another approach is the determination of the total amount of nitrogen present in your sample. Kjeldahl and Dumas can be used to estimate the total nitrogen content, which can be converted into total protein content. This is because each peptide bond contains a nitrogen atom, although some AAs have an additional nitrogen atom in their structure (e.g., arginine and histidine). A nitrogen-to-protein (NTP)-conversion factor is required, and this is calculated from the total amount of AAs present in the sample (Shuuluka et al. 2013; Akgul et al. 2015; Angell et al. 2016; Biancarosa et al. 2017). If insoluble proteins are present, it is not possible to use the colorimetric assays to determine the total protein content because the peptide bonds and/or AAs are hardly accessible and are more or less contained in the matrix. In these cases, Kjeldahl or Dumas method is the most suitable one. Be aware that you should be careful to compare the outcome of the different protein assays with each other as the methods are based on different principles and as such can result in different values (Barbarino and Lourenco 2005). For total proteins, especially Kjeldahl is most often used, whereas for soluble protein Lowry is the preferred protein method. On the other hand, Dumas (total protein) and Bradford (soluble protein) are also suitable, and their advantage is the short assay time. All methods described in this chapter are applicable to all seaweed species regardless of its color (red, green, or brown) or structural characteristics. Furthermore, despite the satisfying efficiency these methods exhibit, they also feature weaknesses that should also be taken into consideration (Table 13.1).

13.3 Materials

13.3.1 Kjeldahl method: Total protein determination

- *Sulfuric acid*: 95%–97% sulfuric acid.
- *Kjeltab*: Kjeltabs Type CK, each tablet contains: 3.5 g K_2SO_4 + 0.4 g $CuSO_4{\cdot}5H_2O$.
- *Boric acid*: Boric acid is dissolved in a solution of 1%. To prepare the solution, 50 g boric acid is dissolved in 5 L distilled water.
- *Sodium hydroxide*: 33% NaOH with 16.5 g L^{-1} sodium thiosulfate.
- *Hydrochloric acid*: 0.1 M (0.1 N) HCl.

13.3.2 Dumas method: Total protein determination

- Oxygen gas for combustion (free of nitrogen).
- Column material for removal of nonnitrogen-containing substances.

Table 13.1 Screening of the main advantages and disadvantages of the methods described in this chapter

Methods	Advantages	Disadvantages
Kjeldahl	Suitable for both dry and liquid samples Measures small amount samples, if needed Measures total proteins (soluble and insoluble)	Time-consuming Measures total nitrogen including nonprotein nitrogen Generates a lot of waste
Dumas	Suitable for both dry and liquid samples Measures total proteins (soluble and insoluble)	Measures total nitrogen, including nonprotein nitrogen A drying step must be performed before the analysis step of the method
Amino acids determination	Accurate measurement	Time-consuming The acid hydrolysis denatures tryptophan and renders very difficult for the determination of cysteine
NTP calculation	Gives more accurate protein measurements It only targets protein nitrogen	It is dependent on the quality of the amino acid analysis previously performed The NTP is highly dependent on the nitrogen-containing components present in the sample Its value can vary between seaweed species, growth conditions, and growth phase at which the seaweed has been harvested

(*Continued*)

Table 13.1 ***(Continued)*** Screening of the main advantages and disadvantages of the methods described in this chapter

Methods	Advantages	Disadvantages
Lowry	Selective measurement given that it only determines soluble proteins Measures/detects very small amount of protein present in the sample	Risk of no color formation because of the interference of other substances such as chelating agents The formation of purple color instead of blue is due to the presence of reducing agents or thiol in the sample The formation of precipitate is because of the presence of surfactant in the sample
Bradford	Fast and simple assay Measures very small amount of protein present in the sample The color is not affected by the presence of reducing agents such as DTT and β-mercaptoethanol	Detergents such as SDS and/or basic conditions affect the color The proteins should contain arginine and/or aromatic residues as the dye binds to these amino acids to form the color The calculation of the amount of proteins is dependent on the type of standard protein used in the assay

- Carrier gas, often helium (free of nitrogen).
- Stock chemical containing a known amount of nitrogen, for instance methionine.
- The stock chemical is lacking nitrogen, for example, cellulose.
- Aluminum or tin cups.
- (Freeze-) drying equipment.
- Grinding equipment (e.g., mortar and pestle).
- Analytical balance, preferably accurate to 0.1 mg.
- Combustion oven >900°C.
- Polishing column.
- Thermal conductivity detector.

13.3.3 *Amino acid determination*

- Analytical grade *o*-phthalaldehyde (OPA) and ethanethiol for derivatization of AA.
- Analytical grade 9-fluorenylmethyl chloroformate (FMOC) for derivatization of Proline.
- Boric acid, potassium chloride, sodium hydroxide (50%) for buffer A.
- Phosphoric acid for the injection diluent.
- Hydrogen chloride and phenol for the acid hydrolysis.
- Disodium phosphate, tetraborate, and sodium azide for eluents.
- Norleucine (internal standard) and AAS18 for the preparation of standards and any additional AA not included in AAS18.
- Column: Acuity UPLC BEH C18 1.7 μm, 2.1 × 150 mm.

13.3.4 *Nitrogen-to-protein-conversion factor determination*

- No specific materials are needed to conduct the calculation of the NTP conversion factor.

13.3.5 *Lowry method: Soluble protein determination*

- Two bovine serum albumin (BSA) ampules (2 mg mL^{-1} each).
- Spectrophotometer cuvettes.
- MilliQ water.
- Folin–Ciocalteau reagent.
- Preprepared modified Lowry reagent containing sodium tartrate ($Na_2C_4H_4O_6$) in an alkaline sodium carbonate buffer (Na_2CO_3), cupric sulfate ($CuSO_4$), and potassium iodide (KI).
- All reagents and BSA ampules can be found in the modified Lowry assay kit (Thermo Fisher Scientific).
- Automatic Pipettors.

13.3.6 Bradford method: Soluble protein assay

- 0.01% (w/v) Coomassie Brilliant Blue G-250 in 4.7% (w/v) ethanol and 8.5% (w/v) phosphoric acid. The reagents can also be obtained already prepared (e.g., Sigma-Aldrich or Thermo Scientific Pierce).
- 0–1.4 mg mL^{-1} BSA.
- Automatic pipettors

13.4 Experimental procedures

13.4.1 Kjeldahl method: Total protein determination

It was first developed in 1883 by the Danish scientist Johan Kjeldahl. The method measures the total amount of organic nitrogen present in the biomass or the extracts. The amount of nitrogen determined will be then converted into the amount of protein using the NTP-conversion factor. Please note that the protocol described applies to a semiautomated system (Gerhardt Analytical Systems, Königswinter, Germany).

13.4.1.1 Digestion

1. Switch on the heater and set the temperature at 420°C.
2. Weigh the sample in the tubes and add a Kjeltabs tablet and 9 mL H_2SO_4 (*Notes 1–3*).
3. Once the heater is at the desired temperature, insert the tubes and put the exhaust system on the top.
4. Switch on the scrubber and water the scrubber.
5. Leave the tubes for 50 min in the heater.
6. Switch off the scrubber and lift the rack with the tubes.
7. Leave the exhaust system on top until no further fumes come out.
8. Add 75 mL distilled water in each tube (maximum cooling time is 10 min).
9. *The samples are ready for the next step*: distillation.

13.4.1.2 Distillation

1. Switch on the VAPODEST 45 and the water.
2. Double check if the vessels are full with distilled water, boric acid, and sodium hydroxide.
3. Put a tube with 75 mL water in the machine and an empty Erlenmeyer. This tube is only to warm up the machine.
4. Select the distillation program on the screen and press enter to start measuring.
5. Wait for 5 min, and the sample is ready.
6. Put a new tube with water and an empty Erlenmeyer to determine the blank value.

7. Repeat this step until the value will be 0.12–0.15 M HCl (titration).
8. Now tubes with the protein samples can be measured.
9. After all samples are tested, activate the cleaning system.

13.4.1.3 Titration

1. Always start with calibrating the electrode by selecting the calibration mode.
2. Set the temperature at 25°C.
3. First, calibrate with the pH 7 buffer, then calibrate with the pH 4 buffer.
4. After the calibration step is over, select Kjeldahl method and start with titration.
5. Record the amount of HCl (mL) titrated.

13.4.1.4 Calculation

Use the following equation:

$$\text{Protein (mg)} = (\text{mL} - \text{blank}) \times 0.1 \times 14 \times KF$$

0.1 is the concentration of HCl used during titration. 14 is the molecular weight of nitrogen; *KF* is a Kjeldahl factor or the so-called NTP-conversion factor; according to Akgul et al. (2015), the universal NTP for seaweeds is 5.

13.4.2 Dumas method: Total protein determination

The method was first proposed by the French chemist Jean-Baptiste Dumas in 1826. It is based on the analysis of total nitrogen in a sample, just like in Kjeldahl's method. Similarly, Dumas' method needs a NTP-conversion factor. The determination of this factor is explained in more detail in Sections 13.4.3 and 13.4.4.

13.4.2.1 Procedure

1. Dry the samples completely before analysis, to guarantee complete combustion (*Note 4*). However, applying very high temperatures is discouraged to prevent evaporation of ammonium or nitrous gas. Usually, lyophilization is used, and if this is not possible, low-temperature drying (60°C–80°C) is to be allowed.
2. Make a calibration curve with the stock chemical containing nitrogen from 1 to 20 mg and should cover the detection range.
3. Weigh the samples in aluminum or tin cups folded shut to prevent sample loss or adsorption of water from the air. For optimal accuracy,

weigh the dried samples in an amount so that the expected nitrogen content falls in the same detection range as the calibration curve.

4. Add a negative or positive control every 20 samples to guarantee data quality. This is done with a chemical containing a known amount of nitrogen (like the calibration chemical) and one without any nitrogen.
5. Individually, add the samples to the combustion oven. The samples are burnt at 900°C in the presence of oxygen to achieve total combustion, after which the detector reports an increase in conductivity over time.
6. Quantify the conductivity peak surface, and calculate the amount of nitrogen with the calibration curve.
7. Convert total nitrogen to protein by using a conversion factor, as explained in Sections 13.4.3 and 13.4.4.

13.4.3 Amino acid determination

This method describes the determination of proteinogenic AAs by acid hydrolysis and automated precolumn derivatization (Meussen et al. 2014). Free AAs can be determined without the performance of acid hydrolysis. The AA composition will be determined by UPLC; after that, the protein-conversion factor can be calculated by using the NTP-conversion factor, which is further explained in Section 13.4.4.

13.4.3.1 Preparation of reagents

1. *FMOC reagents*: Place 5 mg FMOC in a 2 ml vial.
2. Add 2 mL acetonitrile.
3. *OPA reagents*: Place 40 mg OPA in a 10 mL vial and add 2.0 mL of buffer A, 8.0 mL methanol, and 50 μl ethanethiol (*Note 5*).
4. *Buffer A*: Place 1.29 g boric acid and 1.49 g potassium chloride into 50 mL MilliQ water, adjust pH to 10.5 with sodium hydroxide (50%).
5. 5 mL HCl/phenol reagents (1%).
6. *Injection diluent*: Place 48.5 mL of eluent A and 1.5 mL phosphoric acid (85%) into a 50 mL flask.

13.4.3.2 Preparation of eluents

1. Eluent A consists of 10 mM disodium phosphate, 10 mM sodium tetraborate, and 2 mM sodium azide, pH 7.8.
2. Eluent B consists of Methanol–Acetonitrile–MilliQ water (20:60:20).

13.4.3.3 Preparation of standards

1. *Standard 4A (additional AAs)*: In a 50 mL flask, weigh in the additional AAs not included in AAS18. Ensure that the concentration is 2.5 mM for each AA.

2. *AA all stock*: Mix 300 µl of standard AAS18, 300 µl of standard 4A, and 900 µl MilliQ water in an Eppendorf safe-lock tube. Final concentration is 0.5 mM; make a dilution series from 0.5 to 0.01 mM.
3. Internal standard: Weigh 14 mg of norleucine into a 50 mL flask, and fill it up with MilliQ water. Final concentration is 0.4 mM.

13.4.3.4 *Acid hydrolysis*

1. Weigh in 0.2 mg of (freeze-)dried sample in analysis vials and into a reaction vial, add 500 µL of HCl/phenol solution into the reaction vial (N.B.: not in the analysis vials!). Remove air by flushing with nitrogen.
2. Place the reaction vial into the preheated oven at 110°C for 24 h (*Notes 6–8*).
3. After the hydrolysis, remove the reaction vials and let them cool down to room temperature.
4. Remove any excess HCl/phenol solution under vacuum for 2 h.
5. Place the content of the analysis vial into a 50 mL tube with 1 mL internal standard, 4 mL of methanol, and 5 mL of MilliQ water.
6. Before analysis, filter the solution over a 0.2 µm filter into an analysis vial.
7. Dilute samples if necessary, the concentration of each AA must be between 0.01 and 2 mM, and the total AA amount may not exceed 10 mM.

13.4.3.5 *Analysis*

1. Set the column temperature at 50°C.
2. UV_1 set at wavelength 338 nm.
3. UV_2 set at wavelength 263 nm.
4. Flow gradient (Table 13.2).

13.4.4 *Nitrogen-to-protein-conversion factor determination*

13.4.4.1 *Procedure*

1. The NTP-conversion factor is the ratio of proteinogenic protein measured by the total nitrogen content.
2. To obtain the total proteinogenic protein content, this is measured by quantifying and summing up the AAs obtained by the AA determination (see Section 13.4.3) of your sample.
3. Often cysteine and tryptophan are excluded, and the remaining 18 AAs are used (Angell et al. 2016).
4. The amount of protein is expressed as % (w/w) dry matter content.
5. Total nitrogen content can be measured by Kjeldahl (Section 13.4.1) or Dumas method (Section 13.4.2). Here, the amount of total nitrogen

Table 13.2 Flow gradient

Time (min)	Flow (mL min^{-1})	Eluent A (%)	Eluent B (%)
0.00	0.40	80.00	20.00
6.98	0.40	36.00	64.00
10.00	0.40	23.40	76.60
10.20	0.40	0.00	100.00
12.49	0.40	0.00	100.00
12.72	0.40	97.70	2.30
15.00	End	–	–

is expressed as % (w/w) dry matter. The NTP conversion factor is calculated by equation A.

$$\text{Nitrogen-to-protein conversion factor} = \frac{\%\ (\text{w/w})\ \text{total protein}}{\%\ (\text{w/w})\ \text{total nitrogen}} \quad (A)$$

6. If the NTP conversion factor is known (*Note 9*), then you can calculate the amount of proteinogenic protein in your sample by multiplying the nitrogen content of your sample (obtained by Kjeldahl or Dumas) with the NTP conversion factor.

13.4.5 *Lowry method: Soluble protein determination*

The method was first published in 1951 by the American biochemist Oliver H. Lowry. It is a standard method mainly used to determine the amount of soluble proteins. Due to its reliability, Lowry method is considered as the most cited method of all time.

13.4.5.1 *Preparation of the standard solution*

1. While preparing the calibration curve, dissolve the BSA well in glassware. As dissolving needs some time, only shake gently afterward. Stirring will result in foaming.
2. Prepare 10 samples with a concentration range from 0 to 1500 μg mL^{-1} by using MilliQ water as the diluent.

13.4.5.2 *Preparation of Folin–Ciocalteau reagent*

1. Prepare 1 N Folin–Ciocalteau by diluting the reagent (2 N) 1:1 with MilliQ water.
2. Use the prepared Folin–Ciocalteau on the same day, because it is unstable on time.

13.4.5.3 *Test tube protocol*

1. Pipette 0.2 mL of each standard previously prepared.
2. Wait exactly 15 s for each sample, and then add 1 mL modified Lowry reagent to each test tube (*Note 10*).
3. Vortex for 5 s, then cover all tubes and incubate for 10 min (*Notes 11 and 12*).
4. Exactly after 10 min incubation (for each sample) add 100 µL of the prepared (1 N) Folin–Ciocalteau reagent (*Notes 13 and 14*).
5. Set the spectrophotometer at 750 nm.
6. Zero the spectrophotometer with the blank sample that contains only MilliQ water (0 µg mL^{-1} (Protein)).
7. Plot the calibration curve for each concentration prepared and try to get $R^2 = 0.99$.
8. Use the calibration curve to determine the protein concentration of the unknown samples.

13.4.6 *Bradford method: Soluble protein assay*

The Bradford method was first developed by the American scientist, Marion Mckinley Bradford. It is a fast and sensitive method to determine the level of soluble proteins. When the Bradford reagent (Coomassie Brilliant Blue G-250) binds to the protein, the color changes from red to purple, and the absorption maximum changes from 495 to 595 nm. The increase in absorption at 595 nm is monitored (Bradford 1976).

13.4.6.1 *Procedure*

1. Pipette 30 µL sample or standard into a 1 mL plastic cuvette (*Note 15*).
2. Add 1.0 mL Bradford reagents and mix well (*Note 16*).
3. Measure the absorbance at 595 nm.
4. Plot the calibration curve (at least five different BSA concentrations in duplicate) to determine the protein concentration of the unknown samples (*Note 17*).

13.5 Notes

Note 1: To keep the measurement within the appropriate range, there must be between 10 and 100 mg seaweed protein in the sample.

Note 2: It is possible to use liquid samples in this method. Therefore, the digestion part of the protocol has to be changed by following the next three steps:

a. Incubate the sample for 1 h at 180°C.
b. Then incubate the same sample for 1 h at 280°C.
c. Finally, incubate the same sample for 50 min at 420°C.

Note 3: Another alternative is that the liquid sample can be put in the tube and dried at 105°C overnight. Afterward, it can be treated as a solid sample.

Note 4: For seaweeds, inhomogeneity can be an issue, which could partially be solved by (freeze-) drying a larger batch of material and grinding the dried material.

Note 5: Make sure to add ethanethiol inside the fume hood.

Note 6: Not all the AAs present in the seaweed will survive the hydrolysis process.

Note 7: Asparagine is converted into aspartic acid and glutamine into glutamic acid.

Note 8: Tryptophan is completely destroyed, and it is impossible to accurately determine cysteine concentrations in acid hydrolyzed samples.

Note 9: Be aware that the NTP conversion factor for Kjeldahl and Dumas slightly differ as both have another principle to determine the total nitrogen content.

Note 10: Use a new pipette for each reagent and sample, do not use the same pipette for all the samples and reagents.

Note 11: While adding the reagents, it is essential to respect the waiting time between each sample.

Note 12: Respecting the incubation time is important.

Note 13: If the solution is not turning blue, it means that there is interference from undesired components in the sample such as chelating agents.

Note 14: If precipitate forms while adding the reagents, this implies that the samples contain a surfactant or potassium ions.

Note 15: Dissolve the BSA well in glassware. Dissolving needs some time, so only shake gently afterward. Stirring will result in foaming.

Note 16: Coomassie Brilliant Blue G-250 stains glass or quartz cuvettes; the use of disposable polystyrene cuvettes is therefore recommended.

Note 17: Basic conditions and detergents influence the outcome negatively.

Acknowledgment

The authors would acknowledge MacroFuels project. This project has received funding from the European Union's Horizon 2020 research and innovation program under grant agreement No 654010.

References

Akgul, R., Kizilkaya, B., Akgul, F. et al. 2015. Crude protein values and amino acid profiles of some red algae collected from Canakkale. *Indian J. Geo. Mar. Sci.* 44: 527–530.

Angell, A.R., Mata, L., de Nys, R. et al. 2016. The protein content of seaweeds: A universal nitrogen-to-protein conversion factor of five. *J. Appl. Phycol.* 28: 511–524.

Angell, A.R., Paul, N.A., and de Nys, R. 2017. A comparison of protocols for isolating and concentrating protein from the green seaweed *Ulva ohnoi*. *J. Appl. Phycol.* 29: 1011–1026.

Barbarino, E. and Lourenco, S.O. 2005. An evaluation of methods for extraction and quantification of protein from marine macro- and microalgae. *J. Appl. Phycol.* 17: 447–460.

Biancarosa, I., Espe, M., Bruckner, C.G. et al. 2017. Amino acid composition, protein content, and nitrogen-to-protein conversion factors of 21 seaweed species from Norwegian waters. *J. Appl. Phycol.* 29: 1001–1009.

Black, W.A.P. 1950. The seasonal variation in weight and chemical composition of the common British Laminariaceae. *J. Mar. Biol. Assoc. UK* 29: 45–72.

Bradford, M.M. 1976. Rapid and sensitive method for quantitation of microgram quantities of utilizing principle of protein-dye binding. *Anal. Biochem.* 72: 248–254.

Fleurence, J. 1999. Seaweed proteins: Biochemical, nutritional aspects and potential uses. *Trends. Food. Sci. Tech.* 10: 25–28.

Fujiwara-Arasaki, T., Mino, N., and Kuroda, M. 1984. The protein value in human nutrition of edible marine algae in Japan. *Hydrobiologia* 116: 513–516.

Gerber, P.J., Steinfeld, H., Henderson, B. et al. 2013. *Tackling Climate Change Through Livestock: A global Assessment of Emissions and Mitigation Opportunities. Food and Agriculture Organization of the United Nations (FAO)*, Rome, Italy.

Jung, K.A., Lim, S.R., Kim, Y. et al. 2013. Potentials of macroalgae as feedstocks for biorefinery. *Bioresour. Technol.* 135: 182–190.

Meussen, B.J., Zeeland, A.N.T., Bruins, M.E. et al. 2014. A fast and accurate UPLC method for analysis of proteinogenic amino acids. *Food Anal. Method* 7: 1047–1055.

Patarra, R., Paiva, L., Net, A.I. et al. 2011. Nutritional value of selected macroalgae. *J. Appl. Phycol.* 23: 205–208.

Shuuluka, D., Bolton, J.J., and Anderson, R.J. 2013. Protein content, amino acid composition and nitrogen-to-protein conversion factors of *Ulva rigida* and *Ulva capensis* from natural populations and *Ulva lactuca* from an aquaculture system, in South Africa. *J. Appl. Phycol.* 25: 677–685.

Steinfeld, H., Gerber, P., Wassenaar, T. et al. 2006. Livestock's long shadow: Environmental issues and options. *Food and Agriculture Organization of the United Nations (FAO)*, Roman, Italy.

Vilg, J.V. and Undeland, I. 2017. pH-driven solubilization and isoelectric precipitation of proteins from the brown seaweed *Saccharina latissima*-effects of osmotic shock, water volume and temperature. *J. Appl. Phycol.* 29: 585–593.

chapter fourteen

Comprehensive phytohormone quantification in the red alga Pyropia yezoensis *by liquid chromatography–mass spectrometry*

Takakazu Matsuura, Izumi C. Mori, Yoko Ikeda, Takashi Hirayama, and Koji Mikami

Contents

14.1 Introduction

Phytohormones are structurally unrelated compounds, including auxin, abscisic acid (ABA), gibberellins, cytokinins, ethylene, jasmonates, salicylic acid (SA), brassinosteroids, and strigolactones (Kende and Zeevaart 1997; Kamiya 2010). Emerging studies suggest the critical involvement of these phytohormones in the regulation of growth, development, and environmental stress response in plants (Weyers and Paterson 2001; Vanstraelen and Benková 2012). To date, phytohormones are known to exist in seaweeds as in terrestrial plants (Basu et al. 2002; Le Bail et al. 2010; Wang et al. 2014; Mikami et al. 2016), although functions of phytohormones are not entirely understood.

Analysis of phytohormones in algae has a long history, and small molecules identical to phytohormones of terrestrial plants have been detected in a variety of algae (reviewed in Bradley 1991; Evans and Trewavas 1991; Tarakhovskaya et al. 2007; Lu and Xu 2015). As the previous analyses generally focused on individual phytohormones, the overall composition of phytohormones within a given algal species has not been elucidated well. However, efforts have made to develop methods employing gas chromatography–tandem mass spectrogram and liquid chromatography–tandem mass spectrometry (LC–MS/MS) for comprehensive phytohormone analysis in algae, and phytohormone compositions have been characterized in a few algae (Yokoya et al. 2010; Gupta et al. 2011; Wang et al. 2014). In addition, we recently improved comprehensive analysis methods for phytohormones by liquid chromatography–electrospray ionization–tandem mass spectroscopy (LC–ESI–MS/MS) using the red seaweeds *Pyropia yezoensis* and *Bangia fuscopurpurea* (Mikami et al. 2016; Mori et al. 2017). On the basis of our current findings, we describe an updated procedure for comprehensive analysis of phytohormones in *P. yezoenssis* as a model.

14.2 State of the art

For red seaweed samples, we have found that special attention must be paid to two aspects of the experimental procedures. The first is the extraction of phytohormones as the physical characteristics of red alga tissue extracts are peculiar. The second is the choice of deuterium-labeled internal standard for indole-3-acetic acid (IAA), because of contamination of the red algal extracts with an unknown factor possessing a mass-to-charge ratio and retention time in octadecyl silica columns that interferes with the detection of stable isotope-labeled IAAs such as D_2-IAA and D_5-IAA. We have already found that D_7-IAA is compatible for use with crude phytohormone extracts, and thus our latest method utilizes it as the preferred internal standard to quantify IAA in red seaweeds (Mikami et al. 2016). Our findings point to a need for reexamination of comprehensive phytohormone analysis using D_7-IAA in other algae, for which D2-IAA and D_5-IAA were sometimes employed for IAA quantification (see Mikami et al. 2016).

14.3 Materials

14.3.1 Extraction and purification of phytohormones

- *Extraction solvent*: Mixture of acetonitrile (MeCN)/water (80:20) containing 1% (v/v) acetic acid (AcOH). Prepare daily.
- *Internal standard stock solution*: Prepare a stock solution of the extraction solvent containing all four internal standards in amounts shown in Table 14.1. The sources of the internal standards are shown in Table 14.1. Store the internal standard stock solution in a lightproof bottle and replace after approximately six months.
- *Solid-phase extraction cartridges*:
  - Oasis® HLB solid-phase extraction cartridge (1 cm^3, Waters, Milford, MA, USA)
  - Oasis® MCX solid-phase extraction cartridge (1 cm^3, Waters, Milford, MA, USA)
  - Oasis® WAX solid-phase extraction cartridge (1 cm^3, Waters, Milford, MA, USA).
- *Solvents for solid-phase extraction*:
  - MeCN (HPLC grade or higher)
  - Methanol (HPLC grade or higher, MeOH)
  - 0.1 N HCl
  - 0.1 N KOH

Table 14.1 Concentrations of stable isotope-labeled phytohormones in the internal standard stock solution

Compound	Concentration	Source company, Country
D_7-indoleacetic acid[a]	20 ng mL^{-1}	CDN Isotopes, Canada
D_6-isopentenyladenine	1 ng mL^{-1}	OlChemim, Czech Republic
D_4-salicylic acid	200 ng mL^{-1}	OlChemim, Czech Republic
D_6-abscisic acid	20 ng mL^{-1}	OlChemim, Czech Republic

[a] Alternatively, $^{13}C_6$-indoleacetic acid (Cambridge Isotope Laboratories Inc., MA, USA) can be used. D indicates replacement of hydrogen by deuterium.

    - Aqueous solution containing 1% (v/v) AcOH
    - MeCN/water (80:20) containing 1% (v/v) AcOH (extraction solvent)
    - MeCN/water (50:50) containing 5% ammonia
    - MeCN/water (80:20).
- *Reconstitution solution*: aqueous solution containing 0.01% (v/v) AcOH (alternatively, 1% AcOH can be used).
- Liquid nitrogen.
- Pasteur pipettes (146 mm and 229 mm).
- Glass test tubes (13 × 100 mm and 12 × 75 mm).
- Mortars and pestles.
- Disposable round-bottom polypropylene test tubes (14 mL, e.g., Greiner Z617954). Note: must not be made of polystyrene.
- Microcentrifuge tubes (1.5 mL).
- Micropipettes.
- Chemical hood.
- Refrigerated centrifuges to spin 1.5 mL microcentrifuge tubes (e.g., Eppendorf 5418R) and 14 mL polypropylene test tubes (e.g., Eppendorf 5702R).
- Vacuum concentrator (e.g., Mivac quattro concentrator, Genevac).
- Vortex mixer.

14.3.2 Liquid chromatography–mass spectrometry

- Water containing 0.01% (v/v) AcOH.
- MeCN (HPLC grade or higher) containing 0.05% (v/v) AcOH.
- MeOH (HPLC grade or higher) containing 0.2% (v/v) AcOH.
- Aqueous solution containing 0.1% (v/v) formic acid (FA).
- MeCN (HPLC grade or higher) containing 0.1% (v/v) FA.
- Triple quadrupole LC–MS system (Agilent 6410 or equivalent) and associated software (e.g., MassHunter, Agilent).
- A volume of 0.5 mL vials with 250 μL glass inserts (or 96-well polypropylene plates [0.5 mL] with compatible cover caps).

14.4 Experimental procedures

Phytohormones are extracted from fronds into an acidified polar solvent, purified to eliminate interferents by solid-phase extraction, and quantified by LC–MS. The concentration of analytes (endogenous phytohormones) is determined by the internal standard method.

14.4.1 *Extraction*

All procedures should be carried out in a chemical hood:

1. Dilute internal standard stock solution 20 times with the extraction solvent (*Note 1*).
2. Grind lyophilized algal samples (approximately 0.1 g of thalli or protonemata) in liquid nitrogen with mortar and pestle into a fine powder (*Note 2*).
3. Add 4 mL of the extraction solvent containing internal standards to the mortar and incubate at 4°C for 60 min. Briefly, stir the ground tissue in the solvent every 20 min to disperse any aggregates (*Note 3*).
4. Transfer the extract to a polypropylene test tube and centrifuge at 3000 g for 10 min at 4°C.
5. Transfer the supernatant to glass test tube (13 × 100 mm) with a Pasteur pipette (229 mm).
6. Add 4 mL extraction solvent (without the internal standard) to the tube containing the pellet, mix with a vortex mixer, and centrifuge at 3000 g for 10 min at 4°C.
7. Collect the supernatant with a Pasteur pipette (229 mm) and add to the previous one in the same glass test tube (*Note 4*).

14.4.2 *Partial purification of phytohormones with solid-phase extraction*

All procedures should be carried out in a chemical hood. Phytohormones are fractionated into three parts: (1) IAA/ABA, (2) iP, and (3) SA by the following solid-phase extraction steps:

1. Evaporate MeCN in the supernatant with a vacuum concentrator until only the acidic aqueous solution remains.
2. Transfer the acidic aqueous solution to a microcentrifuge tube and centrifuge at 16,000 g for 10 min at 4°C to remove any precipitate that appeared after concentration.
3. Condition Oasis® HLB cartridge with 1 mL of MeCN, 1 mL of MeOH, and 1 mL of 1% AcOH, successively.
4. Load the clear supernatant of the acidic aqueous solution containing extracted phytohormones onto the HLB cartridge with a Pasteur pipette (146 mm, *Note 5*).
5. Wash the cartridge with 1 mL of 1% AcOH aqueous solution.

6. Elute crude phytohormone fraction with 2 mL of MeCN/water (80:20) containing 1% %(v/v) AcOH and collect it in a glass test tube (12 × 75 mm).
7. Add 0.6 mL of 1% AcOH to the eluate and evaporate the MeCN off the eluate with a vacuum concentrator to yield approximately 1 mL of acidic aqueous solution (*Note 6*).
8. Condition Oasis® MCX cartridge with 1 mL of MeCN and 1 mL of MeOH, successively.
9. Activate Oasis® MCX cartridge with 1 mL of 0.1 N HCl and equilibrate with >1 mL of 1% AcOH aqueous solution (*Note 7*).
10. Load the acidic aqueous solution onto the Oasis® MCX cartridge using a 146 mm Pasteur pipette and wash with 1 mL of 1% AcOH aqueous solution.
11. Elute acidic and neutral compounds with 2 mL of MeCN/water (80:20) containing 1% (v/v) AcOH and collect in a glass test tube (12 × 75 mm).
12. After vortex mixing, evaporate a 0.2 mL aliquot to dryness with a vacuum concentrator and prepare for SA quantification by reconstitution in 50 µL of reconstitution solution (Figure 14.1) (*Note 8*). Retain the remaining 1.8 mL of this acidic/neutral eluate for further purification by Oasis® WAX cartridge (as described starting in Step 15).
13. Wash the Oasis® MCX cartridge further with 2 mL of aqueous solution containing 5% ammonia.
14. Elute alkaline compounds from Oasis® MCX cartridge with 2 mL of MeCN/water (50:50) containing 5% ammonia and collect in a glass test tube (12 × 75 mm). Evaporate this alkaline fraction containing iP to dryness with a vacuum concentrator followed by reconstitution in 50 µL of reconstitution solution (Figure 14.1).
15. Add 0.6 mL of 1% AcOH to the acidic/neutral eluate fraction (see Step 12) and evaporate MeCN with a vacuum concentrator to yield approximately 1 mL of aqueous solution.
16. Wet Oasis® WAX cartridge with 1 mL of MeCN and 1 mL of MeOH, successively.
17. Activate Oasis® WAX cartridge with 0.3 mL of 0.1 N KOH and equilibrate with >1 mL of 1% AcOH aqueous solution (*Note 7*).
18. Load the acidic/neutral eluate onto the Oasis® WAX cartridge using a 146 mm Pasteur pipette and wash with 1 mL of 1% AcOH aqueous solution and 2 mL of MeCN/water (80:20), successively.
19. Elute acidic compound fraction containing IAA and ABA with 2 mL of MeCN/water (80:20) containing 1% (v/v) AcOH and collect in a glass test tube (12 × 75 mm).
20. Evaporate the fraction containing IAA and ABA to dryness with a vacuum concentrator followed by reconstitution in 50 µL of reconstitution solution (Figure 14.1).

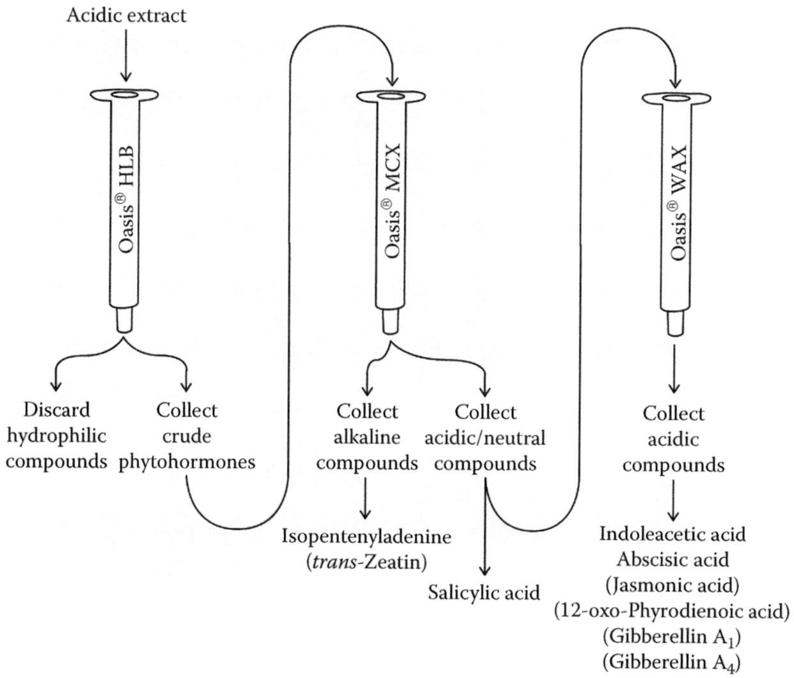

Figure 14.1 Solid-phase extraction procedure. Phytohormones extracted in acidified polar solvent are sequentially processed by Oasis® HLB, Oasis® MCX, and Oasis® WAX cartridges yielding three fractions—alkaline compound fraction, including cytokinins; acidic/neutral compound fraction, including highly acidic salicylic acid; and acidic compound fraction, including indole-acetic acid, abscisic acid, jasmonic acid, 12-oxo-*phtodienoic* acid, gibberellin A_1, and gibberellin A_4.

14.4.3 Liquid chromatography–mass spectrometry analysis

Phytohormone analysis is conducted on an Agilent 1260–6410 Triple quadrupole LC–MS, using a Zorbax Eclipse XDB-C18 column and an XDB-C8 guard column.

1. Set the running parameters. Parameters for LC are shown in Figure 14.2. Parameters for MS are listed in Table 14.2 (*Note 9*).
2. Transfer the reconstituted sample to a vial with 250 μL glass insert or a 96-well plate.
3. Set the vial in the autosampler and run the program.

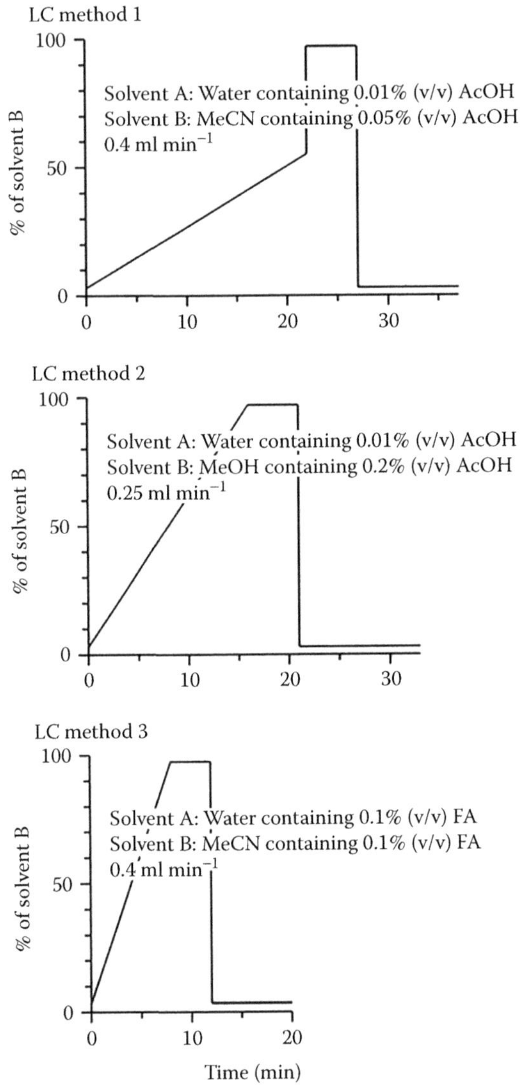

Figure 14.2 Parameters for liquid chromatography. LC method 1 is used for the analysis of indoleacetic acid and abscisic acid. LC methods 2 and 3 are used for analyses of isopentenyladenine and salicylic acid, respectively. The proportion of solvent B increases from 3% to 55% in 22 min in LC method 1. It increases from 3% to 97% in 16 min in LC method 2, and from 3% to 98% in 10 min in LC method 3. After a wash of the column with a high concentration of solvent B for approximately 5 min, the column is washed with 3% solvent B to condition the column. LC, liquid chromatography. AcOH, acetic acid. MeCN, acetonitrile. MeOH, methanol. FA, formic acid. The flow rate of LC methods 1 and 3 is 0.4 mL min^{-1}, and that of LC method 2 is 0.25 mL min^{-1}.

Table 14.2 Parameters for mass spectrometry for quantification of IAA, ABA, iP, and SA

Compound	LC method	Retention time (min)	Time segment (min)	ESI mode	Multiple reaction monitoring transition (*m/z*)	Fragmentor voltage (V)	Collision energy (V)
IAA	1	9.7	9.3–11.8	Positive	176/130	90	15
D_7-IAA					183/135, 136, 137		
ABA	1	12.6	11.8–13.5	Negative	263/153	130	4
D_6-ABA					269/159		
iP	2	12.5	11–16	Positive	204/136	110	11
D_6-iP					210/137		
SA	3	5.6	2–10	Negative	137/93	90	12
D_4-SA					141/97		

See Figure 14.2 for LC methods.

14.4.4 Data analysis

1. Measure the areas of the peaks in the corresponding multiple reaction monitoring for endogenous plant hormones and internal standards.
2. Calculate the amount of plant hormone in the sample and normalize it to the weight of the algal sample.

14.5 Notes

Note 1: The dilution factor can be modified according to the actual concentrations of plant hormones detected in the algal sample.

Note 2: Grinding the tissue with stainless steel beads or tungsten beads can reduce or abolish the detection of some hormones. This may be because of tight binding of hormones to the beads and/or scraping the plastic tube, which increases the surface area of plastic.

Note 3: The extracts may become very viscous and aggregate during the suspension of powdered tissue in the extraction solvent. The extracts will become more fluid as incubation progresses. Stir the extracts in the mortar intermittently to disperse the aggregates.

Note 4: Samples can be stored at −20°C for a week or 4°C for overnight at this step.

Note 5: The volume of the aqueous solution should be adjusted to 1 mL with 1% AcOH, if the volume of the solution is less than that. Hormones may be lost during loading.

Note 6: The addition of 0.6 mL of 1% AcOH aqueous solution prevents the sample from overdrying.

Note 7: Fill the cartridge up to the mouth with 1% AcOH aqueous solution to remove any remaining drops of HCl or KOH on the wall of the cartridge that could affect binding of chemicals to the cartridge.

Note 8: As SA binds to Oasis® WAX cartridge very strongly, the recovery rate is very low. Therefore, it is analyzed by using the eluate from Oasis® MCX cartridge by LC–MS. SA can be eluted by 2% FA/98% MeCN.

Note 9: Parameters for phytohormones that are not detected in *P. yezoensis* but were examined in a previous paper (Mikami et al. 2016) are listed in Table 14.3.

Table 14.3 Parameters for mass spectrometer for quantification of phytohormones measured in Mikami et al. (2016) and 12-oxo-phytodienoic acid (OPDA), which were not detected in gametophytes and sporophytes of *Pyropia yezoensis*

Compound	LC method	Retention time (min)	Time segment (min)	ESI mode	Multiple reaction monitoring transition (*m/z*)	Fragmentor voltage (V)	Collision energy (V)
GA_1	1	8.9	8–9.3	Negative	347/273	160	18
D_2-GA_1					349/275		
JA	1	14.4	13.5–15.7	Negative	209/59	135	11
D_2-JA					211/59		
GA_4	1	17	16–17.3	Negative	331/257	160	20
D_2-GA_4					333/259		
JA-Ile	1	16.2	17.5–20	Negative	322/130	140	14
$^{13}C_6$-JA-Ile					328/136		
OPDA	1	22.4	20–25	Negative	291/165	150	11
D_5-OPDA					296/170		
tZ	2	8.3	5–11	Positive	220/136	100	12
D_5-tZ					225/136, 137		

See Figure 14.2 for LC methods.

References

Basu, S., Sun, H., Brian, L., Quatrano, R.L. and Muday, G.K. 2002. Early embryo development in *Fucus distichus* is auxin sensitive. *Plant Physiol.* 130:292–302.

Bradley, P.M. 1991. Plant hormones do have a role in controlling growth and development of algae. *J. Phycol.* 27:317–321.

Evans, L.V. and Trewavas, A.J. 1991. Is algal development controlled by plant growth substances? *J. Phycol.* 27:322–326.

Gupta, V., Kumar, M., Brahmbhatt, H., Reddy, C.R., Seth, A. and Jha, B. 2011. Simultaneous determination of different endogenetic plant growth regulators in common green seaweeds using dispersive liquid–liquid microextraction method. *Plant Physiol. Biochem.* 49:1259–1263.

Kamiya, Y. 2010. Plant hormones: Versatile regulators of plant growth and development. *Annu. Rev. Plant Biol.* 61:Special Online Compilation.

Kende, H. and J.A.D. Zeevaart. 1997. The five "classical" plant hormones. *Plant Cell* 9:1197–1210.

Le Bail, A., Billoud, B., Kowalczyk, N., Kowalczyk, M., Gicquel, M., Le Panse, S., Stewart, S. et al. 2010. Auxin metabolism and function in the multicellular brown alga *Ectocarpus siliculosus*. *Plant Physiol.* 153:128–144.

Lu, T. and Xu, J. 2015. Phytohormones in microalgae: A new opportunity for microalgal biotechnology? *Trends Plant Sci.* 20:275–282.

Mikami, K., Mori, I.C., Matsuura, T., Ikeda, Y., Kojima, M., Sakakibara, H. and Hirayama, T. 2016. Comprehensive quantification and genome survey reveal the presence of novel phytohormone action modes in red seaweeds. *J. Appl. Phycol.* 28:2539–2548.

Mori, I.C., Ikeda, Y., Matsuura, T., Hirayama, T. and Mikami, K. 2017. Phytohormones in red seaweeds: A technical review of methods for analysis and a consideration of genomic data. *Bot. Mar.* 60:153–170.

Tarakhovskaya, E.R., Maslov, Y.I. and Shishova, M.F. 2007. Phytohormones in algae. *Russ. J. Plant Physiol.* 54:163–170.

Vanstraelen, M. and Benková, E. 2012. Hormonal interactions in the regulation of plant development. *Annu. Rev. Cell Dev. Biol.* 28:463–487.

Wang, X., Zhao, P., Liu, X., Chen, J., Xu, J., Chen, H. and Yan, X. 2014. Quantitative profiling method for phytohormones and betaines in algae by liquid chromatography electrospray ionization tandem mass spectrometry. *Biomed. Chromatogr.* 28:275–280.

Weyers, J.D.B. and Paterson, N.W. 2001. Plant hormones and the control of physiological processes. *New Phytol.* 152:375–407.

Yokoya, N.S., Stirk, W.A., van Staden, J., Novák, O., Turečková, V., Pěnčík, A. and Strnad, M. 2010. Endogenous cytokinins, auxins, and abscisic acid in red algae from Brazil. *J. Phycol.* 46:1198–1205.

chapter fifteen

Total phenolic content and antioxidant capacity analysis of seaweed biomass

Xiaoru Hou, Randi Neerup, and Anne-Belinda Bjerre

Contents

15.1 Introduction

Seaweed is a known source of value-added components such as dietary fibers, proteins, minerals, vitamins, and phenolic compounds. These phenolic compounds are considered as one of the most important classes of natural antioxidants (Machu et al. 2015). A worldwide trend to apply natural antioxidants, instead of synthetic compounds, has gained increasing interest. Main reasons are the possible cause of neurotoxicity, gastritis, and cancer stemming from the degradation of synthetic antioxidants (Nieva-Echevarría et al. 2015). As a result, phenolic compounds extracted from, for example, seaweed are more and more applied to date in, for instance, cosmetics, pharmaceutics, and food/feed ingredients, primarily for their radical scavenging potential and antiaging properties (Zern et al. 2005; Tomas and Kim 2013).

Phenolic compounds are chemical compounds that contain a hydroxyl group (–OH) attached to an aromatic hydrocarbon group (ring of carbons). They are found as simple phenols and polyphenols that contain multiple phenol structural units. Polyphenols are secondary metabolites present in various organisms, including seaweed and higher plants (La Barre et al. 2010). Polyphenols, in general, include phenolic acids, flavonoids, stilbenes, lignans, pigments, and phlorotannins (Chater et al. 2015). Phlorotannins are a major group of phenolic compounds found in brown seaweed and in few species of red seaweed. Phlorotannins are oligo- or polymers of phloroglucinols (PGs) (1,3,5-trihydroxy benzene). More than 150 structures of phlorotannins have been isolated and characterized (Spurr 2014), and the molecular weight varies between 126 and 650 kDa (Koivikko 2008). These compounds can represent up to 20% (Amsler and Fairhead 2006; Koivikko 2008) of the dry weight of brown seaweed.

15.2 State of the art

The content of phenolic compounds in seaweed depends on many factors, for example, seaweed *species*, plant age, position in the plant body, and growing locations and seasons (Marinho-Soriano et al. 2006). The quality and intensity of light, salinity, and nutrient concentration also play an important role for phenolic content in seaweed (Connan et al. 2006). To monitor and optimize the process/strategy of cultivation and harvesting of suitable seaweed feedstocks for extracting phenolic compounds, a quick determination of the phenolic content and the bioactivities (e.g., antioxidant capacity) is of practical importance. Therefore, the current chapter describes suitable methods, which have been modified and adapted specifically to seaweed, for quick determination of the total phenolic content (TPC) (by the Folin–Ciocalteu [FC] assay and Prussian blue [PB] assay) and of the antioxidant capacity (by the DPPH (2,2-diphenyl-1-picrylhydrazyl) assay).

FC assay is the most widely used assay to estimate the TPC in plant extracts. It is a colorimetric method based on electron transfer in an alkaline medium from the phenolic compounds to phosphomolybdic/phosphotungstic acid complexes, resulting in a blue solution, which is then determined spectroscopically at 700–765 nm (Singleton and Rossi 1965). However, the disadvantage is that FC reagent not only reacts with polyphenols but could also react with some other reducing compounds that possibly exist in the crude extracts, and thus skew the results of TPC determination. To increase the specificity for TPC determination, increasing the purification of phenolic compounds extract is one of the possibilities. Another colorimetric method commonly used for determining the TPC is the PB assay (Price and Butler 1977). This method is based on the color-formation reaction between the phenolic compounds and ferricyanide ion $(Fe(CN)_6)^{3-}$. Phenols are oxidized, whereas $Fe(CN)_6^{3-}$ is reduced to the ferrocyanide ion $Fe(CN)_6^{4-}$. $Fe(CN)_6^{4-}$ then reacts with the ferric ion (Fe^{3+}) to form PB metallic complex, $Fe_4[Fe(CN)_6]_3$. Compared with the FC assay, it has the advantages of lower time consumption and higher selectivity (Margraf et al. 2015). The main drawbacks of the original PB assay (Price and Butler 1977) are the formation of precipitates after incubation and an increase in color intensity with time. Graham (1992) developed a stabilized reagent in combination of phosphoric acid (or tetrasodium ethylenediaminetetraacetate) and gum arabic: Phosphoric acid (or tetrasodium ethylenediaminetetraacetate) is used for stopping the reaction by complexing the excess of Fe^{3+} in the mixture, and gum arabic is used for preventing the precipitation by impeding the coalescence of the colloidal particles. However, running a preanalysis to ensure a proper phenolic concentration range is recommended for preventing the potential precipitation of the PB metallic complex, especially when the phenolic content of sample is high. Although the selectivity of phenolic compound is higher in PB assay than in FC assay (Margraf et al. 2015), the purity of the extract still affects the measured result of TPC from PB assay.

The antioxidant capacity of extracted phenolic compounds from seaweed can be determined by, for example, DPPH assay (first developed by Blois 1958). This assay is one of the most widely used assays and has been modified and adapted in many different natural antioxidant studies, because of its simplicity, speed, and sensitivity (Kedare and Singh 2011). $DPPH^{\bullet}$ (free radical) is a stable and commercially available organic nitrogen radical that shows a strong absorption maximum at 517 nm (purple). The color turns from purple to yellow followed by the formation of DPPH by trapping hydrogen from a hydrogen donor (i.e., antioxidant). The antioxidant effect is proportional to the disappearance of $DPPH^{\bullet}$ in test samples and can be spectrophotometrically monitored at 517 nm.

15.3 Materials

15.3.1 Seaweed biomass

Fresh or defrosted seaweed biomass is dried in a 50°C drying oven for two nights (until the biomass dry matter content reaches ~90%) and is then ground and passed through a Ø = 1.0 mm screen.

15.3.2 Equipment

- A UV–vis spectrophotometer with a 96-well flat-bottom microtiter plate reader.
- Moisture content analyzer.
- Thermostatically controlled drying oven (50°C).
- Incubator/water bath.
- Benchtop centrifuge.
- Vortex mixer.
- Magnetic stirrer with temperature controller.
- Analytical balance.
- Single- and multichannel pipette and tips.
- Magnetic stirrer bars.
- Stopwatch.
- Measuring cylinders.
- Volumetric flasks.
- Filter funnels.
- 1.5 and 2 mL plastic microtubes with caps.
- 15/50 mL plastic tubes with screw caps.
- 100 mL autoclavable glass bottles with caps.
- Racks for 1.5–2 mL microtubes and 15/50 mL tubes.
- Disposable 96-well flat-bottom microplate.
- Aluminum foil.

15.3.3 Chemicals

15.3.3.1 Total phenolic content analysis of seaweed biomass

15.3.3.1.1 Folin–Ciocalteu assay

- *FC reagent*: 0.2 N in ultrapure water.
- *Sodium bicarbonate ($NaHCO_3$) solution*: 6% w/v in ultrapure water.
- *Absolute ethanol*: 99% (w/w).
- *PG stock solution*: 2 mg mL^{-1} in absolute ethanol.

15.3.3.1.2 Prussian blue (PB) assay

- *HCl solution*: 0.1 M in ultrapure water.
- *$K_3Fe(CN)_6$*: 0.016 M in ultrapure water.
- *$FeCl_3$*: 0.02 M in 0.1 M HCl.
- *H_3PO_4*: 6.03 M in ultrapure water.
- *Gum arabic solution* (1%): Dilute 1.00 g of gum arabic in a 100 mL volumetric flask with 100 mL ultrapure water. Dissolve by heating it at max 60°C for 30 min. Cool down the solution and adjust with ultrapure water to the 100 mL mark. Filter the solution through a filter paper (20–25 μm).

- *Absolute ethanol*: 99% (w/w).
- *Catechol (1,2-dihydroxybenzene) stock solution*: 150 mg L^{-1} in absolute ethanol.

15.3.3.2 Antioxidant capacity analysis of seaweed biomass: 2,2-Diphenyl-1-picrylhydrazyl assay

- *Absolute ethanol*: 99% (w/w).
- *Ascorbic acid (AA) stock solution*: 100 mg L^{-1} in absolute ethanol.
- *DPPH• (free radical) solution*: 0.1 mM in absolute ethanol. Cover the flask of the prepared solution with aluminum foil.

15.4 Experimental procedures

The procedures for the determination of the TPC (Section 15.4.1) and antioxidant capacity (Section 15.4.2) are described as follows.

15.4.1 Total phenolic content analysis of seaweed biomass

Two quick assays for determination of TPC are described (Sections 15.4.1.1 and 15.4.1.2). FC assay is the most widely used assay for estimating the TPC, whereas PB assay showed the advantages of, for example, lower time consumption (15 min reaction time in PB VS 2 h reaction time in FC, see the procedures in the following sections). Moreover, as being introduced, although the results of TPC from both assays can be skewed by other existing reducing compounds, the selectivities of the assays are different. The two assays can be used as an alternative to each other and to compare the results (Margraf et al. 2015).

15.4.1.1 Folin–Ciocalteu assay

This assay is modified and adapted to seaweed biomass from the previously published methods (Singleton et al. 1998; Velioglu et al. 1998).

1. Measure the dry matter content of the dried, ground, and screened seaweed (preparation described in Section 15.3.1) by using the moisture content analyzer (following the instruction manual of the analyzer you use). Record the moisture content of the sample p (%). Calculate and record the dry matter content of the sample q (%):

$$q = 100\% - p \tag{15.1}$$

2. Mix the dried seaweed with 99% ethanol at a ratio of 1:20 (seaweed biomass [g, dry weight]:ethanol [mL]) in, for example, a 100 mL glass bottle with cap, under continuous stirring at 400 rpm at 22°C ± 1.0°C

for 24 h. Record the weight of the seaweed biomass m (mg) and total volume V (L). Calculate and record the dry weight of seaweed biomass M (mg):

$$M = m \times q \tag{15.2}$$

where q is the recorded dry matter content from step 1.

3. Collect the liquid fraction (i.e., seaweed extract sample) after centrifugation in a benchtop centrifuge at 2200 (×g) for 10 min.
4. Prepare other regents and stock solutions as listed in Section 15.3.3.1.1.
5. *Prepare PG standard series with the range of 20–100 mg L^{-1}*: Dilute the prepared PG stock solution with absolute ethanol, for example, add 0.5, 1.0, 1.5, 2.0, and 2.5 mL of stock solution and then add absolute ethanol to the 50 mL mark of the volumetric flask.
6. If necessary, dilute the seaweed extract sample by ethanol to keep the phenolic content in the extract within the linear range of the concentrations in the standard solution series, that is, 20–100 mg L^{-1}. Record the dilution factor d.
7. Pipette 100 μL each of the below samples to one 2 mL microtube (and make triplicate tubes per sample, as indicated):
   - (Diluted) seaweed extract (×3)
   - Blank (99% ethanol) (×3)
   - *PG standard solution series*: 20 mg L^{-1} (×3), 40 mg L^{-1} (×3), 60 mg L^{-1} (×3), 80 mg L^{-1} (×3), and 100 mg L^{-1} (×3) microtube.

     Label the microtubes by, for example, *seaweed extract, blank, PG standard 20 mg L^{-1}, PG standard 40 mg L^{-1}, PG standard 60 mg L^{-1}, PG standard 80 mg L^{-1}*, and *PG standard 100 mg L^{-1}*.
8. Pipette 750 μL of FC reagent to each of the above microtube and mix well by vortex mixing.
9. Allow the mixture to stand at room temperature (22°C ± 1°C) for 8 min.
10. Pipette 750 μL of 6% $NaHCO_3$ solution to each microtube and mix well by vortex mixing. pH should be approximately 10.
11. Incubate at room temperature (22°C ± 1°C) for 2 h.
12. After incubation, pipette 200 μL of the reacted solution from the marked microtubes, respectively, to a well in the 96-well microtiter plate. Put the plate to a UV–vis spectrophotometer.
13. Record the absorbance at 725 nm ($A_{725\ nm}$). Mean of $A_{725\ nm}$ from the triplicate of each sample (i.e., seaweed extract, blank, and PG standard series) is calculated and used for the following calculations.
14. Plot a PG standard curve (X-axis: TPC [mg L^{-1}], Y-axis: $A_{725\ nm}$). A linear correlation formula is then generated from the curve, that is,

$$Y = kX + b \tag{15.3}$$

15. The concentration of the TPC in the seaweed extract sample, that is, TPC_{SE} (mg L^{-1}) is calculated from the generated formula, based on its measurement of $A_{725\ nm}$:

$$TPC_{SE} = \frac{A_E - A_0}{k} \tag{15.4}$$

where:

TPC_{SE} is the TPC in the extracted sample for measurement
A_E is the absorbance of the seaweed extract sample
A_0 is the absorbance of the blank

16. The TPC in seaweed biomass, that is, TPC_{SW} (%, w/w) is

$$TPC_{SW}(\%) = \frac{TPC_{SE} \times d \times V}{M \times 100} \tag{15.5}$$

where:

d is the (possible) dilution factor recorded in step 6
V (L) and M (mg) are the extraction volume and seaweed biomass dry weight recorded in step 2

15.4.1.2 Prussian blue assay

This assay is modified and adapted to seaweed biomass according to the previously published methods (Price and Butler 1977; Graham 1992).

1. Measure the dry matter content of the dried, ground, and screened seaweed (preparation described in Section 15.3.1) by using the moisture content analyzer (following the instruction manual of the analyzer you use). Record the moisture content of the sample p (%). Calculate and record the dry matter content of the sample q (%) following Equation 15.1.
2. Mix the dried seaweed with 99% ethanol at a ratio of 1:20 (seaweed biomass [g, dry weight]:ethanol [mL]) in, for example, a 100 mL bluecap bottle, under continuous stirring at 400 rpm at 22°C ± 1.0°C for 24 h. Record the weight of the seaweed biomass m (mg) and total volume V (L). Calculate and record the dry weight of seaweed biomass M (mg) following Equation 15.2.
3. Collect the liquid fraction (i.e., seaweed extract sample) after centrifugation in a benchtop centrifuge at 2200 (×g) for 10 min.
4. Prepare other reagents and stock solutions as listed in Section 15.3.3.1.2.
5. Dilute the catechol stock solution with ethanol to prepare the standard solutions within the range of 0.6–15 mg L^{-1}, for example, add 0.2, 0.5, 1.0, 2.0, and 5.0 mL of stock solution, and then add absolute ethanol to the 50 mL mark of the volumetric flask.

6. If necessary, dilute the extracted sample with ethanol with the appropriate dilution factor to keep phenolic content in the diluted extract sample within the linear range of concentrations of the standard solution series, that is, 0.6–15 mg L^{-1}. Record this dilution factor *d*.
7. Pipette 300 μL each of the below samples to one 2 mL microtube (and make triplicate tubes per sample, as indicated):
   - Diluted seaweed extract (×3)
   - Blank (ethanol) (×3)
   - *Standard solution series*: 0.6 mg L^{-1} (×3), 1.5 mg L^{-1} (×3), 3 mg L^{-1} (×3), 6 mg L^{-1} (×3), and 15 mg L^{-1} (×3).

   Label the microtubes by, for example, *seaweed extract, blank, Catechol standard 0.6 mg* L^{-1}, *Catechol standard 1.5 mg* L^{-1}, *Catechol standard 3 mg* L^{-1}, *Catechol standard 6 mg* L^{-1}, and *Catechol standard 15 mg* L^{-1}.
8. Pipette into each of the above microtubes 100 μL of 0.016 M $K_3Fe(CN)_6$ followed by 100 μL of 0.02 M $FeCl_3$ in 0.1 M HCl.
9. Mix the microtubes properly by vortex mixing and incubate the tubes in a water bath at 25°C for 15 min.
10. Pipette 300 μL 6.03 M H_3PO_4 into each of the above-mentioned tubes. Vortex the tubes and incubate the tubes in a water bath at 25°C for 2 min.
11. Pipette 200 μL 1% gum arabic solutions to each of the microtubes and mix it well by vortex mixing.
12. After that, pipette 200 μL of the reacted solution from each of the labeled microtube to a well in the 96-well microtiter plate. Put the plate to a UV–vis spectrophotometer.
13. Record the absorbance at 700 nm ($A_{700\ nm}$). Mean of $A_{700\ nm}$ from the triplicate of each sample (i.e., seaweed extract, blank, and catechol standard series) is calculated and used for the following calculations.
14. Plot a catechol standard curve (*X*-axis: TPC [mg L^{-1}], *Y*-axis: $A_{700\ nm}$). A linear correlation formula is then generated from the curve, that is,

$$Y = aX + c \tag{15.6}$$

15. The concentration of the TPC in the seaweed extract sample, that is, TPC_{SE} (mg L^{-1}) is calculated from the generated formula, based on its measurement of

$$TPC_{SE} = \frac{(A_E - A_0)}{a} \tag{15.7}$$

where:

A_E is the absorbance of the seaweed extract sample
A_0 is the absorbance of the blank

The TPC in seaweed biomass, that is, TPC_{SW} (%, w/w) is then calculated as

$$TPC_{SW}(\%) = \frac{TPC_{SE} \times d \times V}{M \times 100} \tag{15.8}$$

where:

d is the dilution factor recorded in step 6

V (L) and M (mg) are the extraction volume and seaweed biomass dry weight recorded in step 2

15.4.2 *Antioxidant capacity analysis of seaweed biomass: 2,2-Diphenyl-1-picrylhydrazyl assay*

This assay is adapted to seaweed biomass from the previously published methods (Yang et al. 2008; Sanja et al. 2009; Farvin and Jacobsen 2013).

1. Measure the dry matter content of the dried, ground, and screened seaweed biomass (preparation described in Section 15.3.1) by using the moisture content analyzer (following the instruction manual of the analyzer you use). Record the moisture content of the sample p (%). Calculate and record the dry matter content of the sample q (%) following Equation 15.1.
2. Mix the dried seaweed with 99% ethanol at a ratio of 1:20 (seaweed biomass [g, dry weight]:ethanol [mL]) in, for example, a 100 mL glass bottle with cap, under continuous stirring at 400 rpm at 22°C ± 1.0°C for 24 h. Record the weight of the seaweed biomass m (mg) and total volume V (L). Calculate and record the dry weight of seaweed biomass M (mg) following Equation 15.2.
3. Collect the liquid fraction (i.e., seaweed extract sample) after centrifugation in a benchtop centrifuge at 2200 (×g) for 10 min.
4. Prepare other regents and stock solutions as listed in Section 15.3.3.2.
5. *Prepare AA standard*: Mix the prepared AA standard with ethanol to achieve the final concentration of AA standard series in concentration of 1 ~ 10 mg L^{-1}: for example, add 1.0, 2.0, 5.0, 7.5, and 10 mL of stock solution, and then add absolute ethanol to the 100 mL mark of the volumetric flask.
6. Pipette 100 μL each of the below samples to a well in a 96-well microtiter plate (and make triplicate tubes per sample, as indicated):
   - Seaweed extract (×3)
   - Sample control (ethanol) (×3)
   - *Standard solution series*: 1 mg L^{-1} (×3), 2 mg L^{-1} (×3), 5 mg L^{-1} (×3), 7.5 mg L^{-1} (×3), and 10 mg L^{-1} (×3)
7. Pipette 100 μL of the prepared DPPH solution to the above-mentioned wells.

8. Pipette 100 μL seaweed extract + 100 μL ethanol, in triplicate, into extra wells of the microtiter plate, as blank for the seaweed sample.
9. Pipette 100 μL (each of the) standard solution series + 100 μL ethanol, in triplicate, into extra wells of the microtiter plate, as blank for the standard sample.
10. Incubate the microtiter plate for 30 min at room temperature (22°C ± 1.0°C) in darkness.
11. After that, measure the absorbance of each well at 517 nm using a UV–vis microplate spectrophotometer. Mean of the triplicate of each sample (i.e., seaweed extract, sample control, standard series, blank for the seaweed sample, and blank for the standard series) is calculated and used for the following calculations.
12. Antioxidant capacity of the seaweed samples is calculated as
    a. The percentage inhibition (I [%]) of the DPPH radical at 517 nm:

$$\mathrm{I}\,(\%) = \left[1 - \frac{(A_S - A_0)}{A_C}\right] \times 100 \tag{15.9}$$

where:
A_S is the absorbance in the seaweed extract sample
A_0 is the absorbance in the blank for seaweed sample
A_C is the absorbance in the sample control

OR

b. *The AA equivalent antioxidant capacity* (*AEAC*) by the following way:
   – Calculate the percentage inhibition (I%) in each of the AA standard sample:

$$I_{ST}\,(\%) = \left[1 - \frac{(A_{ST} - A_{0ST})}{A_C}\right] \times 100 \tag{15.10}$$

where:
A_{ST} is the absorbance of the standard sample
A_{0ST} is the absorbance in the blank for the corresponding standard sample
A_C is the absorbance in the sample control

   – Plot an AA standard curve (*X*-axis: AA concentration [mg L^{-1}], *Y*-axis: I [%]). A linear correlation formula is then generated from the curve, that is,

$$Y = gX + f \tag{15.11}$$

- AEAC (mg-ascorbic-acid/L) of the seaweed extract sample ($AEAC_{SE}$) is calculated from the above-generated formula, based on its calculated inhibition (%):

$$AEAC_{SE} = \frac{(I_S - f)}{g} \tag{15.12}$$

where I_S is the calculated I (%) of the seaweed extract sample.
- AEAC (g-ascorbic-acid/g-biomass) of the seaweed biomass ($AEAC_{SW}$) is

$$AEAC_{SE} = \frac{(AEAC_{SE} \times V)}{M} \tag{15.13}$$

where V (L) and M (mg) is the extraction volume and seaweed biomass dry weight recorded in step 2.

15.5 *Tips and tricks*

To avoid any possible experiment deviations, for example, small deviation of incubation time or reaction time, always run the blank, sample control (in DPPH assay) and standard series together with the samples to be measured. To avoid precipitation while performing the PB assay, read the absorbance as quickly as possible and strictly follow the addition order of the reagents in the PB assay (Section 15.4.1.2) from step 7 to step 11. Run a pretest to find a proper dilution rate to keep the total phenolic concentration in the seaweed extract sample within the linear range of the standard curve.

Acknowledgment

The current chapter is part of the MacroFuels project. The project has received funding from the European Union's Horizon 2020 research and innovation program under grant agreement No 654010.

References

Amsler, C.D., and V.A. Fairhead. 2006. Defensive and sensory chemical ecology of brown algae. *Adv. Bot. Res.* 43: 1–91.

Blois, M.S. 1958. Antioxidant determinations by the use of a stable free radical. *Nature* 181: 1199–1200.

Chater, P.I., Wilcox, M.D., Houghton, D., and J.P. Pearson. 2015. The role of seaweed bioactives in the control of digestion: Implications for obesity treatments. *Food Funct.* 6: 3420–3427.

Connan, S., Delisle, F., Deslandes, E., and E.A. Gall. 2006. Intra-thallus phlorotannin content and antioxidant activity in *Phaeophyceae* of temperate waters. *Bot. Mar.* 49: 39–46.

Farvin, K.H.S., and C. Jacobsen. 2013. Phenolic compounds and antioxidant activities of selected species of seaweeds from Danish coast. *Food Chem.* 138: 1670–1681.

Graham, H.D. 1992. Stabilization of the Prussian blue color in the determination of polyphenols. *J. Agric. Food Chem.* 40: 801–805.

Kedare, S.B., and R.P. Singh. 2011. Genesis and development of DPPH method of antioxidant assay. *J. Food Sci. Technol.* 48: 412–422.

Koivikko, R. 2008. Brown algal phlorotannins: Improving and applying chemical methods. PhD thesis. University of Turku. Accessed May 8, 2017: http://www.doria.fi/handle/10024/36054

La Barre, S., Potin, P., Leblanc, C., and L. Delage. 2010. The halogenated metabolism of brown algae (*Phaeophyta*), its biological importance and its environmental significance. *Mar. Drugs* 8: 988–1010.

Machu, L., Misurcova, L., Ambrozova, J.V., Orsavova, J., Mlcek, J., Sochor, J., and T. Jurikova. 2015. Phenolic content and antioxidant capacity in algal food products. *Molecules* 20: 1118–1133.

Margraf, T., Karnopp, A.R., Rosso, N.D., and D. Granato. 2015. Comparison between Folin–Ciocalteu and Prussian blue assays to estimate the total phenolic content of juices and teas using 96-Well microplates. *J. Food Sci.* 80: 2397–2403.

Marinho-Soriano, E., Fonseca, P.C., Carneiro, M.A.A., and W.S.C. Moreira. 2006. Seasonal variation in the chemical composition of two tropical seaweeds. *Bioresour. Technol.* 97: 2402–2406.

Nieva-Echevarría, B., Manzanos, J.M., Goicoechea, E., and M.D. Guillén. 2015. 2,6-Di-Tert-Butyl-Hydroxytoluene and its metabolites in foods. *Compr. Rev. Food Sci. Food Saf.* 14: 67–80.

Price, M.L., and L.G. Butler. 1977. Rapid visual estimation and spectrophotometric determination of tannin content of *Sorghum grain*. *J. Agric. Food Chem.* 26: 1268–1273.

Sanja, S.D., Sheth, N.R., Patel, N.K., Patel, D., and B. Patel. 2009. Characterization and evaluation of antioxidant activity of *Portulaca oleracea*. *Int. J. Pharm. Pharmac. Sci.* 1: 74–84.

Singleton, V.L., and J.A. Rossi. 1965. Colorimetry of total phenolics with phosphomolybdic—Phosphotungstic acid reagents. *Am. J. Enol. Vitic.* 16: 144–158.

Singleton, V.L., Rudolf, O., and R.M. Lamuela-Raventós. 1998. Analysis of total phenols and other oxidation substrates and antioxidants by means of Folin–Ciocalteu reagent. *Meth. Enzymol.* 299: 152–178.

Spurr, H.I. 2014. Extraction, separation, and purification of polyphenols, polysaccharides and pigments from British seaweed for high-value applications. PhD thesis. University of Leeds. Accessed May 8, 2017: http://etheses.whiterose.ac.uk/id/eprint/8636

Tomas, N.V., and S.K. Kim. 2013. Beneficial effects of Marine algal compounds in cosmeceuticals. *Mar. Drugs* 11: 146–164.

Velioglu, Y.S., Mazza, G., Gao, L., and B.D. Oomah. 1998. Antioxidant activity and total phenolics in selected fruits, vegetables, and grain products. *J. Agric. Food Chem.* 46: 4113–4117.

Yang, J., Guo, J., and J. Yuan. 2008. *In vitro* antioxidant properties of rutin. *LWT-Food Sci. Technol.* 41: 1060–1066.

Zern, T.L., and M.L. Fernandez. 2005. Cardioprotective effects of dietary polyphenols. *J. Nutr.* 135: 2291–2294.

chapter sixteen

Extraction of phycocyanin and phycoerythrin pigments

Stewart William Beattie, Michèle Morançais, Paul Déléris, Joël Fleurence, and Justine Dumay

Contents

16.1 Introduction

Phycobiliproteins (PBP) are an oligomeric water-soluble algal protein family constituting the main light-harvesting chromoproteins of photosystem II in Rhodophyta and Cyanobacteria. PBPs function as light-capture and transfer systems, comprising up to 50% of water-soluble proteins in Rhodophyta (Dumay et al. 2014) and 60% in Cyanobacteria (Cohen-Bazire and Bryant 1982). These exist on the outer thylakoid membrane in the stroma and are largely responsible for the reflected colors in Rhodophyta (red algae), Cyanophycaea (blue–green algae), Cryptomonads and some Pyrrophyceae, as aggregates of phycobilisomes (Cohen-Bazire and Bryant 1982, Isailovic et al. 2006). These PBPs function predominantly to absorb sunlight in the wavelength range of 470–660 nm, allowing for the species exploiting them to survive at very low light intensities, mainly at depths in the water where chlorophyll-*a* becomes inefficient in light absorption. PBPs are generally split into four groups according to their absorbance maxima, existing as a complex of cores and rods (Figure 16.1). The quantum efficiency with regards to energy transfer between PBPs has been quoted as being close to 100%, with Phycoerythrin (PE) being located at the tip of the rod, Phycocyanin (PC) in the center, and Allophycocyanin (APC) comprising the core that then transfers energy to photosystem II (Figure 16.2) (Galland-Irmouli et al. 2000).

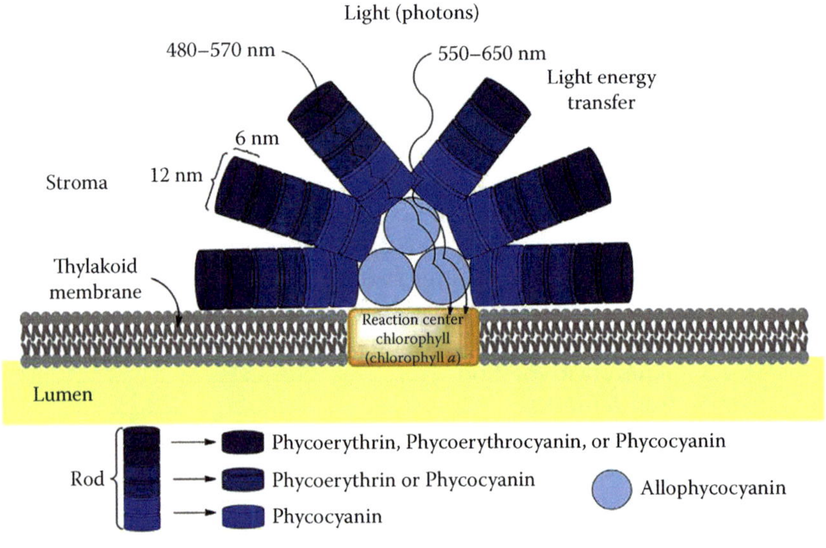

Figure 16.1 Structure of a single phycobilisome situated on the thylakoid membrane. (From Dumay, J. et al., Chapter 11. Phycoerythrins: Valuable proteinic pigments in red seaweeds, in Jacquot, J.P., Gadal, P., Bourgougnon, N. (Eds.), *Advances in Botanical Research* Vol. 71, pp. 321–343, 2014.)

Figure 16.2 Schematic of quantum energy transfer from the tip of a phycobilis.

Certain species of Cyanobacteria, such as those in the genus *Spirulina*, have been found to contain a rich and inexpensive source of C-PC (Bennett and Bogorad 1973, Bhaskar et al. 2005), and it has been posited that the accessory water-soluble PBPs in seaweeds have originated from an endosymbiotic relationship with Cyanobacteria before their evolution into plastids (Dawes 2016).

PBPs are categorized depending on their source, for example, Rhodophyta contain R-Phycocyanin (R-PC) whereas C-Phycocyanin (C-PC) originates from Cyanobacteria and Cryptophyta and B-Phycocyanin (B-PC) from Bangiophycaea, a class of Rhodophyta (Dumay and Morançais 2016). R-PC is classed as a surviving intermediate species from the past that contributes to the transfer of energy from PE to PC and is utilized in modern applications in a similar fashion to R-phycoerythrin (R-PE) as fluorescent markers, natural nontoxic dyes, and as an immunoregulator (Glazer and Hixson 1975, Zhu et al. 2015).

16.2 State of the art

The discovery of phycobilisomes role as the main light-harvesting chromoproteins in the aforementioned phyla can be attributed to Theodor Wilhelm Engelmann in 1883, who concluded that the projection of light on an *Oscillatoria* filament caused an accumulation of aerotactic bacteria in spectral regions where oxygen was involved, which corresponded to the previously stated absorbance regions. This accumulation occurs in photosystem II that is the first protein complex involved in the photosynthesis of oxygen in plants and is the last step in the quantum energy transfer chain situated on the surface of the thylakoid membrane (Figure 16.3).

These PBPs are composed of α and β subunits; however, R-PE (Phycoerythrin of Rhodophyceae) and B-PE (Phycoerythrin of Bangiales) are unique in the sense that they are the only PBPs containing an extra γ subunit (Isailovic et al. 2004, Niu et al. 2006, Sun et al. 2009), which ultimately boosts the pigment structures stability as the extra subunit acts like a rivet to the two central, face-to-face $\alpha\beta_3$-trimers (Glazer and Hixson 1977, Hilditch et al. 1991, Dumay et al. 2014).

The extraction of PBPs is influenced by numerous external factors, with significance given to the biomass-solvent ratio, cellular disruption method, nature of solvent, and extraction time (Abalde et al. 1998, Reise et al. 1998). Furthermore, the efficacy of each method will vary according to species, regardless of PBP intended for extraction (Moraes et al. 2011).

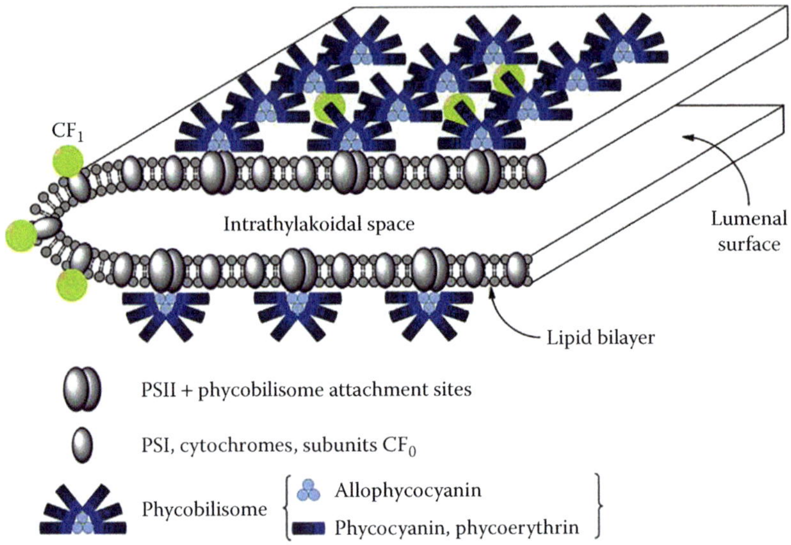

Figure 16.3 Phycobilisomes and their associated photosystems as situated on the thylakoid membrane. (From Dumay, J. et al., Chapter 11. Phycoerythrins: Valuable proteinic pigments in red seaweeds, in Jacquot, J.P., Gadal, P., Bourgougnon, N. (Eds.), *Advances in Botanical Research* Vol. 71, pp. 321–343, 2014.)

This efficacy also depends on the number of steps taken to complete the extraction as it is well known that a greater number of steps will result in a lower final product yield (Kula et al. 1982).

The purity of the extracted PE and PC will ultimately determine which use it is best suited for. Low purity (0.7) will be used as a food colorant as it is nontoxic and noncarcinogenic (Doke 2005). High purity (3.9) can be considered reactive grade, and purity (>3.2) will be implemented into processes involving analyzing fluorescent energy transfer as PBPs have a notably large stokes shift, making then easily detectable through spectroscopy (Rito-Palomares et al. 2001, Patil and Raghavaroa 2007, Dumay et al. 2015).

16.3 Materials

16.3.1 Algal harvesting

PBPs could be recovered from both seaweeds and microalgae.

For seaweeds (*Note 1*)

1. Collect the samples in the shore or from seaweed farms (*Note 2*).
2. Remove epiphytes when required.
3. Rinse thoroughly with tap water and spin dry the sample.
4. Store until use.

16.3.2 *Algal conditioning*

Studies analyzed in the course of writing this chapter have found several phycobiliprotein removal experiments conducted on wet and dry algal biomass. Several studies are reporting different techniques associated with the drying process (Table 16.1) that ultimately affected the purity and yield of PBPs from macroalgae (Sarada et al. 1999, Doke 2005, Oliveira et al. 2009, Chaiklahan et al. 2011, Moraes et al. 2011, Prabakaran and Ravindran 2013).

Studies utilizing these techniques with *Spirulina* sp. have reported differing values of loss for PC yield, with a 5% loss from sun-dried biomass (Chaiklahan et al. 2011), a 5%–7% loss when dried with air circulation (Doke 2005) and a 21% loss with cross-flow drying (Oliveira et al. 2009). One conclusion from Sarada et al. (1999) also detailed that although the temperature differed between drying techniques, it was not a major influencer on PC yield. Regardless of the drying technique in this study (spray, cross-flow, and sun drying), the PC yield remained at a loss of 45% ± 1% when PC recovery from wet biomass is classed as 100%; therefore, it is recommended to perform experiments on fresh wet biomass, either cultivated, harvested or purchased, to optimize yield and purity of PBPs. Purity was also affected in this study as PC recovered from dry samples demonstrated an absorbance peak at 652 nm, indicating that APC was also present within the sample. This could be because of the fact that phycobilisomes are constituted of an APC core surrounded by PC rods tipped with PE, which suggests that heat from drying methods could cause a

Table 16.1 Biomass drying techniques and methodologies

Drying technique	Method
Sun-drying	Wet biomass in a beaker can be placed directly in the sun assuming exterior ambient temperatures exceed 35°C (ancestral method).
Air-drying	Wet biomass can be placed in an air circulation system at 25°C for 1 h in the shade. This can also be performed using a cross-flow system by implementing perpendicular airflows (Oliveira et al. 2009).
Spray-drying	Wet biomass solution is sprayed into a chamber containing a hot gas to eliminate moisture content and leave a powdered form of the initial solution behind (Sarada et al. 1999).
Freeze-drying	Wet biomass can be rapidly frozen before being subjected to a vacuum to remove the water content to preserve it (Nguyen et al. 2016).

decrease in purity and yield because of the peripheral position of the PC and excessive penetration of the phycobilisome by high temperatures (Glazer 1984, Sarada et al. 1999).

Therefore, prefer to carry PBP extraction on dry algae material by the use of one of the techniques.

16.3.3 *Buffer preparation*

Acetate buffer 50 mM pH 5 (1 L) (*Notes 3 and 4*)

- Homogenize 4.32 g of sodium acetate trihydrate (97%) and 1.05 mL of 99% acetic acid in 250 mL of distilled water.
- Set pH at 5 using pH meter.
- Adjust pH if required with the addition of 97% sodium hydroxide (NaOH).
- Complete to 1 L with ultrapure water and store at 4°C for up to one month.

Phosphate buffer 20 mM pH 7.1 (1 L)

- Dissolve 1.58 g of $Na_2HPO_4 \cdot 2H_2O$ and 1.54 g of $NaH_2PO_4 \cdot H_2O$ in 1 L of distilled water.
- Set pH at 7.1 and adjust if necessary before storing at 4°C for up to one month.

16.4 *Experimental procedures*

16.4.1 *Phycobiliproteins extraction*

16.4.1.1 *Algal treatment*

1. Homogenization of cells must also occur before experimentation either manually by grinding, cryo-grinding or mechanically assisted, for example, with pressure 200–400 kg cm^{-2} (Patil and Raghavaroa 2007) (*Note 5*).
2. Biomass can also be frozen or ground with liquid nitrogen and homogenized in the presence of acid-washed neutral sand/ diatomaceous earth (at a ratio of 5:1 respectively) and a weakly acidic buffer at a ratio of 1 g biomass: 20 mL 20 mM sodium phosphate (pH 6.8–7) buffer (Galland-Irmouli et al. 2000, Fleurence 2003, Moraes et al. 2011, Su et al. 2014, Nguyen et al. 2016).

16.4.1.2 *Maceration*

This technique is arguably the most widely employed phycobiliprotein extraction protocol, attributable to low economic impact, short extraction time, and easiness of setup.

1. Homogenize and suspend triplicate masses of samples under agitation (150 rpm) at a ratio of 100 mL water/buffer to 1 g dry weight/7.5 g wet weight biomass (Dumay et al. 2013).
2. Conduct in darkness and allow cell lyses to occur over a 20 min period at 4°C for Cyanobacteria and Rhodophyta.
3. Centrifuge samples at 25,000 g at 4°C for 20 min before filtering and measuring rigorously the supernatant obtained (weight and volume) before spectrophotometric analysis.

16.4.1.3 Acid extraction

Acid extraction must be performed on wet biomass and using different concentrations of acid—both organic and inorganic (*Note 6*).

1. *Inorganic acid extraction*: Treat biomass with HCl (2.0–10 M) at a ratio of 5 g wet biomass:1 mL acid under agitation (150 rpm).
2. Maintain at room temperature in darkness for 24 h (Moraes et al. 2011).
3. Organic extraction has a parallel methodology except HCl is replaced by an organic acid such as acetic acid (2.0–10 M).
4. Clarify by centrifuging at 25,000 g at 4°C for 20 min before filtration, measuring the supernatant obtained (weight and volume) and analyzing spectrophotometrically.

16.4.1.4 Ultrasonic treatment

1. Homogenize and suspend biomass in distilled water/buffer (*Note 7*).
2. With/without the presence of glass pearls (ratio of 1 g biomass:1 g glass pearls), place biomass in an ultrasonic bath at 50 kHz for 40 min to break up cells (Medeiros et al. 2008).
3. Alternatively, sonicate for 6 h at 35 kHz following the procedure of Le Guillard et al. (2015).
4. Centrifuge samples at 25,000 g at 4°C for 20 min, weigh, determine volume, and analyze supernatant by spectrophotometry.

16.4.1.5 Enzymatic hydrolysis

The use of such treatment is well documented (Fleurence 2003, Denis et al. 2009, Dumay et al. 2013, Harnedy and Fitzgerald 2013, Dumay et al. 2015, Le Guillard et al. 2015, Le Guillard et al. 2016, Nguyen et al. 2016) and requires a background knowledge of the cell-wall composition of the subject species to select the appropriate enzymes for cell-wall degradation (Table 16.2) (*Note 8*). The cell walls of Rhodophyta are rich in xylans, making this phylum generally best suited to enzymatic hydrolysis by xylanolytic enzymes obtainable from fungal sources

Table 16.2 Species, study, appropriate enzyme, and optimum conditions for enzymatic hydrolysis (*Note 11*)

Species	Study	Appropriate enzyme	pH + Temp (°C)
Gracilaria verrucosa	Fleurence et al. 1995	• κ carrageenase	pH 6.5–6.8 at 45
		• Cellulase	pH 3.8 at 50
		• Xylanase	pH 5 at 55
		• β-agarase	pH 6–6.5 at 55
Grateloupia turuturu	Denis et al. 2009	• Cellulase	pH 3.8 at 50
		• Agarase	pH 6–6.5 at 55
		• ι carrageenase	pH 6.5–6.8 at 45
		• κ carrageenase	pH 6.5–6.8 at 45
Palmaria palmata	Dumay et al. 2013	• Xylanase	pH 5 at 40
Mastocarpus stellatus	Nguyen et al. 2016	• Xylanase	pH 5 at 55

(Kloareg and Quatrano 1988, Fleurence et al. 1995, Fleurence 2003, Harnedy and Fitzgerald 2013) (*Note 9*).

1. Homogenize biomass before weighing out equal masses of a desired number or replicates for experimentation.
2. Introduce equivalent biomass and buffer values at a ratio of 100 mL water/buffer (e.g., acetate buffer 50 mM, pH 5) to 1 g dry weight/7.5 g wet weight biomass (Dumay et al. 2013).
3. Maintain in darkness under agitation (700 rpm) at specific enzyme optimum temperature and allow 6 h extraction before centrifugation at 25,000 g for 20 min at 4°C (Dumay et al. 2013).
4. Filtrate supernatant to remove large fragments before analyzing by spectrophotometry (*Note 10*).

16.4.2 *Phycobiliprotein purification*

16.4.2.1 *Adsorption bed chromatography*

1. Expanded-bed adsorption column is loaded with STREAMLINE™ phenyl-Sepharose and equilibrated with ammonium sulphate (0.05 M) solution along with an expanded absorbent (*Note 12*).
2. Solution is then pumped into the column in an upward direction at 15 mL min^{-1} at room temperature (Wang 2002, Bermejo et al. 2003, Niu et al. 2006).
3. Ammonium sulphate (0.1 M) solution or sodium acetate solution (0.05 M) can then be used to wash and elute PBPs in a downward direction. PBPs are then collected as a solution as the STREAMLINE™ phenyl-Sepharose has acted as a vector for PBP capture, whereas cell debris and particulate matter has been separated and transported upward (Wang 2002, Niu et al. 2006).

16.4.2.2 Ammonium sulphate precipitation

1. Collected eluates can be fractionated with ammonium sulphate (e.g., 30, 50, and 60% weight/volume) to form precipitates.
2. Precipitates can then be dissolved in a neutral buffer such as sodium phosphate (0.05 M) before dialysis against the same buffer diluted 1000 times over the course of 4 h at 4°C (Zhang and Chen 1999, Bhaskar et al. 2005).
3. Centrifuge at 10,000 g for 30 min at 4°C.
4. Resuspend precipitate in an acetate buffer (0.05 M, pH 5) and centrifuge again using the same settings (Patel et al. 2005, Leema et al. 2010). Repetitive steps and allowing for overnight dialysis can both improve purity, but they will reduce yield due to an increased number of steps.

16.4.2.3 Ion-exchange column

1. Eluates from absorption bed chromatography are eluted with ammonium sulphate solution (e.g., 0.05, 0.1, and 0.2 M) and distilled water before being applied to the ion-exchange column.
2. Pump or pour through the ion-exchange column in a downward direction in the presence of a negatively charged ion exchanger (e.g., STREAMLINE™ Q-Sepharose).
3. Wash with a neutral buffer such as sodium phosphate (0.05 M) to remove loosely bound particles and excess debris, while eluting the PBP from the ion exchanger until the effluent is clear (Wang 2002, Niu et al. 2006, Munier et al. 2015).
4. Elute PBPs by creating a gradient using sodium chloride (0.15 M) and phosphate buffer (0.05 M) before collecting PBPs.

16.4.2.4 Hydroxyapatite chromatography

Hydroxyapatite is a form of naturally occurring calcium that can be created in the form of a neutral solution by combining certain chemicals.

1. Equilibrate hydroxyapatite column with 10 mM phosphate buffer containing 20 mM of sodium chloride in the downward direction before adding eluates (Niu et al. 2006).
2. Eluates from absorption bed chromatography are eluted with ammonium sulphate solution (e.g., 0.05, 0.1, and 0.2 M) and distilled water before being applied to the hydroxyapatite column.
3. Calcium chloride ($CaCl_2$) (2 M) and potassium phosphate (KH_2PO_4) (2 M) can be mixed at a rate of 5 mL min^{-1} using a magnetic stirrer at a ratio of 1:1.1 mL, respectively, to trigger the hydroxyapatite precipitation.
4. As previously mentioned, PBPs are sensitive to pH, therefore, the pH of the hydroxyapatite solution must be maintained at a value of 9 by adding potassium hydroxide (KOH) (2 M) solution.

5. Sodium phosphate (1 mM) buffer containing sodium chloride (NaCl) (20 mM) can be used to wash and remove fine particles from the precipitate before collecting the eluate (Niu et al. 2006, Niu et al. 2007).

16.4.2.5 *Aqueous two-phase extraction*

1. Predetermined quantities of a hydrophilic polymer such as polyethylene glycol (PEG) and salt (e.g., phosphate) are added to a known quantity of crude extract of PBP (*Note 13*).
2. This is termed a polymer-salt extraction, and this results in the PBPs being concentrated in a phase predominantly comprising water (80%–95%) and an increased concentration of a phase-forming component such as PEG (Johansson et al. 1998, Rito-Palomares 2004, Mensi et al. 2014).
3. Vortex mixing or magnetic stirring at room temperature for an hour will then allow for phase separation to occur (Patil and Raghavaroa 2007). This method presents several advantages over polymer–polymer systems as it is cost-efficient, widely used, and applied currently with a stable range of pH 6–9 (Rito-Palomares 2004).

16.4.3 *Analytical method*

As PBPs are segregated on the basis of their absorbance properties, therefore spectrophotometry is the preferred method to analyze supernatant and to identify different absorbance maxima (Table 16.3).

Bennett and Bogorad 1973: *PC*, *APC*, and *PE* concentration (expressed as mg mL^{-1}) can also be calculated using equations that correct for spectral overlaps of the pigments with previous studies detailing their implementation (Bennett and Bogorad 1973, Beer and Eshel 1985, Sarada et al. 1999, Patel et al. 2005, Moraes et al. 2011, Prabakaran and Ravindran 2013).

Equations are required to convert absorbance data into phycobiliprotein concentrations.

$$C-PC\,(\mathrm{mg\,ml^{-1}}) = \frac{A_{615} - 0.474(A_{652})}{5.34} \qquad (16.1)$$

Table 16.3 PBP spectrometry values

PBP	Absorbance maxima (nm)
PE	540–570
PEC	560–600
PC	610–620
APC	650–655

Source: Glazer, A. N. et al., *Proc. Nat. Acad. Sci.*, 73, 428–431, 1976.

$$C - APC\,(\mathrm{mg\,ml^{-1}}) = \frac{A_{652} - 0.208(A_{620})}{5.09} \tag{16.2}$$

$$C - PE\,(\mathrm{mg\,ml^{-1}}) = \frac{[A_{652} - 2.41(PC) - 0.849(APC)]}{9.6} \tag{16.3}$$

$$R - PE\,(\mathrm{mg\,ml^{-1}}) = [(A_{565} - A_{592}) - (A_{455} - A_{592}) \times 0.2] \times 0.12 \tag{16.4}$$

$$R - PC\,(\mathrm{mg\,ml^{-1}}) = [(A_{618} - A_{645}) - (A_{592} - A_{645}) \times 0.15] \times 0.15 \tag{16.5}$$

Silveira et al. (2007): The following equation can be used to calculate the extraction yield of PBP:

$$\mathrm{Yield} = \frac{(\mathrm{PBP\ concentration}) * \mathrm{solvent\ volume\ (mL)}}{\mathrm{Dry\ biomass\ (g)}} \tag{16.6}$$

Purity: Spectrophotometrically measuring the ratio of A_{615}/A_{280} will give a purity index for *PC* (Silveira et al. 2007) whereas the ratio A_{565}/A_{280} will give a purity index for *PE* (Beer and Eshel 1985).

Purification factor: As stated by Antelo et al. (2010).

$$\mathrm{Purification\ factor} = \frac{EP_{\mathrm{P}}}{EP_{\mathrm{C}}} \tag{16.7}$$

where:

EP_{P} is the Extract purity
EP_{C} is the Extract purity of crude extract

Further information and more detailed protocols can be found using the references cited for each process.

16.5 Concluding remarks

Several methods of PBP extraction and purification have been reported with many research projects on this topic having found disparate results with regards to PBP extraction optimization because of the distinct veracity differences between protocols when it comes to PBP extraction. This alone would suggest that to industrialize the aforementioned protocols, the requirement for an efficient, universally acceptable PBP extraction procedure capable of processing large amounts of selected biomass at minimal cost and with environmental concern is necessary. Indeed, all studies cited in this chapter have produced low yield and purity of PBPs if considered from a commercial perspective. The extraction and purification steps combined with the requirement for highly specialized equipment and the meticulous preparation of biomass to obtain such a small amount of final product has severely limited the applications available to PBP and the resulting commercialization of extraction procedures. Centrifugal recovery is a step required in most protocols because of the requirement of the separation of cell debris and product. This step is efficient as no biomass is lost through the process. Nonetheless, from a commercial perspective, it is expensive and energy inefficient when the large volumes of biomass required for industrial production are considered. However, use of less-efficient equipment could hinder the purity of the final product. Therefore, it is imperative that fundamental research be carried out on larger scales to optimize and improve the production and downstream processes of PBP extraction and purification that will ultimately be the solution to achieving supernatant with high content and purity of PBPs. Research will soon be revolving around protocol optimization that will allow for these industrial parameters to become more realistic, permitting the commercialization of an industry dedicated to the extraction and purification of these highly specialized and favorable pigments in the near future that will perhaps expand their applicability and accessibility due to continuous development of modern technology and medicine.

16.6 Notes

Note 1: PBPs content could significantly vary according to environmental factors. Be aware that samples collected during hot and sunny season (it depends of the hemisphere) will not provide as high PBPs yields as those collected during the cold season.

Note 2: Seaweed life cycle could also interfere with the extraction yield. For more accurate results, separate thalli.

Note 3: Depending on the extraction method chosen (especially for enzymatic assisted ones), you can use different buffers according to pH required. Be sure that the buffer chosen fit with the extraction method (enzyme accuracy as example).

Note 4: When PBPs extraction is compared, ensure that the buffer chosen for all the methods is the same (same pH, same ionic strength, and same composition). If not, discuss about the interferences led by the buffer differences.

Note 5: Anionic polysaccharides present within the cell walls of seaweeds make extraction of metabolites challenging and rather inefficient (Denis et al. 2009). The discharge of these highly valuable and useful PBPs from cells can be directly related to cell rupture, but small algae such as *Spirulina* sp. (300–500 μm diameter) are resistant to this in the sense that they have multilayered cell walls, making extraction difficult (Moraes et al. 2011). Regardless of the extraction method, the pH of the system must remain between pH 3 and 10 as this is the range in which PBPs remain stable (Patil and Raghavarao 2007).

Note 6: It should be noted that the method of inorganic extraction by using hydrochloric acid (HCl) can cause release PBPs to precipitate immediately after extraction (Moraes et al. 2011) and, furthermore, negatively impact purity as chlorophyll-*a* will also be extracted by HCl that can be identified by an absorption peak at 663 nm (Doke 2005).

Note 7: This method has been found to be one of the most efficient extraction procedures, especially with the presence of glass beads that were found to increase the yield by 76× when compared with ultrasonic treatment without glass pearls (Moraes et al. 2011), probably due to the fact that as thallus disintegration promotes the release of PBPs from cells, the presence of glass beads most likely catalyzed this process with extra friction.

Note 8: The biochemical composition of Rhodophyta varies greatly according to species, although, as a general assumption, it can be said that red seaweed cell walls comprise a matrix of polysaccharides with a small proportion of proteins and cellulose (Fleurence 2003). Results from previous studies have shown that pH and temperature also have an impact on the enzymatic extraction of proteins and pigments (Fleurence 2003).

Note 9: Enzymes also have nonlinear specific activity values that change with respect to temperature and incubation time; therefore, hydrolysis should be carried out in glass bioreactors under controlled temperature, agitation, and light obscurity to assert the maximum conditional controls.

Note 10: The use of ultrasound to assist in enzymatic hydrolysis has also been tested and can be applied by using *Ultrasonic treatment* or by using a Sonitube® (Le Guillard et al. 2016).

Note 11: Other techniques to extract PBPs exist (e.g., freezing & thawing) but are not applied as much as those stated in this chapter due to increased labor/economic costs or decreased purity and yield efficiency.

Note 12: This method is usually performed in a specialized piece of equipment such as a STREAMLINE™ column that is also resistant to blockage caused by the excess amount of sticky polysaccharides released after cell homogenization (Wang 2002, Niu et al. 2006). This ultimately causes an equilibrium between particle sedimentation and upward flow and a particle size gradient increasing from top to bottom (Hjorth 1999, Wang 2002, Bermejo et al. 2006, UMSC).

Note 13: Commonly implemented as an alternative in purification systems in chemical industries (Hatti-Kaul 2001, Antelo et al. 2010, Ratanapongleka 2010, Goja et al. 2013, Duarte et al. 2015), this process is attractive because of its low processing time and high yield with an efficiency based on the choice of which system is used, for example, polymer–polymer, polymer–salt (Manirafasha et al. 2016).

References

Abalde, J., Betancourt, L., Torres, E., Cid, A., Barwell, C. 1998. Purification and characterisation of phycocyanin from the marine cyanobacterium *Synechococcus* sp. IO9201. *Plant Science* 136: 109–120.

Antelo, F. S., Anschau, A., Costa, J. A. V., Kalil, S. J. 2010. Extraction and purification of C-phycocyanin from *Spirulina platensis* in conventional and integrated aqueous two-phase systems. *Journal of the Brazilian Chemical Society* 21: 5.

Beer, S. and Eshel, A. 1985. Determining phycoerythrin and phycocyanin concentrations in aqueous crude extracts of red algae. *Australian Journal of Marine and Freshwater Research* 36: 785–792.

Bennett, A. and Bogorad, L. 1973. Complimentary chromatic adaption in a filamentous blue-green alga. *The Journal of Cell Biology* 58(2): 419.

Bermejo, R., Acién, F. G., Ibanez, M. J., Fernandez, J. M., Molina, E., Alvarez-Pez, J. M. 2003. Preparative purification of B-phycoerythrin from the microalga *Porphyridium cruentum* by expanded-bed adsorption chromatography. *Journal of Chromatography B* 790(1–2): 317–325.

Bermejo, R., Felipe, M. A., Talavera, E. M., Alvarez-Pez, J. M. 2006. Expanded bed adsorption chromatography for recovery of phycocyanins from the microalga *Spirulina platensis*. *Chromatographia* 63(1–2): 59.

Bhaskar, S. U., Gopalaswamy, G., Raghu, R. 2005. A simple method for efficient extraction and purification of C-phycocyanin from *Spirulina platensis* Geitler. *Indian Journal of Experimental Biology* 43: 277–279.

Chaiklahan, R., Chirasuwan, N., Loha, V., Tia, S., Bunnag, B. 2011. Separation and purification of phycocyanin from *Spirulina* sp. using a membrane process. *Bioresource Technology* 102(14): 7159–7164.

Cohen-Bazire, G. and Bryant, D. A. 1982. Chapter 7. Phycobilisomes: Composition and Structure. In: Carr, N. G. and Whitton, B. A. (Eds.). *The Biology of Cyanobacteria*, pp. 143–190. Oxford, UK: Blackwell.

Dawes, C. 2016. Chapter 4. Macroalgae systematics. In: Levine, I. and Fleurence, J. (Eds.). *Seaweed in Health and Disease Prevention*, pp. 107–138. London, UK: E-Book, Elsevier.

Denis, C., Morançais, M., Gaudin, P., Fleurence, J. 2009. Effect of enzymatic digestion on thallus degradation and extraction of hydro soluble compounds from *Grateloupia turuturu*. *Botanica Marina* 52: 262–267.

Doke, J. M. Jr. 2005. An improved and efficient method for the extraction of phycocyanin from *Spirulina* sp. *International Journal of Food Engineering* 1(5): 2.

Duarte, A. W. F., Lopes, A. M., Molino, J. V. D., Pessoa, A., Sette, L. D. 2015. Liquid-liquid extraction of lipase produced by psychrotrophic yeast *Leucosporidium scottii* L117 using aqueous two-phase systems. *Separation and Purification Technology* 156(2): 215–225.

Dumay, J., Clément, N., Morançais, M., Fleurence, J. 2013. Optimisation of hydrolysis conditions of *Palmaria palmata* to enhance R-phycoerythrin extraction. *Biosource Technology* 131: 21–27.

Dumay, J. and Morançais, M. 2016. Chapter 9. Proteins and pigments. In: Levine, I. and Fleurence, J. (Eds.). *Seaweed in Health and Disease Prevention*, pp. 275–308. London, UK: E-book, Elsevier.

Dumay, J., Morançais, M., Munier, M., Le Guillard, C. L., Fleurence, J. 2014. Chapter 11. Phycoerythrins: Valuable proteinic pigments in red seaweeds. In: Jacquot, J. P., Gadal, P., and Bourgougnon, N. (Eds.). *Advances in Botanical Research*, Vol. 71, pp. 321–343.

Dumay, J., Morançais, M., Nguyen, H. P. T., Fleurence, J. 2015. Chapter 5. Extraction and purification of R-phycoerythrin from marine red algae. In: Stengel, D. B. and Connan, S. (Eds.). *Natural Products from Marine Algae (NPMA). Methods in Molecular Biology*, p. 1308.

Engelmann, T. W. 1883. Bacterium photometricum. *Archive for the Whole Physiology of Humans and Animals* 30(1): 95–124.

Fisheries and Aquaculture Statistics (FAO). 2014. *FAO yearbook*. Rome, Italy: Food and Agriculture Organisation of the United Nations.

Fleurence, J. 2003. R-phycoerythrin from red macroalgae: Strategies for extraction and potential application in biotechnology. *Applied Biotechnology, Food Science and Policy* 1(1): 1–6.

Fleurence, J., Massiani, L., Guyader, O., Mabeau, S. 1995. Use of enzymatic cell wall degradation for improvement of protein extraction from *Chondrus crispus*, *Gracilaria verrucosa*, and *Palmaria palmata*. *Journal of Applied Phycology* 7: 393.

Galland-Irmouli, A. V., Pons, L., Luçon, M. et al. 2000. One-step purification of R-phycoerythrin from the red macroalga Palmaria palmata using preparative polyacrylamide gel electrophoresis. *Journal of Chromatography B: Biomedical Sciences and Applications* 739(1): 117–123.

Glazer, A. N. 1984. A Macromolecular complex optimised for light energy transfer. *Biochimica et Biophysica Acta* 768: 29–51.

Glazer, A. N., Apell, G. S., Hixson, C. S., Bryant, D. A., Rimon, S., Brown, D. M. 1976. Biliproteins of cyanobacteria and rhodophyta: Homologous family of photosynthetic accessory pigments. *Proceedings of the National Academy of Science* 73: 428–431.

Glazer, A. N. and Hixson, C. S. 1975. Characterisation of R-phycocyanin. Chromophore content of R-phycocyanin and C-phycoerythrin. *Journal of Biological Chemistry* 250: 5487–5495.

Glazer, A. N. and Hixson, C. S. 1977. Subunit structure and chromophore composition of rhodophytan phycoerythrins. *Porphyridium cruentum* B-phycoerythrin and b-phycoerythrin. *Journal of Biological Chemistry* 252(1): 32–42.

Goja, A. M., Tang, H., Cui, M., Li, C. 2013. Aqueous two-phase extraction advances for bioseparation. *Journal of Bioprocessing and Biotechnology* 4: 1–8.

Gupta, A. and Sainis, J. K. 2010. Isolation of C-phycocyanin from *Synechococcus* sp. *Journal of Applied Phycology* 22: 231–233.

Harnedy, P. A. and Fitzgerald, R. J. 2013. Extraction of protein from the macroalga *Palmaria palmata*. *LWT-Food Science and Technology* 51(1): 375–382.

Hatti-Kaul, R. 2001. Aqueous two-phase systems. *Molecular Biotechnology* 19: 269–277.

Hilditch, C. M., Balding, P., Jenkins, R., Smith, A. J., Rogers, L. J. 1991. R-phycoerythrin from the macroalga *Corallina officinais* (Rhodophyceae) and application of a phycofluor probe for detecting sugar-binding sites on cell membranes. *Journal of Phycology* 3(4): 345–354.

Hjorth, R. 1999. Expanded bed adsorption: Elution in expanded bed mode. *Bioseparation* 8: 1–9.

Isailovic, D., Li, H. W., Yeung, E. S. 2004. Isolation and characterisation of R-phycoerythrin subunits and enzymatic digests. *Journal of Chromatography* 1051: 119–130.

Isailovic, D., Sultana, I., Phillips, G. J., Yeung, E. S. 2006. Formation of fluorescent proteins by the attachment of phycoerythrobilin to R-phycoerythrin alpha and beta apo-subunits. *Analytical Biochemistry* 358: 38–50.

Johansson, H. O., Karlstrom, G., Tjerneld, F., Haynes, C. A. 1998. Driving forces for phase separation and partitioning in aqueous two-phase systems. *Journal of Chromatography B* 711: 3–17.

Kloareg, B. and Quatrano, R. S. 1988. Structure of the cell walls of marine algae and ecophysical functions of the matric polysaccharides. *Oceanography and Marine Biology—An Annual Review* 26: 259–315.

Kula, M. R., Kroner, K. H., Hustedt, H. 1982. Purification of enzymes by liquid-liquid extraction. *Advances in Biochemical Engineering* 24: 73–118.

Le Guillard, C., Berge, J. P., Donnay-Moreno, C. et al. 2016. Soft liquefaction of the red seaweed *Grateloupia turuturu* Yamada by ultrasound-assisted enzymatic hydrolysis process. *Journal of Applied Phycology* 28(4): 2575–2585.

Le Guillard, C., Dumay, J., Donnay-Moreno, C. et al. 2015. Ultrasound-assisted extraction of R-phycoerythrin from *Grateloupia turuturu* with and without enzyme addition. *Algal Research* 12: 522–528.

Leema, J. T. M., Kirubagaran, R., Vinithkumar, N. V., Dheenan, P. S., Karthikayulu, S. 2010. High value pigment production from *Arthrospira (Spirulina) platensis* cultured in seawater. *Bioresource Technology* 101(23): 9221–9227.

Manirafasha, E., Ndikubwimana, T., Zeng, X., Lu, Y., Jing, K. 2016. Phycobiliprotein: Potential microalgae derived pharmaceutical and biological reagent. *Biochemical Engineering Journal* 109: 282–296.

Medeiros, F. O., Alves, F. G., Lisboa, C. R., Martins, D. S., Burkert, C. A. V. K., Kalil, S. J. 2008. Ondas ultrasonicas e perolas de vidro: Um novo metodo de extracao de B-galactosidase para uso em laboratorio. *Quimica Nova, Brazil* 31(2): 336.

Mensi, F., Ksouri, J., Seale, E., Romdhane, M. S. 2014. Optimising R-phycoerythrin purification from the red algae *Gracilaria verrucosa* in polymer-salt aqueous two-phase system response surface methodology. *Bulletin de l'institut National des Sciences et Technologies de la Mer* 41: 63–74.

Moraes, C. C., Sala, L., Cerveira, G. P., Kalil, S. J. 2011. C-phycocyanin extraction from *Spirulina platensis* wet biomass. *Brazilian Journal of Chemical Engineering* 28: 1.

Munier, M., Morançais, M., Dumay, J., Jaouen, P. 2015. One-step purification of R-phycoerythrin from the red edible seaweed *Grateloupia turuturu*. *Journal of Chromatography* 992: 23–29.

Nguyen, H. P. T., Morançais, M., Fleurence, J., Dumay, J. 2016. *Mastocarpus stellatus* as a source of R-phycoerythrin: Optimisation of enzyme assisted extraction using response surface methodology. *Journal of Applied Phycology* 29(3): 1563–1570.

Niu, J. F., Wang, G. C., Lin, X. Z., Zhou, B. C. 2007. Large-scale recovery of C-phycocyanin from *Spirulina platensis* using expanded bed adsorption chromatography. *Journal of Chromatography B* 850: 267–276.

Niu, J. F., Wang, G. C., Tseng, C. K. 2006. Method for large-scale isolation and purification of R-phycoerythrin from red alga *Polysiphonia urceolata*. *Protein Expression and Purification* 49: 23–31.

Oliveira, E. G., Rosa, G. S., Moraes, M. A., Pinto, L. A. A. 2009. Characterisation of thin layer drying of *Spirulina platensis* utilising perpendicular air flow. *Bioresource Technology* 100: 1297–1303.

Patel, A., Mishra, S., Pawar, R., Ghosh, P. K. 2005. Purification and characterisation of C-phycocyanin from cyanobacterial species of marine and freshwater habitat. *Protein Expression and Purification* 40(2): 248–255.

Patil, G. and Raghavaroa, K. S. M. S. 2007. Aqueous two phase extraction for purification of C-phycocyanin. *Biochemical Engineering Journal* 34(2): 156–164.

Prabakaran, P. and Ravindran, A. D. 2013. Efficacy of different extraction methods of phycocyanin from *Spirulina platensis*. *International Journal of Research in Pharmacy and Life Science* 1(1): 15–20.

Ratanapongleka, R. 2010. Recovery of biological products in aqueous two phase systems. *International Journal of Chemical Engineering and Applications* 1: 191–198.

Reise, A., Mendes, A., Lobo-Fernandes, L., Empis, J. A., Novais, J. M. 1998. Production, extraction and purification of phycobiliproteins from *Nostoc* sp. *Bioresource Technology* 66: 181–187.

Rito-Palomares, M. 2004. Practical application of aqueous two-phase partition to process development for the recovery of biological products. *Journal of Chromatography B* 807(1): 3–11.

Rito-Palomares, M., Nunez, L., Amador, D. 2001. Practical application of aqueous two-phase systems for the development of a prototype process for C-phycocyanin recovery from *Spirulina maxima*. *Journal of Chemical Technology and Biotechnology* 76: 1273–1280.

Sarada, R., Pillai, M. G., Ravishankar, G. A. 1999. Phycocyanin from *Spirulina* sp. Influence of processing of biomass on phycocyanin yield, analysis of efficacy of extraction methods and stability studies on phycocyanin. *Process Biochemistry* 34: 795–801.

Silveira, S. T., Burkert, J. F. M., Costa, J. A. V., Burkert, C. A. V., Kalil, S. J. 2007. Optimisation of phycocyanin extraction from *Spirulina platensis* using factorial design. *Bioresource Technology* 98(8): 1629.

Su, C. H., Liu, C. S., Yang, P. C., Syu, K. S., Chiuh, C. C. 2014. Solid-liquid extraction of phycocyanin from *Spriulina platensis*: Kinetic modelling of influential factors. *Separation and Purification Technology* 123: 64–68.

Sun, L., Wang, S., Gong, X., Zhao, M., Fu, X., Wang, L. 2009. Isolation, purification and characteristics of R-phycoerythrin from a marine macroalga *Heterosiphonia japonica*. *Protein Expression and Purification* 64: 146–154.
Wang, G. 2002. Isolation and purification of phycoerythrin from red alga *Gracilaria verrucosa* by expanded-bed-adsorption and ion-exchange chromatography. *Chromatographia* 56(7–8): 1–5.
Zhang, Y. M. and Chen, F. 1999. A simple method for efficient separation and purification of C-phycocyanin and allophycocyanin from *Spirulina platensis*. *Biotechnology Techniques* 13: 601–603.
Zhu, L. P., Gan, S. G., Lv, A. J. 2015. Chapter 51. Efficient purification and active configuration investigation of R-phycocyanin from *Polysiphonia urceolata*. In: Zhang, T. C. and Nakajima, M. (Eds.). *Advances in Applied Biotechnology*, Vol. 332, pp. 489–496.

chapter seventeen

Quantification and localization of reactive oxygen species in marine macrophytes

Manoj Kumar, Loretto Contreras-Porcia, Nirali M. Kumar, and Peter J. Ralph

Contents

17.1 Introduction

Marine macrophytes (including seaweeds and seagrasses) are the assemblage of diverse groups of phototrophic marine plants and are considered marine ecological engineers. Both provide a range of ecologically and economically valuable biological services; however, they are under constant threat from a wide range of anthropogenic stressors, climate variation, invasive species, and pathogens and are often exposed to low- and high-tidal cycle on daily bases (Contreras-Porcia et al. 2011; Kumar et al. 2011a, b). The relative abundance, survivability, and distribution of marine macrophytes in such environments are principally determined by their tolerance abilities to diverse stresses. Any adverse effect on marine macrophytes because of fluctuating and stressful environmental conditions can directly or indirectly lead to perturbations at higher trophic levels and can eventually affect the integrity and sustainability of marine ecosystems. The climatic variations and anthropogenic stressors in their ecological niche trigger a series of physiological processes and accumulation of reactive oxygen species (ROS), including singlet oxygen ($^1O_2$), superoxide radicals ($O_2^{\cdot-}$), hydrogen peroxide (H_2O_2), or hydroxyl radicals ($OH^{\cdot}$).

The cell organelles such as chloroplast, mitochondria, and peroxisomes are vital organelles involved in highly oxidizing metabolic activities with intense electron flow rate and are the major source of ROS production. Furthermore, microsomes (e.g., cytochrome P450 and diamine oxidase), peroxisomes (e.g., enzymes involved in fatty acid oxidation), and some enzymes in the plasma membrane (e.g., NADPH oxidase and lipooxygenase) have been identified as a potential source of ROS generator (Kumar et al. 2011b, 2014; Wojtala et al. 2014). ROS are produced directly by excitation of O_2 and subsequent formation of singlet oxygen, or by the transfer of one, two, or three electrons to O_2 or by protonation, which results in the formation of $O_2^{\cdot-}$, H_2O_2, or $OH^{\cdot}$ radicals (Bischof and Rautenberger 2012). Furthermore, $O_2^{\cdot-}$, if protonated, can form the perhydroxyl radical ($HO_2^{\cdot}$) and can also react with reactive nitrogen species (RNS), that is, nitric oxide radical ($NO^{\cdot}$, signaling free radical) to form more toxic peroxynitrite (OON^-) radical. Under physiological conditions, both ROS and RNS can act as signal molecules to regulate cellular metabolism. They also affect the transmission of signals across cells and can be considered second messengers during growth, differentiation, and cell death (Wojtala et al. 2014). Although many ROS and RNS, while acting as a signal molecule, modify gene expression and modulate the activity of specific defense proteins (Kumar et al. 2014), high concentrations of ROS and/or RNS can disrupt the cellular redox homeostasis, and can lead to oxidative burst exacerbating cellular damages. In this context, marine macrophytes express a battery of antioxidant enzymes and nonenzymatic antioxidants such as ascorbate, glutathione, tocopherol, carotenoids, phenolics, and flavonoids, which are of immense importance and play a crucial role in controlling cellular ROS levels, redox status, and cellular repair, thereby minimizing the risk of detrimental effects resulting from ROS-mediated oxidative stress.

To attribute a cell-signaling event in response to stress, ROS estimation is essential to detect and characterize these species accurately as an indicator of oxidative stress. ROS are highly reactive and exhibit very short lifetimes that vary from nanoseconds to seconds. For example, $OH^{\cdot}$ reacts with most of the organic compounds and has a lifetime of about 10 ns. $O_2^{\cdot-}$ exhibits a low steady-state concentration of around 10^{-10} M in different cells or organelle types (Gardner 2002). The half-life of $O_2^{\cdot-}$ depends on its concentration. At a concentration of 10 μM, $O_2^{\cdot-}$ exhibits a half-life time of 0.2 ms in water, whereas at a lower concentration of 1 μM, half-life rises to 20 ms. $^1O_2$ exhibits a lifetime of 2.7 μs (Steffen-Heins and Steffens 2015).

17.2 State of the art

Several analytical approaches, including electron paramagnetic resonance (EPR, including spin trap and spin probe), chemiluminescence, fluorescence, and spectrophotometry have been used to detect ROS in land and marine plants (Contreras-Porcia et al. 2009, 2011; Kalyanaraman et al. 2012; Kumar et al. 2011a, b, 2012; Dikalov and Harrison 2014; Steffen-Heins and Steffens 2015; Strittmatter et al. 2016, and references therein). Nevertheless, it is very challenging to monitor ROS successfully in biological systems because of their significantly low concentration, the enzymatic defense systems, and the different compartments of the living cell. ROS detection methods are based on a unique biochemical reaction mechanism with merits and demerits—describing them is out of the scope of this article; however, interested readers can refer to other articles (Kalyanaraman et al. 2012; Dikalov and Harrison 2014; Winterbourn 2014; Steffen-Heins and Steffens 2015, and references therein). Here we describe the most common methods used in marine macrophytes to quantify ROS using fluorescence probes such as dihydroethidium (DHE) and dichloro-dihydrofluorescein diacetate (DCFH–DA), and their localization using stains such as nitro blue tetrazolium (NBT, for $O_2^{\cdot-}$) and 3,3′-diaminobenzidine (DAB, for H_2O_2).

17.2.1 ROS detection reaction mechanism using dihydroethidium and dichlorodihydrofluorescein diacetate

Dihydroethidium (DHE) and its mitochondrion-targeted form mito-SOX (a triphenyl-phosphonium cation conjugated to DHE) is a widely

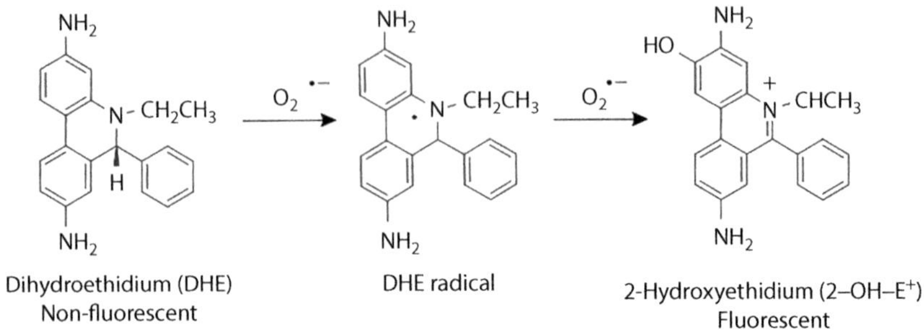

Figure 17.1 Formation of DHE superoxide-specific product 2-hydroxyethidium (2–OH–E+).

used fluorescence probe for detecting cellular and mitochondrial $O_2^{\bullet-}$ (Dikalov and Harrison 2014). The red fluorescence formed from the two-electron oxidation product 2–OH–E+ is equated to intracellular $O_2^{\bullet-}$ (Figure 17.1). However, this approach is limited, because DHE can also form ethidium (E+) and other dimers of HE (HE–HE, HE–E+, E+–E+) by nonspecific redox reactions. The fluorescent spectra of ethidium and 2–OH–E+ overlap that make it difficult to distinguish such products by using confocal microscopy or other fluorescence-based microscopic assays to measure only 2–OH–E+ accurately. Therefore, careful use of specific wavelengths of excitation might allow separation of these signals, so that confocal imaging or other fluorescence-based approaches could be used. One should take extra efforts to accurately measure the 2–OH–E+ using HPLC-fluorescence or LC–MS-based analytical methods.

DCFH–DA is commonly used for detecting intracellular H_2O_2 and oxidative stress. This probe is cell permeable, which first gets intracellularly hydrolyzed by esterase into the DCFH carboxylate anions, and subsequently its oxidation results in the formation of a fluorescent product dichlorofluorescein (Figures 17.2 and 17.3). This product can

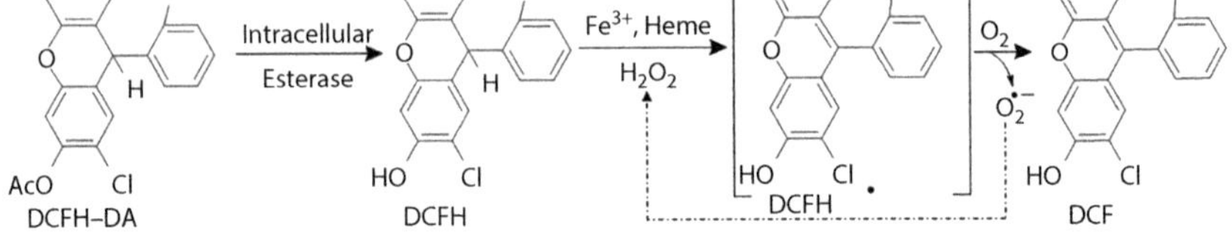

Figure 17.2 DCFH–DA oxidative reaction mechanism for H_2O_2 detection.

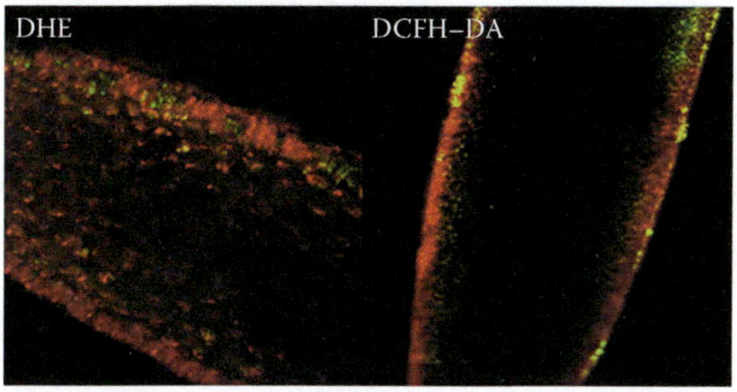

Figure 17.3 ROS burst visualized in *Lessonia spicata* under copper stress (100 μg L^{-1}) by fluorescence microscopy using DHE and DCFH–DA probes. (Adapted from Contreras-Porcia, L. et al., *Aquat. Toxicol.*, 94, 94–102, 2009.)

be monitored by either fluorescent-based analytical methods, microscopy, or by flow cytometry. Despite the immense popularity of these methodologies, it cannot be reliably used to solely measure the intracellular H_2O_2 in a cell. It is highly essential to keep in mind the following limitations when using DCFH–DA probe for accurate interpretation of data: (1) DCFH does not directly react with H_2O_2; (2) several one-electron oxidizing species (including $OH^{\cdot}$, HOCl, and several ROS) oxidize DCFH to DCF; (3) DCF can actually produce H_2O_2 through the reaction of DCF radical with the oxygen, leading to artifactual amplification of fluorescence signal intensity; (4) redox active metals, cytochrome c, and heme peroxidases can also catalyze DCFH oxidation (Kalyanaraman et al. 2012).

17.2.2 *Histochemical localization mechanism using nitro blue tetrazolium and DAB staining*

Cell type-specific ROS have been detected using histochemical approaches. Localization and quantification of H_2O_2 using DAB in different cell compartments are possible, wherein DAB reacts with H_2O_2 in a peroxidase-catalyzed reaction resulting in an oxidized insoluble brown precipitate. For localization of $O_2^{\cdot-}$ in a plant tissue, the nitrosubstituted aromatic compound NBT is useful, wherein NBT specifically reacts with superoxide ions, and oxidized NBT forms a purple/blue formazan precipitate (Figure 17.4). Such techniques have recently been implemented in ROS localization in seaweeds and seagrasses (Figure 17.5) (Kumar et al. 2011a, b, 2012; Strittmatter et al. 2016).

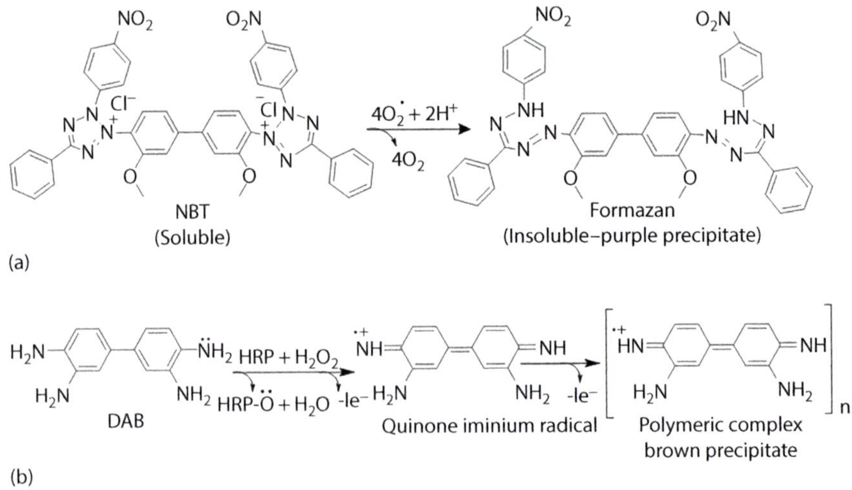

Figure 17.4 (a) NBT and (b) DAB oxidative reaction mechanism for ROS detection.

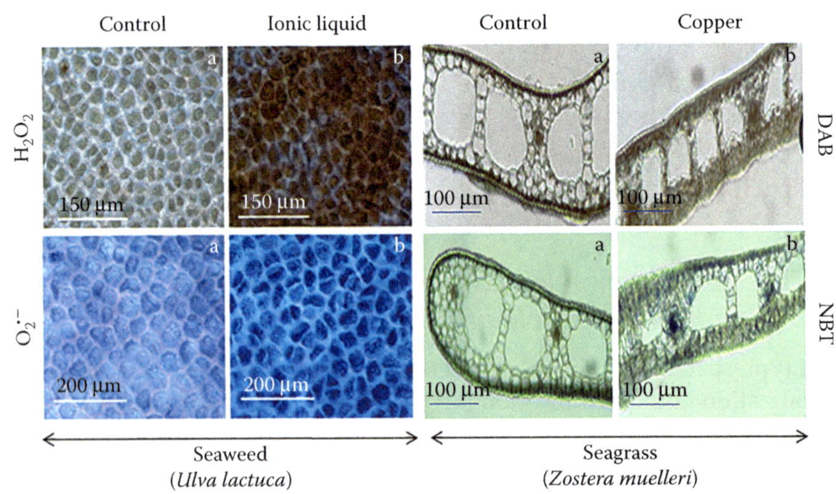

Figure 17.5 Histochemical localization of ROS in seaweeds and seagrasses exposed to toxic ionic liquid [C12mim]Br (0.14 mM for four days) and copper (250 mM; seven days). (From Kumar, M. et al., *Chem. Res. Toxicol.*, 24, 1882–1890, 2011b.)

17.3 Material

17.3.1 Reactive oxygen species visualization by microscopy

- 0.2 μm-filtered natural seawater (NSW)
- 10 μM DCFH–DA (Calbiochem, UK) prepared in filtered seawater
- 100 μM DHE (Molecular Probes, Invitrogen, Australia) prepared in filtered seawater
- Fluorescence microscopy coupled with camera
- Razor blade
- Glass slide
- Cover slips

17.3.2 Reactive oxygen species quantification using dihydroethidium and dichlorodihydrofluorescein diacetate

- 10 μM DCFH–DA (Calbiochem, UK) prepared in filtered seawater
- DCF standard curve prepared in filtered seawater
- 100 μM DHE (Molecular Probes, Invitrogen, Australia) prepared in filtered seawater
- 40 mM Tris–HCl buffer pH 7.0 (for DCF determination) and 7.5 (for 2–OH–E^+ determination)
- Liquid nitrogen
- Amber glasses or bottles (50–100 mL)
- Porcelain mortar and pestle
- Spectrofluorometer
- Room-temperature centrifuge
- Agitator plate

17.3.3 ROS histochemical localization

- 0.2 μm-filtered NSW
- Vacuum infiltration unit (desiccator connected to vacuum pump)
- Glass Petri dishes (90 × 30 mm)
- Nitroblue tetrazolium chloride (Sigma-Aldrich)
  - Prepare 0.1% (w/v) NBT solution in NSW (filtered and autoclaved) in an amber bottle
- Diaminobenzidine tetrahydrochloride (Sigma-Aldrich)
  - Prepare 0.1% (w/v) DAB solution (pH 3.8) in NSW in an amber bottle
- Water bath
- Absolute ethanol

- 50% glycerol
- Razor blade
- Glass slide
- Coverslips
- Stereomicroscope coupled with camera

17.4 Experimental procedures

17.4.1 Reactive oxygen species visualization by microscopy

1. Cut by hand the seaweed thalli or seagrass leaves from control and treatments samples.
2. Incubate the samples at room temperature (20°C–25°C) with DCFH–DA or DHE solution for 30 min or 1 h under gentle shaking.
3. Rinse the tissue in filtered sea waters four times.
4. Blot-dry the tissue gently.
5. Put the samples quickly on glass slip.
6. Select an excitation wavelength of 480–488 nm and an emission wavelength of 525–590 nm.

17.4.2 ROS quantification using DHE and DCFH–DA

1. Incubate 1–2 g fresh weight of the samples at room temperature (20°C–25°C) in DCFH–DA or DHE solution for 30 min or 1 h under gentle shake.
2. Rinse the tissue at room temperature in filtered seawaters four times, and blot-dry the tissue gently.
3. Freeze the tissue in liquid nitrogen until reaching a fine powder using porcelain mortar and pestle.
4. Homogenize the tissue in 5–10 mL of Tris–HCL.
5. Centrifugate at 20,000 g for 10–15 min, recover the supernatant, and put immediately on ice.
6. The fluorescence determination of the extract must be carried out immediately, using an excitation wavelength of 480–488 nm and an emission wavelength of 525–590 nm.
7. The ROS values will be expressed as nmol DCF per gram of fresh tissue and nmol 2–OH–E^+ per gram of fresh tissue.

17.4.3 Histochemical localization for reactive oxygen species detection

1. Excise the seaweed thalli or seagrass leaves from control and treatments samples. For example, thalli of seaweed (*Ulva lactuca*) exposed to ionic liquid ([C12mim]Br; 0.14 mM) for four days and leaves of

seagrass *Zostera muelleri* exposed to copper (250 mM) for seven days were used for ROS histochemical localization.

2. Cut the thalli/leaves using a razor blade in 1 × 1 cm size and immerse in 25 mL of 0.1% NBT or 0.1% DAB solution. For negative control, supplement the NBT solution with 10 mM sodium azide.
3. Vacuum infiltrates the leaves by building a vacuum (~50–60 kPa for 2 min). Release the vacuum gently and repeat the procedure two to three times until the thalli/leaves are completely infiltrated.
4. Incubate the infiltrated samples at room temperature for 1–2 h under room light.
5. After incubation, drain off the staining solution and transfer the thalli/leaves samples in a Falcon tube (50 mL) containing 30 mL of absolute ethanol for chlorophyll removal.
6. Remove the chlorophyll for proper visualization of the stain by series of ethanol washes while heating the samples in a boiling water bath for 10–15 min (or more if necessary, with intermittent shaking, add fresh ethanol in every wash).
7. Cool and transfer the samples in 50% glycerol to capture the images using a stereomicroscope by keeping the samples on a slide.
8. Hand-cut sections can also be prepared from stained samples after removal of chlorophyll and can be fixed subsequently in 50% glycerol for imaging.
9. Superoxide ions react with NBT and appear as blue formazan. Hydrogen peroxide reacts with DAB and appears as a brown precipitate (Figure 17.5).

17.5 Tips and tricks

17.5.1 Reactive oxygen species visualization

- The incubation time depends on the tissue thickness. It is recommended to use thin cuts.
- The rinsing depends on the tissue.
- Both DCFH–DA and DHE solutions must be prepared freshly before use.
- The excitation and emission wavelengths must be adjusted according to the samples.

17.5.2 Reactive oxygen species quantification

- Both DCFH–DA or DHE solutions must be prepared freshly before use.
- The supernatant for quantification must be clear without rests of tissue.

- The DHE standard curve should be recorded along with the samples.
- Will be using the extinction coefficient of 2–OH–E^+ for quantification depending on the excitation wavelength.
- The excitation and emission wavelengths must be adjusted according to the samples. Use additional controls as the tissue without probes.

17.5.3 Reactive oxygen species histochemical localization

- For the negative control, the NBT solution will be supplemented with 10 mM sodium azide before the infiltration.
- NBT solution can also be prepared in 50 mM potassium phosphate buffer (pH 6.5–7.5). However, it is preferred to use NSW for marine samples.
- A low pH of 3.8 is necessary for proper solubilization of DAB. Set the pH by using 0.1 N HCl.
- Both NBT and DAB staining solutions must be prepared freshly before use.
- It is important to release the vacuum gently during vacuum infiltration to enable better infiltration of leaves.
- Perform the step of chlorophyll removal by boiling the mixture in a fume hood to prevent the spread of bad odor.

Acknowledgment

Author MK is grateful to the Australian Research Council for awarding him Discovery Early Career Research Award (DECRA Fellowship, DE150100461-2015). Author LC-P is grateful to FONDECYT 1170881 and DI-1245-16/R (Universidad Andrés Bello).

References

Bischof, K. and Rautenberger, R. 2012. Seaweed responses to environmental stress: Reactive oxygen and antioxidative strategies. In C. Wiencke, and K. Bischof (Eds.), *Seaweed Biology. Novel Insights into Ecophysiology, Ecology and Utilization Ecological Studies* (Vol. 219, pp. 109–132). Heidelberg, Germany: Springer.

Contreras-Porcia, L., Mella, D., Moenne, A., and Correa, J.A. 2009. Differential responses to copper-induced oxidative stress in the marine macroalga *Lessonia nigrescens* and *Scytosiphon lomentaria* (Phaeophyceae). *Aquat. Toxicol.* 94: 94–102.

Contreras-Porcia, L., Thomas, D., Flores, V., and Correa, J.A. 2011. Tolerance to oxidative stress induced by desiccation in *Porphyra columbina* (Bangiales, Rhodophyta). *J. Exp. Bot.* 62: 1815–1829.

Dikalov, S.I. and Harrison, D.G. 2014. Methods for detection of mitochondrial and cellular reactive oxygen species. *Antioxid. Redox Signal.* 20: 372–382.

Gardner, P.R. 2002. Aconitase: Sensitive target and measure of superoxide. *Meth. Enzymol.* 349: 9–23.

Kalyanaraman, B., Darley-Usmar, V., Davies, K.J., Dennery, P.A., Forman, H.J., Grisham, M.B., Mann, G.E., Moore, K., Roberts, L.J., and Ischiropoulos, H. 2012. Measuring reactive oxygen and nitrogen species with fluorescent probes: challenges and limitations. *Free Radic. Biol. Med.* 52(1): 1–6.

Kumar, M., Bijo, A.J., Baghel, R.S., Reddy, C.R.K., and Jha, B. 2012. Selenium and spermine alleviate cadmium induced toxicity in the red seaweed *Gracilaria dura* by regulating antioxidants and DNA methylation. *Plant Physiol. Biochem.* 51: 129–138.

Kumar, M., Gupta, V., Trivedi, N., Kumari, P., Bijo, A.J., Reddy, C.R.K., and Jha, B. 2011a. Desiccation induced oxidative stress and its biochemical responses in intertidal red alga *Gracilaria corticata* (Gracilariales, Rhodophyta). *Environ. Exp. Bot.* 72:194–201.

Kumar, M., Kumari, P., Reddy, C.R.K., and Jha, B. 2014. Salinity and desiccation induced oxidative stress acclimation in seaweeds. In N. Bourgougnon (Ed.), *Sea Plants* (Vol. 71, pp. 91–123). San Diego, CA: Academic Press Publisher, Elsevier.

Kumar, M., Trivedi, N., Reddy, C.R.K., and Jha, B. 2011b. Toxic effects of imidazolium ionic liquids on the green seaweed *Ulva lactuca*: Oxidative stress and DNA damage. *Chem. Res. Toxicol.* 24(11): 1882–1890.

Steffen-Heins, A. and Steffens, B. 2015. EPR spectroscopy and its use in planta—a promising technique to disentangle the origin of specific ROS. *Front. Environ. Sci.* 3: 15.

Strittmatter, M., Grenville-Briggs, L.J., Breithut, L., Van-West, P., Gachon, C.M.M., and Küpper, F.C. 2016. Infection of the brown alga *Ectocarpus siliculosus* by the oomycete *Eurychasma dicksonii* induces oxidative stress and halogen metabolism. *Plant Cell Environ.* 39(2): 259–271.

Winterbourn, C.C. 2014. The challenges of using fluorescent probes to detect and quantify specific reactive oxygen species in living cells. *Biochim. Biophys. Acta: General Subjects* 1840(2): 730–738.

Wojtala, A., Bonora, M., Malinska, D., Pinton P., Duszynski, J., and Wieckowski, M.R. 2014. Methods to monitor ROS production by fluorescence microscopy and fluorometry. *Meth. Enzymol.* 542: 243–262.

chapter eighteen

Metabolomics of intra- and extracellular metabolites from micro- and macroalgae using GC–MS and LC–MS

Constanze Kuhlisch, Gianmaria Califano, Thomas Wichard, and Georg Pohnert

Contents

18.1 Introduction

Metabolomics techniques aim to comprehensively extract, quantify, and evaluate metabolites from a given organism or community and have developed into an indispensable tool in life sciences (Allwood et al. 2011; Aldridge and Rhee 2014) (for more information, refer to a recent special issue in *Current Opinion in Chemical Biology*) (Schroeder and Pohnert 2017). Different approaches have been brought forward that allow us to answer a multitude of questions about the physiology of an organism and to generate new hypotheses about its response to stress (Goulitquer et al. 2012; Krug and Müller 2014). Among others, metabolomics allows us to map changes in primary metabolism that reflect the regulation of biochemical pathways to categorize samples by using metabolic fingerprinting techniques or to identify metabolites that are regulated in stress or interaction situations by comparative metabolic profiling. Especially, comparative metabolomics is suitable for the generation of hypotheses about the role of primary and secondary metabolites (Lee and Fiehn 2008; Kuhlisch and Pohnert 2015). Most commonly, mass spectrometric techniques are used for the recording of metabolomics data. With complex samples, mass spectrometry (MS) can be coupled to efficient separation techniques such as ultra performance liquid chromatography (UPLC) and gas chromatography (GC). By using such hyphenated analytical techniques, the profiling of hundreds of metabolites within minutes is feasible. Fundamental prerequirements for successful metabolomics investigations are the experiment planning, sample preparation, extraction, data recording, and the statistical evaluation of the results. In this contribution, we describe a very robust and validated protocol for the generation of metabolic data from marine organisms. Initially developed for the investigation of the diatom *Skeletonema marinoi*, the protocol was successfully adapted to the investigation of different algal taxa (Nylund et al. 2011; Vidoudez and Pohnert 2012; Mausz and Pohnert 2015), including flagellated gametes and thalli of the green macroalga *Ulva mutabilis* and water samples for exometabolomic profiling (Alsufyani et al. 2017). Even profiling of entire plankton communities is feasible without the need for further alterations (Kuhlisch et al. submitted), and it can be predicted that only minor adaptations to the protocol are required to make it suitable for a broad range of other (marine) organisms. By introducing a few additional experimental steps, the method that was initially developed for the investigation of the endometabolome (i.e., the intracellular metabolites) can also be used to monitor the exometabolome (i.e., the metabolites released into the surrounding seawater; see Barofsky et al. 2009 and Alsufyani et al. 2017). For high-quality metabolomics, the sample preparation and handling are decisive, and we thus place here attention to fully document and describe the workflow. We also introduce one selected workflow for data analysis of comparative metabolomics datasets.

18.2 State of the art

In the past, qualitative and quantitative metabolite investigations were conducted by targeted chemical analyses of certain metabolites or metabolite classes (Gravot et al. 2010; Dittami et al. 2011). However, along with the improvement of hyphenated analytical techniques, the simultaneous and untargeted metabolic profiling of a broader range of substance classes including amino acids, organic acids, sugars, and fatty acids became feasible as reviewed in Dunn (2008). Complex biological extracts are thereby separated by, for example, GC, LC, or capillary electrophoresis, followed by their subsequent analysis using MS. Various available mass analyzers such as quadrupole, time of flight (TOF), and the Orbitrap provide a range of instruments with variable sensitivity, mass resolution, and mass accuracy. Besides LC–MS, one of the most common analytical platforms for metabolic profiling is GC–MS because of its high chromatographic resolution, sensitivity, and availability of reference libraries (Dunn 2008). Many experimental steps for GC–MS metabolic profiling were established in terrestrial plants and transferred to mammals and microbes (Fiehn 2008). Thus, the first protocols in plant sciences included plant-specific protocols for enzyme quenching by liquid nitrogen, cell homogenization, and metabolite extraction in hot methanol before the separation of polar metabolites, evaporation for derivatization, GC–MS analysis, data processing, and statistical analysis. A two-step derivatization with optimized methoxylation conditions and *N*-methyl-*N*-trimethylsilyl-trifluoroacetamide (MSTFA) as silylation reagent was introduced, and DB-5ms columns were recommended for separation (Fiehn et al. 2000a, b; Roessner et al. 2000). Isotopically labeled primary metabolites (Fiehn et al. 2000a) and the polyol ribitol (Roessner et al. 2000) were introduced as internal standards (ISs) for semiquantitative metabolite analysis. In the following, several steps of the initial protocols were improved (Lisec et al. 2006) or adjusted to the needs of other study systems (Winder et al. 2008), especially for particularly error-prone steps such as metabolic quenching (Álvarez-Sánchez et al. 2010), or cell extraction, for which a one-phase-solvent mixture was proposed by Gullberg et al. (2004), who also sorted out the effect of oximation time and temperature during derivatization. Later, GC–TOF–MS systems came in focus, which offer faster scan times compared with quadrupole-MS (Wagner et al. 2003). Furthermore, data processing and analysis strategies were evaluated, for example, handling of multiple derivatization products (Kanani and Klapa 2007), appropriate data-scaling methods (van den Berg et al. 2006), or the choice of uni- and multivariate statistical analyses (Saccenti et al. 2014).

Within the field of marine sciences, metabolomics approaches are still at their advent (Goulitquer et al. 2012). Metabolic profiling of microalgae was first established for the freshwater microalga *Chlamydomonas reinhardtii* (Bölling and Fiehn 2005), followed by the introduction of protocols for the investigation of the marine diatoms *Phaeodactylum tricornutum* (Allen et al. 2008), *Skeletonema marinoi* (Vidoudez and Pohnert 2012), and *Thalassiosira pseudonana* (Bromke et al. 2013). With *Gracilaria vermiculophylla* and *Gracilaria chilensis* (Nylund et al. 2011; Weinberger et al. 2011), the first marine macroalgae were profiled followed by the brown algae *Ectocarpus siliculosus*, *Laminaria digitata*, and *Lessonia spicata* (Ritter et al. 2014; Ritter et al. 2017). Some efforts have already been made to adjust experimental procedures to the needs of macroalgae. Thus, the robust cell wall structures were extracted after flash freezing in liquid nitrogen and thorough grinding with mortar and pestle. Prior to this, epibionts were removed gently if thalli were collected in the field. A challenge that has not been addressed thus far is the high morphological and chemical diversity of the diverse macroalgal taxa and life-cycle stages.

In contrast to the profiling of endometabolites of marine algae, rather few investigations exist, which cover exometabolites (Barofsky et al. 2009; Gillard et al. 2013; Becker et al. 2014; Longnecker et al. 2015). The methodology in this field is currently developing as reviewed in Minor et al. (2014). Main challenges are the separation of cells and surrounding medium without cell leakage and the extraction of dissolved metabolites. Thus, cells should be removed, for example, by gentle filtration over sand or GF/C filters (Barofsky et al. 2009; Alsufyani et al. 2017). The most common extraction method for dissolved metabolites in seawater is solid-phase extraction (SPE) as it is fast, simple, and removes the salt load of marine samples that would otherwise interfere with subsequent analytical processes. Metabolite coverage depends both on the choice of the SPE cartridge adsorbent and eluting solvents. Styrene divinylbenzene phases showed highest recoveries thus far, and we propose CHROMABOND® Easy cartridges due to their high recovery of m/z-retention time pairs (Barofsky et al. 2009).

In the following protocol, we focus on sample generation for GC–MS analysis in addition to LC–MS and give a brief overview of processes in data treatment. Different sampling and extraction strategies for the diverse macroalgal life stages are presented. We also introduce a protocol for profiling the exometabolome of micro- and macroalgae.

18.3 Materials

18.3.1 Solvents

- Acetone (certified ACS, Fisher Chemical)
- Chloroform (for HPLC, HiPerSolv CHROMANORM® [≥99.8%, filtered through 0.2 μm, packed under N_2, stabilized with 0.6% ethanol], VWR)
- Ethanol (gradient grade for LC, LiChrosolv®; ≥99.9%, filtered through 0.2 μm, Merck)
- Hexane (for GC, SupraSolv® [≥98%], Merck; stored over 4 Å molecular sieve)
- Methanol (for HPLC, Chromasolv®Plus [≥99.9%], Sigma-Aldrich)
- Pyridine (for HPLC, Chromasolv®Plus [≥99.9%], Sigma-Aldrich; stored over 4 Å molecular sieve under argon)
- Tetrahydrofuran (THF; for HPLC, HiPerSolv CHROMANORM® [≥99.7%, filtered through 0.2 μm, packed under N_2, not stabilized], VWR; stored under argon)
- Water (for HPLC, Chromasolv®Plus [filtered through 0.2 μm], Sigma-Aldrich)
- Extraction solution (see Table 18.1 in Section 18.3.4)
- Column elution solution (see Table 18.2 in Section 18.3.4)
- Internal standard (IS) solution (see Table 18.3 in Section 18.3.4)
- Methoxyamine hydrochloride (98%, Sigma-Aldrich; hygroscopic and thus stored in a desiccator under argon and vacuum-dried before use)
- Retention time index (RI) solution (see Table 18.4 in Section 18.3.4)
- MSTFA (1 mL vials, Macherey–Nagel)

18.3.2 Equipment

- Fume hood
- Lab coat
- Safety goggles
- Chemical-resistant gloves (nitril or latex, dependent on solvent)
- Eppendorf® pipettes (1000, 100, and 10 μL, recently checked and calibrated)

- Filtration pump system with vacuum control and Nalgene® tubing
- DURAN® filtering flasks (1 L, KECK™ assembly, Erlenmeyer shape) with rubber seals (Figure 18.1)
- DURAN® dismountable filter holders (ø 50 mm, 250 mL top manifold, sintered filter discs [Porosity 4], FKM seals, PP outlet funnel) (Figure 18.1a)
- Tweezers
- 25 mL glass beaker
- TissueLyser II (QIAGEN, max. speed: 30 frequencies s^{-1})
- Lyophilizer
- Polytetrafluoroethylene (PTFE) tubing (ø 0.8 mm, each ~30 cm) (Figure 18.1b)
- Vortex mixer
- Ultrasonic bath
- Centrifuge (with temperature control, up to 30,000 × g)
- Vacuum evaporation facility (desiccator) and vacuum diaphragm pump (down to 7 mbar, with vacuum control, connected to argon)
- Precision balance (±0.1 mg)

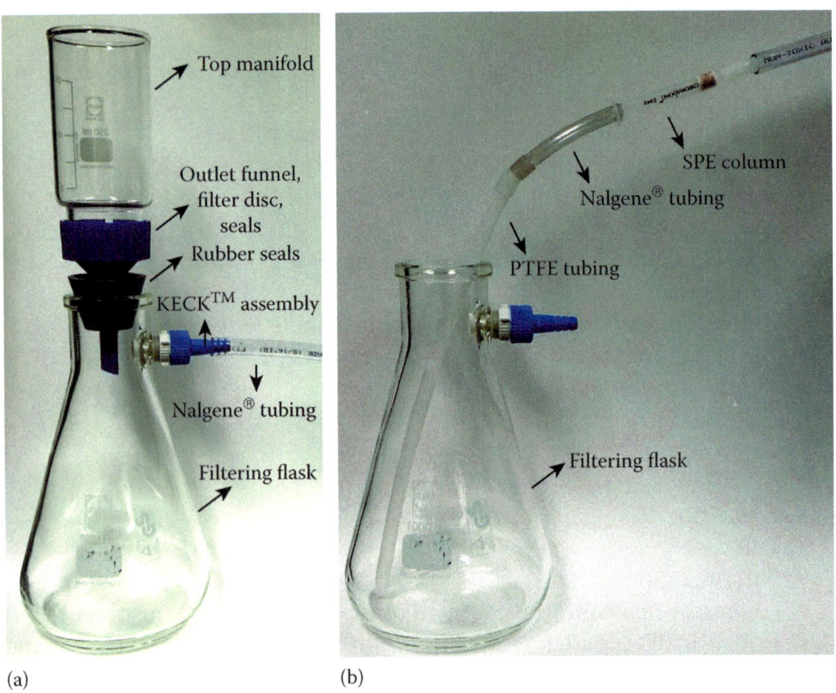

Figure 18.1 Setup of filtration system (a) and filtrate extraction system (b).

- Hamilton® glass syringes (100 µL)
- Heating block (for 1.5 mL vials)

18.3.3 Consumables

- Eppendorf® pipette tips (blue, yellow, and white)
- Eppendorf® centrifuge tubes (1.5 mL, 2 mL)
- Whatman® glass microfiber filters (ø 47 mm, grade GF/C)
- SPE columns CHROMABOND® Easy (3 mL, 200 mg, Macherey–Nagel)
- Metal beads (ø 3 mm, stainless steel)
- Centrifuge tubes (15 mL)
- Glass screw neck vials both 1.5 and 4 mL, the 1.5 mL vials have to fit GC–MS or LC–MS autosamplers
- Glass inserts for 1.5 mL vials (200 µL) with metal springs compatible with autosamplers
- Screw caps for 1.5 and 4 mL vials with septa (PTFE coated)
- Liquid nitrogen

18.3.4 Solution recipes

The following solutions have to be prepared in advance (Tables 18.1 through 18.4):

Table 18.1 Extraction solution

Composition	Final ratio (v:v:v)	Quantity to add (for 10 mL final solution)
Methanol	1	2 mL
Ethanol	3	6 mL
Chloroform	1	2 mL

Note: Prepare daily and precool at −20°C before use. During cell sampling, store on ice to ensure an ice-cold solution for extraction. The composition is optimized for diatom cells (*Skeletonema marinoi*) and was suitable for the extraction of other marine organisms but can be adjusted to ensure optimal metabolite extraction if using organisms with specific mechanical or chemical properties.

Table 18.2 Column elution solution

Composition	Final ratio (v:v:v)	Quantity to add (for 20 mL final solution)
Methanol	1	10 mL
THF	1	10 mL

Note: Store in inert containers (ideally glass sealed with PTFE caps). The composition can be adjusted to ensure optimal elution of the SPE column.

Table 18.3 Internal standard (IS) solution

Composition	Final concentration	Quantity to add (for 1 mL final solution)
Ribitol or $^{13}C_6$-sorbitol	4 mM	100 µL of stock solution at 40 mM
Water	–	900 µL

Note: Prepare a 40 mM stock solution by dissolving 24.9 mg ribitol (>99%, Sigma-Aldrich) in water. Store stock and working solution at −20°C in inert containers (ideally glass sealed with PTFE caps). Thaw before use. The polyol ribitol serves as IS in this protocol as it has not been reported from marine algae thus far. However, it has been shown in bacteria, fungi, plants, and green microalgae (Bieleski 1982). Here, isotopically labeled ISs are recommended such as $^{13}C_6$-sorbitol (Bölling and Fiehn 2005). Both the final concentration and composition of the IS can be adjusted with regard to the experiment.

Table 18.4 Retention time index (RI) solution

Composition	Final concentration	Quantity to add (for 1 mL final solution)
Decane, $C_{10}H_{22}$	1 mM	10 µL of stock solution at 100 mM
Pentadecane, $C_{15}H_{32}$	1 mM	10 µL of stock solution at 100 mM
Nonadecane, $C_{19}H_{40}$	1 mM	10 µL of stock solution at 100 mM
Docosane, $C_{22}H_{46}$	1 mM	10 µL of stock solution at 100 mM
Octacosane, $C_{28}H_{58}$	1 mM	100 µL of stock solution at 10 mM
Dotriacontane, $C_{32}H_{66}$	1 mM	100 µL of stock solution at 10 mM
Hexatriacontane, $C_{36}H_{74}$	0.5 mM	100 µL of stock solution at 5 mM
Hexane	–	660 µL

Note: Prepare stock solutions for *n*-alkanes from C_{10} to C_{36} (all >99%, Sigma-Aldrich) in hexane. Stock and working solutions are stored at −20°C and thawed before use. The RI solution is used to calculate system-independent retention times, which is, for example, necessary for comparison with externally measured reference compounds and thus metabolite identification. Final concentration and composition can be adjusted with regard to the experiment.

18.3.5 Instrumental setup

Common GC–MS (and LC–MS) instruments can be used. The described procedure can be easily adjusted to the available instrumentation.

GC–MS instrument: Agilent 6890N gas chromatograph, equipped with Agilent 7683B autosampler, coupled to Waters® Micromass GCT Premier™ mass spectrometer (orthogonal acceleration time-of-flight MS).

Injection parameters: 1 µL injection volume, split/splitless injector at 300°C, splitless to split 10 mode (can be adapted to sample and instrument properties), deactivated Agilent split liner (4 × 6.3 × 78.5 mm inner ø × outer ø × length) with glass wool.

GC parameters: Constant helium (5.0) flow at 1 mL min^{-1}, Agilent J&W DB-5ms column (30 m, 0.25 mm internal diameter, 0.25 μm film thickness, 10 m Duraguard precolumn), oven temperature program: 60°C for 1 min, ramp to 310°C at 15°C min^{-1}, 310°C for 10 min.

MS parameters: Electron impact source (70 eV) at 300°C, scan rate of 5 scans s^{-1}, dynamic range extension mode, resolution of >6000 at *m*/*z* 501.97.

18.4 Experimental procedures

In the following section, all steps from sampling to statistical analysis are described in detail (Figure 18.2). Preceding steps such as the design of the experimental setup are not in focus of this protocol. It is, however,

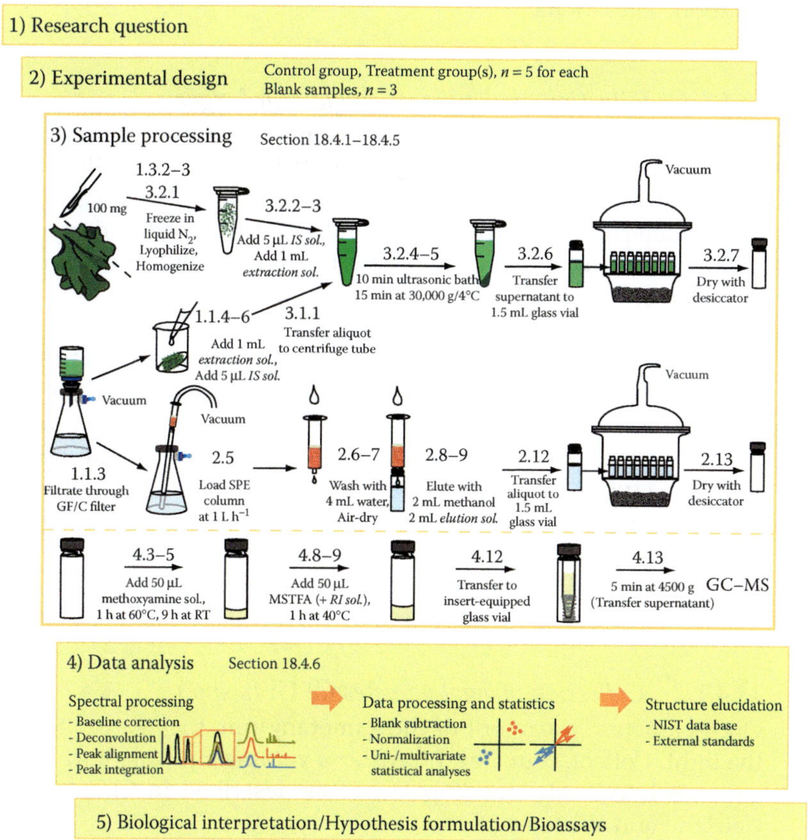

Figure 18.2 Metabolomics workflow for the analysis of intra- and extracellular metabolites of micro- and macroalgae by GC–MS. All numbers refer to steps within the protocol.

recommended to set up at least five biological replicates. Further, the sampling of blanks that should undergo identical treatment as the biological samples is essential for the later identification of contaminants.

18.4.1 *Sampling and metabolic quenching*

The presented protocol was originally developed for metabolite profiling of phytoplankton samples. Therefore, this section describes several strategies for collecting algal biomass depending on its origin, including a sampling of single-celled microalgae (Section 18.4.1.1), of single-cell stages of macroalgae (Section 18.4.1.2), and of the whole thalli (Section 18.4.1.3). As metabolites underlie diurnal fluctuations, the sampling time should be identical for all replicates. Until metabolic *quenching* (arresting the metabolic activity), all steps have to be conducted as rapidly as possible to prevent metabolic alterations.

18.4.1.1 *Filtration of planktonic single-celled algae*

1. Gently shake the culture to homogeneously distribute the cells.
2. Collect aliquots for cell counting (*Note 1*), bioassays, or collection of other metadata if needed.
3. Filter a determined culture volume (*Note 2*) under reduced pressure (~600 mbar) through a GF/C filter. The filter should not run dry. Keep the filtrate at 4°C until SPE (see Section 18.4.2.).
4. Immediately transfer the wet filter to a 25 mL glass beaker.
5. Immediately quench the metabolism by quickly resuspending the cells in 1 mL cold (–10°C) extraction solution. Therefore, rinse cells as far as possible off the filter by repeatedly pipetting the extraction solution over the filter. Transfer the suspension into a 1.5 mL centrifuge tube.
6. Add 5 μL IS solution and vortex 10 s.
7. Store the sample on ice.
8. Repeat steps 3–7 for all replicates. Continue with step 4 in Section 18.4.1.3.

18.4.1.2 *Collection of algal gametes (of* Ulva *spp.)*

1. Once gametes swarm out of the gametangium, they gather toward the light. Collect, count, and transfer a volume equivalent to 5×10^6 gametes into 2 mL centrifuge tubes as described in Chapter 9 by Califano and Wichard (2018).
2. *Optional*: Free swimming gametes can be sampled by centrifugation (*Note 3*).

3. Once the gametes are attached to the 2 mL centrifuge tube wall or are pelleted after centrifugation, remove the growth medium with a pipette, and immediately freeze the sample in liquid nitrogen. Because of the low recovery rate of entire cells from surfaces, the tube in which the gametes were grown should also be used for extraction. Continue with step 4 in Section 18.4.1.3.

18.4.1.3 Collection of algal thalli

1. Gently clean the thalli. First, scrape off epiphytes with a scalpel and then wash three times with autoclaved (filtered) seawater.
2. Collect about 100 mg fresh weight of a specific tissue (e.g., blade tissue or rhizoid, cut with a scalpel) in a 2 mL centrifuge tube and immediately freeze in liquid nitrogen.
3. Remove remaining water by lyophilization at 0.001 mbar at –50°C until completely dry.
4. *Optional*: Samples can be stored at –20°C for a few days and –80°C for several weeks.
5. Samples are ready for extraction (see Section 18.4.3).

18.4.2 Solid-phase extraction of extracellular metabolites

The extraction of extracellular metabolites from the filtrates can be done in parallel for all replicates. Therefore, randomize the samples.

1. Condition a CHROMABOND® Easy column directly before use. Therefore, pipette 4 mL methanol onto the column and let it flow by gravity into a waste vial.
2. Wash the column with 4 mL water. Let it flow by gravity into a waste vial.
3. Take out all filtrates from the fridge (step 3 in Section 18.4.1.1) and place them on ice.
4. Connect the PTFE tube in line with the column. Place the PTFE tube into the filtration flask containing the filtrate and connect the column with the vacuum system.
5. Pass the sample slowly through the column at a flow rate of ~1 L h^{-1}.
6. Disconnect the column from the PTFE tube and wash the column with 4 mL water.
7. Air-dry the column with the vacuum system. Dry column adsorbent is bright red again.
8. Elute the column by gravity with 2 mL methanol into a 4 mL glass vial.
9. Elute in the second step with 2 mL column elution solution into the same vial.

10. Add 5 μL IS solution, close the vial, and vortex for 10 s.
11. *Optional:* Samples can be stored at –20°C for a few days and at –80°C for several weeks.
12. Transfer for each sample an aliquot of 1.5 mL into a 1.5 mL glass vial. For LC–MS see *Note 4*.
13. Evaporate to dryness under vacuum using a desiccator (*Note 5*). Reduce the pressure from ambient pressure to 0 mbar considering solvent boiling points (*Note 6*).
14. Vent the desiccator slowly with dry air or argon (*Note 7*) and immediately close all vials.
15. Samples are ready for derivatization (see Section 18.4.4).

18.4.3 Extraction of intracellular metabolites

The extraction of intracellular metabolites can be carried out in parallel for all replicates. Therefore, randomize the samples. Depending on the cell-wall morphology of the studied organism, different cell disruption methods might be suitable of which two are described in this section.

Optional: Thaw samples from step 4 in Section 18.4.1.3 (or directly use samples from step 5 in Section 18.4.1.3).

18.4.3.1 Cell disruption by ultrasound treatment

1. Vortex the samples for 30 s. Transfer an aliquot equivalent to ~5×10^7 cells into a new 1.5 mL centrifuge tube (*Note 2*).
2. Add extraction solution to reach an adequate cell-to-solvent ratio that is ideally equivalent to a cell density of ~5×10^5 cells μL^{-1} solvent. Continue with step 4 in Section 18.4.3.2.

18.4.3.2 Cell disruption with a bead mill

1. Place the centrifuge tubes in a precooled (–80°C) TissueLyser II tube support. Add two metal beads per tube, close again, and disrupt the cells using the TissueLyser II for 30 s at a frequency of 30 s^{-1} (*Note 8*). During this procedure, the sample will remain frozen.
2. Add 5 μL IS solution.
3. Add 1 mL extraction solution to each sample and vortex vigorously to homogenize the sample and allow a more uniform extraction.
4. Place the samples for 10 min into an ultrasonic bath.
5. Centrifuge the samples at 30,000 × g for 15 min at 4°C.
6. Transfer the debris-free supernatants into 1.5 mL glass vials. For LC–MS analysis see *Note 4*.

7. Evaporate to dryness under vacuum using a desiccator (*Note 5*). Reduce the pressure from ambient pressure to a 0 mbar setting considering solvent boiling points (*Note 6*).
8. *Optional*: Traces of salt from the medium will precipitate as crystals that restrain water molecules. Thus, keep the pressure setting at 0 mbar for an additional hour to ensure an entirely dry sample.
9. Vent the desiccator slowly with dry air or argon (*Note 7*) and immediately close all vials.
10. Samples are ready for derivatization (see Section 18.4.4).

18.4.4 Two-step-derivatization for GC–MS analysis

To analyze a broad range of metabolites by GC–MS, the volatility and thermostability of some substance classes need to be enhanced by derivatization. In a first step, ketones and aldehydes are derivatized to oximes, and then functional groups such as –OH, $-NH_2$, –SH, or –COOH as present in, for example, sugars, fatty acids, or amino acids are chemically derivatized by silylation. Directly use the dried samples from step 15 in Section 18.4.2 or step 10 in Section 18.4.3.2.

11. Prepare the methoxyamine solution immediately before derivatization. Weigh 20 mg dried methoxyamine hydrochloride in a 1.5 mL glass vial.
12. Add 1 mL pyridine, close the vial, and ensure complete dissolution by sonication in an ultrasonic bath for at least 5 min.
13. Pipette 50 µL methoxyamine solution to each sample (a maximum of 20 samples is recommended, *Note 9*, and immediately close the vials. Vortex 60 s to redissolve the extract.
14. Incubate at 60°C for 1 h.
15. Subsequently incubate at room temperature for 9 h (up to 16 h, *Note 10*).
16. Prepare the silylation solution. Therefore, remove a new vial of MSTFA from the fridge, and let it warm up to room temperature. Thaw and vortex the RI solution.
17. Add with a glass syringe 40 µL RI solution to 1 mL MSTFA and vortex the vial. Rinse the syringe with hexane after use.
18. Add with a glass syringe 50 µL silylation solution to each sample. Prevent any cross-contamination between samples. Rinse the syringe with acetone after use.
19. Incubate at 40°C for 1 h (*Note 11*).
20. Let the samples cool down to room temperature.
21. *Optional*: In the case of condensation along the glass, briefly centrifuge the vials (~5 s).
22. Transfer each sample into a glass insert and close the vial.

23. Centrifuge all samples at 8000 × g for 5 min using 15 mL centrifuge tubes padded with a cloth as vial mounting. In the case of precipitate formation, transfer the supernatant into a new glass insert.
24. Analyze the batch of samples immediately (*Note 9*) by GC–MS (see Section 18.4.5.).

18.4.5 GC–MS analysis

1. Directly use the derivatized samples from step 24 in Section 18.4.4.
2. Analyze each batch of samples in a random order to diminish, for example, a potential systematic effect of increasing time delay between silylation and injection. For analysis details, see the *Instrumental setup.*
3. Run air injections before, in between, and after each batch to check for contaminations.
4. Use a new glass liner every 21 injections or if air injections indicate contamination.
5. For data analysis see Section 18.4.6.

18.4.6 Data analysis for GC–MS data

In the following protocol, a canonical analysis of principal coordinates (CAP, Anderson and Willis 2003) is used to investigate metabolic alterations. In comparison with other common multivariate statistical analyses, CAP is less sensitive to hidden correlations within the dataset (McCune et al. 2002), which are a direct result from metabolites generating multiple peaks, for example, due to different levels of derivatization. Alternatively, a range of online platforms is available offering different data processing and analysis strategies. These include MetaboAnalyst (Xia et al. 2009; see Figure 18.3e), XCMS Online (Tautenhahn et al. 2012; see Figure 18.3d), and Workflow4Metabolomics (Giacomoni et al. 2015). The following workflow is also applicable to LC–MS data after adaptation of the peak detection procedure (Alsufyani et al. 2017).

1. Conduct a background-noise correction for each chromatogram using, for example, the CODA tool of MassLynx 4.1 (Waters, MCQ = 0.8, Smoothing window = 5).
2. Convert the raw data files into NetCDF (.cdf), for example, by using the Databridge tool of MassLynx 4.1.
3. Deconvolute all chromatograms with AMDIS 2.71 (NIST, http://chemdata.nist.gov/, 2012) with the following parameters: minimum match factor = 30, type of analysis = simple, low/high m/z = auto,

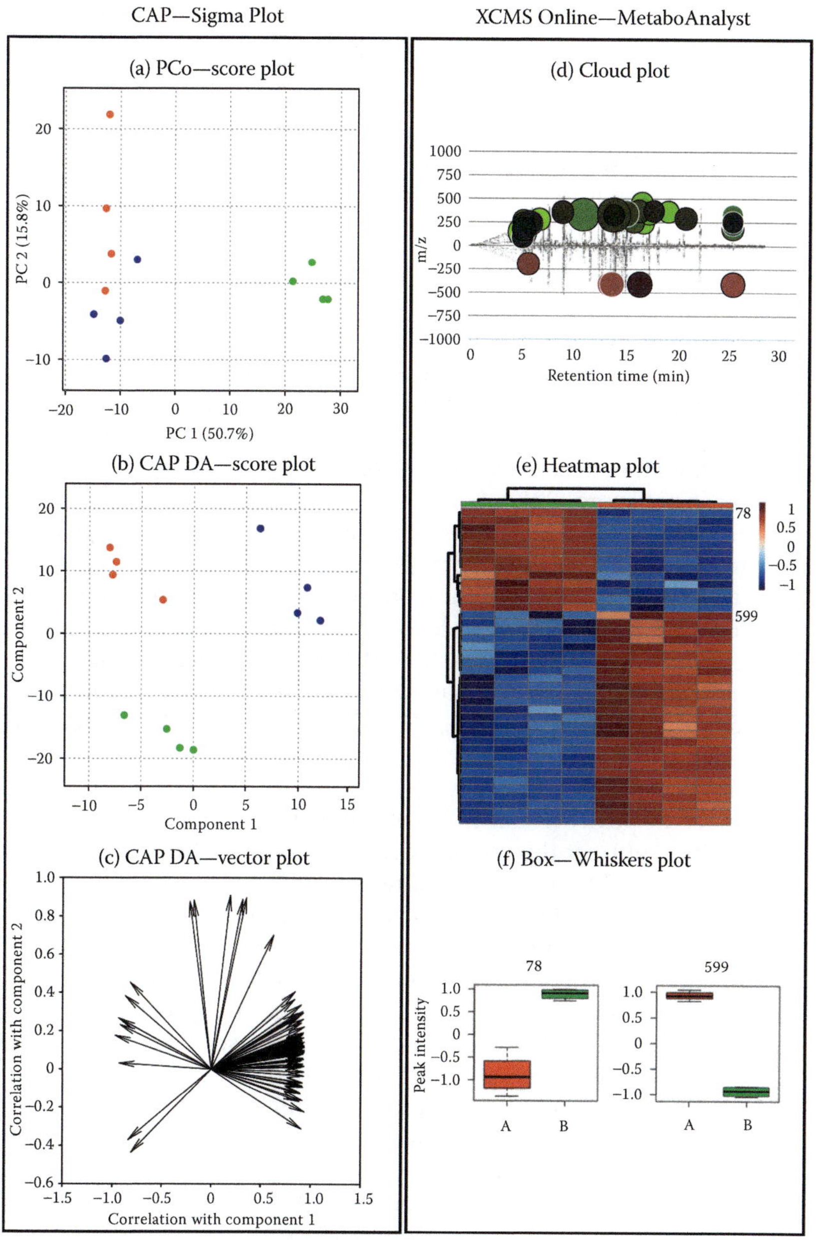

Figure 18.3 Statistical data analysis approaches for untargeted metabolomics datasets. A combination of unsupervised (a) and supervised (b, c) multivariate analyses with univariate analyses (d) is recommended. Selected metabolites of interest can be further described with heat maps (e) or box plots (f). *(Continued)*

Figure 18.3 **(Continued)** All plots were generated from a dataset that was obtained by GC–MS analysis of the macroalga *Ulva mutabilis* with either three (a–c) or two (d–f) treatment groups ($n = 4$). Score plots of (a) principal coordinate analysis and (b) CAP show the separation of sample groups. The vector plot of the CAP (c) and the *t*-test-based cloud plot (d, from XCMS Online) indicate characteristic metabolites. These selected metabolites can be visualized semiquantitatively with a heat map (e, from MetaboAnalyst) or box plots (f, from XCMS Online).

instrument type = quadrupole, component width = 32, omitted m/z = 147, 176, 193, 207, adjacent peak subtraction = 2, resolution = low, sensitivity = medium, shape requirement = low, column bleed = 207. Run one batch job each for extra- and intracellular samples.

4. Integrate the deconvoluted peaks with MET-IDEA 2.08 (http://bioinfo.noble.org, 2012). Select the chromatogram with the highest number of deconvoluted components as ion file and the following parameters for peak integration: chromatography = GC, average peak width = 0.08, minimum peak width = 0.5, maximum peak width = 2, peak start/stop slope = 1.5, adjusted retention time accuracy = 0.25, peak overload factor = 0.9, MS = TOF, mass accuracy = 0.1, mass range = 0.3, lower mass limit = 100, ion per component = 1, exclude ion list = 73, 147, 281, 341, 415. In the output file, the peak area of each variable as described by model ion (m/z) and retention time (min) is listed for each sample.
5. Import the peak area file in Excel (Microsoft® Office, 2010). For each variable, subtract the median area of all blanks from each sample. Remove variables with a resulting negative peak area.
6. Normalize the data. Divide each peak area by the peak area of the IS of the same sample. For intracellular metabolites, normalize to extracted biomass (e.g., fresh weight) as well. For extracellular metabolites, normalize to the sum of all peak areas within one sample.
7. Export the dataset as a text file for further statistical analysis. Sample and variable descriptors have to be omitted.
8. Perform a CAP with CAP12 (Anderson and Willis 2003, http://www.esapubs.org/archive/ecol/E084/011/suppl-1.htm) by using the following parameters: transformation = none, standardization = none, distance measure = Bray–Curtis dissimilarity, discriminant analysis mode, number of principal coordinates axes chosen by the program, 999 random permutations test.
9. Display the resulting sample coordinates as score plot and the metabolites as loading plot, for example, with SigmaPlot 13.0 (Systat Softwares). Evaluate the principal coordinate analysis score plot (Figure 18.3a) to detect sample outliers. Evaluate the CAP score plot

(Figure 18.3b) for correct sample groups. Use the correlation values of the original variables with the CAP axes to generate a vector plot (Figure 18.3c) and select highly correlated variables. Describe these variables by their m/z and t_R values (see Section 18.4.6, step 4).
10. Compare the mass spectra of all highly correlated variables as extracted by AMDIS (see Section 18.4.6, step 4) with spectral libraries using MS Search 2.0 (NIST, http://chemdata.nist.gov/, 2005). A common commercial database is the NIST library. Other, noncommercial alternatives are available which often focus on specific metabolites, for example, the Golm-library for plant metabolites. Document the quality of spectral comparison with the *R*-match value.
11. Compare the retention times of all highly correlated variables with externally measured reference compounds by calculating nonlinear retention indices (van den Dool and Kratz 1963). Follow Fiehn et al. (2008) for metabolomics standards initiative-compliant identification.

18.5 Notes

For a successful untargeted metabolomic approach, the personal security and sample protection against contaminants and chemical reactions with H_2O and O_2 are essential. It is thus recommended to wear personal protective equipment and work under fume hoods whenever necessary. Moreover, read the material safety data sheets (MSDS) of all reagents and follow the recommendations for correct disposal. As biological samples are rather complex, avoid any contamination by humans or lab equipment (e.g., fatty acids, plasticizers). It is thus advisable to wear gloves, use glass or PTFE whenever possible, and rinse lab equipment with solvents before use. For the same reason, blanks should be carried through the whole procedure. As the derivatization is moisture sensitive, all involved chemicals and samples have to be dry, which is especially challenging for salt water samples. Samples should be exposed to air as rarely as possible. Several metabolites are labile. Therefore, work should be carried out without longer interruptions.

Note 1: For subsequent data normalization, it is important to determine the amount of extracted biomass, for example, as cell count or fresh weight. For a quantitative comparison or differential screening, it is best to collect the same biomass among samples wherever possible.

Note 2: The sample volume should provide ~5×10^7 algal cells of a cell volume of ~100 μm^3 as tested for the diatom *Skeletonema marinoi*. With ~12 pg C $cell^{-1}$ (Menden-Deuer and Lessard 2000), this cell number equals ~0.6 mg C. For other organisms, the required cell numbers have to be determined in test runs or indirectly estimated on the basis of the amount of carbon per cell.

Note 3: Alternatively, algal cells can be sampled by centrifugation. Cold solvent quenching before centrifugation is necessary to prevent metabolic alterations during centrifugation (Bölling and Fiehn, 2005).

Note 4: At this stage, the samples can be directly used for LC–MS analysis as described in Barofsky et al. (2009) and Alsufyani et al. (2017) for extracellular extracts and in Barofsky et al. (2010) for cell extracts. Therefore, transfer an aliquot of 100 μL into insert-equipped glass vials and measure in one batch in randomized order.

Note 5: All solvents and water residues have to be evaporated before the derivatization. This can be achieved with a desiccator but also under a flow of nitrogen. Any sample contamination should, however, be prevented.

Note 6: Pressure reduction results similarly to heating in solvent boiling. In a desiccator, however, a boiling delay can occur. To avoid any loss of extract because of this uncontrolled boiling, reduce the pressure stepwise. Select the pressure steps at the boiling points of the different solvents as listed elsewhere.

Note 7: To vent the desiccator with dry air, mount a $CaCO_3$-filled column at the air inflow. Alternatively, argon can be used as shielding gas as it is denser than air and inert.

Note 8: The speed (or the frequency) of the grinding has to guarantee cell-wall disruption in a short period of time (before thawing) and has to be optimized for the respective tissue in test runs.

Note 9: The batch size is limited to maximum 20 samples because of the instability of the silylated samples. If, for example, standing in the autosampler of a GC–MS instrument at room temperature, decomposition would be substantial if more than 20 samples in a row would be measured (Kanani et al. 2008). As one GC–MS run lasts about 30 min, 20 samples can be analyzed within ~10 h. Biological replicates should be randomized within one batch.

Note 10: The oximation depends both on reaction temperature and time. To reduce the decomposition of, for example, sucrose and increase derivatization efficiency, an oximation for 1 h at 60°C and 16 h at room temperature was proposed by Gullberg et al. (2004). However, to be able to work up two batches per day, the latter time is reduced to 9 h without significant losses. With the addition of MSTFA, the oximation reaction is quenched because of silylation of the reactive amine group of methoxyamine.

Note 11: MSTFA is a strong trimethylsilyl group (TMS; $Si[CH_3]_3$) donor, and thus a common silylation reagent. During the silylation reaction, the active hydrogen of functional groups such as –OH, –NH, $-NH_2$ –SH, and –COOH are replaced by a TMS group. As each functional group has different reaction kinetics, a heating time of 1 h compromises silylation efficiency and required reaction times.

Acknowledgments

We gratefully thank Alina Hera for her help in the *Ulva mutabilis* analysis. Furthermore, we acknowledge Franziska Speck for testing protocol comprehensibility in the lab and Kathleen Thume for proofreading the manuscript. The current work was supported by the Jena School of Microbial Communication (CK, TW), by the Collaborative Research Center 1127 "Chemical Mediators in complex Biosystems" and CRC 1067 "AquaDiva" (TW, GP) and by the European Union's Horizon 2020 research and innovation program under the Marie Sklodowska-Curie grant agreement No. 642575 (GC, TW, and GP).

References

Aldridge, B. B., and K. Y. Rhee. 2014. Microbial metabolomics: Innovation, application, insight. *Curr. Opin. Microbiol.* 19:90–96.

Allen, A. E., J. Laroche, U. Maheswari et al. 2008. Whole-cell response of the pennate diatom *Phaeodactylum tricornutum* to iron starvation. *Proc. Nat. Acad. Sci.* 105 (30):10438–10443.

Allwood, J. W., R. C. De Vos, A. Moing et al. 2011. Plant metabolomics and its potential for systems biology research background concepts, technology, and methodology. *Methods Enzymol.* 500:299–336.

Alsufyani, T., A. Weiss, and T. Wichard. 2017. Time course exo-metabolomic profiling in the green marine macroalga *Ulva* (Chlorophyta) for identification of growth phase-dependent biomarkers. *Mar. Drugs* 15 (1):14. doi: 10.3390/md15010014.

Álvarez-Sánchez, B., F. Priego-Capote, and M. D. Luque de Castro. 2010. Metabolomics analysis II. Preparation of biological samples prior to detection. *TrAC, Trends Anal. Chem.* 29 (2):120–127.

Anderson, M. J., and T. J. Willis. 2003. Canonical analysis of principal coordinates: A useful method of constrained ordination for ecology. *Ecology* 84 (2): 511–525.

Barofsky, A., P. Simonelli, C. Vidoudez et al. 2010. Growth phase of the diatom *Skeletonema marinoi* influences the metabolic profile of the cells and the selective feeding of the copepod *Calanus* spp. *J. Plankton Res.* 32 (3):263–272.

Barofsky, A., C. Vidoudez, and G. Pohnert. 2009. Metabolic profiling reveals growth stage variability in diatom exudates. *Limnol. Oceanogr.: Methods* 7:382–390.

Becker, J., P. Berube, C. Follett et al. 2014. Closely related phytoplankton species produce similar suites of dissolved organic matter. *Front. Microbiol.* 5 (111). doi: 10.3389/fmicb.2014.00111.

Bieleski, R. L. 1982. Sugar alcohols. In *Plant Carbohydrates I: Intracellular Carbohydrates,* (Eds.) F. A. Loewus and W. Tanner, pp. 158–192. Berlin, Germany: Springer.

Bölling, C., and O. Fiehn. 2005. Metabolite profiling of *Chlamydomonas reinhardtii* under nutrient deprivation. *Plant Physiol.* 139 (4):1995–2005.

Bromke, M. A., P. Giavalisco, L. Willmitzer, and H. Hesse. 2013. Metabolic analysis of adaptation to short-term changes in culture conditions of the marine diatom *Thalassiosira pseudonana*. *PLoS One* 8 (6):e67340.

Califano, G., and T. Wichard. 2018. Preparation of axenic cultures in *Ulva* (Chlorophyta). In *Protocols for Macroalgae Research*, (Eds.) B. Charrier, T. Wichard, and C.R.K. Reddy, Chapter 9. Boca Raton, FL: CRC Press, Taylor & Francis Group.

Dittami, S. M., A. Gravot, D. Renault et al. 2011. Integrative analysis of metabolite and transcript abundance during the short-term response to saline and oxidative stress in the brown alga *Ectocarpus siliculosus*. *Plant Cell Environ.* 34 (4):629–642.

Dunn, W. B. 2008. Current trends and future requirements for the mass spectrometric investigation of microbial, mammalian and plant metabolomes. *Phys. Biol.* 5 (011001).

Fiehn, O. 2008. Extending the breadth of metabolite profiling by gas chromatography coupled to mass spectrometry. *TrAC-Trend. Anal. Chem.* 27 (3):261–269.

Fiehn, O., J. Kopka, P. Dormann, T. Altmann, R. N. Trethewey, and L. Willmitzer. 2000a. Metabolite profiling for plant functional genomics. *Nat. Biotechnol.* 18 (11):1157–1161.

Fiehn, O., J. Kopka, R. N. Trethewey, and L. Willmitzer. 2000b. Identification of uncommon plant metabolites based on a calculation of elemental compositions using gas chromatography and quadrupole mass spectrometry. *Anal. Chem.* 72 (15):3573–3580.

Fiehn, O., G. Wohlgemuth, M. Scholz et al. 2008. Quality control for plant metabolomics: Reporting MSI-compliant studies. *Plant J.* 53 (4):691–704.

Giacomoni, F., G. Le Corguille, M. Monsoor et al. 2015. Workflow4Metabolomics: A collaborative research infrastructure for computational metabolomics. *Bioinformatics* 31 (9):1493–1495.

Gillard, J., J. Frenkel, V. Devos et al. 2013. Metabolomics enables the structure elucidation of a diatom sex pheromone. *Angew. Chem. Int. Edit.* 52 (3):854–857.

Goulitquer, S., P. Potin, and T. Tonon. 2012. Mass spectrometry-based metabolomics to elucidate functions in marine organisms and ecosystems. *Mar. Drugs* 10 (4):849–880.

Gravot, A., S. M. Dittami, S. Rousvoal et al. 2010. Diurnal oscillations of metabolite abundances and gene analysis provide new insights into central metabolic processes of the brown alga *Ectocarpus siliculosus*. *New Phytol.* 188 (1):98–110.

Gullberg, J., P. Jonsson, A. Nordström, M. Sjöström, and T. Moritz. 2004. Design of experiments: An efficient strategy to identify factors influencing extraction and derivatization of *Arabidopsis thaliana* samples in metabolomic studies with gas chromatography-mass spectrometry. *Anal. Biochem.* 331 (2):283–295.

Kanani, H., P. K. Chrysanthopoulos, and M. I. Klapa. 2008. Standardizing GC-MS metabolomics. *J. Chromatogr. B: Anal. Technol. Biomed. Life Sci.* 871 (2):191–201.

Kanani, H. H., and M. I. Klapa. 2007. Data correction strategy for metabolomics analysis using gas chromatography-mass spectrometry. *Metab. Eng.* 9 (1): 39–51.

Krug, D., and R. Müller. 2014. Secondary metabolomics: The impact of mass spectrometry-based approaches on the discovery and characterization of microbial natural products. *Nat. Prod. Rep.* 31 (6):768–783.

Kuhlisch, C., J. Althammer, A. F. Sazhin, H. H. Jakobsen, J. C. Nejstgaard, and G. Pohnert. Metabolomics-derived marker metabolites to characterize *Phaeocystis pouchetii* in natural plankton communities. submitted.

Kuhlisch, C., and G. Pohnert. 2015. Metabolomics in chemical ecology. *Nat. Prod. Rep.* 32 (7):937–955.

Lee, D. Y., and O. Fiehn. 2008. High quality metabolomic data for *Chlamydomonas reinhardtii*. *Plant Methods* 4. doi: 10.1186/1746-4811-4-7.

Lisec, J., N. Schauer, J. Kopka, L. Willmitzer, and A. R. Fernie. 2006. Gas chromatography mass spectrometry-based metabolite profiling in plants. *Nat. Protocols* 1 (1):387–396.

Longnecker, K., M. C. K. Soule, and E. B. Kujawinski. 2015. Dissolved organic matter produced by *Thalassiosira pseudonana*. *Mar. Chem.* 168:114–123.

Mausz, M. A., and G. Pohnert. 2015. Phenotypic diversity of diploid and haploid *Emiliania huxleyi* cells and of cells in different growth phases revealed by comparative metabolomics. *J. Plant Physiol.* 172:137–148.

McCune, B., J. B. Grace, and D. L. Urban. 2002. *Analysis of Ecological Communities*. 2 ed., Vol. 28. Gleneden Beach, Oregon: MjM software design.

Menden-Deuer, S., and E. J. Lessard. 2000. Carbon to volume relationships for dinoflagellates, diatoms, and other protist plankton. *Limnol. Oceanogr.* 45 (3):569–579.

Minor, E. C., M. M. Swenson, B. M. Mattson, and A. R. Oyler. 2014. Structural characterization of dissolved organic matter: A review of current techniques for isolation and analysis. *Environ. Sci. Process Impacts* 16 (9):2064–2079.

Nylund, G. M., F. Weinberger, M. Rempt, and G. Pohnert. 2011. Metabolomic assessment of induced and activated chemical defence in the invasive red alga *Gracilaria vermiculophylla*. *PLoS One* 6 (12):e29359.

Ritter, A., L. Cabioch, L. Brillet-Guéguen et al. 2017. Herbivore-induced chemical and molecular responses of the kelps *Laminaria digitata* and *Lessonia spicata*. *PLoS One* 12 (3):e0173315.

Ritter, A., S. M. Dittami, S. Goulitquer et al. 2014. Transcriptomic and metabolomic analysis of copper stress acclimation in *Ectocarpus siliculosus* highlights signaling and tolerance mechanisms in brown algae. *BMC Plant Biol.* 14 (1):116.

Roessner, U., C. Wagner, J. Kopka, R. N. Trethewey, and L. Willmitzer. 2000. Technical advance: Simultaneous analysis of metabolites in potato tuber by gas chromatography-mass spectrometry. *Plant J.* 23 (1):131–142.

Saccenti, E., H. C. J. Hoefsloot, A. K. Smilde, J. A. Westerhuis, and M. M. W. B. Hendriks. 2014. Reflections on univariate and multivariate analysis of metabolomics data. *Metabolomics* 10 (3):361–374.

Schroeder, F., and G. Pohnert. 2017. Editorial overview: Omics techniques to map the chemistry of life. *Curr. Opin. Chem. Biol.* 36:v–vi.

Tautenhahn, R., G. J. Patti, D. Rinehart, and G. Siuzdak. 2012. XCMS Online: A web-based platform to process untargeted metabolomic data. *Anal. Chem.* 84 (11):5035–5039.

van den Berg, R. A., H. C. Hoefsloot, J. A. Westerhuis, A. K. Smilde, and M. J. van der Werf. 2006. Centering, scaling, and transformations: Improving the biological information content of metabolomics data. *BMC Genomics* 7:142. doi: 10.1186/1471-2164-7-142.

van den Dool, H., and P. D. Kratz. 1963. A generalization of retention index system including linear temperature programmed gas-liquid partition chromatography. *J. Chromatogr.* 11 (4):463–471.

Vidoudez, C., and G. Pohnert. 2012. Comparative metabolomics of the diatom *Skeletonema marinoi* in different growth phases. *Metabolomics* 8 (4):654–669.

Wagner, C., M. Sefkow, and J. Kopka. 2003. Construction and application of a mass spectral and retention time index database generated from plant GC-EI-TOF-MS metabolite profiles. *Phytochemistry* 62 (6):887–900.

Weinberger, F., U. Lion, L. Delage et al. 2011. Up-regulation of lipoxygenase, phospholipase, and oxylipin-production in the induced chemical defense of the red alga *Gracilaria chilensis* against epiphytes. *J. Chem. Ecol.* 37 (7):677–686.

Winder, C. L., W. B. Dunn, S. Schuler et al. 2008. Global metabolic profiling of *Escherichia coli* cultures: An evaluation of methods for quenching and extraction of intracellular metabolites. *Anal. Chem.* 80 (8):2939–2948.

Xia, J. G., N. Psychogios, N. Young, and D. S. Wishart. 2009. MetaboAnalyst: A web server for metabolomic data analysis and interpretation. *Nucleic Acids Res.* 37:W652–W660.

chapter nineteen

Preparative extraction of exometabolites from seaweed surfaces

Florian Weinberger

Contents

19.1 Introduction

The surface chemistry of seaweeds and other marine organisms is of increasing interest in biotechnology and ecology. The molecular components that are released, taken up, or perceived by an alga reflect all its interactions with the biotic and abiotic environment. For this reason, these compounds are often carriers of ecologically or physiologically important functions. In particular, the relationships between algal hosts and compatible and noncompatible epibionts are often (if not always) driven by molecular communication, and they typically involve chemical deterrents or cues. Marine biotechnology increasingly targets such compounds for antifouling applications. At the same time, ecologists studying seaweed–epibiont interactions also aim to identify compounds that are bioactive when they are present on macroalgal surfaces at physiological concentration. In particular, the second goal requires extraction methods that yield surface metabolites and exclude tissue metabolites. Solid-phase extraction (SPE) is available to prepare relatively small samples for sensitive analytical approaches, such as LC–MS or GC–MS (Chapter 20 by Cirri and Pohnert 2017). However, surface metabolites are often required in larger quantities or even in bulk amounts, for example, when they are to be tested for bioactivities or when they need to be purified for less-sensitive analytical approaches, such as NMR spectroscopy. Under such conditions, SPE is usually not affordable, and approaches based on the solvent extraction are required: The seaweed to be extracted is dipped alive into an organic solvent (de Nys et al. 1998). Organic solvents generally have a stronger capacity to dissolve lipids than water. However, a destruction of the outer membranes of epidermal cells—which are primarily composed of lipids—has to be prevented if only surface and not intracellular metabolites are to be extracted. At the same time, the sensitivity of marine macrophytes to organic solvents differs, and specific techniques need to be developed for each target organism. The current chapter provides a general protocol for the development of large-scale preparative surface extraction methods.

19.2 State of the art

Solvent surface-extraction protocols have already been published for brown and red, but not for green macroalgae (Table 19.1). The resistance of any seaweed to an organic solvent depends on (a) the nature of the solvent; (b) the exposure time; and (c) certain traits of the alga, such as the structure of the outer cell wall. Main constituents of the outer cell wall are usually polysaccharides. On account of their polar character, polysaccharides exclude nonpolar solvents more than those that are polar. As a consequence, nonpolar solvents such as hexane are generally tolerated for longer exposure times than dichloromethane or even more polar solvents (Table 19.1). However, the nonpolar character of hexane also implicates that it can only extract nonpolar surface compounds, which may exclude many relevant metabolites (Saha et al. 2012). In most cases, it will be necessary to overcome this limitation, which can often be achieved if hexane is mixed with a more polar solvent and the exposure time is reduced. Thus, the resistance of the target seaweed to various solvents during various exposure times has to be compared to identify an optimal combination of nondetrimental extraction even of less unpolar compounds. The following protocol for the development of an extraction method has so far been successfully employed for brown (Saha et al. 2011) and red (Wang et al. 2016) seaweeds, and for the seagrass *Zostera marina* (Guan, unpublished).

Different approaches have already been used to evaluate the intactness of epidermal cells after solvent treatment (see references in Table 19.1). The following protocol uses Evans Blue to stain damaged cells. Evans Blue cannot cross intact plasma membranes and has a high affinity to proteins, which are much more abundant within cells than in the extraplasmatic space.

Table 19.1 Exposure times to different organic solvents that are tolerated before epidermal cells of seaweeds are disrupted

Species	Solvent	Time (s)	Reference
	(A) Rhodophyta		
Laurencia obtusa	Hexane	60 or less	de Nys et al. (1998)
	DCM, DE, EA, and MeOH	<10	de Nys et al. (1998)
Laurencia rigida	1% or less DCM in hexane	30	Nylund et al. (2007)
	1% DCM in hexane	30 or less	Nylund et al. (2007)
Delisea pulchra	Hexane	40 or less	Nylund et al. (2007)
	DCM, DE, EA, and MeOH	<10	de Nys et al. (1998)
Pterocladia capillacea	Hexane	40 or less	Nylund et al. (2007)
Solieria robusta	Hexane	40 or less	Nylund et al. (2007)
Gracilaria vermiculophylla	25% DCM in hexane	5 or less	Wang et al. (2016)
	(B) Phaeophyta		
Dictyopteris acrostichoides	4% or less DCM in hexane	30	Nylund et al. (2007)
	4% DCM	40 or less	Nylund et al. (2007)
Dilophus marginatus	6% or less DCM in hexane	30	Nylund et al. (2007)
	6% DCM in hexane	40 or less	Nylund et al. (2007)
Taonia atomaria	MeOH	15 or less	Othmani et al. (2016)
	50% MeOH in DCM	20 or less	Othmani et al. (2016)
Fucus vesiculosus	4% DCM in hexane	40 or less	Brock et al. (2007)
	50% MeOH in hexane	5 or less	Lachnit et al. (2010)
	50% MeOH in hexane	10 or less	Saha et al. (2011)
Fucus serratus	50% MeOH in hexane	5 or less	Rickert et al. (2016)

Note: DCM = Dichloromethane, DE = Diethyl ether, EA = Ethyl acetate, MeOH = Methanol.

As a consequence, damaged and intact cells can be easily distinguished by microscopic observation after application of Evans Blue (Weinberger et al. 2005; Saha et al. 2011; Wang et al. 2016).

Once a reliable extraction protocol is available, concentrations of obtained extracts need to be related to the extracted surfaces. The size of these surfaces is strongly dependent on algal morphology. Surface areas (SA) can be determined through analysis of thallus projections (TP), which may be generated with a standard flatbed scanner. Scanning each sample before extraction is usually impossible because the scanning procedure comes along with air exposure, temperature increase, and exposure to intense light, which may affect the extraction result. However, algal SA are—within species and life-cycle stage—to some degree correlated with mass. As a consequence, mean surface/mass ratios can be determined and used to calculate approximate SA from fresh weights.

19.3 Materials

- Algal samples to be extracted should be generally collected and maintained under low-stress conditions. Desiccation, temperature shock, or salinity shock have to be prevented for obvious reasons. If different age cohorts, sexes, or populations are to be studied with the final protocol (see 19.4.3), make sure that they are all investigated during the preparatory steps (19.4.1 and 19.4.2), as they could differ in their sensitivities.
- Sufficient amounts of cooled natural seawater to allow for repeated rinsing and interim maintenance of sample seaweeds (*Note 1*).
- Salad spinner.
- Balance.
- Vent hood.
- Magnetic stirrer.
- Glass beakers (size adapted to the size of seaweeds to be extracted).
- Organic solvent.
- Natural seawater containing 0.05 % w/v of Evans Blue (20 mL per sample).
- Rack with test tubes for incubation during staining (20 mL).
- Large- and medium-sized forceps.
- Stopwatch.
- Microscope with glass slides.
- Flatbed scanner.
- Imaging software (e.g., Fiji, a free software package available for Windows, MacOS, Linux, and other systems [Schindelin et al. 2012]).
- Ruler.
- Filtration devise with glass microfiber filters.
- Evaporator.
- Lyophilizer.

19.4 Experimental procedures

19.4.1 Development of extraction protocol

All extractions should be carried out in a vented hood with clean equipment. Apart from the solvent treatment itself examined, algae need to be maintained at low-stress conditions (*Note 1*).

1. *Water removal*: To remove adhering water, spin an algal individual for 60 s in a salad spinner at full speed. Do not blot with paper or a towel, as this might remove surface compounds.
2. *Extraction*: Dip algae individually for a defined time (stop seconds) into a beaker containing stirred solvent. Tissue cuts should not get into contact with the solvent, as this would result in extraction of tissue metabolites.
3. *Coarse rinsing*: Immediately after performing step 2, throw the alga into a large beaker containing seawater to remove the solvent. Move it vigorously with a large forceps for some seconds.
4. *Fine rinsing*: Immediately after performing step 3, sample a thallus section (approximately 5 cm), and rinse it further in a second beaker containing seawater.
5. *Staining*: Immediately after performing step 4, introduce the thallus section into a vial containing 20 mL of seawater with the addition of Evans Blue (0.05 % w/v). Incubate 30 min at low-stress conditions (darkness, suitable temperature).
6. *Removal of excess dye*: Rinse stained thallus section in fresh seawater.
7. *Microscopic observation*: Take photos and compare the staining result to negative controls (=unextracted algae) and positive controls (=algae placed for 5 min into pure methanol).
8. Repeat step 1 to step 6 with various solvents (*Note 2*) and for different extraction times, always with naïve algal individuals. Treatments are always resulting in the same staining result as negative controls are treatments that do not damage epidermal cells and that may be used in the following steps.

19.4.2 Determination of algal surface/mass ratio

Once a reliable extraction protocol is available, concentrations of obtained extracts usually need to be related to the extracted surfaces. These are strongly dependent on the algal morphology but can be determined from TP as follows:

1. *Selection of algae*: At least 10 individuals should be examined. If groups are to be compared that may potentially differ in their surface/mass ratios (e.g., populations, age cohorts, or sexes), each of these groups should be examined separately.
2. *Water removal*: Remove adhering water by using the same technique as in 19.4.3.1 below (spinning individuals for 60 s in a salad spinner at full speed).
3. *Weighing*: Determine the fresh weight of each alga on balance.
4. *Interim maintenance*: To prevent shrinking or other deformations due to drought, place the algae into individual beakers containing seawater.

5. *Scanning*: Dry the algae with paper, and spread them on a flatbed scanner. Individuals that are too large or exhibit nonflat morphologies can be cut into pieces, which need to be distributed on the scanner without overlap. A scale is placed on the scanner as well. To obtain a good background contrast, cover the scanner with spotlessly clean and bright white material (e.g., an isopor panel). Scan at high resolution and save the obtained image in a standard graphics format.
6. *Image analysis*: Use imaging software to determine the total projection area of each alga.
7. Determine SA by using either Equation 19.1 or 19.4. Most seaweeds exhibit either flat or filamentous morphologies. For thalli with flat morphology, SA can be calculated from TP as

$$SA = 2 \times TP \tag{19.1}$$

In contrast, filaments can be regarded as cylinders, therefore

$$SA = \pi \times \text{thallus diameter} \times \text{thallus length} \tag{19.2}$$

and, because

$$TP = \text{thallus diameter} \times \text{thallus length} \tag{19.3}$$

$$SA = \pi \times TP \tag{19.4}$$

8. *Calculate mean surface/mass-ratios*: For each scanned algal individual, divide SA by the fresh weight obtained in 19.4.2.3, and then, calculate the mean surface/mass ratio.

19.4.3 Extraction

All extractions should be carried out in a vented hood with clean equipment. Prior to the extraction, examined algae need to be maintained at low-stress conditions.

1. *Water removal:* To remove adhering water, spin algal individuals for 60 s in a salad spinner at full speed. Do not blot with paper or a towel, as this might remove surface compounds.
2. *Weighing:* Determine the fresh weight of algal material on balance.
3. *Extraction:* Dip algae individually for a suitable time into a beaker containing stirred suitable solvent (solvent and exposure time were identified following 19.4.1). Tissue cuts should not get into contact with the solvent. Surface extracted algae may be either discarded or conserved for other purposes.

4. *Filtration:* Remove particles from the solvent by immediate filtration through a glass microfiber filter.
5. *Evaporation:* Evaporate the organic solvent in vacuo at a temperature <40°C. In nearly all cases, the sample will also contain a small volume of water that originally adhered to the alga. Samples that are free of organic solvent should be lyophilized prior to storage at –20°C.
6. *Determine extracted SA*: For each sample, multiply the fresh weight obtained in step 19.4.3.2 with the mean surface/mass ratio obtained in step 19.4.2.8.

19.5 Notes

Note 1: To prevent stress, use seawater with a similar salinity and temperature as in the collection site of the algal material.

Note 2: In most cases, nonpolar solvents such as hexane are tolerated best (Table 19.1) and should be tested first. The solvent polarity may then be increased stepwise until exposure times of less than 5 s result in cell damage.

References

Brock, E., G. M. Nylund, and H. Pavia. 2007. Chemical inhibition of barnacle larval settlement by the brown alga *Fucus vesiculosus*. *Mar. Ecol. Progr. Ser.* 337:165–174.

Cirri, E. and G. Pohnert. 2017. Disruption-free solid phase extraction of surface metabolites from macroalgae. In *Protocols for Macroalgae Research*, B. Charrier, T. Wichard, and C.R.K. Reddy, (Eds.) Chapter 20, 311–321. Boca Raton, FL: CRC Press, Taylor & Francis Group.

de Nys, R., S. A. Dworjanyn, and P. D. Steinberg. 1998. A new method for determining surface concentrations of marine natural products on seaweeds. *Mar. Ecol. Progr. Ser.* 162:79–87.

Lachnit, T., M. Wahl, and T. Harder. 2010. Isolated thallus-associated compounds from the macroalga *Fucus vesiculosus* mediate bacterial surface colonization in the field similar to that on the natural alga. *Biofouling* 26 (3):247–255.

Nylund, G. M., P. E. Gribben, R. de Nys, P. D. Steinberg, and H. Pavia. 2007. Surface chemistry versus whole-cell extracts: antifouling tests with seaweed metabolites. *Mar. Ecol. Progr. Ser.* 329:73–84.

Othmani, A., J. F. Briand, M. Aye, M. Molmeret, and G. Culioli. 2016. Surface metabolites of the brown alga *Taonia atomaria* have the ability to regulate epibiosis. *Biofouling* 32 (7):801–813.

Rickert, E., M. Lenz, F. R. Barboza, S. N. Gorb, and M. Wahl. 2016. Seasonally fluctuating chemical microfouling control in *Fucus vesiculosus* and *Fucus serratus* from the Baltic Sea. *Mar. Biol.* 163 (10):203.

Saha, M., M. Rempt, B. Gebser, J. Grueneberg, G. Pohnert, and F. Weinberger. 2012. Dimethylsulphopropionate (DMSP) and proline from the surface of the brown alga *Fucus vesiculosus* inhibit bacterial attachment. *Biofouling* 28 (6):593–604.

Saha, M., M. Rempt, K. Grosser, G. Pohnert, and F. Weinberger. 2011. Surface-associated fucoxanthin mediates settlement of bacterial epiphytes on the rockweed *Fucus vesiculosus*. *Biofouling* 27 (4):423–433.
Schindelin, J., I. Arganda-Carreras, E. Frise et al. 2012. Fiji: An open-source platform for biological-image analysis. *Nat. Meth.* 9 (7):676–682.
Wang, S., G. Wang, F. Weinberger, D. Bian, M. Nakaoka, and M. Lenz. 2016. Anti-epiphyte defences in the red seaweed *Gracilaria vermiculophylla*: Non-native algae are better defended than their native conspecifics. *J. Ecol.* 105 (2): 445–457.
Weinberger, F., G. Pohnert, M. L. Berndt, K. Bouarab, B. Kloareg, and P. Potin. 2005. Apoplastic oxidation of L-asparagine is involved in the control of the green algal endophyte *Acrochaete operculata* Correa & Nielsen by the red seaweed *Chondrus crispus* Stackhouse. *J. Exp. Bot.* 56 (415):1317–1326.

chapter twenty

Disruption-free solid-phase extraction of surface metabolites from macroalgae

Emilio Cirri and Georg Pohnert

Contents

20.1 Introduction

Surface metabolites play a fundamental role in the mediation of interactions on biotic surfaces of, for example, macroalgae, corals, or sponges. Such compounds control settling processes, regulate predator–prey relationships, and mediate infection processes (Wahl 2009, Dobretsov et al. 2013, da Gama et al. 2014). A hallmark of such interactions is the locally much focused action of the compounds in question (Dworjanyn et al. 1999, 2006). The surface concentration of such metabolites often exceeds the concentration in the medium, but it can even exceed concentrations within the tissue as exuded compounds can accumulate in a diffusion-limited laminar boundary layer. Despite their ecological importance, until now, only a few methods allow the estimation of local concentrations of surface metabolites. As a consequence, many investigations on the effect of surface metabolites were based on bioassays with extracts of whole organisms (Hellio et al. 2000, Nylund et al. 2011). Such experiments do not reflect the real ecological relevance of surface-active substances because only metabolites at the surface or near a producer should be considered (Nylund et al. 2007). The determination of metabolites within the laminar boundary layer around an aquatic organism, a thin film of about 100–200 μm that determines the transition between the surface and the surrounding water is thus crucial for experiment planning and evaluation. Thus far, the laminar boundary layer has been studied to determine uptake rates of nutrients (Wheeler 1980) or to model them (Hadley 2014), or it has been investigated in correlation with climate changes and oceans' acidification (Cornwall 2014), but the study of metabolites identity and concentration at macroalgal surface is crucial both from an ecological and an industrial point of view, especially in the framework of antifouling substances research (Bhadury and Wright 2004, Rajan 2016).

20.2 State of the art

Several methods to study the laminar boundary layer have been established during the last 20 years: These include Raman microspectroscopy (Grosser et al. 2012), mass-spectrometry imaging, such as MALDI and DESi (Slaveykova 2009, Andras 2012) and the widely used dipping methods (de Nys et al. 1998, Lachnit et al. 2010). For extraction, algae are immersed in a solvent for a short period, during which the metabolites are partially extracted from the surface. Care has to be taken that the solvent does not damage the cells of the extracted organism. After concentration in vacuum, the extracts can be submitted to analytical methods, such as GC–MS and LC–MS. Dipping methods are really easy to handle and can be used with a lot of different species of macroalgae. By optimizing the solvent (or mixtures of solvents) for extractions, as well as the solvent's volume, these techniques allow extracting a very different kind of metabolites, proving to be flexible and adaptable to specific questions (Chapter 19 by Weinberger 2018). Although useful, dipping methods are problematic as solvent exposure can cause cell lysis, and thereby contamination of the surface extract with intracellular metabolites; these problems depend both on the algal species and on the physiological conditions, as well as on the part of the alga that needs to be extracted. Some algae only tolerate exposure to rather nonpolar solvents such as hexane for few seconds. However, these solvents only cover a very limited range of nonpolar metabolites and do not penetrate surface-associated water. If solvent mixtures containing methanol are employed, massive damage of the algae could be observed, thereby questioning the validity of results. To overcome these limitations,

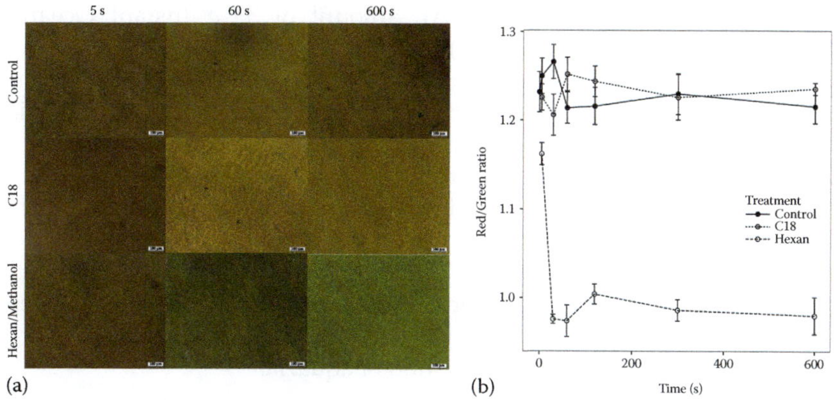

Figure 20.1 Evaluation of surface damage by different extraction methods. (a) Photographs of *F. vesiculosus* surfaces (scale bars 100 µm). Top row: Control after removal from water for 5, 60, and 600 s. Middle row: algae after C18 extraction. Bottom row: after hexane/methanol dipping for the same time spans. (b) Evaluation of cell damage after Evans blue staining by red/green ratio analysis at 5, 30, 60, 120, 300, 600 s exposure to C18 material (gray), hexane/methanol dipping (white) and control (black), ($n = 5 \pm$ SD). (Reprinted with permission from Cirri et al. 2016.)

we developed a new, nondestructive solvent-free and universal method for extracting secondary metabolites from marine macroorganisms (Cirri 2014, Cirri et al. 2016). The method is based on the adsorption of organic metabolites onto C18 extraction sorbent and is not harmful to algal surface cells (Figure 20.1).

The technique has been optimized regarding recovery, reproducibility, and ease of use with the brown macro alga *Fucus vesiculosus* as a model organism. *F. vesiculosus* is a common, well-studied brown alga that can be found on the coasts of the North Sea, the western Baltic Sea, and the Atlantic and Pacific Oceans. However, also the green alga *Caulerpa taxifolia* and the red alga *Gracilaria vermiculophylla* were extracted for proof of concept, demonstrating the universality of the method that can be potentially used for all aquatic macroscopic organisms. Here, a detailed commented protocol is given, based on our publication (Cirri et al. 2016).

20.3 Materials

20.3.1 Fucus vesiculosus

- *Fucus vesiculosus* samples can be collected independently of the season. The alga can be extracted directly after collection in the field or can be kept in aquaria as described below.

- Artificial seawater (ASW): An amount of 33 g Instant Ocean™ (Aquarium Systems, France) per liter of deionized water is stirred at least for 12 h to accomplish a complete dissolution. In our experiment, ASW was diluted with deionized water to half of the initial concentration to reproduce the salinity of the Baltic Sea.
- Nutrient solutions: A volume of 0.22 mL K_2HPO_4 solution (1.79 g K_2HPO_4/200 mL water) and 0.59 mL of a $NaNO_3$ solution (17.99 g $NaNO_3$/200 mL water) were added to 1 L of ASW.
- Glass Aquaria (7 L or more).
- Controlled climate chamber (15°C) under 12 h light/12 h darkness regime (light intensity of 65 µmol m^2 s^{-1}).
- Air pump for constant ventilation of aquaria.

20.3.2 Solid-phase surface extraction method

- Fully end-capped SPE material C18-reversed phase silica gel material (pore size 90 Å, particle dimensions 40–63 µm, Sigma-Aldrich, Deisenhofen, Germany)
- Forceps
- Empty 6 mL polypropylene columns with PE (polyethylene) frits (Macherey-Nagel, Düren, Germany)
- ASW (Artificial seawater as described in 20.3.1 point 2)
- 500 mL spray bottle
- Deionized water
- Methanol (HPLC Grade, Sigma-Aldrich, Germany)
- Plastic Petri dishes (92 × 10 mm)
- Glass or plastic funnels (100 mm diameter)
- Silicone adaptors for Büchner flask
- Büchner flask (500 mL)
- A vacuum pump (ca. 550 mbar) (not essential, *Note 10*)
- 4 mL glass vials with a black cap
- 1.5 mL glass vials with a blue cap
- 200 µL glass inserts and metal springs
- Analytical balance
- Digital camera
- ImageJ (Rasband 1997) or another image-processing software

20.4 Experimental procedures

The following operations in Figure 20.2 are suitable for different species of macroalgae when they are in healthy condition (not wounded or damaged). The method can easily be adapted to other aquatic organisms by changing handling parameters.

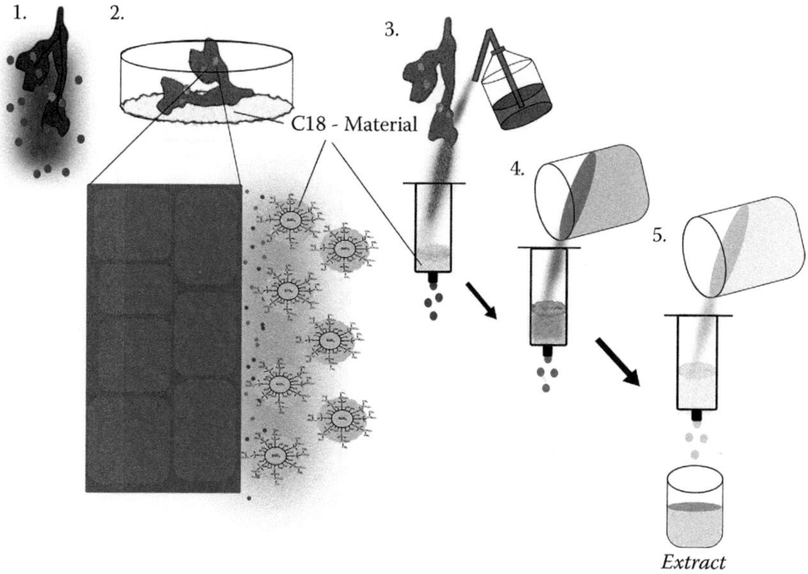

Figure 20.2 Schematic workflow of the C18 method. (1) Algal fronds are removed from the water and left for 2 min to remove excess water by dripping; (2) fronds are transferred to Petri dishes and covered with absorption material; (3) the C18 material is washed off with excess seawater and collected in an empty solid-phase extraction cartridge equipped with a frit; (4) the material is washed with deionized water to remove salts; (5) elution with organic solvents finalizes sample preparation. (Reprinted with permission from Cirri et al. 2016.)

20.4.1 Extraction procedure

1. Spread C18 SPE material (*Notes 1 and 2*) on the small thecae of the Petri dish and weigh it. For a Petri dish with a diameter of 9.2 cm (92 × 10 mm), an amount of 0.5 g of SPE material is sufficient to reach uniform covering of the plate (*Note 3*).
2. Weight the empty polypropylene cartridge (*Note 2*).
3. Take the piece of alga that you desire to extract out of the aquarium (in this case, the fronds of *F. vesiculosus*). Pieces of around 40 cm^2 of this alga were small enough to fit in a 9.2 cm diameter Petri dish and sufficient for generating surface extracts that can be investigated in GC–MS and LC–MS.
4. Hold the alga for 2 min so that excess of water can drop off (*Note 4*).
5. Place the alga inside the dish, close it, and shake it manually for about 10 s, taking care to obtain a homogenous cover of the surface with SPE material.

Figure 20.3 Vacuum setup for solid-phase extraction.

6. Leave the alga in the Petri dish for 60 s (*Note 5*).
7. Take the alga out of the Petri dish using forceps, shaking it gently to remove excess SPE material.
8. Hold the alga with forceps over the funnel connected with the empty polypropylene column and with the Büchner flask (Figure 20.3).
9. Wash the algal surface with ASW from a spray bottle to remove the most of SPE material. The SPE material is collected in an empty 6 mL polypropylene column (*Note 6*).
10. Wash the funnel and the cartridge with the SPE material with excess (ca. 20 mL) MilliQ water to remove salt from the seawater/medium, by taking care that all SPE material is going into the cartridge. While washing, you can apply a gentle vacuum (≈550 mbar) to make the powder settle in the cartridge (0.5–1 mL bed volume) (*Note 7*).
11. Add an appropriate internal standard for recovery calculations.
12. Elute the compounds with methanol (three times 0.5 mL) into a 4 mL glass vial under ambient pressure. Optimal flow rate should be maximum one drop per second (*Note 8*).
13. Remove the solvent under a stream of nitrogen and redissolve the sample in 100 µL of methanol (*Notes 9 and 10*). Transfer the sample into a 250 µL glass insert placed in a 1.5 mL glass vial (or other vials suitable for the autosampler of your own GC or HPLC).

14. Store the sample at –20°C or proceed to measure it with the desired technique (GC–MS or LC–MS, as in, e.g., Cirri et al. 2016, Vidoudez and Pohnert 2011) (Chapter 18 by Kuhlisch et al. 2018).
15. Place the extracted alga on a white, plane surface equipped with a scale bar and take a picture. Calculate the surface of the alga as shown in Chapter 19 by Weinberger (2017) with an image-processing software (in our case, we used the open source software ImageJ, http://imagejnihgov/ij/.) (*Note 11*). For thalli with flat morphology, surface areas (SA) can be calculated as

$$SA = 2 \times TP$$

where TP is the thallus projection.

In contrast, filaments can be regarded as cylinders, therefore

$$SA = \pi \times \text{thallus diameter} \times \text{thallus length;}$$

and, because TP = thallus diameter × thallus length,

$$SA = \pi \times TP.$$

After the C18 material is completely dry, weight the cartridge to calculate the amount of extraction powder used for the experiment.

20.4.2 Data evaluation

1. After LC–MS or GC–MS measurements, integrate both the chromatographic peak area of the internal standard and the substance(s) of interest with the appropriate software associated with your analytical instrument (e.g., Waters® Masslynx or Thermo® Xcalibur).
2. Calculate the ratio between the chromatographic peak area of the internal standard and the substance(s) of interest for a relative quantification. For absolute quantification, the ratio should be compared with an external calibration curve.
3. Normalize the quantity (relative or absolute) to the SA previously calculated with ImageJ.
4. For recovery calculation, normalize the chromatographic peak area of the internal standard (in our experiment, canthaxanthin) to the weight of the C18 material effectively extracted.

20.5 Notes

Security advice: Care should be taken that the dust of the SPE material is not inhaled as aerosols are problematic under prolonged exposure. MeOH is flammable and toxic, thus handling under a fume hood or a good ventilated area is advised.

Note 1: Take care of where you are cutting your alga: During the following step of the method, you have to avoid washing and removing the SPE material that is in contact with damaged tissue in cutting sites. Otherwise, you would risk extracting internal metabolites.

Note 2: Our method development was limited to test a different type of C18 SPE material, but the flexibility of this approach allows extending the range of absorbing material that can be used. For example, material contained in prepackaged SPE cartridges (HLB, StrataX, and Ionic Exchanger) can be used as well. The choice depends on the polarity of the metabolites under consideration. It is not necessary to purchase new empty cartridges: You can use old used cartridges; empty them and clean them before reusing them.

Note 3: The Petri dish dimensions can be changed as desired, and smaller or bigger samples can be used, depending on the availability of biological material and the concentration of surface metabolites. The *Petri dish* method was preferred to other methods because it is faster and assures a uniform and simultaneous covering of the algal surface. A drawback of the method is the loss of a huge percentage of SPE material as only about one-fifth of the powder spread in the Petri dish attached to the algal surface. However, this powder can be reused as it does not get contaminated by exometabolites of the alga. An alternative to the Petri dish method could be to use a small mesh sieve to spread the SPE material over the algal surface. Either plastic or glass funnels can be used. Glass guarantees less chemical contaminations. However, the plastic material allows for stirring and collecting the C18-powder easily, as the silica material tends to stick to the glass surfaces.

Note 4: It is essential that the surface is not getting completely dry: The thin layer of water that remains on the surface (laminar boundary layer) is the environment from which metabolites shall be determined.

Note 5: Different extraction times were tested, and 60 s was chosen as the best compromise between a good interaction and absorption of metabolites with C18 material and an easy recovery of the powder from the alga.

Note 6: The amount of water used for the washing steps is not a crucial parameter, but for more precise absolute quantification of metabolites, this step can also be easily standardized, always by using the same amount of water.

Note 7: Different vacuum setups could be used, as a vacuum manifold (Visiprep™, Supelco, USA) for multiple, simultaneous extractions. However, the use of a vacuum pump is not necessary. In case the vacuum is not applicable, the settlement of the SPE material will require longer time and a larger amount of water to make sure that the absorbing material is properly packed.

Note 8: Our method was optimized for 1.5 mL of methanol as an extraction solvent, but as in all SPE methods, the amount and the polarity of the solvent (or mixture solvents) can be adapted. Other solvents can be applied depending on the used solid phase and the nature of the metabolites which should be extracted. C18 material does not need to be dry before elution; however, each material has its specific treatment to be fully effective. For all these information, also check the recommendations of the manufacturer.

Note 9: If needed or desired, the sample can be stored at this point at –20°C, and the drying step could be performed afterward.

Note 10: The volume of redissolution should be adapted to some surface metabolites present and the sensitivity of the analytical instrument to obtain a response in the linear range. For GC–MS analysis of polar compounds, a derivatization protocol can be used starting from this step. For more information about derivatization, see chapter 18 by Kuhlisch et al. (2017) and Vidoudez and Pohnert (2011).

Note 11: The measurement of the surface is an essential step of this method, but sometimes difficult to achieve, because of the texture and morphology of the alga. Try to get your sample as flat as possible, by using, for example, a glass slide to cover your alga and some weights to keep the alga plane.

Acknowledgments

The authors acknowledge funding the Jena School for Microbial Communication and the International Max Planck Research School on the Exploration of Ecological Interactions with Molecular and Chemical Techniques and the EU Marie Sklodowska-Curie Initial Training Network (ITN) program Algal Microbiome: Friends or Foes (ALFF) for funding.

References

Andras, T.D., Alexander, T.S., Gahlena, A., Parry, R.M., Fernandez, F.M., Kubanek, J., Wang, M.D., and Hay, M.E. 2012. Seaweed allelopathy against coral: Surface distribution of a seaweed secondary metabolite by imaging mass spectrometry. *J. Chem. Ecol.* 38:1203–1214.

Bhadury, P. and Wright, P.C. 2004. Exploitation of marine algae: Biogenic compounds for potential antifouling applications. *Planta* 219 (4):561–578.

Cirri, E. 2014. Development and optimization of an extraction method for the analysis of algal, surface metabolites, Master Thesis, University of Florence.

Cirri, E., Grosser, K., and Pohnert, G. 2016. A solid phase extraction based non-disruptive sampling technique to investigate the surface chemistry of macroalgae. *Biofuling* 32(2):145–153.

Cornwall, E.C., Boyd, P., McGraw, C.M., Hepburn, C.D., Pilditch, C.A., Morris, J.N., Smith, A.M., and Hurd, C.L. 2014. Diffusion boundary layers ameliorate the negative effects of ocean acidification on the temperate coralline macroalga *Arthrocardia corymbosa*. *PLoS One* 9 (9):e109468.

da Gama, B.A.P., Plouguerne, E., and Pereira, R.C. 2014. The antifouling defence mechanisms of marine macroalgae. In: *Advances in Botanical Research—Sea Plants*. Bourgounon, N., Jacquoud, J.P., and Gadal, P. (Eds.), pp. 413–440. Amsterdam, the Netherlands: Elsevier.

de Nys, R., Dworjanyn, S.A., and Steinberg, P.D. 1998. A new method for determining surface concentrations of marine natural products on seaweeds. *Mar. Ecol. Prog. Ser.* 162:79–87.

Dobretsov, S., Abed, R.M.M., and Teplitski, M. 2013. Mini-review: Inhibition of biofouling by marine microorganisms. *Biofouling* 29:423–441.

Dworjanyn, S.A., de Nys, R., and Steinberg, P.D. 1999. Localisation and surface quantification of secondary metabolites in the red alga *Delisea pulchra*. *Mar. Biol.* 133:727–736.

Dworjanyn, S.A., Wright, J.T., Paul, N.A., de Nys, R., and Steinberg, P.D. 2006. Cost of chemical defence in the red alga *Delisea pulchra*. *Oikos* 113:13–22.

Grosser, K., Zedler, L., Schmitt, M., Dietzek, B., Popp, J., and Pohnert, G. 2012. Disruption-free imaging by Raman spectroscopy reveals a chemical sphere with antifouling metabolites around macroalgae. *Biofouling* 28:687–696.

Hadley, S., Wild-Allen, K., Johnson, C., and Macleod, C. 2014. Modeling macroalgae growth and nutrient dynamics for integrated multi-trophic aquaculture. *J. Appl. Phycol.*, 27 (2):901–916.

Hall, S., Wild-Allen, K., Johnson, C., and Macleod, C. 2015. Modeling macroalgae growth and nutrient dynamics for integrated multi-trophic aquaculture. *J. Appl. Phycol.* 27 (2):901–916.

Hellio, C., Bremer, G., Pons, A.M., Le Gal, Y., and Bourgougnon, N. 2000. Inhibition of the development of microorganisms (bacteria and fungi) by extracts of marine algae from Brittany, France. *Appl. Microbiol. Biotechnol.* 54:543–549.

Kuhlisch, C., Califano, G., Wichard, T., and Pohnert, G. 2018. Metabolomics of intra- and extracellular metabolites from micro- and macroalgae using GC-MS and LC-MS. In *Biotechnology Protocols for Macroalgae Research*, Charrier, B., Wichard, T., Reddy, C.R.K. (Eds.), Chapter 18, 281–302. Boca Raton, FL: CRC Press, Taylor & Francis Group.

Lachnit, T., Fischer, M., Kuenzel, S., Baines, J.F., and Harder, T. 2013. Compounds associated with algal surfaces mediate epiphytic colonization of the marine macroalga *Fucus vesiculosus*. *FEMS Microbiol. Ecol.* 84:411–420.

Lachnit, T., Wahl, M., and Harder, T. 2010. Isolated thallus-associated compounds from the macroalga *Fucus vesiculosus* mediate bacterial surface colonization in the field similar to that on the natural alga. *Biofouling* 26:247–255.

Nylund, G.M., Gribben, P.E., de Nys, R., Steinberg, P.D., and Pavia, H. 2007. Surface chemistry versus whole-cell extracts: Antifouling tests with seaweed metabolites. *Mar. Ecol. Prog. Ser.* 329:73–84.

Nylund, G.M., Weinberger, F., Rempt, M., and Pohnert, G. 2011. Metabolomic assessment of induced and activated chemical defence in the invasive red alga *Gracilaria vermiculophylla*. *PLoS One* 6:e29359.

Rajan, R., Selvaraj, M., Palraj, S., and Subramanian, G. 2016. Studies on the anticorrosive & antifouling properties of the *Gracilaria edulis* extract incorporated epoxy paint in the Gulf of Mannar Coast, Mandapam, *India. Prog. Org. Coat,* 90:448-454.

Rasband, W. 1994–2014. ImageJ, US National Institutes of Health, Bethesda, MD. Available from: Accessed December 22, 2017, https://imagej.nih.gov/ij/

Saha, M., Rempt, M., Gebser, B., Grueneberg, J., Pohnert, G., and Weinberger, F. 2012. Dimethylsulphopropionate (DMSP) and proline from the surface of the brown alga *Fucus vesiculosus* inhibit bacterial attachment. *Biofouling* 28:593–604.

Saha, M., Rempt, M., Grosser, K., Pohnert, G., and Weinberger, F. 2011. Surface-associated fucoxanthin mediates settlement of bacterial epiphytes on the rockweed *Fucus vesiculosus. Biofouling* 27:423–433.

Slaveykova, V.I., Guignard, C., Eybe, T., Migeon, H.-N., and Hoffmann, L. 2009. Dynamic NanoSIMS ion imaging of unicellular freshwater algae exposed to copper. *Anal. Bioanal. Chem.* 393:583–589.

Vidoudez, C. and Pohnert, G. 2011. Comparative metabolomics of the diatom Skeletonema marinoi in different growth phases. *Metabolomics* 8(4):654–669.

Wahl, M., Molis, M., Ringelhan, F., Rohnstock, L., and Lenz, M. 2009. Effects of grazing pressure on the induction of anti-herbivory defences in bladderwrack (*Fucus vesiculosus*) in different seasons. *8th International Temperate Reefs Symposium.* Accessed January 12, 2016. Adelaide, Australia.

Weinberger, F. 2018. Preparative extraction of exometabolites from seaweed surfaces. In *Biotechnology Protocols for Macroalgae Research*, Charrier, B., Wichard, T., Reddy, C.R.K. (Eds.), Chapter 19, 303–310. Boca Raton, FL: CRC Press, Taylor & Francis Group.

Wheeler, W.N. 1980. Effect of boundary layer transport on the fixation of carbon by the giant kelp *Macrocystis pyrifera. Mar. Bio.* 56(2):103–110.

section three

Cellular and molecular characterization

chapter twenty one

The immunodetection and in situ *imaging of cell-wall polysaccharides in brown algae*

Amandine Siméon, Delphine Duffieux, Cécile Hervé, Sophie Le Panse, Paul Knox, and Thomas Torode

Contents

21.1 Introduction

Cell walls in brown algae are major cell compartments involved in important biological roles such as algal growth and development, osmotic regulation of cell surfaces, and defense responses. Such a variety of functional properties is supported by dynamic and complex cell-wall architectures. Brown algal cell walls are composed predominantly of polysaccharides together with lower amounts of phenols, proteins, and halide compounds (Deniaud-Bouët et al. 2014). The fucose-containing sulfated polysaccharides and the alginates are the major sets of matrix polysaccharides, which are by far prevalent over crystalline polymers, including cellulose. Structural variants in cell-wall polysaccharides of brown algae are known and these are indicative of diverse functions. So far, most attempts to study cell-wall polymers in brown algae used labor-intensive chemical extractions, with the consequent loss of most spatial and developmental information. The challenge now is to discern features of specific polysaccharide structures, their modulation, and dynamics in cell-wall regions and tissues. One of the best ways to detect and assess the presence of polysaccharides in algal materials is by immunohistochemical techniques, based on the use of tagged proteins with specific recognition capacities (Lee et al. 2011). Currently, most of the molecular probes used to bind specific polysaccharide epitopes in context are monoclonal antibodies. The development of such tools is still in its infancy in brown algae but recent advancements have extended our capabilities to study cell-wall biology in these systems. The current chapter focuses on chemical staining of cell walls and the use of monoclonal antibodies for the specific detection of polysaccharide epitopes in brown algae.

21.2 *State of the art: Imaging cell walls in brown algae*

Brown algae have a unique cell-wall biochemistry not present in other taxa that is organized in a three-dimensional complex architecture and composed predominantly of polysaccharides. The anionic and matrix polysaccharides alginates and fucose-containing sulfated polysaccharides (FCSPs) are by far prevalent over crystalline β-glucans, which include cellulose. Proteins, arabinogalactan proteins (Hervé et al. 2016), phenolic compounds known as phlorotannins, and halide compounds as iodide, are additional compounds found in brown algal cell walls. It is believed that FCSPs interlock the β-glucan scaffold, whereas alginate–phenol linkages are key players in regulating the rigidity of the wall (Deniaud-Bouët et al. 2014). Each of these groups of polysaccharides displays distinct elements of chemical complexity. The arrangement of alginates in three block types has been described in some detail, although it is obvious that a vast continuum of intermediate structures can be found *in planta* (Campa et al. 2004; Hervé et al. 2016). The FCSPs are highly diverse in terms of their degree of sulfation, esterification, methylation, molecular weight, and conformation of sugar residues (Deniaud-Bouët et al. 2014); less heterogeneous FCSPs include the sulfated fucans, which are polymers featuring a backbone based on α-L-fucopyranosyl residues only. All these variants exhibit discrete biological and mechanical properties that still need to be explored in the context of the wall at both the cell and tissue levels.

To date, limited attention has been paid in brown algae to the particular arrangements of the various polymers forming the cell wall and how they are interlinked. Equally important is the fact that the variations are not just between lineages and species, but might also occur between organs, tissues, and microdomains of the wall. Most of the chemical knowledge gained on brown algal cell walls used analyses that require homogenization of large amounts of material, generally including a diversity of cell types. The information of each cell type is lost on analysis. In land plants, the development of immunolocalization techniques has made significant contributions to the exploration of many aspects of the biology of cell-wall polysaccharides (Lee et al. 2011). These methods generally use well-defined monoclonal antibodies that have been developed from immunization procedures with cell-wall–derived materials. This technology has been used for decades in plant biology and is now extending to brown algal cell walls. Recently, specific monoclonal antibodies toward cell-wall polysaccharides of brown algae were generated and characterized. They include the brown algal monoclonal (BAM) antibodies that bind structural variants of FCSPs (BAM1–4) and alginates (BAM6–11) (Torode et al. 2015, 2016). Although the structural features recognized by these antibodies have not always been fully defined, they allow detailed cytological analysis to be performed. They are likely to significantly increase our understanding of the organization and developmental dynamics of the polysaccharides in brown algal cell walls. In optical microscopy, these antibodies can be used in combination to general stains, which are useful for the detection of all cell walls in tissues.

Studies on plant cell-wall polysaccharides indicate that some epitopes might not be readily accessible in all cell walls, because of other polymers masking them (Marcus et al. 2008; Hervé et al. 2009; Xue et al. 2013). The use of dedicated enzymes to remove populations of cell-wall polysaccharides was instrumental in these study processes to reveal epitopes. The situation is likely to be similar in brown algae and the use of specific enzymatic treatments can be a valuable addition to the immunolabeling techniques.

The current chapter discusses the use of specific antibodies in combination with histochemical techniques for the detection of cell-wall polysaccharides in tissues or at cell surfaces in brown algae. A set of protocols is provided for both optical and electron microscopy. Additional guidance is given for the general staining of all cell walls and the enzymatic pretreatment with alginate lyases to remove alginates.

21.3 *Materials*

- Seawater and fresh brown algae.
- Razor blades or Leitz 1320 freezing microtome (Ernst LeitzWetzlar GmbH, Wetzlar, Germany) coupled with Tissue Freezing Medium®

(Leica Biosystems Nussloch GmbH, Germany) to cut algal material for light microscopy.
- Microtome ultracut Leica UCT, diamond knife (Diatome), and Jeol 1400 transmission electron microscope to cut and observe algal material for electron microscopy.
- *Fixative solution*: A good fixative to start with is a 4% (w/v) solution of formaldehyde in 50% (v/v) seawater diluted with distilled water. Alternatively, 2% (w/v) formaldehyde with 0.25% (w/v) of glutaraldehyde (Sigma-Aldrich), 0.001% (w/v) of Tween 80, and 342 mM NaCl in 25 mM PIPES (pH 7.2) can be used.
- Polysine-coated microscopy slides (VWR, France).
- Poly-L-lysine solution 0.1% in water (Sigma-Aldrich).
- Microscope slide cover slips (VWR, France).
- Microscope slide mailers (VWR, France).
- Super PAP hydrophobic pen (Agar Scientific).
- Milk powder (Régilait, France) or bovine serum albumin (Sigma-Aldrich) to be used as blocking agents.
- Molecular probes that recognize algal cell-wall polysaccharides. A range of monoclonal antibodies that specifically bind to sulfated fucans (BAM1 to BAM4) and alginates (BAM6 to BAM11) are available through PlantProbes (www.plantprobes.net). The LM7 monoclonal antibody is classically used to detect pectic homogalacturonan in plants, but it also binds alginates (www.plantprobes.net) as shown in Figure 21.1. These antibodies are derived using rat hybridoma systems, and thus require anti-rat secondary reagents, as outlined below.

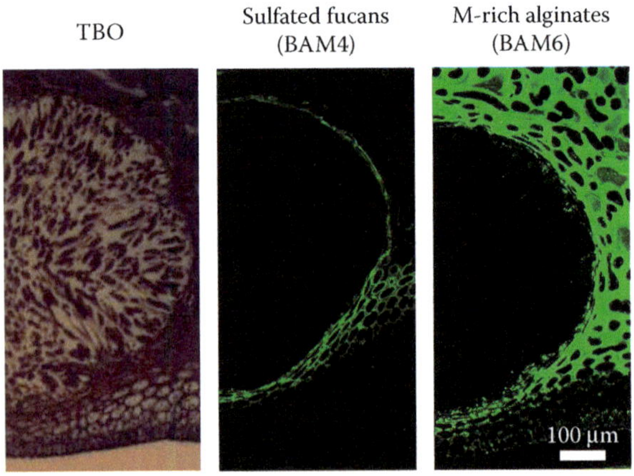

Figure 21.1 Micrographs showing toluidine blue O (TBO) staining of cell walls and indirect immunofluorescence labeling of sulfated fucan and alginate epitopes in equivalent sections of *Fucus serratus* male conceptacles.

- Calcofluor White (fluorescent brightener, Sigma-Aldrich).
- *Toluidine Blue O (Sigma-Aldrich)*: Use 1.3 mM in 0.1 M phosphate buffer (pH 6.8).
- *Ruthenium Red (Sigma-Aldrich)*: Use 0.5 mM in 0.1 M ammonium acetate (pH 6.85).
- Phosphate-buffered saline (PBS) containing 137 mM NaCl, 2.7 mM KCl, 10 mM Na_2HPO_4, and 1.8 mM KH_2PO_4.
- *Alginate lyases*: We use in-house alginate lyases but commercial alginate lyases (Megazyme, Sigma-Aldrich) should give equivalent results.
- *Anti-fade reagents*: Citifluor PBS AF3 (Agar scientific).
- *Secondary antibodies for fluorescence microscopy*: Anti-rat immunoglobulin coupled to a fluorochrome, generally FITC (Sigma-Aldrich) or Alexafluor (Invitrogen).
- *Secondary antibodies for electron microscopy*: We use goat anti-rat IgG ultrasmall with the Silver Enhancement Reagent (both from Aurion). The nano-gold antibody has a low particle diameter, which decreases the steric hindrance as compared to classical conjugates. The postlabeling silver enhancement allows enlarging the average gold particle size by coating the gold particle, greatly enhancing visibility in electron microscopy.
- Absolute ethanol to prepare aqueous solutions (30%–100%) for dehydration procedures.
- LR White resin hard grade, containing 0.5% of the catalyst benzoin methyl ether (Delta microscopies).
- Nonpolar Lowicryl resin HM20 embedding kit.
- Gelatin capsules size 4 for resin embedding (Electron Microscopy Sciences).
- Formvar film on 200 square mesh nickel grids (Delta microscopies) to collect thin sections.
- Uranyl acetate.
- Lead citrate.
- Fish gelatin.
- 0.5 M ammonium chloride in 0.1 M phosphate buffer pH 7.4.

21.4 Experimental procedures

21.4.1 Preparation of algal material and fixation procedures

1. Small filamentous algae such as *Ectocarpus* can be plunged directly into fixative solutions (*Note 1*). To ease the subsequent immunolabeling procedure, the fixed material can be attached to pretreated slides. To do so, soak a Q-tip with 0.1% poly-L-lysine, spread the solution on the slide, and let it dry. Apply the fixed and washed sample on the slide and allow it to dry.

2. Early developmental stages of brown algae such as zygotes and embryos can be directly cultivated on polylysine-coated slides or cover slips before the fixation stage. The slides are maintained in Petri dishes and covered with a drop of seawater and/or immersed after natural adhesion of the algae to the substratum. At selected times the seawater is washed out and replaced by the fixative solution.
3. Samples of reasonable quality for light microscopy can be obtained by sectioning from large samples or stiff materials such as blades from Laminariale species. They are obtained by hand sectioning with a razor blade and directly put in the fixative solution.
4. The use of a microtome allows generating thinner sections of equivalent size. We provide instructions for the use of a Leitz 1320 freezing microtome, but other systems can be used. Fill the sample holder with a Tissue Freezing Medium and place the sample. Coat the entire material with the Tissue Freezing Medium and wait until total freezing before sectioning. Sections of 60 μm are a good starting point for Laminariale stipes. Place sections in the fixative solution.
5. For resin-embedding procedures, pieces of such materials (generally no thicker than 5 mm) are sectioned and placed in fixative solution. Specific recommendations for the embedding and sectioning are explained in Sections 21.4.2. through 21.4.4.
6. Maintain in fixative at 4°C for at least 1 h and for no more than overnight. Fixation duration is increasing with the size and thickness of the sample and must be determined. Fixed and slide-mounted samples can be stored for a reasonable time period (up to one month) before being used for optical microscopy. To do this, we maintain the slides in a fourfold dilution of fixative solution in microscope slide mailers.
7. Fixed samples should be washed twice in buffer without fixative before labeling.

21.4.2 Embedding protocol for London resins White resin

The London resins (LR) are hydrophilic acrylic resins that tolerate the water remaining in samples. The LR White resin can be used at room temperature (RT), 4°C, or at low temperature (–20°C). A dehydration performed at a low temperature protects most antigenic sites from denaturation. We carry out our protocol at 4°C until the end of the dehydration step and at RT during infiltration.

1. Fix samples.
2. Rinse the samples three times for 10 min with 0.5 M ammonium chloride in 0.1 M phosphate buffer pH 7.4 to quench the free aldehyde groups.

3. Dehydrate for 10 min with 30% ethanol followed by 10 min with 50% ethanol, then 3 × 15 min with 70% ethanol, 3 × 15 min with 90% ethanol, and finally 3 × 15 min with absolute ethanol.
4. Infiltrate the specimens with a 3:1 mixture of ethanol and LR White resin for 1 h.
5. Infiltrate the specimens with a 1:1 mixture of ethanol and LR White resin for 1 h.
6. Infiltrate the specimens with a 1:3 mixture of ethanol and LR White resin overnight.
7. Infiltrate the specimens with pure LR White resin. Repeat three times every 3 h and incubate overnight.
9. Transfer the samples to gelatin capsules. Fill the capsules with 100% LR White resin and place the caps. Make sure to exclude any oxygen as this will inhibit polymerization.
10. Allow polymerization either at 37°C for five days (to minimize the heat effect on a reduction of the antigenicity) or at 60°C for two days.

21.4.3 *Embedding protocol for Lowicryl resin*

Lowicryls are acrylate–methacrylate mixtures that polymerize to form cross-linked saturated carbon chain structures (Kellenberg et al. 1980; Carlemalm et al. 1982; Acétarin et al. 1986). These resins can be used at low temperatures to improve the preservation of molecular structures and antigenicity. A low viscosity facilitates the rapid penetration of the resin into the sample at low temperature. Two groups of Lowicryl resins are available: (1) the polar (K4M and K11M) and (2) the nonpolar (HM20 and HM23) Lowicryl resins. After dehydration at low temperature the solvent is replaced by the resin and the polymerization is achieved using UV light.

1. Fix and rinse samples at 4°C as explained earlier.
2. Transfer to 30% ethanol on ice for 30 min.
3. Continue dehydration as follows in precooled solvent solutions:
   - 50% ethanol from 0°C to –20°C, 2 × 30 min.
   - 70% ethanol from –20°C to –35°C, 2 × 45 min.
   - 90% ethanol from –35°C to –35°C, 2 × 45 min.
   - Absolute ethanol from –35°C to –35°C, 2 × 1 h.
4. Replace the solvent by the resin with a 1:2 mixture of ethanol and Lowicryl resin for 2 h, followed by a 2:1 mixture overnight.
5. Infiltrate with the resin as follows:
   - Pure Lowicryl HM20 resin at –35°C for 1 h, repeat 2 times.
   - Pure Lowicryl HM20 resin at –35°C overnight.
   - Pure Lowicryl HM20 resin at –35°C for one day and overnight.
6. Induce polymerization under long waves UV at –35°C for 48 h.

7. Store the samples at room temperature. Lowicryl blocks can be stored for several years but should be protected against the humidity uptake.

21.4.4 *Sectioning of resin-embedded samples*

1. These instructions relate to the use of Leica Ultracut UCT.
2. Prepare glass knives.
3. Trim the block using the razor blade and then more precisely using the glass knife of the microtome.
4. For electron microscopy Lowicryl HM20 or LRWhite sections of a ~70 nm thickness can be cut at room temperature with a diamond knife.
5. Rinse the nickel grids with ethanol and allow them to dry onto filter paper within glass Petri dishes. Collect sections on grids.

21.4.5 *Imaging algal cell walls using general stains*

General stains are not specific enough to ascertain the exact nature of the targeted polymer but they offer a quick and easy alternative to indicate all cell walls in sections. Calcofluor White (CW) binds to a variety of β-glucans that include cellulose and fluoresces under UV excitation. CW poorly binds cell walls in brown algae, and other stains are preferentially used such as ruthenium red (RR) or toluidine blue O (TBO), as shown in Figure 21.1. RR reveals carboxylated polysaccharides (alginates). TBO binds to polysaccharides containing carboxyl or sulfated groups (alginates, fucans) and putative sulfated polyphenols (*Note* 2).

1. *For CW staining*: Incubate the sample with 0.015% CW solution diluted in distilled water for no more than 5 min in darkness. Wash abundantly with PBS. Mount slide and examine in fluorescence microscopy.
2. *For RR staining*: Incubate the sample with RR dye solution during 10 min at room temperature. Wash five times with 0.1 M ammonium acetate (pH 6.85). Mount slide and examine in light microscopy.
3. *For TBO staining*: Incubate the sample with TBO dye solution during 5–10 min, at room temperature. Wash five times with 0.1 M phosphate buffer (pH 6.8). Mount slide and examine in light microscopy.

21.4.6 *Immunolabeling for fluorescence microscopy*

1. Use the hydrophobic pen to draw a circle around the slide-mounted sample
2. Block nonspecific binding sites by incubation with PBS/MP for at least 45 min at room temperature (RT) or overnight at 4°C

3. Wash three times with PBS during 5 min
4. Incubate with primary antibody at an appropriate dilution in PBS/MP (generally 1:10 for hybridoma cell culture supernatant and 1:1000 for purified antibodies) for at least 1 h at RT (*Note 3*)
5. Wash three times with PBS during 5 min
6. Incubate with secondary antibody conjugated to a fluorochrome (generally FITC or AlexaFluor) diluted at 1:100 in PBS/MP for at least 1 h at RT in darkness (*Note 4*)
7. Wash three times with PBS during 5 min
8. Incubate with either CW or TBO as above-mentioned
9. Wash abundantly with PBS and mount slides with an antifading solution

21.4.7 Immunolabeling for electron microscopy

1. Grids with sections facing down are incubated 15 min at RT on a droplet (at least 20 μL) of phosphate buffer pH 7.4 containing glycine to block free aldehyde groups (*Note 5*).
2. Block nonspecific binding with 0.5% bovine serum albumin (BSA) and 0.2% fish gelatin (optional) in 0.1 M phosphate buffer.
3. Incubate with primary antibody at an appropriate dilution (generally between 1 μg mL^{-1} to 10 μg mL^{-1}, which is equivalent to a 10–100-fold dilution of a hybridoma cell culture supernatant) in 0.1 M phosphate buffer containing 0.5% of BSA in a moist chamber overnight.
4. Wash seven times with phosphate buffer containing 0.5% of BSA.
5. Incubate with secondary antibody (goat anti-rat IgG ultrasmall coupled to 1.6 nm gold particle) diluted at 1:200 in phosphate buffer containing 0.5% of BSA.
6. Wash seven times with phosphate buffer containing 0.5% of BSA.
7. Incubate with the silver enhancement reagents during 40 min.
8. Wash as before with phosphate buffer and then extensively in distilled water.
9. The sections are contrasted using heavy metal. Float grids on a drop of saturated uranyl acetate for 5–10 min. Wash on drops of distilled water and air dry. Incubate on drops of lead citrate for 3 min in the presence of NaOH pellets to reduce carbon dioxide. Examine under an electron microscope. This facilitates the detection of epitopes with high resolution as shown in Figure 21.2.

21.4.8 Pretreatments before immunolabeling

To date, most demonstrated cases of cell-wall polysaccharide epitope masking are in plants. Similar cases of epitope masking can be explored

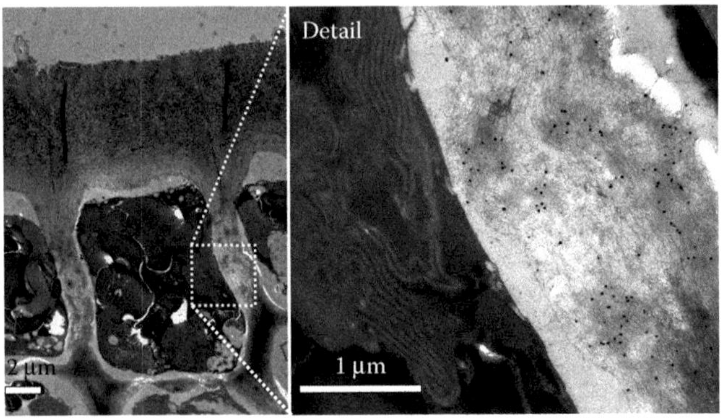

Figure 21.2 Micrographs showing immunogold electron microscopy labeling of alginate in a section of *Saccharina latissima* tissue. The black dots in cell walls feature the labeling.

in brown algae by degradation of alginates through the use of commercially available alginate lyases.

1. Block nonspecific binding sites by incubation of the slide-mounted sample with PBS/MP for at least 45 min.
2. Wash three times with PBS during 5 min.
3. Incubate with alginate lyase (10 µg/mL) in Tris–NaCl buffer for 1 h at RT. Alternatively, most alginates can be solubilized using a 2.8 mM sodium carbonate solution for 1 h at RT (*Note 6*).
4. Wash with three changes of PBS.
5. The material is now ready for immunolabeling as detailed in Sections 21.4.5 through 21.4.7.

21.5 Notes

Note 1: Compared to plant or animal cells, cells from marine species are hyperosmotic and particular care should therefore be taken to maintain the fixative in iso-osmolarity with those cells, that is, 1100 mOsm.

Note 2: The TBO staining method can be used as a postlabeling procedure during which it can be very useful in quenching the autofluorescence emitted by the algal cells.

Note 3: Controls are essential to validate the results—omitting the primary antibody is a good control to determine any nonspecific binding.

Note 4: Specific care should be taken to avoid photobleaching of the fluorochromes. The labeling procedure is performed in darkness and the light exposure under the microscope should be kept to a minimum.

Note 5: The upper side of the grids (*i.e.*, without sections) must be kept clean and dry during all steps.

Note 6: Application of polysaccharide-degrading enzymes to materials not fixed to a glass slide is likely to result in separation of cells and may cause degradation of samples.

Acknowledgments

We acknowledge funding from the French National Research Agency with regard to the IDEALG program with reference ANR-10-BTBR-04.

References

Acétarin, J.D., Carlemalm, E., and Villiger, W., 1986. Developments of new Lowicryl resins for embedding biological specimen of even lower temperatures. *J. Micros.* 143, 81.

Campa, C., Holtan, S., Nilsen, N., Bjerkan, T.M., Stokke, B.T., and Skjak-Braek, G., 2004. Biochemical analysis of the processive mechanism for epimerization of alginate by mannuronan C-5 epimerase AlgE4. *Biochem. J.* 381 (1), 155–164.

Carlemalm, E., Garavito, R.M., and Villiger, W., 1982. Resin development for electron microscopy and an analysis of embedding at low temperature. *J. Micros.* 126, 123.

Deniaud-Bouët, E., Kervarec, N., Michel, G., Tonon, T., Kloareg, B., and Hervé, C., 2014. Chemical and enzymatic fractionation of cell walls from fucales: Insights into the structure of the extracellular matrix of brown algae. *Ann. Bot.* 114 (6), 1203–1216.

Hervé, C., Rogowski, A., Gilbert, H.J., and Knox, J.P., 2009. Enzymatic treatments reveal differential capacities for xylan recognition and degradation in primary and secondary plant cell walls. *Plant J.* 58 (3), 413–422.

Hervé, C., Siméon, A., Jam, M., Cassin, A., Johnson, K.L., Salmeán, A.A., Willats, W.G.T., Doblin, M.S., Bacic, A., and Kloareg, B., 2016. Arabinogalactan proteins have deep roots in eukaryotes: Identification of genes and epitopes in brown algae and their role in fucus serratus embryo development. *New Phytologist* 209 (4), 1428–1441.

Kellenberg, E., Carlemalm, E., Villiger, W., Roth, J., and Garavito, R.M., 1980. *Low Denaturation Embedding for Electron Microscopy of Thin Sections*. Chemische Werke Lowi GmbH, Waldkraiburg, Germany, F.R.G, pp. 1–59.

Lee, K.J.D., Marcus, S.E., and Knox, J.P., 2011. Cell wall biology: Perspectives from cell wall imaging. *Mole. Plant* 4 (2), 212–219.

Marcus, S.E., Verhertbruggen, Y., Hervé, C., Ordaz-Ortiz, J.J., Farkas, V., Pedersen, H.L., Willats, W.G.T., and Knox, J.P., 2008. Pectichomogalacturonan masks abundant sets of xyloglucan epitopes in plant cell walls. *BMC Plant Biol.* 8, 60.

Torode, T.A., Marcus, S.E., Jam, M., Tonon, T., Blackburn, R.S., Hervé, C., and Knox, J.P., 2015. Monoclonal antibodies directed to fucoidan preparations from brown algae. *PLoS One* 10 (2), e0118366.

Torode, T.A., Siméon, A., Marcus, S.E., Jam, M., LeMoigne, A.M., Duffieux, D., Knox J.P., and Hervé, C., 2016. Dissection of cell wall assembly dynamics during early embryogenesis in the brown alga Fucus using monoclonal antibodies. *J. Exp. Bot.* 67 (21), 6089–6100.

Xue, J., Bosch, M., and Knox, J.P., 2013. Heterogeneity and glycan masking of cell wall microstructures in the stems of *Miscanthus* × *giganteus,* and its parents *M. sinensis* and *M. sacchariflorus. PLoS One* 8 (11), e82114.

chapter twenty two

Atomic force microscopy based analysis of cell-wall elasticity in macroalgae

Thomas Torode, Marina Linardic, J. Louis Kaplan, and Siobhan A. Braybrook

Contents

22.1 *Introduction*

22.1.1 *Biomechanical characterization of macroalgal cell walls*

The three independent multicellular macroalgae lineages (red, green, and brown) show conserved adaptations to reducing drag, improving buoyancy, and displaying remarkable developmental plasticity in their body forms. As walled organisms, the structure, shape, and mechanical properties of macroalgae are dependent on the cell wall—a complex hydrated composite composed predominantly of polysaccharides. The emergent cell-wall mechanical properties are reliant on the composition, abundance, and architecture of the components of the cell wall, which can be modulated by the alga. The challenge now is to discern how microscale mechanical properties of cell walls contribute to the formation of functionalized body forms and macroscale physical properties, which allow macroalgae to grow in a range of aquatic environments. Macroalgae grow in a variety of diverse and demanding ecological niches: lentic (lakes), lotic (rivers), eulittoral (intertidal), sublittoral (submerged), and epipelagic (free-floating). Atomic force microscopy (AFM) has recently been used to study cell-wall mechanics in plants but remains in its infancy for macroalgae research. The current chapter focuses upon the application of AFM techniques to macroalgal cell walls, with specific reference to unique challenges faced by algal researchers.

22.2 State of the art

22.2.1 Mechanical characterization of macroalgal cell walls

In comparison with terrestrial plants, macroalgae do not need to resist gravity to grow. The body forms of most macroalgae have therefore evolved to be flexible and pliable to survive forces in an aqueous environment. This has resulted in a body unique tissue organization: a tightly packed outer meristoderm surrounding a more loosely organized medulla. The medulla cells, which are not in close association with neighboring cells, feature large extracellular spaces between cell walls (Torode et al. 2015). The general cell-wall composition of macroalgae features a reduction in the abundance of load-bearing cellulose microfibrils, and a larger portion of hydrogel matrix polysaccharides. Furthermore, the cell walls of algae are generally thicker than those of terrestrial plants.

In plants, the use of AFM-based techniques has been used to determine tissue and cell mechanics (Milani et al. 2011; Peaucelle et al. 2011; Fernandes et al. 2012; Sampathkumar et al. 2014) and localized changes in cell walls because of exogenous application and endogenous expression of cell-wall modifying agents (Braybrook and Peaucelle 2013). To date, there has been limited application of AFM-based techniques toward macroalgae, focusing on the early life stages of embryos of *Fucus serratus* (Torode, Linardic, Braybrook, unpublished), and early sporophytes of *Ectocarpus siliculosus* (Tesson and Charrier 2014). Both these samples are ideal for AFM analysis because of their relatively small size and the self-adhesive properties of their cells to the underlying substratum. However, they represent a small fraction of the immense variety of morphology within the macroalgae. Within this chapter, we will provide examples of excised thalli and cultured zygotes.

The current chapter builds upon the expanding field of AFM-based techniques used to study plant cell walls (Braybrook 2015) and will provide technical directions to optimize experiments for macroalgae. The relatively recent advent of AFM techniques in plant sciences, and the comparative ease of transferable AFM-based skills, equipment, and software, allows a rapid and successful application of AFM techniques to macroalgae research.

AFM-based indentation, as presented here, allows one to analyze the elasticity of a biological sample. The AFM tip is used to indent into the material, up to a maximum force, and is then withdrawn. Data are collected throughout the process in the form of force and piezo-height (Figure 22.1a; indentation and retraction portions of curve). The elasticity of the material can be evaluated in two ways: by fitting a tangent to

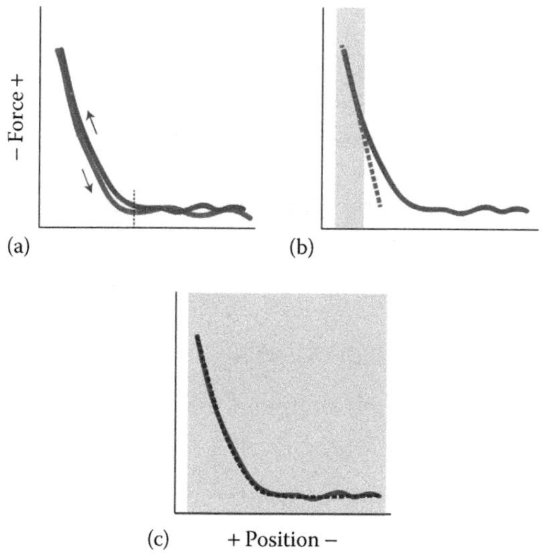

Figure 22.1 An illustration of AFM-based indentation data and analysis. AFM-based indentation experiments can measure elastic responses within the algal cell-wall material. (a) A single indentation event consists of an approach (darker curve) in which the tip moves toward the sample until it reaches contact (dashed line) with the sample surface. The indentation then commences, and force can be seen to increase. The retraction of the tip from the sample (gray curve) can also be tracked. The indentation portion of the curve may be fit by using two methods, both of which yield measures of elasticity. (b) A tangent (dashed orange line) fit to the upper portion of the indentation curve (light-gray box) yields a measure of stiffness (N m^{-1}). (c) The whole indentation curve (dotted gray box) may be used to fit a Hertzian contact model (dark-black dashed line), yielding an elastic/indentation modulus (Pa or N m^{-2}). The cases for use of each method are discussed at the end of the current chapter.

the indentation curve at the upper range of indentation (Figure 22.1b), or by fitting a Hertzian-contact model to the whole indentation curve (Figure 22.1c). In the method below, we will outline how one acquires such data and chooses to analyze it.

22.3 Materials

- Seawater (artificial or natural)
- Fresh algae
- Razor blades/scalpel
- Forceps
- Petri dishes
- Microscope slides
- Gene Frame Seals (Thermo Scientific, 11570294)
- 1% (w/v) low-melt agarose in seawater
- AFM Cantilevers: 0.5 and 3 N m^{-1} stiffness; 10 nm, 1 and 10 μm tips (Windsor Scientific, UK)
- Two-component epoxy resin
- Silica beads of various diameters (e.g., 5, 10, 50 μm; www.microspheres-nanospheres.com)
- Atomic force microscope with top-view optics (JPK NanoWizard3, JPK AG, Berlin)

22.4 Experimental procedures

22.4.1 Preparation of algal material for atomic force microscopy analysis

Samples for AFM analysis must first be mounted such that the sample does not move, is accessible to the cantilever, remains hydrated, and is not physically restricted in a way that would alter the mechanical properties of the sample region. We have found that macroalgal samples, other than those which self-adhere to glass, are best mounted in low-melt agarose upon slides.

1. Small pieces (~0.5 cm^2) of macroalgal material can be excised from the alga and maintained in seawater prior to their mounting for AFM analysis.
2. Prepare the slide. Attach a Gene Frame to the slide by peeling off the backing and squaring with the top/bottom slide edges. This creates a reservoir for the seawater, keeping the sample hydrated and containing the liquid.

3. Melt the agarose (1.5% [w/v] with seawater) and place ~250 μL (adjust volume depending on sample number and size) onto the center of the slide (within the Gene Frame), allow to partially set, and then settle the excised algal piece on top of the agar (*Note 1*). Partial setting allows the gel to have structure enough to hold the sample but still enough pliability to form a good connection and fit the sample in place.
4. Place the algae sample near to top of the agarose mound to ensure relative flatness of the sample and minimize the chance of agarose covering the sample.
5. Flood the reservoir with seawater.
6. Early stages of embryogenesis and sporophyte germination can be studied without additional mounting support due to the natural adhesive properties of the developing alga. To do this, settle fertilized eggs/spores upon desired substratum (microscope slides) that can be loaded into the AFM. When ready to start the AFM-indentations, place a Gene Frame around the samples which forms a reservoir of seawater to maintain hydration.

22.4.2 *Choosing a force for indentation*

The indentation depth is important for the experiment, but dependent on the biological question being asked, and some prior knowledge of the cell wall and mucilage thickness can help the experimenter choose a good indentation depth. For example, the outer meristoderm cell wall of *Ulva flexuosa* is ~25 μm with a ~5 μm mucilage layer (Messyasz et al. 2013); in *Saccharina latissima*, the wall thickness is ~2 μm with a mucilage layer ~6 μm (Chapter 21 by Siméon et al.). In these examples, the experimenter should choose their depth on the basis of their target (mucilage or wall and mucilage). To determine the force of indentation that yields the desired indentation depth, a range of forces should be tested. Lower forces result in smaller indentation depths, which can lead to error and overestimation of elastic properties, whereas larger forces can generate indentation depths far greater than the wall—thereby measuring internal tissue properties, instead of the desired target mechanical properties alone.

1. Mount a macroalgal sample onto agar (as mentioned above).
2. Run a series of force spectroscopy experiments, using a range of indentation forces. In our hands, by using a 3 N m^{-1} cantilever, a force range of 100–1500 nN is suitable (Figure 22.2). Force-spectroscopy experiments are simply a handful of point-AFM-indentations, as opposed to the grid-based force maps used later. Upper and lower bounds for this range of forces can be found from evaluation of

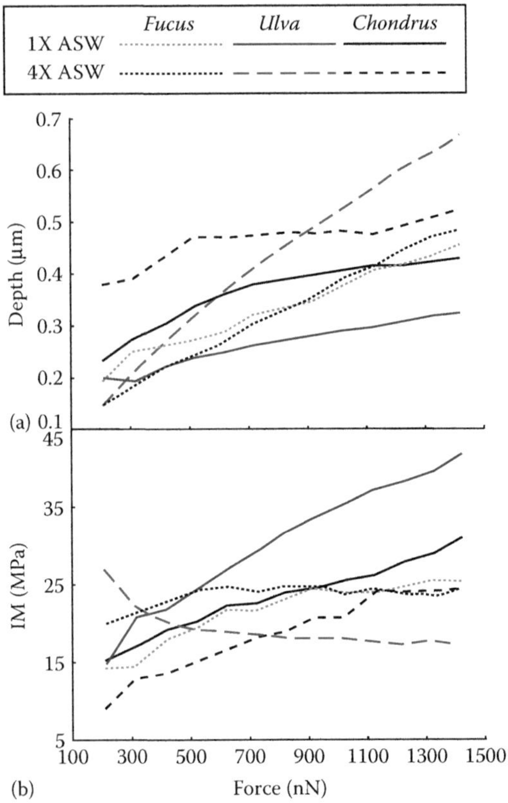

Figure 22.2 Influence of indentation force and media upon indentation depth and modulus in three macroalgal species. *Fucus*, *Ulva*, and *Chondrus* samples were subjected to a set of force spectroscopy experiments at a range of forces in 1X (solid) and 4X (dashed) artificial seawater (ASW). Resulting indentation depths (a) allow the experimenter to determine an appropriate force for further experiments and to examine the linearity of the material response. Resulting indentation (b) moduli (IM) allow the experimenter to assess the linearity of the material response as well, which should be considered alongside knowledge of cell-wall geometry. Examining sensitivity of both results to increased ASW concentration allows for the consideration of sample sensitivity to evaporation of indentation media during the experiment; *Fucus* is relatively insensitive, whereas *Ulva* is very sensitive.

force/depth plots, shown by AFM testing software, of initial tests at several reasonable forces (e.g., somewhere in range above) and allowing a generous margin of error for potential sample heterogeneity.

3. Analyze the data to extract the indentation depths and moduli. This can be achieved using the AFM manufacturer's software

(here JPK Data Processing, JPK AG, DE) or by custom-built code. In general, a Hertz fit to the indentation phase of the force indentation curve will yield an indentation modulus (IM; an approximation of the Young's modulus; MPa). In cases of extreme adhesion, a tangential fit yielding a stiffness value (N m^{-1}) may be more appropriate (Braybrook 2015).

4. Plot force by indentation depth and force by IM. Use the first plot to determine which forces yielded an indentation depth appropriate for the question and sample being tested: as an example, a depth that is <5% of the meristoderm cell depth. Use the second plot to assess whether your chosen force(s) are within a linear range, in which the IM change is relatively flat as force increases. This is ideal but may not always be achievable (Figure 22.2, *Fucus* vs. *Ulva* or *Chondrus*). You should choose a force for your experiments, which yields the desired indentation depth, based on these data.

22.4.3 Influence of osmoticum upon macroalgal mechanics

For the AFM-based mapping of elasticity in any given sample, we recommend using a grid of AFM-indentations, or force maps. Large, high-resolution force-mapping can take several hours to finish. In this time, water loss through evaporation may be detrimental to the sample, altering its mechanical properties. This is particularly important when using seawater in which evaporation of water can lead to rapid accumulation of salts and an altered osmoticum. Therefore, we recommend determining the influence of hypersaline conditions upon samples before initiating force scan experiments. This is variable between the species of macroalgae we have tested in 1X and 4X artificial seawater: *Fucus* shows no difference; *Chondrus* shows a similar trend, but variation in absolute values; and *Ulva* shows a large variation in mechanical properties (Figure 22.2). We recommend testing the sensitivity of your sample, so appropriate adjustments can be made if needed.

1. Mount a macroalgal sample onto agar, flood with 1X sea water.
2. Find a location on the sample that is easily identifiable in the future (take a screen shot of the location if the AFM is mounted with a camera).
3. Run a series of force-spectroscopy experiments, using the range of indentation forces used previously (see above).
4. Remove the microscope slide from the AFM machine, and swap the media for a stronger concentration (*Note 2*), leaving 1 h to allow diffusion of salts.
5. Repeat the series of force-spectroscopy experiments, using the range of indentation forces.

6. Compare the resultant indentation depths and IMs for the two concentrations of media.
7. Assess whether your sample is sensitive to changes in salinity. If it is, we recommend that the solution be exchanged often during the experiment. This may be achieved by making an initial waterline mark, and topping up to this during the experiment with distilled water. The changes seen here may be due to complex phenomenon: turgor pressure may be affected (as in plants) resulting in a lower measured IM. In the case of *Ulva*, in which the cell wall is over 20 μm thick (Messyasz et al. 2013) and a maximum indentation of 0.5 μm is applied, it is unlikely that turgor was influencing the measurements in 1X seawater; the material properties of the wall, and its hydrogel matrix, may change due to both changes in salt concentration and dehydration. It is likely that this is the dominant phenomenon seen in Figure 22.2 for *Ulva* and *Chondrus*.

22.4.4 Selection of tip size

Due to the large variety in shape and form within the macroalgae, it is important to determine a suitable tip size for use in experiments. Tip size influences the possible resolution of force mapping experiments, and it can also affect both the values of IM obtained and the tissue depth from which those data originate (larger beads deform larger areas of tissue) (*Note 3*).

1. Mount a macroalgal sample onto agar, flood with 1X seawater.
2. Find a location on the sample that is easily identifiable, this will aid in relocation of the position for analysis (take a screen shot of the location if the AFM has a camera).
3. Run a force-mapping experiment at the determined force. Force-mapping experiments allow for a positional map of both sample topography and IM/stiffness to be produced. This is helpful in determining where you are in a sample and what features you may be observing.
4. Change cantilever tip to another size and repeat your scan as above, ensuring that the indentation force remains constant. It is often difficult to return to the exact position after switching tips, use the image from step 2 to help you relocate.
5. Analyze the data, and plot the resultant IM/stiffness as a force map.
6. An ideal tip size will give good spatial resolution of tissue features. We recommend using a 10 nm tip as a starting point due to affordability and excellent resolution. Figure 22.3 demonstrates how tip size affects both the IM and resolution of the force map.

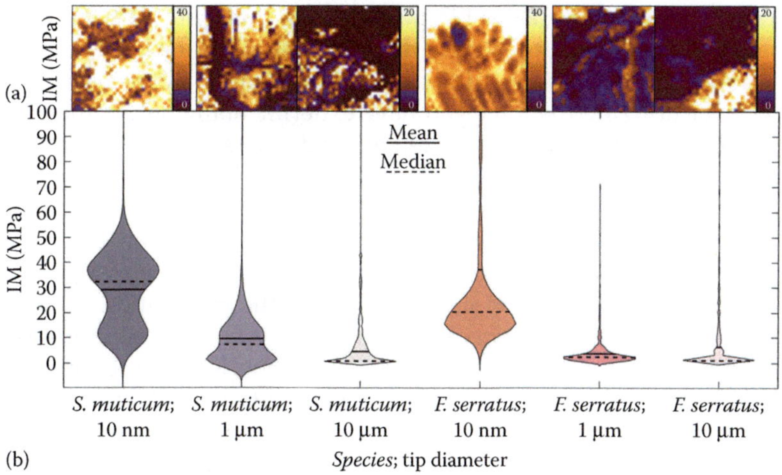

Figure 22.3 The influence of AFM tip size on spatial resolution and indentation modulus. Two species of brown alga, *F. serratus* and *S. muticum*, were tested using three tip diameters but constant indentation force (10 nm, 1 μm, 10 μm; 500 nN). Maps of indentation modulus (IM; 10 × 10 μm grids) are shown in (a). Smaller tip sizes yield better *XY* resolution of surface structures and larger tip sizes give spatially averaged values. (b) IM values vary with tip size and this may alter comparability of datasets between experiments. The experimenter should be consistent and clear on the tip size chosen.

22.4.5 *Performing atomic force microscopy based elasticity measurements*

As mentioned earlier, elasticity data in complex samples are best acquired in a force map or grid. This provides positional and topographical information about the sample that can help the experimenter to gather data from the appropriate structures or features of interest. A detailed protocol for plant AFM-based elasticity measurements is available as a reference for details of tip-calibration and data analysis (Braybrook 2015). In this final section, we will outline a basic elasticity experiment and highlight challenges unique to macroalgal samples.

1. Mount a macroalgal sample onto agar, flood with 1X seawater.
2. Place the sample-containing dish under the AFM head (with an appropriate, calibrated tip).
3. Navigate to the sample and area you wish to test. Note that several bio-AFMs allow the operator to choose the grid size and resolution of a force map. Depending on your biological question and resolution need, you may range from a 10 × 10 μm grid to a 100 × 100 μm

grid with 8 × 8–128 × 128 points along that grid. The larger and denser your grid, the longer the experiment but the better the resolution. This decision requires some optimization by the user.

4. After utilizing the force you have experimentally determined, start acquiring data in your force-map grid. Pay attention to the force–position curves produced by the AFM during the run (most bio-AFMs allow real-time observation of each curve as it is produced); macroalgal samples can suffer from adhesion due to their coatings of extracellular mucilage. When strong adhesion is present, as visualized by a negative force at the start of the indentation curve and end of the retraction curve (Figure 22.4), it is best to use a stiffness fit for elasticity determination. When adhesion is only present on the retraction curve, one can safely fit a Hertz-type model to the indentation curve to obtain an IM. Figure 22.5 shows the topography of a *Fucus* zygote rhizoid alongside the IM and stiffness maps obtained through analysis.
5. Occasionally, some samples will appear very soft, indenting much more than expected. In our experience, this results from one of two

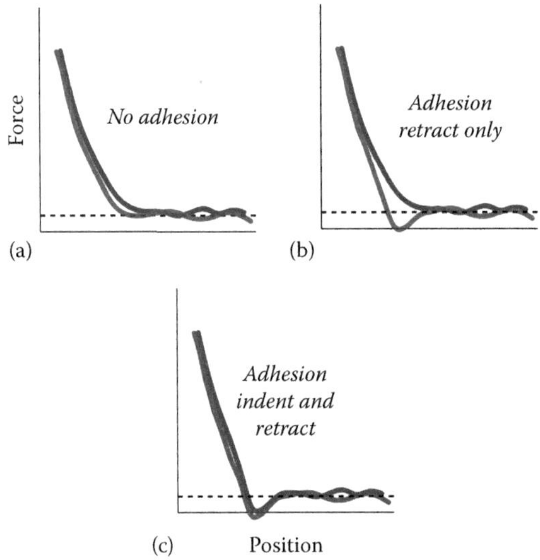

Figure 22.4 (a) The effect of sample adhesion on AFM-based indentation experiments. In an ideal situation, there is no adhesion between the sample and the tip. Neither the approach (dark gray) nor retraction (gray) curves result in values that dip below the baseline noncontact force (dashed line). (b) Biological samples often do show adhesion that may result in the tip being stuck on the sample during retraction or (c) additionally being pulled down onto the surface early. When adhesion is present, stiffness analysis is more appropriate.

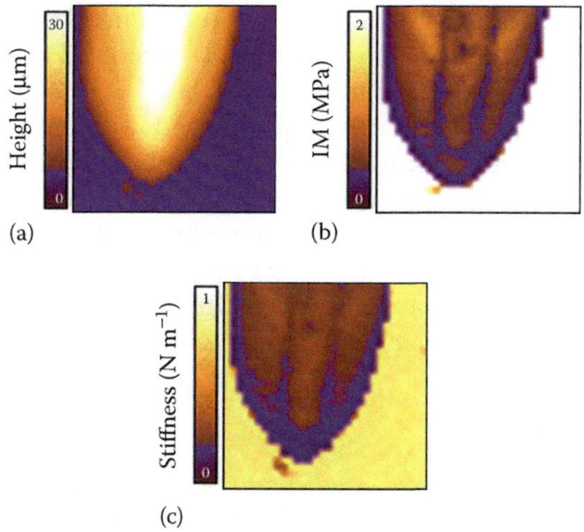

Figure 22.5 An example of data obtained from AFM-based indentation tests. Force maps of a rhizoid of *F. serratus*, showing (a) sample height; (b) indentation modulus (IM, MPa); and (c) stiffness (N m^{-1}). Indentation modulus and stiffness show similar qualitative trends. The sample height map may be used to filter out nonsample data (e.g., the glass surface here, low height and high IM, and stiffness).

scenarios: (1) either the sample has become unmounted in the agarose or is covered in agarose, or (2) there is an inordinate amount of mucilage on the sample. The latter can be helped by thoroughly rinsing your samples before mounting, the former requires remounting of the sample.

6. When extracting data from AFM maps (i.e., for comparison of regions), we recommend plotting values on the basis of height of the sample and extracting the IM values as a secondary value. This removes unconscious bias from selection of data. In Figure 22.5a, the lowest part of the topographic map is the glass slide (also the highest IM and stiffness values), and this could be used as a height filter threshold.

22.5 Notes

Note 1: When orientating samples for AFM analysis, consider the topographical structure of the algae in relation to angle and approach of the cantilever. The cantilever requires access to the sample—disrupting this access through ridges/obstacles will limit the capacity to perform the experiment. Obtaining an unobstructed sampling will require familiarization with the AFM you are using.

Note 2: When swapping solutions, we recommend placing the entire microscope slide (and therefore sample) inside a Petri dish and gently flooding with desired media.

Note 3: Manufactured tips can be purchased at a 10 nm diameter, and these are the most common types of probes. Specially made 1 μm tips can be ordered from most companies but will be more expensive. Tipless cantilevers are considerably cheaper than buying cantilevers premounted with larger (>1 μm) beads. Beads of a variety of sizes can be bought and mounted directly onto tipless cantilevers through two-component epoxy resin glue.

Acknowledgments

Funding for the work presented here was provided by the Gatsby Charitable Foundation (GAT3396/PR4; T.A.T. and S.A.B). M.L. was supported by the R. Lewin and F.E. Fritsch Prize Studentship in Phycology. J.L.K. was supported by the George and Lillian Schiff Studentship in Engineering. We thank our colleagues in the Phycomorph Network for their support and collaboration (COST Action FA1406; http://www.phycomorph.org/).

References

Braybrook, S.A. 2015. Chapter 13—Measuring the elasticity of plant cells with atomic force microscopy. *Methods in Cell Biology* 125: 237–254.

Braybrook, S.A., and Peaucelle, A. 2013. Mechano-chemical aspects of organ formation in arabidopsis thaliana: The relationship between auxin and pectin. *PLoS One* 8(3): e57813.

Fernandes, A.N., Chen, X.Y., Scotchford, C.A., Walker, J., Wells, D.M., Roberts, C.J., Everitt, N.M. 2012. Mechanical properties of epidermal cells of whole living roots of *Arabidopsis thaliana*: An atomic force microscopy study. *Physical Review E* 85: 1–8.

Messyasz, B., Czerwik-marcinkowska, J., Massalski, A., Uher, B., Rybak, A., Szendzina, L., Pikosz, M. 2013. Morphological and ultrastructural studies on *Ulva flexuosa* subsp. Pilifera (Chlorophyta) from Poland. *Acta Societatis Botanicorum Poloniae* 82: 157–163.

Milani, P., Gholamirad, M., Trass, J., Arnéodo, A., Boudaoud, A., Argoul, F., Hamant, O. 2011. *In vivo* analysis of local wall stiffness at the shoot apical meristem in *Arabidopsis* using atomic force microscopy. *The Plant Journal* 67: 1116–1123.

Peaucelle, A., Braybrook, S.A., Le Guillou, L., Bron, E., Kuhlemeier, C., Höfte, H. 2011. Pectin-induced changes in cell wall mechanics underlie organ initiation in Arabidopsis. *Current Biology* 21: 1720–1726.

Sampathkumar, A., Krupinski, P., Wightman, R., Milani, P., Berquand, A., Boudaoud, A., Hamant, O., Jönsson, H., Meyerowitz, E.M. 2014. Subcellular and supracellular mechanical stress prescribes cytoskeleton behavior in Arabidopsis cotyledon pavement cells. *eLife* 3: e01967.

Tesson, B., and Charrier, B. 2014. Brown algal morphogenesis: Atomic force microscopy as a tool to study the role of mechanical forces. *Frontier in Plant Science* 5: 1563.

Torode, T.A., Marcus, S.E., Jam, M., Tonon, T., Blackburn, R.S., Hervé, C., Knox, J.P. 2015. Monoclonal antibodies directed to fucoidan preparations from brown algae. *PLoS One* 10(2): e0118366.

chapter twenty three

Dynamic and microscale mapping of cell growth

Case of *Ectocarpus* filament cells

Hervé Rabillé, Bernard Billoud, Elodie Rolland, and Bénédicte Charrier

Contents

23.1 Introduction

Macroalgal cells are surrounded by a stiff cell wall that prevents any relative cell displacement inside the multicellular thallus (Mirabet et al. 2011; Charrier et al. 2012). As a consequence, their development and morphogenesis rely mainly on the pattern of cell division and cell elongation, which impacts the proper final shape (Kuchen et al. 2012; Bassel et al. 2014). Cell growth and morphogenesis imply mechanical deformations, which depends on molecular signaling pathways impacting the mechanical properties of the cellular components and/or act through force generation (Boudaoud 2010; Uyttewaal et al. 2010; Robinson et al. 2013). Before getting an integrated understanding of morphogenetic mechanisms, geometrical deformations of cells and tissues must be characterized and quantified (Mirabet et al. 2011). Moreover, characterization of cell or tissue mechanical properties (i.e., deformability) is needed to assess the physical and structural basis of deformations and to understand how such deformations are regulated (Geitmann 2006; Routier-Kierzkowska and Smith 2013). Deformations can be measured in response to external or internal force variations (i.e., turgor) and allow one to get insight into the possible biological mechanisms regulating the mechanical deformations (Peters and Bernstein 1997). Deformation can be linked to the localization of cellular factors such as cell-wall components, delivered vesicles, or elements of the cytoskeleton.

Many studies monitored the displacement of natural or artificial surface markers regularly or randomly scattered at the cell surface during growth and morphogenesis. Such studies were mainly conducted in green land plants (Richards and Kavanagh 1943; Erickson and Sax 1956; Hejnowicz and Brodzki 1960; Schnyder et al. 1987; Kutschera and Briggs 1988; Shaw et al. 2000; Menand et al. 2007; Armour et al. 2015; Yanagisawa et al. 2015), green algae (Green 1965b; Chen 1973; Métraux et al. 1980; Hejnowicz et al. 1977; Serikawa and Mandoli 1998; Von Dassow et al. 2001), and fungi (Castle 1958; Soll and Herman 1983; Soll et al. 1985; Staebell and Soll 1985; Bartnicki-Garcia et al. 2000; Abenza et al. 2015) but were much less attempted on other groups of macroalgae (Waaland et al. 1972; Kataoka 1982).

To our knowledge, no kinetic studies of cell or tissue growth pattern were conducted in brown algae (*Phaeophyceae*), whereas this class of macroalgae displays interesting specific features, among which are its great morphological diversity and its phylogenetic position (Charrier et al. 2012; Bogaert et al. 2013). In the current chapter, we describe a protocol to observe the fine pattern of cell-surface deformations during growth or in response to turgor variations (osmotic shocks) in the sporophytic filaments of *Ectocarpus siliculosus*, a model species for the brown algae. Cell surface was marked with commercial fluorescent microspheres that get stuck on the cell wall, and deformations were observed by epifluorescence or confocal laser-scanning microscopy. Instructions for the image analysis and calculation of cell-wall deformations are also provided. In the future, such protocol could be adapted on morphologically more complex, multicellular brown algal thalli to provide a more accurate description of the volume changes potentially experienced during the lifetime of these organisms.

23.2 State of the art

During the twentieth century, surface-marking experiments have been extensively used to follow the deformations of tissue and cell surface during the development of animal (Cox and Peacock 1978), plant, and fungal organisms (see references in the introduction). The principle was to follow, in the course of time, the displacement rate of markers regularly or randomly scattered at the surface of an organ. Doing so allowed one to localize the growing region, its local growth rate, and potential growth anisotropy. In land plants, most kinematic studies were focused on roots (Erickson and Sax 1956; Hejnowicz and Brodzki 1960), leaves (Richards and Kavanagh 1943; Kemp 1980; Schnyder et al. 1987, 1990), and stem growth (Kutschera and Briggs 1988). Determining the rate of distance increase between markers or between markers and a reference point provided segmental growth rates (Peters and Bernstein 1997). More complex calculation allowed one to calculate the *growth field* or *field of displacement vectors* that can provide many different kinematic, mechanical, and physiological information regarding the growing organ (Silk and Erickson 1979; Silk 1984). These numerical approaches are beyond the scope of the present protocol, which focuses on the analysis of cellular growth and morphogenesis.

Many studies were focused on individual cell morphogenesis. The aim was to observe the local growth dynamics, and in some cases, to assess the mechanical properties of the extracellular matrix (i.e., the plant and algal cell walls). Reaching these objectives required that (1) the surface of the cell be labeled with fine markers, to allow a resolution at the subcellular level; and (2) the relative displacement of markers be followed by time-lapse microscopy and their trajectory be recorded, traced, and analyzed.

To fulfill these requirements, different kinds of surface markers were used. In rare cases, items such as dust or cell fragments naturally occurring at the cell surface were considered (Suzaki and Williamson 1985), but in most studies artificial markers deposited on the cell surface were required. Their nature depends on the chemical composition of the cell wall and the growth medium. Small grains were released close to the cells (by pipetting), to which they bound randomly through ionic, weak chemical interactions with the cell wall. First studies using giant-celled characean algae such as *Nitella* or *Acetabularia* used anionic resin particles (Green 1965a, b; Chen 1973; Hejnowicz et al. 1977; Métraux et al. 1980; Kataoka 1982) or carbon particles (Waaland et al. 1972; Serikawa and Mandoli 1998; Von Dassow et al. 2001), whereas Castle (1958) used starch grain on the tip-growing sporangiophore of *Phycomyces*. In other eumycetes, polystyrenes beads (Soll et al. 1985; Staebell and Soll 1985) or carbon particles (Bartnicki-Garcia et al. 2000) were used. In all these studies, the displacements of surface markers were followed by time-lapse optical microscopy, sometimes with special devices using horizontal objectives to observe cells growing vertically (Castle 1958; Green 1965b; Chen 1973; Métraux et al. 1980). Most recent studies on fungal or land plant tip-growing cells used fluorescent surface markers such as commercial fluorescent microspheres (Shaw et al. 2000; Dumais et al. 2004; Menand et al. 2007; Abenza et al. 2015; Yanagisawa et al. 2015) or droplets of fluorescent paint (Armour et al. 2015), and deformation was monitored using epifluorescence or confocal scanning microscopy.

After the time-lapse acquisition, images must be extracted, and individual particle trajectories are mapped to quantify cell-surface deformations. In the first studies, pictures were imprinted, and morphometric measurements were done manually on imprinted images (Castle 1958; Green 1965b; Kataoka 1982). Use of image analysis software allowed more rapid image normalization (for size and growth axis orientation) and analysis (Soll et al. 1985; Staebell and Soll 1985; Bartnicki-Garcia et al. 2000; Shaw et al. 2000; Mine and Okuda 2003; Menand et al. 2007; Abenza et al. 2015; Armour et al. 2015; Yanagisawa et al. 2015).

Several methods for the calculation of the deformation were employed. The simplest studies used surface-marking experiments only to locate and measure the extent of the growth site in individual cells (Soll et al. 1985; Serikawa and Mandoli 1998; Menand et al. 2007). Bartnicki-Garcia et al. (2000) finely traced the trajectory of markers on the growing tip and compared them with theoretical trajectories predicted by various growth models. However, most studies, especially those dedicated to tip-growing cells, calculated the relative elemental growth rate as described in Peters and Bernstein (1997). This approaches the irreversible and ideal extension rate of an infinitely small elemental piece of cell wall in two perpendicular directions (and thus the growth anisotropy) as a function of its position along the cells (generally relatively to a point of reference) (Castle 1958; Green 1965b; Chen 1973; Hejnowicz et al. 1977; Shaw et al. 2000; Von Dassow et al. 2001; Dumais et al. 2004; Abenza et al. 2015). Some more complex, computational approaches also exist (see, e.g., Armour et al. 2015) but are out of the scope of this chapter.

23.3 *Materials*

23.3.1 *Reagents*

- Seawater.
- Ethanol 70% (v:v).

A stock solution of fluorescent microspheres. We used FluoSpheres™ amine, 0.2 µm, red (F8763, Molecular Probes). In the rest of the chapter, these surface markers will be referred to as *fluospheres*. Concentration of stock solution is 2% (w:v) in distilled water. Working solution of fluospheres is prepared by diluting the stock solution to 0.1%–0.01% (w:v) in seawater (Natural Sea Water [NSW] or artificial seawater [ASW]). The two solutions must be kept at 4°C in the dark. Fluospheres tend to agglomerate and to accumulate at the bottom of the tube, so the solution must be vigorously vortexed or sonicated before use. Keep the solution sterile if possible.

23.3.2 Ectocarpus *culture*

The material necessary for the standard cultivation of *Ectocarpus* was described previously (Le Bail and Charrier 2013). To carry out the protocol, additional material is required.

Sterile glass-bottom Petri dishes or sterile glass bottom multiwell plates (×6 or ×24 wells) for tissue culture. If cultivating *Ectocarpus* on these materials is unsuccessful, culture it on 22 × 22 mm sterile coverslips (that must be autoclaved in an oven at 180°C for 2 h) loaded in conventional Petri dishes (55 or 90 mm in diameter). Then prepare *homemade* glass bottom Petri dishes from this material according to the protocol described in Section 23.4.1.2.

23.3.3 *Equipment and software*

- Blu-tack®.
- Double-faced adhesive tape.
- Parafilm® strips.
- Waterproof silicone glue.
- Fine forceps, needle, and scalpel (or box cutter).
- Empty Petri dishes, which bottom is lined with Parafilm®.
- 1000 and 100 µL pipettors.
- Inverted epifluorescence or confocal laser-scanning microscope, equipped with a camera and a motorized stage. The whole system must be piloted by software that allows recording multiple positions on the stage and in time-lapse mode. We used a DMI6000-inverted epifluorescence videomicroscope (Leica) or TCS SP5

Acousto-Optical Beam Splitter (AOBS)–inverted confocal microscope (Leica) equipped with the LAS AF (v2.2.1, Leica) software.
- White LED lamp programmed on a 14:10 h day/light cycle.
- Confined, air-conditioned culture cabinet enclosing the microscope to maintain a temperature compatible with *Ectocarpus* cultivation.
- Hot air pistol, which temperature should be around 600°C–700°C.
- Metal perforator used to make holes in the bottom of the 55 mm polystyrene Petri dishes. This perforator is held at the extremity of a wood stick and heated up to about 600°C–700°C.

23.4 Experimental procedures

23.4.1 Preparation of living algal material for time-lapse microscopy

23.4.1.1 Gametes release on glass-bottom Petri dish, multiwell plate or free coverslips

A detailed protocol for the production of *Ectocarpus* filaments and gametes is described in Le Bail and Charrier (2013). Specifically for this chapter, the instructions are as follows:

1. Prepare *Ectocarpus* sporophytic filaments by loading gametes in sterile glass-bottom Petri dishes or coverslips as described in Le Bail and Charrier (2013).
2. Once the young filaments initiate branching (usually after 2 weeks), they are ready to be used for the time-lapse experiment.

23.4.1.2 Preparation of "homemade" glass bottom Petri dishes

Cultivation of *Ectocarpus* in Petri dishes equipped with a glass bottom is necessary for accurately monitoring the cell-surface markers under the microscope. Difficulties were met when trying to grow *Ectocarpus* filaments in glass-bottom Petri dishes or multiwell plates bought from commercial suppliers (*Note 1*). Alternatives consist in preparing *homemade* glass bottom Petri dishes. The protocol involves sticking glass coverslips with *Ectocarpus* filaments germinated on it below conventional Petri dishes (typically 55 mm in diameter) previously drilled with a hole in their center.

1. Two weeks before, make filaments grow on glass coverslips according to the procedure described in Section 23.4.1.1.
2. By using a metal perforator heated with a hot air gun (adequate temperature: 600°C–700°C), make a hole at the center of the bottom of empty 55 mm wide Petri dishes (*Note 2*). Ideally, the hole must be between 10 and 12 mm in diameter (*Note 3*).

3. Under a sterile laminar flow hood, spray the Petri dishes (and their lid) copiously with 70% ethanol and let them dry overnight under the UV light of the hood.
4. The next day, keep the dishes under the laminar flow hood. Once they are completely dry, cut 22 × 22 mm square pieces of double-faced adhesive tape (i.e., of the size of a coverslip). Stick each piece below the bottom of a drilled dish (*Note 4*), centering it on the hole of the dish.
5. By using a fine sterilized scalpel or cutter, cut the central part of each piece of the adhesive tape facing the hole of the dish.
6. Open the culture dishes bearing the inoculated coverslips. Remove most of the filaments on the surface of the coverslip with fine forceps, except a few at the very center.
7. Remove the coverslip from its Petri dish using forceps and a fine needle. Gently dry the coverslips with a paper towel, except at the center of the top face, where a few number of filaments must be kept (*Note 5*). Be sure to let a small drop of seawater at this place to protect those filaments from desiccation. This drop must have a diameter lower than that of the central hole of the drilled dish.
8. Put the coverslip on the lid of a 50 mL Falcon tube. Then stick it right below a drilled dish with a piece of double-faced adhesive tape. Make sure that the square-shaped piece of double-faced adhesive tape is right on the coverslip, and that the central filaments within their drop of seawater end up in the hole of the dish. Press gently for about 10 s on the internal face of the dish, around the central hole (*Note 6*).
9. Immediately fill the dish with seawater and seal it with Parafilm®.
10. Paste the border of the coverslip (i.e., at the junction with the bottom of the dish) with silicone glue using the tip of a pipettor cone (*Note 7*).
11. Place the dish under standard culture conditions for one night. Make the dish straddle two inverted empty dishes or Falcon tube lids to have the bottom face (with silicone glue) not in contact with the shelf of the culture cabinet.
12. One day later, when the silicone glue is dry, the *homemade* glass-bottom Petri dishes are ready to be used for time-lapse experiments.

23.4.2 Cell-surface deformation monitoring using time-lapse microscopy

High magnification (×1000 or ×2000) is required to finely observe the position and displacement of fluospheres at the surface of *Ectocarpus* cells. Moreover, the focus plane must be thin enough to distinguish fluospheres in the different planes of the cell. For these reasons, we recommend confocal microscopy (*Note 8*). We used a TCS SP5 AOBS–inverted confocal microscope (Leica) or a DMI6000-inverted video microscope, which are both controlled by a LAS AF (v2.2.1, Leica) software.

23.4.2.1 *Prior preparation of the equipment*

The microscope devices must be switched on and ready to use before filaments labeling with fluospheres.

1. Select the appropriate lasers. As we used red fluospheres (F8763, excitation-emission maximums: 580/605), we used a DPSS 561 nm laser.
2. Select the appropriate excitation wavelength and appropriate detection range (Photomultiplier Tubes [PMT] module) according to the characteristic of the fluospheres used. Activate the *bright-field* PMT that will reconstruct a virtual bright-field image of the cell. Both fluorescent and bright-field pictures must be acquired at every acquisition (*Note 9*).

23.4.2.2 *Cell-surface labeling with fluorescent microspheres*

1. Open a glass bottom Petri dish (commercial or homemade, cf Section 23.4.1.2) or multiwell plate, with filaments germinated inside. Remove most of the culture medium and let the filaments immersed into some drops of the medium.
2. Just above the filaments, add a small volume (20–50 µL) of fluosphere solution (*Note 10*). Pipette the fluosphere solution up and down repeatedly and rapidly on the filaments for 1 min (*Note 11*). Take care not to tear off the filaments from the glass surface.
3. Add some milliliters of seawaters completed with 10 µL/mL of Provasoli's enrichment medium (ASWp or NSWp) and shake the dish gently and manually to wash out the exceeding fluospheres not bound to the filament surface.
4. Remove the washing medium and fill the dish/well with fresh seawater. Seal with Parafilm® stripes. Mount it immediately under the microscope previously prepared for time-lapse experiments (Section 23.4.2.1).

23.4.2.3 *Measuring irreversible deformations during growth (plastic strain)*

In this section, we describe how to monitor cell-surface deformation during growth or morphogenesis. The example here is the tip-growth of *Ectocarpus* sporophyte vegetative filaments. The seawater volume in which filaments are incubated (~15 mL in a 55 mm Petri dish, or 2–5 mL in a well of a multiwell plate) is enough to allow normal growth without renewal of the medium during a few days.

1. As soon as the filaments have been labeled with fluorescent markers (see Section 23.4.2.2), mount the dish on the appropriate object holder. If there is any slack between the dish and the object holder, stuck it by inserting small pieces of Blu-tack® or modeling clay.
2. Mount the object-holder on the motorized stage of the inverted microscope and locate an interesting individual at the surface of the glass bottom (×100 immersion objective is recommended).
3. Select the cells of interest: Adequate cells must be located just above the glass surface and with their growth axis parallel to it (i.e., focus plane) (*Note 12*). Mark the position of each cell (*x*/*y*/*z* coordinates) (*Note 13*). From this moment, the dish must not be moved from its location on the motorized stage. If the imaging software allows it, optimize the experiment by monitoring several cells in different wells of a multiwell plate, thus testing different growth conditions in the same experiment (*Note 14*). Save the set of positions for the next time point.
4. Once all the cells with valuable labeling have been selected and marked, initiate the time-lapse and acquire the first time point (t_0).
5. Before acquiring pictures at the next steps (t_n), check that the positions were conserved (*Note 15*). Acquire a new series of pictures and repeat the procedure as many times as necessary.
6. When the time-lapse is completed, export the pictures. Figure 23.1 shows an example of time-lapse series of pictures produced by a growing apical cell, showing displacement of fluospheres restricted to the dome.

23.4.2.4 Measuring deformations induced by hypo- or hypertonic shock (mainly elastic deformations)

To acquire data about the in-plane elasticity of the cell wall, deformation of the cell surface in response to osmotic shocks can be monitored. That will also provide information about the mechanical anisotropy of the cell wall, that is, the different stretching potentials in one direction compared with the perpendicular one.

1. Prepare the microscope and all the necessary devices as described in Section 23.4.2.1. As the deformations take place on very short term, culture cabinet and LED lamp are not required (*Note 16*).
2. After filament labeling with fluospheres (Section 23.4.2.2), add a small amount of seawater (osmolarity ~1100 mOsm L^{-1}) in the dish. Do not seal it with Parafilm®.
3. Mount the dish under the microscope without its lid. Record the position (*x*/*y*/*z*) of a set of interesting cells. To ensure reliable acquisitions of very fast cell response to osmotic shocks, we recommend

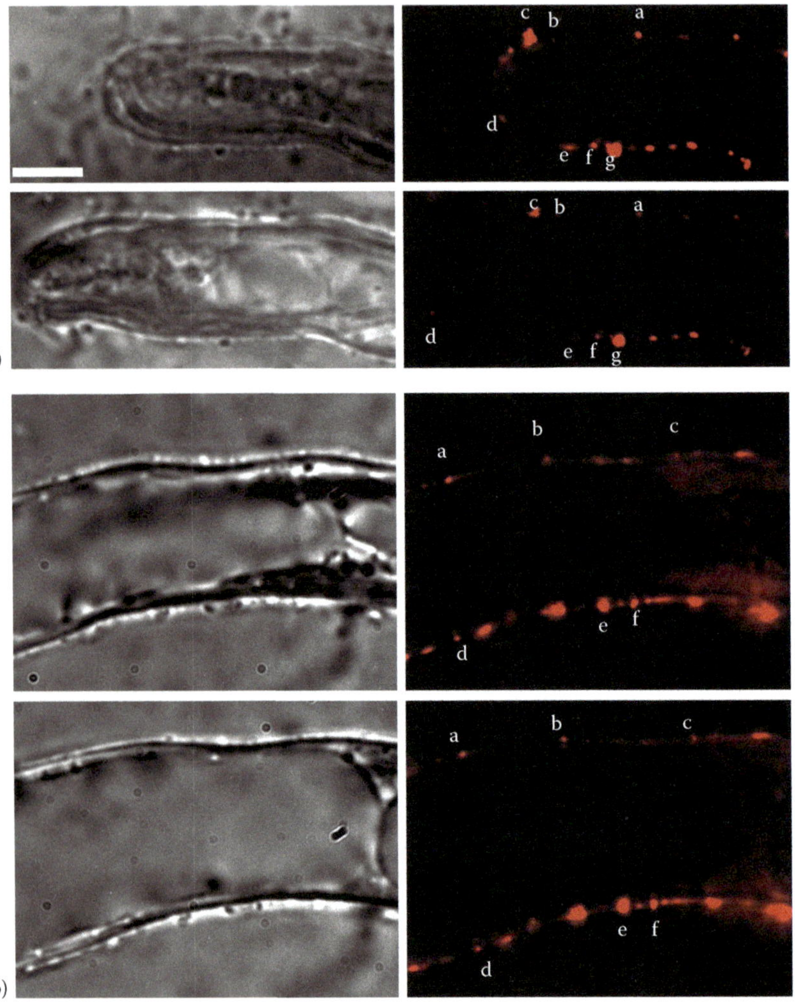

(a)

(b)

Figure 23.1 Example of surface deformation pattern of *Ectocarpus* cells observed by surface labeling with fluospheres, and using time-lapse fluorescent microscopy. Some individual markers (or marker aggregates) are labeled with letters to display their respective displacement. (a) Cell-wall strain observed during growth of one *Ectocarpus* apical cell. Time elapsed between the first (top) and the second (bottom) acquisitions was 4 h. Note the displacement of marker *d* compared with markers *c* and *e*; (b) example of cell-wall deformation in an *Ectocarpus* (subapical) cell in response to a hypotonic shock (transfer from NSW ~1100 mOsm L^{-1}, to half-strength NSW ~550 mOsm L^{-1}, resulting in cell inflation). Images show the cell just before (top) and ~1 min after transfer into the diluted medium (bottom). Note the significant circumferential displacement of markers *a*, *b*, and *c* compared with the position of markers *d*, *e*, and *f*, whereas the meridional displacement is negligible. (left) bright-field image; (right) corresponding fluorescent picture. Bar = 5 μm for all pictures.

not selecting more than 4–5 positions, and no more than 1–2 focus planes per position (*Note 17*).

4. Acquire the first series of pictures corresponding to the initial state of the cells (t_0 = normal osmolarity) (*Note 18*).
5. By using a pipette, take off most of the seawater from the dish.
6. Immediately after, fill the dish with a large volume (from 2 to 15 mL, according to the volume available) of hypotonic or hypertonic seawater (*Note 19*). Then, correct each position that will probably be modified in the *z*-axis because of the change of medium (*Note 20*).
7. After a short time (elastic stretching of *Ectocarpus* cell wall takes about 1 min), acquire a second series of pictures at the same positions. This second series will correspond to the stable, inflated (after hypotonic shocks), or shrunk (after hypertonic shocks) state of the cells.
8. Export the pictures. Figure 23.1 shows a cell from an *Ectocarpus* filament, surface deformation of which was monitored after a hypotonic shock in half-strength seawater.

23.4.3 *Picture analysis: Quantification of cell-surface marker displacement*

This final section provides some instructions for image analysis, measurement, and quantification of surface-marker displacement and for corresponding calculation of significant geometric or mechanical parameters.

1. For each time point and each position, export one bright-field picture of the median plane of the cell and one or several fluorescent pictures of one or several planes of the cell showing fluospheres.
2. Process images using dedicated software such as GIMP (Smith and Joost 2012; available at https://www.gimp.org). First, if time-lapse facilities do not allow one to faithfully superpose successive images, this should be done manually by using translations and/or rotations, taking static markers outside the cell as landmarks. On each image, draw a 1-pixel large line onto the cell wall, making the cell contour. For convenience, this line should be drawn in color if the microscope picture is black-and-white, and/or drawn on a separate layer. Similarly, denote each marker by a pixel in another color.
3. From each cell contour, compute a cubic spline, thus creating one spline per image. Smoothing should be carefully adjusted to avoid effects at pixel-size scale, yet keeping local contour curvature. Similarly, build a trajectory for each marker, by joining all marker pixels of the same color in the time series, leading to one spline per marker. The degree of each spline should be three, but if less than four-time steps are available, then the degree should be $n - 1$ for n images.

4. Consider each intersection between a cell contour and a marker trajectory (each represented by its spline). From these, it is possible to compute the mean velocities of each point during one interimage duration, and the angle between the two curves (deduced from the first derivatives at the intersection point). These raw data can be processed by using standard statistical analysis software, such as R Core Team (2016).

Software designed to perform steps 3 and 4 can be obtained upon request to the authors.

23.5 Notes

Note 1: In most cases, we observed a bloom of bacteria invading the whole glass bottom, rapidly killing and consuming all the early *Ectocarpus* filaments.

Note 2: This step is dangerous and must be carried out in adequate safety conditions. The metal perforator must be held at the extremity of a handle made of an isolating material (e.g., wood) to prevent any burning injury.

Note 3: The hole must be much lower in size compared with the size of the coverslip (22 × 22 mm).

Note 4: To efficiently stick the double-faced adhesive tape on the exterior face of the dish bottom, turn the dish upside-down without its lid, on the lid of a 50 mL Falcon tube (the top face of the Falcon lid must be in contact with the internal face of the bottom of the dish). Then press the adhesive tape on the exterior face of the dish bottom.

Note 5: Clean and dry the peripheral border of the coverslip to ensure a good sticking of the coverslip under the double-faced adhesive tape (cf the next step). Indeed, *Ectocarpus* and bacteria growing on the surface of the coverslip seem to deposit some substances that make the surface more hydrophilic. Cleaning the surface with a paper towel will thus prevent the central drop of seawater from spreading on the coverslip.

Note 6: You can press on the internal face of the Petri dish around the central hole using a 50 mL Falcon tube without its lid and turned upside-down.

Note 7: Be careful to keep the center of the coverslip clean of any trace of glue, especially the area facing the central hole. This area must be perfectly clean when high-magnification observations are planned. Make sure that the silicone glue forms a thin bulge all around the periphery of the coverslip, which does not diffuse to the center of the coverslip. It would otherwise impede the movement of the thin high-magnification objective.

Note 8: It is also possible to use epifluorescence video microscope. Note that, on both types of microscope, the exposition of *Ectocarpus* filament to strong white or fluorescent light has deleterious effects on the cell physiology. In our experiment, we observed that less than half of the apical cells continue to grow after time-lapse initiation.

Note 9: Bright-field pictures are necessary for drawing the contour of the cell required for picture analyses.

Note 10: The fluospheres tend to aggregate and sink at the bottom of the tube. Vortex or sonicate the solution just before using it. The concentration of the working fluospheres solution must be adjusted according to the algal material used and to the required density of fluospheres on the cell surface. Make some tests before starting a time-lapse. We used a 0.1% (w:v) working concentration.

Note 11: The rapid flow of fluospheres solution around the filament will enhance the sticking of fluospheres on the surface of the cell.

Note 12: Observing the cell in a plane passing through the symmetry axis will facilitate the geometrical analysis of fluosphere displacement.

Note 13: Each position is given a number. When needed, depending on the software used, clicking on the appropriate number on the dropdown list allows one to reach any position on the coverslips rapidly. Each position can be easily reassigned, suppressed, or renamed.

Note 14: If working on a multiwell plate with oil objective, it may be necessary to put back the objective when switching from one well to another. In this situation, the number of well usable for one time-lapse would be limited, and the time-lapse would have to be conducted manually.

Note 15: If the time interval between each acquisition is large (i.e., 12 h or more), it might be preferable to switch off all the devices except the air-conditioned cabinet and the LED lamps between each acquisition to avoid any extra heat produced by the equipment. Make sure that the positions were maintained before any subsequent acquisition.

Note 16: However, if the environment around the microscope is too hot (due to the activity of the camera or another device), the use of an air-conditioned cabinet would be required.

Note 17: At every chosen position, all the cells to be analyzed should be more or less in the same meridional plane. For this purpose, you should select filaments lying on the glass surface. Nonetheless, note that the physical adherence of cells to the glass surface could impact the deformability of cells.

Note 18: When positions are unstable, acquire each position one by one.

Note 19: The large volume of hypo/hypertonic medium will exceed by far that of the remaining initial medium. Therefore, the impact of the latter on the final osmolarity is negligible.

Note 20: The addition of several milliliters of medium in the dish will increase the total weight on the object-holder and the stage. This can lower all the positions by several μm and thereby require correction of the *z*-position before the acquisition.

References

Abenza, J. F., E. Couturier, J. Dodgson, J. Dickmann, A. Chessel, J. Dumais, and R. E. C. Salas. 2015. Wall mechanics and exocytosis define the shape of growth domains in fission yeast. *Nature Commun* 6. Nature Publishing Group: 8400. doi:10.1038/ncomms9400.

Armour, W. J., D. A. Barton, A. M. Law, and R. L. Overall. 2015. Differential growth in periclinal and anticlinal walls during lobe formation in arabidopsis cotyledon pavement cells. *Plant Cell* 27 (9): 2484–2500. doi:10.1105/tpc.114.126664.

Bartnicki-Garcia, S., C. E. Bracker, G. Gierz, R. López-Franco, and H. Lu. 2000. Mapping the growth of fungal hyphae: Orthogonal cell wall expansion during tip growth and the role of turgor. *Biophys J* 79 (5): 2382–2390. doi:10.1016/S0006-3495(00)76483-6.

Bassel, G. W., P. Stamm, G. Mosca, P. Barbier de Reuille, D. J. Gibbs, R. Winter, A. Janka, M. J. Holdsworth, and R. S. Smith. 2014. Mechanical constraints imposed by 3D cellular geometry and arrangement modulate growth patterns in the arabidopsis embryo. *Proc Natl Acad Sci USA* 111 (23): 8685–8690. doi:10.1073/pnas.1404616111.

Bogaert, K. A., A. Arun, S. M. Coelho, and O. De Clerck. 2013. Brown algae as a model for plant organogenesis. In *Plant Organogenesis: Methods and Protocols*, I. De Smet (Ed.), Vol. 959, pp. 97–125. New York: Springer Science+Business Media. doi:10.1007/978-1-62703-221-6.

Boudaoud, A. 2010. An introduction to the mechanics of morphogenesis for plant biologists. *Trends Plant Sci* 15 (6): 353–360. doi:10.1016/j.tplants.2010.04.002.

Castle, E. S. 1958. The topography of tip growth in a plant cell. *J Gen Physiol* 41: 377–383.

Charrier, B., A. Le Bail, B. de Reviers, A. L. Bail, and B. D. Reviers. 2012. Plant proteus: Brown algal morphological plasticity and underlying developmental mechanisms. *Trends Plant Sci* 17 (8): 468–477. doi:10.1016/j.tplants.2012.03.003.

Chen, J. C. W. 1973. The kinetics of tip growth in the nitella rhizoid. *Plant Cell Physiol* 14 (4): 631–640.

Cox, R. W., and M. A. Peacock. 1978. The velocity field of growing ear cartilage. *J Anat* 126 (Pt 3): 555.

Dumais, J., S. R. Long, and S. L. Shaw. 2004. The mechanics of surface expansion anisotropy in *Medicago Truncatula* root hairs. *Plant Physiol* 136 (2): 3266–3275. doi:10.1104/pp.104.043752.

Erickson, R. O., and K. B. Sax. 1956. Elemental growth rate of the primary root of zea mays. *Proc Natl Acad Sci USA* 100 (5): 487–498.

Geitmann, A. 2006. Experimental approaches used to quantify physical parameters at cellular and subcellular levels. *Am J Bot* 93 (10): 1380–1390. doi:10.3732/ajb.93.10.1380.

Green, P. B. 1965a. Anion-exchange resin spheres as marking material for wet cell surfaces. *Exp Cell Res* 40: 195–196.

Green, P. B. 1965b. Pathways of cellular morphogenesis. A diversity in nitella. *J Cell Biol* 27 (2): 343–363. doi:10.1083/jcb.27.2.343.
Hejnowicz, Z., and P. Brodzki. 1960. The growth of root cells as the function of time and their position in the root. *Acta Soc Bot Pol* 29 (4): 625–644. doi:10.5586/asbp.1960.032.
Hejnowicz, Z., B. Heinemann, and A. Sievers. 1977. Tip growth: Patterns of growth rate and stress in the chara rhizoid. *Zeitschrift für Pflanzenphysiologie* 81 (5): 409–424. doi:10.1016/S0044-328X(77)80176-1.
Kataoka, H. 1982. Colchicine-induced expansion of vaucheria cell apex. Alteration from isotropic to transversally anisotropic growth. *Bot Mag Tokyo* 95 (3): 317–330. doi:10.1007/BF02488543.
Kemp, D. R. 1980. The location and size of the extension zone of emerging wheat leaves. *New Phytol* 84: 729–737.
Kuchen, E. E., S. Fox, P. Barbier de Reuille, R. Kennaway, S. Bensmihen, J. Avondo, G. M. Calder et al. 2012. Generation of leaf shape through early patterns of growth and tissue polarity. *Science* 335 (111): 1092–1095. doi:10.1126/science.1245938.
Kutschera, U., and W. R. Briggs. 1988. Growth, in vivo extensibility, and tissue tension in developing pea internodes. *Plant Physiol* 86 (1): 306–311.
Le Bail, A., and B. Charrier. 2013. Culture methods and mutant generation in the filamentous brown algae *Ectocarpus Siliculosus*. In *Plant Organogenesis: Methods and Protocols*, I. De Smet (Ed.), Vol. 959, pp. 235–245. New York: Springer Science+Business Media New York.
Menand, B., G. Calder, and L. Dolan. 2007. Both chloronemal and caulonemal cells expand by tip growth in the moss *Physcomitrella Patens*. *J Exp Bot* 58 (7): 1843–1849. doi:10.1093/jxb/erm047.
Métraux, J.-P., P. A. Richmond, and L. Taiz. 1980. Control of cell elongation in *Nitella* by endogenous cell wall pH gradients. *Plant Physiol* 65 (2): 204–210.
Mine, I., and K. Okuda. 2003. Extensibility of isolated cell walls in the giant tip-growing cells of the Xanthophycean alga *Vaucheria Terrestris*. *Planta* 217 (3): 425–435. doi:10.1007/s00425-003-0993-8.
Mirabet, V., P. Das, A. Boudaoud, and O. Hamant. 2011. The role of mechanical forces in plant morphogenesis. *Ann Rev Plant Biol* 62: 365–385. doi:10.1146/annurev-arplant-042110-103852.
Peters, W. S., and N. Bernstein. 1997. The determination of relative elemental growth rate profiles from segmental growth rates. *Plant Physiol* 113: 1395–1404.
R Core Team. 2016. *R: A Language and Environment for Statistical Computing*. Vienna, Austria: R Foundation for Statistical Computing. (https://www.R-project.org/)
Richards, O. W., and A. J. Kavanagh. 1943. The analysis of the relative growth gradients and changing form of growing organisms: Illustrated by the tobacco leaf. *Am Nat* 77 (772): 385–399. doi:10.1086/281140.
Robinson, S., A. Burian, E. Couturier, B. Landrein, M. Louveaux, E. D. Neumann, A. Peaucelle, A. Weber, and N. Nakayama. 2013. Mechanical control of morphogenesis at the shoot apex. *J Exp Bot* 64 (15): 4729–4744. doi:10.1093/jxb/ert199.
Routier-Kierzkowska, A. L., and R. S. Smith. 2013. Measuring the mechanics of morphogenesis. *Curr Opin Plant Biol* 16 (1): 25–32. doi:10.1016/j.pbi.2012.11.002.
Schnyder, H., C. J. Nelson, and J. H Coutts. 1987. Assessment of spatial distribution of growth in the elongation zone of grass leaf blades. *Plant Physiol* 85 (1): 290–293. doi:10.1104/pp.85.1.290.

Schnyder, H., S. Seo, I. F. Rademacher, and W. Kühbauch. 1990. Spatial distribution of growth rates and of epidermal cell lengths in the elongation zone during leaf development in *Lolium Perenne* L. *Planta* 181 (3): 423–431. doi:10.1007/BF00195897.

Serikawa, K. A., and D. F. Mandoli. 1998. An analysis of morphogenesis of the reproductive whorl of *Acetabularia Acetabulum*. *Planta* 207 (1): 96–104. doi:10.1007/s004250050460.

Shaw, S. L., J. Dumais, and S. R. Long. 2000. Cell surface expansion in polarly growing root hairs of *Medicago Truncatula*. *Plant Physiol* 124 (3): 959–970. doi:10.1104/pp.124.3.959.

Silk, W. K. 1984. Quantitative descriptions of development. *Ann Rev Plant Physio* 35 (1): 479–518. doi:10.1146/annurev.arplant.35.1.479.

Silk, W. K., and R. O. Erickson. 1979. Kinematics of plant growth. *J Theor Biol* 76 (4): 481–501. doi:10.1016/0022-5193(79)90014-6.

Smith, J., and R. Joost. 2012. *GIMP for Absolute Beginners*. Berkeley, CA: Apress.

Soll, D. R., and M. A. Herman. 1983. Growth and the inducibility of mycelium formation in *Candida Albicans*: A single-cell analysis using a perfusion chamber. *J Gen Microbiol* 129 (9): 2809–2824.

Soll, D. R., M. A. Herman, and M. A. Staebell. 1985. The involvement of cell wall expansion in the two modes of mycelium formation of *Candida Albicans*. *J Gen Microbiol* 131 (9): 2367–2375.

Staebell, M., and D. R. Soll. 1985. Temporal and spatial differences in cell wall expansion during bud and mycelium formation in *Candida Albicans*. *J Gen Microbiol* 131 (6): 1467–1480. doi:10.1099/00221287-131-6-1467.

Suzaki, T., and R. E. Williamson. 1985. Euglenoid movement in *Euglena Fusca*: Evidence for sliding between *Pellicular Strips*. *Protoplasma* 124 (1–2): 137–146. doi:10.1007/BF01279733.

Uyttewaal, M., J. Traas, and O. Hamant. 2010. Integrating physical stress, growth, and development. *Curr Opin Plant Biol* 13 (1): 46–52. doi:10.1016/j.pbi.2009.10.004.

Von Dassow, M., G. M. Odell, and D. F. Mandoli. 2001. Relationships between growth, morphology and wall stress in the stalk of *Acetabularia Acetabulum*. *Planta* 213 (4): 659–666. doi:10.1007/s004250100538.

Waaland, S. D., J. R. Waaland, and R. Cleland. 1972. A new pattern of plant cell elongation: Bipolar band growth. *J Cell Biol* 54: 184–190.

Yanagisawa, M., A. S. Desyatova, S. A. Belteton, E. L. Mallery, J. A. Turner, and D. B. Szymanski. 2015. Patterning mechanisms of cytoskeletal and cell wall systems during leaf trichome morphogenesis. *Nat Plants* 1 (3): 15014. doi:10.1038/nplants.2015.14.

chapter twenty four

Actin fluorescent staining in the filamentous brown alga Ectocarpus siliculosus

Hervé Rabillé, Maria Koutalianou, Bénédicte Charrier, and Christos Katsaros

Contents

24.1 Introduction

Brown algae (Phaeophyceae) are a very interesting group of macroalgae that acquired a complex multicellularity independently of land plants or animals (Bogaert et al. 2013). In the past 10 years, genetic and molecular studies of fundamental processes governing brown algal development have benefited from the selection of *Ectocarpus siliculosus* as a new model species for this group, because of its very simple filamentous body plan and of the easiness of its cultivation (Peters et al. 2004; Charrier et al. 2008; Le Bail and Charrier 2013).

A major component involved in brown algal morphogenesis is the cytoskeleton, consisting of microtubules (MTs) and actin filaments (AFs). The role of the cytoskeleton in the brown algal morphogenesis has repeatedly been underlined in several studies, particularly after the development of immunofluorescence techniques (Katsaros et al. 2006, for literature).

Contrary to higher plants, brown algae bear a unique and complex cell wall, which caused problems in the localization of cytoskeleton, when protocols described for plant cells were followed. Therefore, it became necessary to develop modified protocols suitable for brown algae (Coelho et al. 2012a).

As in land plants (Viridiplantae), fungi and animals (Opistokonts), AFs are thought to play a very important role in cell and tissue development in brown algae. Previous works showed their importance in the polarization processes in fucoid zygotes (Hable et al. 2003; Bisgrove and Kropf 2004; Hable and Kropf 2005; Bogaert et al. 2013), branching in filamentous gametophytes (Varvarigos et al. 2004), cytokinesis (Karyophyllis et al. 2000a; Bisgrove and Kropf 2004; Nagasato et al. 2010), and growth and morphogenesis of tip-growing apical cell of *Sphacelaria rigidula* (Karyophyllis et al. 2000b; Katsaros et al. 2002, 2003, 2006).

In some of these studies, the fine structure of AF networks was determined by immunolabeling. AF networks can display various organizations, and to understand their role in cellular processes such as polarization, intracellular trafficking or growth, knowledge of its 3D organization within the cell is required (Chebli et al. 2013).

When the transformation is possible, an actin-binding protein probe labeled with GFP can be used to reveal the actin organization in living cells (Wilsen et al. 2006; Meijer et al. 2014), but this technique is not available in brown algae yet. Immunolocalization on fixed material represents a good alternative to GFP tagging, with the additional advantage of revealing the majority of the AF structures present in a cell in a single experiment, contrary to *in vivo* GFP-tagged probe that can only recognize some specific AF configurations (Wilsen et al. 2006).

As AFs are very delicate structures, moreover sensitive to the fixation procedure (Katsaros et al. 2006), a modified protocol has been developed for brown algal cells. The detailed steps of this alternative protocol using Rhodamine–Phalloidin labeling are described in this chapter. It has proven to be successful in *Sphacelaria* and *Dictyota*, but the version given has been adapted for *Ectocarpus siliculosus*, the model brown algal species for which protocols for immunolabeling of cytoplasmic proteins has already been established (Coelho et al. 2012a, b).

24.2 State of the art

Actin fluorescent staining has been tried in all the major plant lineages, including land plants (Traas et al. 1987; Kakimoto and Shibaoka 1987a; Sonobe and Shibaoka 1989; Doris and Steer 1996; Lovy-Wheeler et al. 2005), fungi (Heath et al. 2000), microalgae (Harper et al. 1992; Pflügl-Haill et al. 2000; Hawkins et al. 2003), and macroalgae (Garbary et al. 1992; Katsaros et al. 2006). A fluorescent probe conjugated to phalloidin, a toxic bicyclic peptide first extracted from the fungus *Amanita phalloides* (Barden et al. 1987), is used to label AFs, although several anti-actin antibodies are also used.

Most of AF staining techniques on plant cells were first established for terrestrial plants (Perdue and Parthasarathy 1985; Kakimoto and Shibaoka 1987a, b; Parthasarathy 1987; Traas et al. 1987; Sonobe and Shibaoka 1989; Heslop-Harrison and Heslop-Harrison 1991; Doris and Steer 1996; Wasteneys et al. 1997; Vitha et al. 2000; Lovy-Wheeler et al. 2005; Dyachok et al. 2010). One of the major inputs from these works on land plant immunocytochemistry was the use of *m*-maleimidobenzoyl *N*-hydroxysuccinimide ester (commonly abbreviated as MBS) as an actin-stabilizing reagent added before the fixation steps (Sonobe and Shibaoka 1989; Doris and Steer 1996). Another finding was the necessity to partially digest the cell wall (CW) to allow the fluorescent probe to get access to the cytoplasm. In this case, the enzymes used were cellulases and pectolyases (Lloyd et al. 1979; Kakimoto and Shibaoka 1987b; Sonobe and Shibaoka 1989; Vitha et al. 2000; Szechyńska-Hebda et al. 2006; Dyachok et al. 2010).

In green macroalgae, very few studies of AF organization were conducted, except in the Characean algae (*Chara, Nitella*) and some coenocytic algae such as *Acetabularia* and *Caulerpa*. As in land plants, the fixation procedure was commonly carried out with classical aldehyde compounds, but sometimes it could be omitted, and AFs were labeled directly *in vivo* (Tewinkel et al. 1989; Sampson and Pickett-Heaps 2001). Curiously, the CW digestion step was omitted in most studies. In some cases, simple cell permeabilization appeared sufficient (Tewinkel et al. 1989; Braun and Wasteneys 1998; Sampson and Pickett-Heaps 2001), whereas in others, methods to get through the CW barrier were used (e.g., freeze-chattering of rhizoid and protonemata of characean algae; Braun and Wasteneys 1998). Moreover, in the giant coenocytic green alga *Acetabularia*, the CW showed to be insensitive to enzymatic treatments (Sawitzky et al. 1996), and for this species and other giant coenocytic algae, AF staining was conducted on microdissected cells (Menzel 1987) or cytoplasm extruded from its cell wall (La Claire II 1989; Mine et al. 2001).

AF staining procedure was also developed for red macroalgae in the 1990s (McDonald et al. 1993). Red macroalgae possess a complex CW (Garbary et al. 1992; Popper et al. 2011) and so several enzyme cocktails have been used, including cellulase alone (Reis et al. 2013), a mix of cellulase and snail gut extract (McDonald et al. 1993), or β-glucuronidase (Garbary et al. 1992; Garbary and McDonald 1996). However, the CW digestion step had to be adapted to each species (McDonald et al. 1993), sometimes leading to damaged cell ultrastructures, which raises some doubts about the observed AF organization. As in green macroalgae, some studies reported successful AF labeling in red macroalgae without CW digestion steps (Kim and Kim 1999; Kim et al. 2001; Wilson et al. 2003).

On brown macroalgae (*Phaeophyceae*), few AF staining studies have been conducted until now, and most of them were only focused on fucoid zygotes during polarization and germination (Brawley and Robinson 1985; Kropf et al. 1989; Alessa and Kropf 1999; Hable et al. 2003). In the earliest studies, protocols did not include any CW digestion step, but only cell permeabilization using compounds such as saponin or Triton X-100 after fixation (Brawley and Robinson 1985; Kropf et al. 1989; Alessa and Kropf 1999). However, fucoid zygotes are free naked cells, directly in contact with the external medium, in which the cell wall is just beginning to form (Bisgrove and Kropf 2001). Therefore, this simple procedure would not be as effective for other cell types, especially for cells encased in a tissue. Moreover, *Phaeophyceae* have a very complex cell wall, the composition of which strongly differs from that of land plants or other macroalgae. As mentioned in the introduction, land plants and brown algae differ strongly in their CW composition and structure, cellulose, hemicellulose, arabinogalactan, and callose being the only polymers that the two groups have in common (Popper et al. 2011; Deniaud-Bouët et al. 2014; Hervé et al. 2016). The two major components of the brown algal cell wall are alginates and sulfated fucans that are specific to this class of macroalgae (Mabeau and Kloareg 1987; Popper et al. 2011). Brown algal cell wall also contains proteins, halogenated and sulfated phenolic molecules (named phlorotannins), and halides (Deniaud-Bouët et al. 2014). As a consequence, the digestion steps necessary to permeabilize the cell must be adapted to these particular molecules.

In 2000, Karyophyllis and colleagues adapted several protocols from previous studies on pollen tubes (Sonobe and Shibaoka 1989) and red algae (Garbary et al. 1992) to improve AF staining on the apical cells of *Sphacelaria rigidula* (Karyophyllis et al. 2000a, b). This protocol improved AF staining by introducing MBS as an AF-stabilizing agent before chemical fixation. In addition, CW is digested by a complex mixture of enzymes, including cellulase, β-glucuronidase, and sulfatase (both from abalone acetone powder), pectinase, hemicellulase (both from macerozyme product, Onozuka R-10, Yakult), laminarase, and xylanase (both from driselase product, Sigma) (Karyophyllis et al. 2000a). Using Rhodamine–Phalloidin or anti-actin antibody to tag AFs and with some modifications, this new protocol allowed the visualization of the fine AF organization in several species and cell types of *Phaeophyceae,* including apical and other vegetative cells of *Sphacelaria rigidula* (Karyophyllis et al. 2000b), and *Dictyota dichotoma* (Katsaros et al. 2002), polarizing cells of *Macrocystis pyrifera* gametophytes (Varvarigos et al. 2004, 2007) and again in fucoid zygotes (Hable et al. 2003; Bisgrove and Kropf 2004). Recently, it has been successfully applied to *Ectocarpus* sporophyte cells, and the different steps are described in this chapter.

24.3 Materials

24.3.1 Chemicals

- DMSO 100%.
- Glycerol 100%.
- Cell-wall digesting enzymes:
  - Cellulase (Onozuka R-10, Yakult).
  - Hemicellulase (H2125, Sigma).
  - Driselase (D8037-1G, Sigma).
  - Macerozyme (Onozuka R-10, Yakult).
  - Pectinase (P2401, Sigma).
  - Alginate lyase (A1603, Sigma).
- *m*-maleimido benzoic acid *N*-hydroxy succinimide ester (MBS, M2786, Sigma).
- Natural or artificial seawater (ASW) (to wash free-floating filaments before prefixation. *Must be autoclaved and kept sterile, if possible*).
- Paraformaldehyde (PFA, powder).
- *p*-phenylenediamine (PDA).
- Poly-L-lysine (1 mg mL^{-1}).
- Rhodamine–Phalloidin (Sigma or Biotium) or AlexaFluor568–Phalloidin (in general, AlexaFluor568 is more suited for plant material than Rhodamine–Phalloidin).
- Triton X-100.

24.3.2 Solutions and recipes

24.3.2.1 Preparation of Rhodamine–Phalloidin (R415) or AlexaFluor568–Phalloidin (A12380, life technology)

Dilute the content of a whole tube (300 U) into 1.5 mL of methanol, following the manufacturer's directions. The final concentration is about 6.6 μM (=200 U mL^{-1}). Store at −20°C in the dark.

The instructions for the preparation of buffers and other solutions are indicated in Tables 24.1 through 24.10.

24.3.2.2 *Preparation of buffers*

Table 24.1 Microtubule stabilizing buffer (MTB) modified for actin

Composition	Final concentration	Quantity to add (for 1 L final)
Pipes	50 mM	15.12 g
EGTA	5 mM	1.318 g
$MgSO_4$	5 mM	1.902 g
KCl	25 mM	1.86 g
NaCl	4% (w/v)	40 g
PVP (Polyvinylpyrrolidone 25)	2.5% (w/v)	25 g
DTT ((−)-1,4-dithio-L-threitol)	1 mM	0.154 g
Pure water	–	qsp 1 L

Note: Adjust pH to 7.4. Store at 4°C.

Table 24.2 Phosphate-buffered saline (PBS)

Composition	Final concentration	Quantity to add (for 1 L final)
NaCl	137 mM	8.01 g
KCl	0.7 mM	0.052 g
Na_2HPO_4	5.1 mM	0.72 g
KH_2PO_4	1.7 mM	0.22 g
Distilled water	–	qsp 1 L

Note: Adjust pH to 7.4. Store at 4°C.

Table 24.3 Stock solution of MBS (*m*-maleimido benzoic acid *N*-hydroxy succinimide ester)

Composition	Final concentration	Quantity to add (for 1 mL final)
MT-buffer	–	980 μL
DMSO	2%	20 μL
MBS	100 mM	0.0943 g

Note: Store in the dark at −20°C. As the product is better preserved as a powder (at −20°C), prepare preferably a small volume of stock solution just before a series of experiments.

Table 24.4 Actin-stabilization solution

Composition	Final concentration	Quantity to add (for 1 mL final)
MTB	–	975 μL
Triton X-100	0.2%	2 μL
DMSO	2%	20 μL
MBS	300 μM	3 μL of a stock solution at 100 mM

Note: Prepare fresh solution just before use.

Table 24.5 Fixation solution

Composition	Final concentration	Quantity to add (for 10 mL final)
MT buffer	–	10 mL
PFA (Paraformaldehyde)	4%	0.4 g

Note: Distribute in 1 mL aliquots and store at −20°C.

Table 24.6 Cell wall lysis buffer

Composition	Final concentration	Quantity to add (for 2 mL final)
PBS:MTB 1:1	–	1946 μL
Triton X-100 (pure)	0.2% (v:v)	4 μL
Rhodamine – (or AlexaFluor568) – Phalloidin	0.17 μM	50 μL of stock solution at 6.6 μM
Cellulase	2% (w:v)	40 mg
Hemicellulase	2% (w:v)	40 mg
Driselase	1% (w:v)	20 mg
Macerozyme	1% (w:v)	20 mg
Pectinase	0.5% (w:v)	10 mg

Note: Just before use, prepare the lysis buffer (with PBS, MTB, Triton X-100, and Rh-Ph). Stir and dissolve the enzymes in it, and centrifuge for 5 min to eliminate the precipitate. Adjust the pH to 5.5 (this is important for the activity of enzymes). The given composition of the cell wall digestion solution is only indicative and must be adjusted for each type of material. We also recommend trying using alginate–lyases (concentration 1%–2% w:v), because alginate is a common cell wall compound in brown algae.

24.3.2.3 Other solutions

Table 24.7 Extraction solution

Composition	Final concentration	Quantity to add (for 1 mL final)
DMSO	5%	50 μL
Triton X-100	3%	30 μL
PBS	–	920 μL

Note: Prepare fresh solution just before use.

Table 24.8 Actin-staining solution

Composition	Final concentration	Quantity to add (for 300 μL final)
Rhodamine–Phalloidin	0.33 μM	15 μL of stock solution at 6.6 μM
PBS:MTB	–	285 μL

Note: Rhodamine–Phalloidin can be replaced by AlexaFluor568–Phalloidin (A12380, Life Technology), that is thought to give better results on plant cells. In both cases, prepare the staining solution just before use. Store on the ice and protect from light.

Table 24.9 DNA staining solution

Composition	Concentration	Quantity to add (for 10 mL final)
PBS	–	9.9 mL
Hoechst 33258 (94403, Sigma)	10 μg mL^{-1}	100 μL of a 1 mg mL^{-1} stock solution

Note: Store at 4°C.

Table 24.10 Mounting solution

Composition	Concentration	Quantity to add (for 1 mL final)
Glycerol	33.3%	800 μL
PBS	66.6%	400 μL
PDA (*p*-phenylenediamine, P6001, Sigma)	0.2% (w:v)	0.02 g

Note: Store at 4°C, in the dark. The solution gets darker with time.

24.3.3 *Equipment*

1.54 mL Eppendorf® tubes.
15- and 50 mL plastic tubes.
Fine forceps.
Glass coverslips (sterilized, if possible).
Microscope slides (sterilized, if possible).
Parafilm®.
Pipettes and tips (10/100/1000 μL range).
Plastic or glass Petri dishes.
Plastic Pasteur pipette.

24.4 *Experimental procedures*

24.4.1 *Preparation of poly-L-lysine-coated coverslips*

This is for free-floating filaments. They must be prepared at least one day before the staining experiment (*Note 1*).

1. Soak coverslips in nitric acid for at least several hours.
2. Wash the coverslips with distilled water, then with acetone, and again with water. Dispose them one-by-one on a paper towel, and let them dry until all drops of water have disappeared.
3. Using a cotton bud spread a solution of 1 mg mL^{-1} poly-L-lysine on the whole surface of coverslips.
4. Let the coverslips dry at room temperature (RT). The poly-L-lysine solution is thick and dries very slowly, so it is better to prepare the coverslips at least one day in advance. Drying step can be speeded up by putting the coverslips under a chemical hood.

24.4.2 *Preparation of* Ectocarpus *samples*

24.4.2.1 *Filaments grown on coverslips*

The preparation of *Ectocarpus* algal material is described in Chapter 23 by Rabillé et al.

1. On the bottom of the lid of an empty Petri dish, place a round piece of Whatman paper (that must cover all the Petri dish's bottom surface). Press a small piece of cotton wool on the rim of the lid. Add a small volume of water on the Whatman paper and the piece of cotton wool to have enough humidity. Remove any exceeding water. Finally, dispose a large piece of Parafilm® on the Whatman paper (*Note 2*).

2. Using forceps, takeoff the coverslips with the filaments grown on them from the bottom of the culture dish. Wipe the coverslips' bottom and edge with a paper towel.
3. Put the coverslips on the surface of the Parafilm® with the filaments upside. For all the subsequent steps, dispose a small volume of each solution on the surface of the coverslips and close the moist chamber for incubation.

24.4.2.2 *Free-floating filaments*

1. Gently crap or takeoff the filaments from the medium with forceps, and stack them in a pile. Quickly wash the pile of filaments by gently shaking it in clean sea water.
2. Put a dense pile of filaments into an Eppendorf tube. The following steps up to the enzyme digestion will take place in the tube. When changing the liquids by a pipette, be careful of not losing or damaging the filaments (*Note 3*).
3. After the cell wall lysis step, spread the filaments on the surface of poly-L-lysine-coated coverslips (or slides) with forceps. Separate every tuft of filaments from one another (*Note 4*).
4. Let some medium evaporate, keeping the coverslips damp. The rest of the protocol is then proceeded as for filaments that are directly grown on coverslips. Alternatively, the procedure can be continued in Eppendorf tube.

24.4.3 *Actin staining procedure*

1. Incubate the filaments with the actin-stabilization solution (300 μM MBS, 2% DMSO, 0.1% Triton X-100 in MTB, pH 7,4) for 30 min in dark at room temperature (RT) (*Note 5*).
2. Wash 3 × 10 min with MTB, at RT (this step might be omitted, *Note 6*).
3. Fix the filament with 4% PFA in MTB for 40 min at RT (*Note 5*).
4. Wash 3 × 10 min in PBS:MTB 1:1 (v:v), at RT.
5. Incubate the filaments with the cell wall lysis solution (2% cellulase, 2% hemicellulase, 1% driselase, 1% macerozyme, 0.5% pectinase, 0.2% Triton X-100, 0.17 μM Rhodamine–Phalloidin, pH 5.5) for 15–20 min in dark, at RT.
6. Wash 3 × 10 min in PBS:MTB 1:1, in the dark at RT (*Note 7*).
7. Optional: Incubate the filaments in the extraction solution (2% DMSO in MT buffer) for 10 min, in the dark, RT (*Note 5*).
8. Wash 3 × 10 min in PBS:MTB 1:1, in the dark at RT.
9. Incubate the filaments in actin-staining solution (0.66 μM Rh–Ph in PBS:MT 1:1) in the dark for at least 1 h or overnight at 4°C or at RT (*Note 5*).
10. Wash 3 × 10 min in PBS, in the dark at RT.

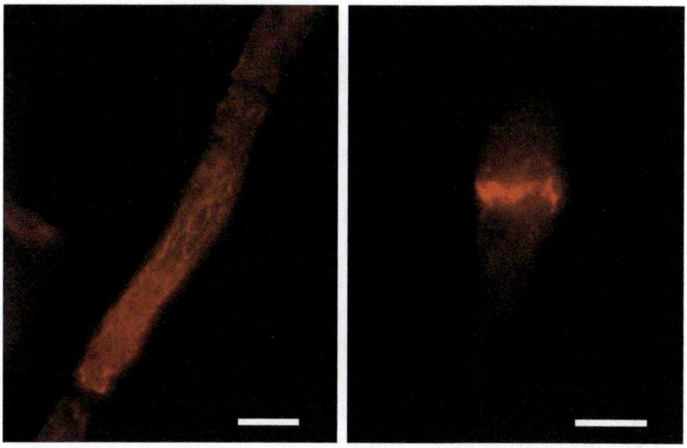

(a) (b)

Figure 24.1 Organization of actin cytoskeleton in vegetative cells of *Ectocarpus* filaments labeled with Rhodamine–Phalloidin, as seen under epifluorescence microscopy. Long bundles of actin filaments are observed in intermediate (elongated) cell during interphase (a) and during cytokinesis (b), in which actin filaments are densely packed at the cytokinetic diaphragm plane. Scale bars represent 10 μm on each picture.

11. Stain the nucleus in 10 μg mL^{-1} Hoechst 33258 in PBS for 10 min, in the dark at RT.
12. Wash 3 × 10 min in PBS, in the dark at RT.
13. Mount in 0.2% phenylenediamine in glycerol:PBS 2:1. The slides are now ready to be observed under the fluorescent microscope. Two examples of actin-staining pattern observed in *Ectocarpus* sporophyte vegetative cells are shown in Figure 24.1.

24.5 Notes

Note 1: Alternatively, slides can be used instead of coverslips.

Note 2: By turning over the Petri dish on its lid, you create a moist chamber in which the coverslips will be incubated. The moisture avoids the evaporation of the medium deposited on the surface of the coverslips, which is important especially for long incubation times. The Parafilm® is hydrophobic and prevents the medium to flow over the surface of the coverslips.

Note 3: Avoid sucking filaments into the pipette tip when removing the medium from the tube while manipulating as gently as possible. If needed, slowly centrifuge the tube to take the filaments at the bottom of the tube. Avoid centrifuging filaments until they have been fixed.

Note 4: Do not pressure the filaments with the forceps, just spread them around very gently, because *Ectocarpus* filaments are easily damageable. Do not leave large piles of entangled filaments, which prevent proper staining of the filament, subsequently interfering with the observation. If needed, spread the filaments under a dissecting microscope using a fine needle.

Note 5: Incubation duration might require adjustments, depending on the amount and nature of the treated material.

Note 6: The first washing step can be omitted. Alternatively, remove half quantity of the first solution above and complete with the solution of PFA (8%) without washing step.

Note 7: Spread free-floating filaments on poly-L-lysine-coated coverslips (or slides) as described in Section 24.4.1. The rest of the protocol is identical for both types of sample.

Acknowledgments

This research was cofinanced by the National and Kapodistrian University of Athens (program "Kapodistrias"), the Roscoff Biology Station (Centre National de la Recherche Scientifique et Université Pierre et Marie Curie) and the COST Action Phycomorph FA 1406.

References

Alessa, L., and D. L. Kropf. 1999. F-Actin marks the rhizoid pole in living *Pelvetia compressa* zygotes. *Development* 126 (1): 201–209.

Barden, J. A., M. Miki, B. D. Hambly, and C. G. dos Remedios. 1987. Localization of the phalloidin and nucleotide-binding sites on actin. *Eur J Biochem* 162 (3): 583–588.

Bisgrove, S. R., and D. L. Kropf. 2001. Cell wall deposition during morphogenesis in fucoid algae. *Planta* 212 (5–6): 648–658.

Bisgrove, S. R., and D. L. Kropf. 2004. Cytokinesis in brown algae: Studies of asymmetric division in fucoid zygotes. *Protoplasma* 223 (2–4): 163–173.

Bogaert, K. A., A. Arun, S. M. Coelho, and O. De Clerck. 2013. Brown algae as a model for plant organogenesis. In *Plant Organogenesis: Methods and Protocols*, I. De Smet (Ed.), 959: 97–125. New York: Springer Science+Business Media.

Braun, M., and G. O. Wasteneys. 1998. Distribution and dynamics of the cytoskeleton in graviresponding protonemata and rhizoids of characean algae: Exclusion of microtubules and a convergence of actin filaments in the apex suggest an actin-mediated gravitropism. *Planta* 205 (1): 39–50.

Brawley, S. H., and K. R. Robinson. 1985. Cytochalasin treatment disrupts the endogenous currents associated with cell polarization in fucoid zygotes: Studies of the role of F-actin in embryogenesis. *J Cell Biol* 100 (4): 1173–1184.

Charrier, B., S. M. Coelho, A. Le Bail, T. Tonon, G. Michel, P. Potin, B. Kloareg, C. Boyen, A. F. Peters, and J. M. Cock. 2008. Development and physiology of the brown alga *Ectocarpus siliculosus*: Two centuries of research. *New Phytol* 177 (2): 319–332.

Chebli, Y., J. Kroeger, and A. Geitmann. 2013. Transport logistics in pollen tubes. *Molecular Plant* 6 (4): 1037–1052.
Coelho, S. M., D. Scornet, S. Rousvoal, N. Peters, L. Dartevelle, A. F. Peters, and J. M. Cock. 2012a. Immunostaining of *Ectocarpus* cells. *Cold Spring Harb Protoc* 2012 (3): 369–372.
Coelho, S. M., D. Scornet, S. Rousvoal, N. Peters, L. Dartevelle, A. F. Peters, and J. M. Cock. 2012b. Isolation and regeneration of protoplasts from *Ectocarpus*. *Cold Spring Harb Protoc* 2012 (3): 361–364.
Deniaud-Bouët, E., N. Kervarec, G. Michel, T. Tonon, B. Kloareg, and C. Hervé. 2014. Chemical and enzymatic fractionation of cell walls from fucales: Insights into the structure of the extracellular matrix of brown algae. *Ann Bot* 114 (6): 1203–1216.
Doris, F. P., and M. W. Steer. 1996. Effects of fixatives and permeabilisation buffers on pollen tubes: Implications for localisation of actin microfilaments using phalloidin staining. *Protoplasma* 195: 25–36.
Dyachok, J., C.-M. Yoo, K. Palanichelvam, and E. B. Blancaflor. 2010. Sample preparation for fluorescence imaging of the cytoskeleton in fixed and living plants roots. In *Cytoskeleton - Methods and Protocols*, R.H. Gavin (Ed.), 54: 157–169. New York: Springer Science+Business Media New York.
Garbary, D. J., A. R. McDonald, and J. G. Duckett. 1992. Visualization of the cytoskeleton in red algae using fluorescent labelling. *New Phytol* 120: 435–444.
Garbary, D. J., and A. R. McDonald. 1996. Fluorescent labelling of the cytoskeleton in *Ceramium strictum* (Rhodophyta). *J Phycol* 32 (1): 85–93.
Hable, W. E., and D. L. Kropf. 2005. The Arp2/3 complex nucleates actin arrays during zygote polarity establishment and growth. *Cell Motil Cytoskel* 61 (1): 9–20.
Hable, W. E., N. R. Miller, and D. L. Kropf. 2003. Polarity establishment requires dynamic actin in fucoid zygotes. *Protoplasma* 221 (3–4): 193–204.
Harper, J., D. McCurdy, M. Sanders, and J. Salisbury. 1992. Actin dynamics during the cell cycle in *Chlamydomonas reinhardtii*. *Cell Motil Cytoskel* 126: 117–126.
Hawkins, E. K., J. J. Lee, and M. Correia. 2003. Polar localization of filamentous actin in cells of the scale-forming alga *Pleurochrysis* sp. *Protoplasma* 220 (3–4): 233–236.
Heath, I. B., G. Gupta, and S. Bai. 2000. Plasma membrane-adjacent actin filaments, but not microtubules, are essential for both polarization and hyphal tip morphogenesis in *Saprolegnia ferax* and *Neurospora crassa*. *Fungal Genet Biol* 30 (1): 45–62.
Hervé, C., A. Siméon, M. Jam, A. Cassin, K. L. Johnson, A. A. Salmeán, W. G. T. Willats et al. 2016. Arabinogalactan proteins have deep roots in eukaryotes: Identification of genes and epitopes in brown algae and their role in *Fucus serratus* embryo development. *New Phytol* 209 (4): 1428–1441.
Heslop-Harrison, J., and Y. Heslop-Harrison. 1991. The actin cytoskeleton in unfixed pollen tubes following microwave-accelerated DMSO-permeabilisation and TRITC-phalloidin staining. *Sex Plant Reprod* 4: 6–11.
Kakimoto, T., and H. Shibaoka. 1987a. A new method for preservation of actin filaments in higher plant cells. *Plant Cell Physiol* 28 (8): 1581–1585.
Kakimoto, T., and H. Shibaoka. 1987b. Actin filaments and microtubules in the preprophase band and phragmoplast of tobacco cells. *Protoplasma* 140 (2–3): 151–156.
Karyophyllis, D., C. Katsaros, and B. Galatis. 2000. F-Actin Involvement in apical cell morphogenesis of *Sphacelaria rigidula* (Phaeophyceae): Mutual alignment between cortical actin filaments and cellulose microfibrils. *Eur J Phycol* 35 (2): 195–203.

Karyophyllis, D., C. Katsaros, I. Dimitriadis, and B. Galatis. 2000. F-actin organization during the cell cycle of *Sphacelaria rigidula* (Phaeophyceae). *Eur J Phycol* 35 (1): 25–33.

Katsaros, C., D. A. Karyophyllis, and B. D. Galatis. 2002. Cortical F-actin underlies cellulose microfibril patterning in brown algal cells. *Phycologia* 41 (2): 178–183.

Katsaros, C., D. Karyophyllis, and B. Galatis. 2003. F-actin cytoskeleton and cell wall morphogenesis in brown algae. *Cell Biol Int* 27 (3): 209–210.

Katsaros, C., D. Karyophyllis, and B. Galatis. 2006. Cytoskeleton and morphogenesis in brown algae. *Ann Bot* 97 (5): 679–693.

Kim, G. H., T. A. Klotchkova, B. C. Lee, and S. H. Kim. 2001. FITC-phalloidin staining of F-actin in *Aglaothamnion oosumiense* and *Griffithsia japonica* (Rhodophyta). *Bot Mar* 44 (5): 501–508.

Kim, S., and G. H. Kim. 1999. Cell-cell recognition during fertilization in the red alga, *Aglaothamnion oosumiense* (Ceramiaceae, Rhodophyta). In *Sixteenth International Seaweed Symposium*, pp. 81–89. Dordrecht, the Netherlands: Springer.

Kropf, D. L., S. K. Berge, and R. S. Quatrano. 1989. Actin localization during *Fucus* embryogenesis. *Plant Cell* 1 (2): 191–200.

La Claire II, J. W. 1989. Actin cytoskeleton in intact and wounded coenocytic green algae. *Planta* 177: 47–57.

Le Bail, A., and B. Charrier. 2013. Culture methods and mutant generation in the filamentous brown algae *Ectocarpus siliculosus*. In *Plant Organogenesis: Methods and Protocols*, I. De Smet (Ed.), 959: 235–245. New York: Springer Science+Business Media.

Lloyd, C. W., A. R. Slabas, A. J. Powell, G. MacDonald, and R. A. Badley. 1979. Cytoplasmic microtubules of higher plant cells visualised with anti-tubulin antibodies. *Nature* 279: 239–241.

Lovy-Wheeler, A., K. L. Wilsen, T. I. Baskin, and P. K. Hepler. 2005. Enhanced fixation reveals the apical cortical fringe of actin filaments as a consistent feature of the pollen tube. *Planta* 221 (1): 95–104.

Mabeau, S., and B. Kloareg. 1987. Isolation and analysis of the cell walls of brown algae: *Fucus spiralis, F. ceranoides, F. vesiculosus, F. serratus, Bifurcaria bifurcata* and *Laminaria digitata*. *J Exp Bot* 38 (9): 1573–1580.

McDonald, A. R., D. J. Garbary, and J. G. Duckett. 1993. Rhodamine-phalloidin staining of F-actin in rhodophyta. *Biotech Histochem* 68 (2): 91–98.

Meijer, H. J. G., C. Hua, K. Kots, T. Ketelaar, and F. Govers. 2014. Actin dynamics in *Phytophthora infestans*; rapidly reorganizing cables and immobile, long-lived plaques. *Cell Microbiol* 16 (6): 948–961.

Menzel, D. 1987. The cytoskeleton of the giant coenocytic green alga *Caulerpa* visualized by immunocytochemistry. *Protoplasma* 139 (2–3): 71–76.

Mine, I., K. Okuda, and D. Menzel. 2001. Poly(A)+ RNA during vegetative development of *Acetabularia peniculus*. *Protoplasma* 216 (1–2): 56–65.

Nagasato, C., A. Inoue, M. Mizuno, K. Kanazawa, T. Ojima, K. Okuda, and T. Motomura. 2010. Membrane fusion process and assembly of cell wall during cytokinesis in the brown alga *Silvetia babingtonii* (Fucales, Phaeophyceae). *Planta* 232 (2): 287–298.

Parthasarathy, M. V. 1987. *In situ* localization of actin filaments in higher plant cells using fluorescent probes. *Plant Mol Biol Rep* 5 (1): 251–259.

Perdue, T. D., and M. V. Parthasarathy. 1985. In situ localization of F-actin in pollen tubes. *Eur J Cell Biol* 39 (1): 13–20.

Peters, A. F., D. Marie, D. Scornet, B. Kloareg, and J. M. Cock. 2004. Proposal of *Ectocarpus siliculosus* (Ectocarpales, Phaeophyceae) as a model organism for brown algal genetics and genomics. *J Phycol* 40 (6): 1079–1088.

Pflügl-Haill, M., L. Vidali, J. W. Vos, P. K. Hepler, and U. Lütz-Meindl. 2000. Changes of the actin filament system in the green alga *Micrasterias denticulata* induced by different cytoskeleton inhibitors. *Protoplasma* 212: 206–216.

Popper, Z. A., G. Michel, C. Hervé, D. S. Domozych, W. G. T. Willats, M. G. Tuohy, B. Kloareg, and D. B. Stengel. 2011. Evolution and diversity of plant cell walls: From algae to flowering plants. *Ann Rev Plant Biol* 62: 567–590.

Reis, V. M., L. S. Oliveira, R. M. F. Passos, N. B. Viana, C. Mermelstein, C. Sant'Anna, R. C. Pereira et al. 2013. Traffic of secondary metabolites to cell surface in the red alga *Laurencia dendroidea* depends on a two-step transport by the cytoskeleton. *PLoS One* 8 (5): e63929.

Sampson, K., and J. D. Pickett-Heaps. 2001. Phallacidin stains the kinetochore region in the mitotic spindle of the green algae *Oedogonium* spp. *Protoplasma* 217 (4): 166–176.

Sawitzky, H., J. Willingale-Theune, and D. Menzel. 1996. Improved visualization of F-actin in the green alga *Acetabularia* by microwave-accelerated fixation and simultaneous FITC-phalloidin staining. *Histochem J* 28 (5): 353–360.

Sonobe, S., and H. Shibaoka. 1989. Cortical fine actin filaments in higher plant cells visualized by rhodamine-phalloidin after pretreatment with m-maleimidobenzoyl N-hydroxysuccinimide ester. *Protoplasma* 148: 80–86.

Szechyńska-Hebda, M., M. Wędzony, E. Dubas, H. Kieft, and A. Van Lammeren. 2006. Visualisation of microtubules and actin filaments in fixed BY-2 suspension cells using an optimized whole mount immunolabelling protocol. *Plant Cell Rep* 25 (8): 758–766.

Tewinkel, M., S. Kruse, H. Quader, D. Volkmann, and A. Sievers. 1989. Visualization of actin filament pattern in plant cells without pre-fixation. A comparison of differently modified phallotoxins. *Protoplasma* 149: 178–182.

Traas, J. A., J. H. Doonan, D. J. Rawlins, P. J. Shaw, J. Watts, and C. W. Lloyd. 1987. An actin network is present in the cytoplasm throughout the cell cycle of carrot cells and associates with the dividing nucleus. *J Cell Biol* 105 (1): 387–395.

Varvarigos, V., B. Galatis, and C. Katsaros. 2007. Radial endoplasmic reticulum arrays co-localize with radial F-actin in polarizing cells of brown algae. *Eur J Phycol* 42 (3): 253–262.

Varvarigos, V., C. Katsaros, and B. Galatis. 2004. Radial F-actin configurations are involved in polarization during protoplast germination and thallus branching of *Macrocystis pyrifera* (Phaeophyceae, Laminariales). *Phycologia* 43 (6): 693–702.

Vitha, S., F. Baluska, M. Braun, J. Samaj, D. Volkmann, and P. W. Barlow. 2000. Comparison of cryofixation and aldehyde fixation for plant actin immunocytochemistry: Aldehydes do not destroy F-actin. *Histochem J* 32 (8): 457–466.

Wasteneys, G. O., J. Willingale-Theune, and D. Menzel. 1997. Freeze shattering: A simple and effective method for permeabilizing higher plant cell walls. *J Microsc* 188 (1): 51–61.

Wilsen, K. L., A. Lovy-Wheeler, B. Voigt, D. Menzel, J. G. Kunkel, and P. K. Hepler. 2006. Imaging the actin cytoskeleton in growing pollen tubes. *Sex Plant Reprod* 19 (2): 51–62.

Wilson, S. M., J. A. West, and J. D. Pickett-Heaps. 2003. Time-lapse videomicroscopy of fertilization and the actin cytoskeleton in *Murrayella periclados* (Rhodomelaceae, Rhodophyta). *Phycologia* 42 (6): 638–645.

chapter twenty five

Cryofixation of brown algae for transmission electron microscopy

Chikako Nagasato, Christos Katsaros, and Taizo Motomura

Contents

25.1 Introduction

Chemical fixation involves the immersion of samples in a solution that inactivates and stabilizes cellular contents. Although the general protocol is the same, modifications are required to give optimal results (Huang and Yeung 2015). The chemical fixation procedure is complicated, but it has been a well-accepted method, giving good-quality images. The main steps in chemical fixation are as follows: (a) fixation using aldehyde fixatives (glutaraldehyde, paraformaldehyde) to cross-link proteins and osmium tetroxide (OsO_4) to preserve lipids in membranes and lipoproteins; (b) dehydration using an organic solvent (ethanol, acetone); and (c) embedding samples in resins (Epon, Spurr, LR White, Lowcryl, etc.).

In most cases, the chemical fixation protocols used for land plants and animals are inadequate to apply to brown algae. Therefore, special modifications should be developed for this group (Clayton and Beaks 1983, Hallam and Luff 1988). One of the characteristics of brown algae that should be considered during preparation of transmission electron microscopy (TEM) samples is that the cell wall contains alginates and fucoidans in addition to cellulose (Kloareg and Quatrano 1988). These polysaccharides flow out into the sample preparation buffers in conventional chemical fixation methods. For example, alginate becomes soluble under low concentrations of calcium in the buffer. Swollen cell walls caused by low calcium buffer induce detachment of the plasma membrane from the cell wall. Therefore, phosphate buffer is inadequate for preparation of brown algal cells, because calcium binds to phosphate. Thus, many researchers use cacodylate buffer with 0.1%–1% calcium or seawater for the fixation buffer. On the other hand, mucilaginous polysaccharides in the cell wall of brown algae disturb penetration of fixatives and resins into the tissue, and as a result, it is difficult to make ultrathin sections with satisfactory preservation of the cell structures. Phenolic compounds in physodes, which are membrane-bounded structures with osmiophilic or electron-dense material, also cause a problem for chemical fixation in brown algae. Leakage of the phenolics into the cytoplasm affects fixation of samples for TEM, usually resulting in negative contrast images of the cell structure. To trap the phenolics from physodes, caffeine is usually added to the aldehyde fixative and washing buffer (Clayton and Beaks 1983). Components of the fixative and selection of the buffer become crucial factors affecting TEM images of brown algae.

Physical fixation is achieved by *cryofixation*, composed of rapid freezing (10^4–10^5 K s^{-1}) to the temperature of liquid nitrogen (–196°C) or below and subsequent freeze substitution. All metabolic activities in the cell are instantly stopped by cryofixation, which preserves the native status of cells. However, cryofixation is not always successful in all samples. In cryofixation, the most important thing is to prevent formation of ice crystal in cells. One problem with cryofixation is that there is a limitation in the area or depth of sample preserved, as it is difficult to achieve high cooling to produce amorphous ice instead of ice crystals. This depth-limit is usually a few micrometers. After cryofixation, frozen cellular water should be substituted with an organic solvent under low temperature (–180°C) (Gilkey and Staehelin 1986). There are three methods in cryofixation: immersion, metal contact, and high-pressure freezing. In immersion freezing, specimens are plunged into liquid propane. This method is simple and handy, but freezing depth is less than 10–20 μm. In the metal contact and high-pressure freezing methods, rapid freezing is performed by liquid helium or liquid nitrogen using expensive instrumentation. The metal contact method permits a well-preserved area with a depth up to 20 μm. High-pressure freezing guarantees high-quality images of 600 μm thickness from the surface (McDonald et al. 2007). In the current chapter, the immersion-freezing method is described, as applied to brown algae.

25.2 State of the art

The first application of cryofixation and freeze substitution for TEM samples in brown algae was in fucoid zygotes of *Phyllospora comosa* and *Hormosira banksii* (Schoenwaelder and Clayton 2000). Subsequently, observations of ultrastructure using this method expanded to several species of brown algae, that is, zygotes of *Scytosiphon lomentaria* (Nagasato and Motomura 2002) and *Silvetia babingtonii* (Nagasato et al. 2010); apical cells of *Halopteris congesta* (Katsaros et al. 2009), *Sphacelaria rigidula* (Katsaros et al. 2009), and *Dictyota dichotoma* (Terauchi et al. 2012); and plurilocular sporangia of *Ectocarpus siliculosus* (Fu et al. 2013). The images in these reports indicate that detachment of the plasma membrane from the cell wall does not occur and membranous structures are well preserved, showing minimal appearance of artifacts and reflecting their original structure (Figure 25.1). In the case of brown algae, the immersion-freezing method preserves cell structure to a depth of more than 20 μm from the cell surface, but the success rate is less than 10%. In land plants,

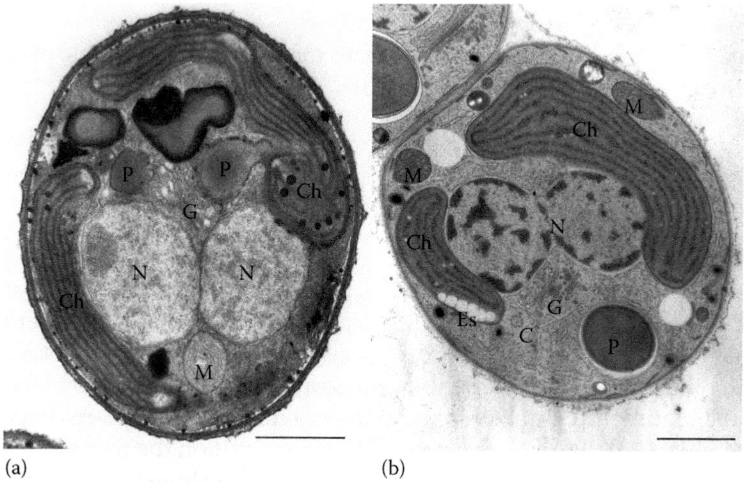

Figure 25.1 Comparison of ultrastructural images between chemical and cryofixation methods. (a) Chemical fixation; (b) cryofixation. Centriole (C), chloroplast (Ch), eye spot (Es), Golgi body (G), mitochondrion (M), nucleus (N), pyrenoid (P). Scale bars, 1 μm.

cryoprotectants, such as sucrose, are used to suppress the formation of extracellular ice crystals (Seguí-Simarro 2015). In brown algae, cryoprotectants have not been used thus far.

25.3 Materials

25.3.1 Reagents

Liquid nitrogen
Liquid propane
Acetone
Dry ice
2% Osmium tetraoxide (OsO_4) in acetone
Spurr Low Viscosity Embedding Kit (Polysicences, Inc., Warrington, PA, USA)

25.3.2 Equipment

Filter paper
Thin, acetone-resistant support film
Gold wire loop (5–10 mm diam.) with formvar film
Metal caps (appropriate for liquid propane and liquid nitrogen, for example, small metal milk jug around 10 mL)
Polypropylene tube (50 mL)
Cryogenic tubes (2 mL)
Glass vial (2 or 4 mL)
Small styrene containers (appropriate for holding metal caps, polypropylene tubes, and cryogenic vials)

25.4 Experimental procedures

25.4.1 Preparation of material

A thin, acetone-resistant support film, similar to that used for acrylamide gel electrophoresis, or a formvar film on gold wire loops (5–10 mm diameter) are used as sample carriers into the cryogen (Figure 25.2a). Motile cells bearing flagella, such as gametes and zoospores, are collected by centrifugation. Floating thalli are chopped with a razor blade into pieces less than 5 mm wide. Cell pellets and chopped thalli are put on the formvar-coated gold wire loop. To induce germination after germlings of reproductive cells are fixed, motile reproductive cells are inoculated on the thin film attached to a Petri dish with adhesive tape. Just before cryofixation, the film with attached germlings is cut into a right triangle to easily distinguish the front and back.

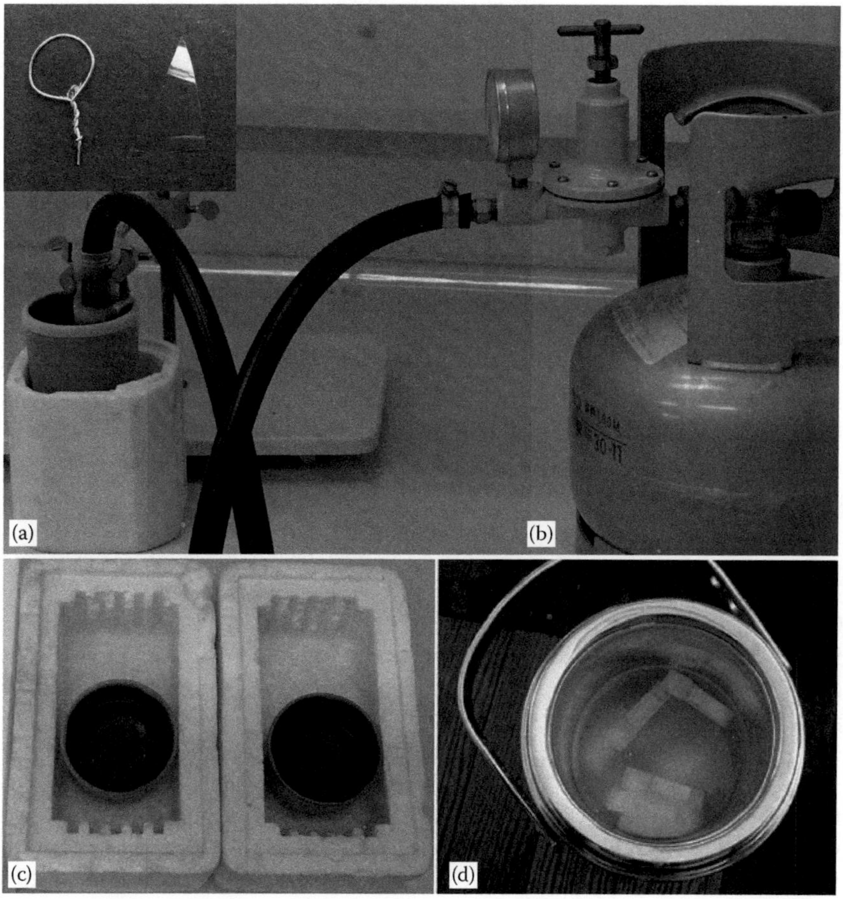

Figure 25.2 Preparation for immersion freezing. (a) Sample carriers. A formvar film on gold wire loops and a thin support film with acetone tolerance; (b) making liquid propane in the fume hood. Place a hose connected to a propane gas cylinder (right) into a precooled 50 mL polypropylene tube; (c) liquid propane and liquid nitrogen poured into separate metal caps on the liquid nitrogen bath; and (d) dry-ice acetone to keep cryogenic vials under –80°C.

25.4.2 Cryogen and substitution fixative

To make liquid propane, a hose connected to a propane gas cylinder is placed in a 50 mL polypropylene tube, which is cooled in a liquid nitrogen bath (Figure 25.2b) (*Note 1*). About 5 mL of liquid propane is poured into the small metal cap (Figure 25.2c), which has been precooled in a liquid nitrogen bath. Liquid nitrogen is then decanted into another metal cap. Instead of propane, ethane or a mixture of propane and ethane has also been used as cryogens (Ryan 1992). However, ethane has not been used as a cryogen for brown algae.

The substitution fixative is adjusted by adding OsO_4 to acetone to a final concentration of 2%. It is then transferred into a 2 mL cryogenic vial and cooled in a liquid nitrogen bath.

25.4.3 Cryofixation

Excess culture medium is quickly drained with filter paper before immersion into liquid propane, taking care to avoid plasmolysis. The specimen's carrier is held with forceps and rapidly plunged into the metal cap filled with liquid propane. Quick immersion of the sample into liquid propane is a key step for obtaining quality images in cryofixation, as it prevents the growth of ice crystals. After 10 s, the sample is transferred into liquid nitrogen and then preserved into precooled (<–80°C) 2% osmium–acetone substitution fixative.

25.4.4 Freeze substitution

Cryogenic vials (2–3 mL volume) containing samples in OsO_4 are kept in dry-ice acetone (–80°C) or a deep freezer for two days (Figure 25.2d) (*Note 2*). At this low temperature, frozen water in the cells is gradually substituted with acetone. After two days, the cryogenic vials are moved to –20°C for 2 h, then at 4°C for 2 h, and finally, at room temperature for 1 h. Osmium tetroxide fixes phospholipids and proteins as the temperature rises. The specimen carriers are taken out of the cryogenic vials and washed with acetone several times at room temperature. Samples are then removed from these carriers with forceps. If samples adhere to the thin film, the film is kept intact until the last step. After washing with acetone, samples are transferred into glass vials (2 or 4 mL).

25.4.5 Infiltration and embedding of samples into resin

For conventional ultrastructural observations of biological specimens, Epon 812 and Spurr's resins are frequently used. Spurr's shows much lower viscosity and has a faster infiltration into cells than Epon 812. In brown algae, Spurr's is preferably used as the embedding resin. Spurr's Low Viscosity Embedding Kit includes 3,4-epoxycyclohexylmethyl 3,4-epoxycyclohexanecarboxylate (ERL 4221), diglycidyl ether of polypropylene glycol (DER 736), nonenyl succinic anhydride, and dimethylaminoethanol (DMAE) as an accelerator. According to the manufacture's protocol, these resin media are compounded. It is possible to preserve the resin mixture without DMAE at –20°C for several months, but it should return to room temperature before using it. To perform the procedure, the resin is first diluted to 10% with acetone. The resin concentration is then increased 10% every half day until it reaches 100%. A rotary shaker is used to facilitate the infiltration of resin into cells. Samples in the vials are exchanged with fresh 100% resin two times, and they are then infiltrated with 100% resin overnight. New resin containing DMAE is added to the samples and is kept for half a day. Finally, samples are embedded in the resin on dishes of aluminum foil, and polymerization is carried out in an oven at 60°C overnight.

In the case of film with adhering germlings, the resin pedestal must be prepared beforehand with the same resin using aluminum-foil dishes. The film is placed upside down with a little resin on the polymerized resin pedestal. After polymerization, excess resin surrounding the film is trimmed, the film removed with forceps, and the samples remain on the flat resin disc (*Note 3*).

25.5 Notes

Note 1: Liquid propane is explosive when mixed with oxygen for a long time, and, therefore, each of the above steps should be carried out in a fume hood. After finishing these steps, liquid propane is immediately discarded outdoors. The filter paper is torn off by hand before use to allow the promotion of absorption of water from the small sample. Cool the cryogenic vials containing substitution medium before use. Leave the cryogenic vials on liquid nitrogen until half of the liquid freezes.

Note 2: Osmium tetroxide is highly hazardous. When handling osmium tetroxide, wear gloves and carry out all steps in the fume hood. To avoid contamination of freezers and refrigerators with OsO_4, put the cryogenic vials in a metallic can with a cover and seal it with plastic tape.

Note 3: Vials containing samples immersed in more than 90% resin should be put in a plastic container with silica gel to control humidity.

25.6 Concluding remarks

Cryofixation is beneficial in obtaining high-quality ultrastructural images compared with conventional chemical fixation, although there are limitations. In the current chapter, a procedure of immersion freezing was described in detail. This method can preserve the cell structure at 10–20 μm depth from the cell surface. An additional advantage of this method is that special equipment is not necessary. To obtain high-quality images across a wide range of objectives, high-pressure freezing is applied. In this case, specimens are subjected to a pressure of 2100 bars, which depresses the melting point of water and limits the formation of destructive ice crystals. This method preserves cell structures up to 600 μm in depth. In brown algae, tissue from *Fucus distichus* and huge apical cells from *Halopteris congesta* have been studied using this method (Nagasato et al. 2015, Nagasato et al. 2017). High-quality ultrastructural images were impossible to obtain using immersion-freezing methods. The disadvantages of high-pressure freezing methods are that the instrument is expensive, and a lot of liquid nitrogen is used. However, almost all samples were well preserved and gave informative images using this technique.

Cryofixation is also more suitable for immunofluorescence and immunoelectron microscopy than chemical fixation, because antigen-recognition regions against the target molecules are preserved in cryofixation but are not with chemical fixation (Guo and Huang 2015). In brown algae, cryofixation has been applied using both procedures (Fu et al. 2014, Fu et al. 2016). To obtain structures close to the native state of the cells, cryofixation should be employed in cytological procedures.

References

Clayton, M. N., and Beakes, G. W. 1983. Effects of fixatives on the ultrastructure of physodes in vegetative cells of *Scytosiphon lomentaria* (Scytosiphonaceae, Phaeophyta). *Journal of Phycology* 19: 4–16.

Fu, G., Nagasato, C., Ito, T., Müller, D. G., and Motomura, T. 2013. Ultrastructural analysis of flagellar development in plurilocular sporangia of *Ectocarpus siliculosus* (Phaeophyceae). *Protoplasma* 250: 261–272.

Fu, G., Nagasato, C., Oka, S., Cock, J. M., and Motomura, T. 2014. Proteomics analysis of heterogeneous flagella in brown algae (*stramenopiles*). *Protist* 165: 662–675.

Fu, G., Nagasato, C., Yamagishi, T., Kawai, H., Okuda, K., Takao, Y., Horiguchi, T., and Motomura, T. 2016. Ubiquitous distribution of helmchrome in phototactic swarmers of the *stramenopiles*. *Protoplasma* 253: 929–941.

Gilkey, J. C., and Staehelin, L. A. 1986. Advances in ultrarapid freezing for the preservation of cellular ultrastructure. *Journal of Electron Microscopy Technique* 3: 177–210.

Guo, F., and Huang, B. Q. 2015. Immunogold labeling for electron microscopy: Strategy and problem solving. In *Plant Microtechniques and Protocols*, E. C. T. Yeung, C. Stasolla, C. Sumner, and B. Q. Huang (Eds.), pp. 225–249. Springer International Publishing, Switzerland.

Hallam, N. D., and Luff, S. E. 1988. The fixation and infiltration of larger brown algae (*Phaeophyta*) for electron microscopy. *British Phycological Journal* 23: 337–346.

Huang, B. Q., and Yeung, E. C. 2015. Chemical and physical fixation of cells and tissues: An overview. In *Plant Microtechniques and Protocols*, E. C. T. Yeung, C. Stasolla, C. Sumner, and B. Q. Huang (Eds.), pp. 23–43. Springer International Publishing, Switzerland.

Katsaros, C., Motomura, T., Nagasato, C., and Galatis, B. 2009. Diaphragm development in cytokinetic vegetative cells of brown algae. *Botanica Marina* 52: 150–161.

Kloareg, B., and Quatrano, R. S. 1988. Structure of the cell walls of marine algae and ecophysiological functions of the matrix polysaccharides. *Oceanogry and Marine Biology, An Annual Review* 26: 259–315.

McDonald, K. L., Morphew, M., Verkade, P., and Müller-Reichert, T. 2007. Recent advances in high-pressure freezing: Equipment and specimen loading methods. In *Methods in Molecular Biolology*, Vol. 369, *Electron Microscopy: Methods and Protocols*, 2nd Ed., J. Kuo (Ed.), pp. 143–173. Humana Press, Totowa, NJ.

Nagasato, C., Inoue, A., Mizuno, M., Kanazawa, K., Ojima, T., Okuda, K., and Motomura, T. 2010. Membrane fusion process and assembly of cell wall during cytokinesis in the brown alga, *Silvetia babingtonii* (Fucales, Phaeophyceae). *Planta* 23: 287–298.

Nagasato, C., and Motomura, T. 2002. Ultrastructural study on mitosis and cytokinesis in *Scytosiphon lomentaria* zygotes (Scytosiphonales, Phaeophyceae) by freeze-substitution. *Protoplasma* 219: 140–149.

Nagasato, C., Tanaka, A., Ito, T., Katsaros, C., and Motomura, T. 2017. Intercellular translocation of molecules via plasmodesmata in the multiseriate filamentous brown alga, *Halopteris congesta* (Sphacelariales, Phaeophyceae). *Journal of Phycology* 53: 333–341.

Nagasato, C., Terauchi, M., Tanaka, A., and Motomura, T. 2015. Development and function of plasmodesmata in zygotes of *Fucus distichus*. *Botanica Marina* 58: 229–238.

Ryan, K. P. 1992. Cryofixation of tissues for electron microscopy: A review of plunge cooling methods. *Scanning Microscopy International* 6: 715–743.

Schoenwaelder, M. E., and Clayton, M. N. 2000. Physode formation in embryos of *Phyllospora comosa* and *Hormosira banksii* (Phaeophyceae). *Phycologia* 39: 1–9.

Seguí-Simarro, J. M. 2015. High-pressure freezing and freeze substitution of in vivo and in vitro cultured plant samples. In *Plant Microtechniques and Protocols*, E. C. T. Yeung, C. Stasolla, C. Sumner, and B. Q. Huang (Eds.), pp. 117–134. Springer International Publishing, Switzerland.

Terauchi, M., Nagasato, C., Kajimura, N., Mineyuki, Y., Okuda, K., Katsaros, C., and Motomura, T. 2012. Ultrastructural study of plasmodesmata in the brown alga *Dictyota dichotoma* (Dictyotales, Phaeophyceae). *Planta* 236: 1013–1026.

chapter twenty six

Probing the subcellular topography of seaweeds

Transmission electron microscopy, immunocytochemistry, and correlative light microscopy

Sandra C. Raimundo and David S. Domozych

Contents

26.1 Introduction

Macroscopic marine algae or *seaweeds* comprise a diverse range of multicellular photosynthetic eukaryotes that are classified as brown-, red-, and green algae. Taxonomically, green seaweeds (Chlorophyta) belong to the green lineage in the eukaryotic tree of life (Viridiplantae), and are evolutionarily closer to land plants than other seaweed taxa (Becker and Marin 2009, Leliaert et al. 2012); red seaweeds (Rhodophyta) are a distinct and ancient eukaryotic algal lineage (Yoon et al. 2004), whereas brown seaweeds (Phaeophyceae) form an independent lineage, which is neither related to the other algal groups nor to land plants (Patterson 1999). Marine algae exhibit a wide array of morphological forms ranging from giant kelps with distinctive thallus parts, such as holdfast, stipe, and blade, to small filamentous or sheet-like phenotypes. Marine seaweeds are important primary producers and some are keystone species in particular habitats where they constitute the base of nutrient cycles and ultimately define biotic diversity. Many seaweeds form extensive beds in intertidal and subtidal zones of rocky shores, providing the physical infrastructure for distinct ecosystems (Bold and Wynne 1978, Lobban and Harrison 1994, Graham and Wilcox 2000). Although humans have historically used many of these algae for food, fertilizers, and gelling agents, seaweeds are now taking on even greater economic significance in the biofuel, pharmaceutical, and bioremediation industries (Cardozo et al. 2007, Wargacki et al. 2012, Enquist-Newman et al. 2014, Cunha and Grenha 2016). Yet, the basic biology of most seaweeds, including fundamental physiological processes related to growth, development, and reproduction is poorly resolved and lags significantly behind what is known about land plants. At the cell biology level, though some seaweed species have been studied as model organisms for specific biological phenomena, for example, adhesion in *Ulva* sp. (Callow and Callow 2006) and polarity establishment in fertilized eggs in *Fucus* sp. (Hervé et al. 2016, Torode et al. 2016), little is known about the vast majority of taxa. Recently, new and valuable molecular studies have emerged and are providing keen insight into the evolution, phylogenetic relationships, and physiological events of various marine algae (Cock et al. 2010, Collén et al. 2013). Similar endeavors focused at the cellular level will be required to help translate this wealth of new molecular data. The technological foundation of many cell biology studies is microscopy. This includes light microscopy (LM)-based imaging such as those employing immunofluorescence labeling and fluorescent protein transformants, and transmission electron microscopy (TEM)-based analyses. LM and TEM imaging may also be coupled in correlative microscopy studies to provide complementary data and different topological perspectives. For example, in immunocytochemical investigations, immunofluorescence-based imaging yields overview maps of specific epitope distribution in various tissues and organs, whereas complementary TEM-based immunogold labeling provides high-resolution epitope location within the constituent cells therein. One of the most critical aspects of microscopy based imaging involves the proper preparation of the sample whereby experimental probing takes place with least alteration/destruction of the specimen.

Preparation of seaweeds for microscopy based analyses is a major challenge (Hallam and Luff 1988). Large seaweeds containing cell walls with gel-forming polysaccharides pose significant problems for the researcher attempting to prepare samples without unwanted extractions, incomplete preservation of cell–tissues, and/or generation of misleading artifacts. Although no method has been shown to be the perfect solution for preparing these algae for microscopy, there are fundamental strategies that provide practical starting points. This chapter describes a set of techniques that can be used for fixing and embedding seaweeds in plastic for subsequent sectioning and observation for TEM and LM microscopy.

26.2 State of the art

26.2.1 General considerations

For the majority of cellular studies, specimens need to be fixed, dehydrated, embedded in plastic, and ultimately sectioned. One fixation strategy entails cryofixation whereby cells/tissues are rapidly frozen in a cryogen (e.g., liquid propane) at temperatures below –180°C. This is usually followed by a freeze substitution of the tissue at –80°C or lower with various chemical fixatives (e.g., formaldehyde, glutaraldehyde, and osmium tetroxide) dissolved in ethanol, methanol, or acetone, depending on the embedding medium of choice. After substitution, the cells/tissues are infiltrated with a plastic embedding medium and precisely positioned in a capsule or mold. UV-based or heat polymerization follows to yield a solid plastic block containing the specimen. The initial freezing of cells and tissues must be rapid to avoid the production of damaging ice crystals. However, the depth of a good cryofixation is often limited because of the inadequate means of rapid freezing (e.g., freeze plunging of tissue segments or spray freezing of cells into a cryogen) and/or the size of the tissue to be frozen. Seaweeds with thick walls and high amount of mucilaginous intercellular polysaccharides are especially difficult specimens for cryofixation, and therefore studies have been limited to small structures such as single-celled zygotes or embryos (Nagasato and Motomura 2002, Nagasato et al. 2010, 2014). An improvement of the freezing quality using high-pressure units and/or coupled with cryomicrotomy and cryo-microscopy is possible but is often prohibitively expensive, therefore limited to a few laboratories. Most laboratories employ chemical-fixation protocols followed by dehydration and embedding in a wide assortment of embedding agents, especially plastics. Often, this strategy provides very good results for both structural and specific labeling studies (e.g., immunolabeling). Likewise, the judicious choice of the right fixatives and embedding chemicals can yield materials suitable for both TEM and LM imaging, that is, correlative microscopy. In this chapter, a basic protocol for chemical fixation followed by plastic embedding that can be used both for TEM or LM is provided. Experimental adjustments of specific stages described here especially when dealing with difficult specific specimens are encouraged to maximize results.

26.2.2 Protocol considerations

To obtain optimal results when chemically fixing seaweeds, three specific features need to be carefully considered. First, particular conditions must be maintained during fixation to prevent shrinkage, extraction of cellular material, and/or production of artifacts. Specifically, the fixation buffer should be adjusted to pH 7.8 or matching the pH of the habitat from which the seaweed was collected. Osmotic conditions that match those of the seaweed's habitat must also be carefully considered. An osmometer is very helpful to discern osmotic conditions, although the use of natural seawater as the fixation buffer is often satisfactory. Second, choosing the fixative must take into account the goals of the microscopy-based experiment. Formaldehyde is a fixative that rapidly penetrates tissues and most often provides good quality for LM imaging. However, for TEM, either it is not used or is combined in a *cocktail* with other fixatives such as glutaraldehyde. Known for its superb cross-linking activity during fixation, glutaraldehyde is the preferred fixative for TEM studies and typically yields superior subcellular ultrastructural preservation. Glutaraldehyde, though, may interfere with the immunolabeling especially when using antibodies that recognize protein epitopes. To address this problem and remove any excess of glutaraldehyde, sections fixed with this chemical can be treated with ammonium chloride before labeling. Nonetheless, glutaraldehyde causes little or no problem when using antibodies that recognize epitopes present in polysaccharides (which constitutes most of the commercially available monoclonal antibodies [mAbs] used in plant and algal cell wall research). Another fixation concern may occur if one wishes to use osmium tetroxide, a chemical that binds to membrane lipids and significantly enhances contrast for TEM imaging. However, osmium tetroxide should not be used in immunogold labeling or with common plastics; for example, London Resin (LR) White resin, and should be removed from the tissues by preincubating the sections in 3%–5% (v/v) H_2O_2 before immunolabeling. It is also important to note that many of the newer TEMs have energy filter lenses that allow a superb contrast of nonosmicated specimens and thereby eliminate the need to use osmium tetroxide in the fixation process. Third, the choice of the embedding chemical (e.g., plastic) is also critical, depending on the ultimate goal of the microscopy study. Some plastics are polymerized with heat (e.g., epoxy plastics such as Spurr's Low-Viscosity Embedding Mixture) that may potentially cause alterations to the antigens recognized by mAbs. Other agents may be polymerized with UV at a wide range of temperatures. Many laboratories employ LR White as it can be either UV-polymerized (room temperature [RT], 4°C or –20°C) or heat polymerized (60°C). Other plastics can be UV polymerized at very low temperature (e.g., –40°C or lower in the case of Lowicryl® embedding media) but require special equipment and are often a source of noxious vapors.

26.3 Materials

The following materials represent some of the resources required for processing seaweeds for microscopic studies.

26.3.1 Fixation, dehydration, and embedding

All chemicals described in the list below are available from diverse commercial sources (e.g., Electron Microscopy Sciences [EMS], USA; Agar Scientific, UK; Sigma-Aldrich®, USA). Vacuum-sealed glass ampules with specific concentrations of fixatives (Figure 26.1a) are recommended as they are relatively inexpensive, easy to use, and safer to dispose. Both natural and artificial seawater are appropriate to prepare the fixation buffer. If a different buffer (e.g., sodium cacodylate) is chosen, high purity deionized water (e.g., 18 MΩ resistivity) should be used. All fixations should be performed under a fume hood, and all personnel should wear gloves and protective eye gear. Institutional and government-based safety regulations for hazardous waste disposal should be followed (*Note 1*). The materials needed include the following:

- Seawater (natural or artificial).
- Styrofoam™ container with ice.
- 9 cm diameter plastic Petri dish.
- Tweezers.
- Glass scintillation vials.
- Disposable pipettes.
- Bell-jar connected to a vacuum pump or vacuum filtration unit.
- Single-edge blades of carbon steel.
- Laboratory shaker.
- Paraformaldehyde 16% (w/v), EM grade.
- Glutaraldehyde 25% EM grade distillation purified.
- Osmium tetroxide 4% (w/v) aqueous solution.

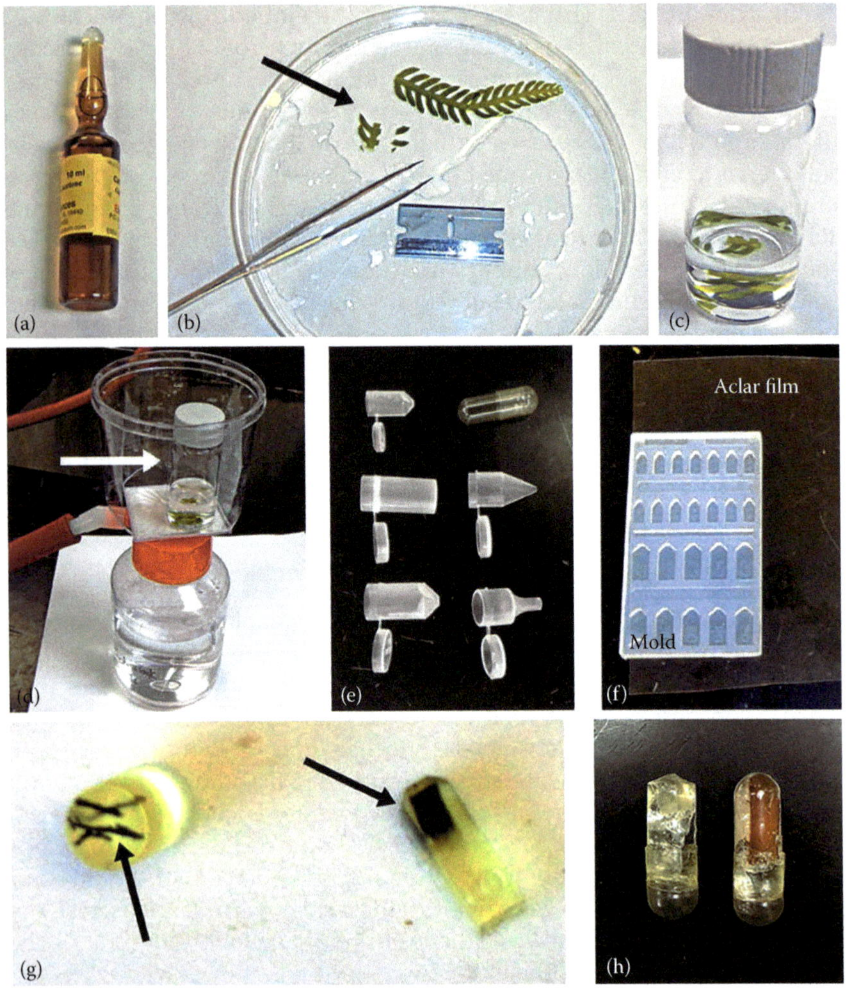

Figure 26.1 (a) Vacuum-sealed glass ampule of glutaraldehyde. (b) On collection of the seaweeds from the field, pieces are cut arrow and placed in a Petri dish containing the fixative solution. (c) After cutting smaller pieces, these are placed in glass scintillation vials containing the fixative solution. (d) Filtration unit attached to a sink aspirator can be used to remove air trapped in the samples. (e) A variety of *BEEM*® and gelatin capsules can be used for polymerization. (f) Molds can also be chosen to polymerize samples, although if using LR White embedding medium, Aclar® film must be applied to seal the molds. (g) Samples prepared for TEM have a black appearance, whereas samples prepared for LM often look transparent, or retain some of the original color of the sample.

- Acetone or ethanol (depending on the embedding medium of choice).
- Fixative solution: 2% (v/v) paraformaldehyde with 1% (v/v) glutaraldehyde in seawater or buffer of choice. Prepare fresh and keep the solution in a box with ice.
- Postfixative: 1% (v/v) osmium tetroxide in ice cold seawater or buffer of choice. Prepare freshly before use.
- LR White embedding medium-grade acrylic resin.
- Low-viscosity Embedding Media Spurr's kit.
- Plastic cup with paper lid (for Spurr's preparation).
- Dehydrating series solutions: If using LR White, prepare solutions of 10% (v/v), 30% (v/v), 50% (v/v), 70% (v/v), 90% (v/v), and 100% (v/v) of ethanol in seawater. If using Spurr's, prepare the same concentrations, but use acetone in seawater.

26.3.2 *Polymerization*

- BEEM® embedding capsules, gelatin capsules, or molds for flat embedding
- Aclar® 33c embedding film (if using LR White)
- Tweezers
- Laboratory UV lamp (for LR White medium)
- Laboratory oven (for Spurr's medium)

26.3.3 *Sectioning*

- Ultramicrotome such as the Leica EM UC6 ultramicrotome
- Histo-diamond knife (e.g., DiATOME, Hatfield, PA, USA)
- Single-edge blades of carbon steel
- High precision and ultrafine tweezers
- Microscope slides
- Formvar-coated copper, nickel, or gold grids
- Chloroform
- Squeeze lab bottle with dH_2O
- Glass pipette with a plug of cotton inside
- Filter paper ripped in pieces
- *Eyelash* tool: Use glue or nail polish to attach an eyelash or a fine hair of an artist's fine brush to a toothpick handle
- Grid storage box

26.3.4 *Uranyl acetate/Lead citrate staining*

- Sodium hydroxide pellets.
- Water without dissolved CO_2: Boil dH_2O in a glass bottle for 4–6 min in a microwave. Cap and let it cool down overnight before use.
- Squeeze lab bottle with dH_2O.

- Disposable pipettes.
- Uranyl acetate 1% (w/v) solution.
- Lead citrate 1% (w/v) solution.
- High precision and ultrafine tweezers.
- 100 mL glass beakers.
- Glass Petri dish (100 mm diameter).
- Filter paper ripped in pieces.
- Silicone staining plates.
- Grid storage box.

26.3.5 *Immunogold labeling*

- PBST: Add 10 mL of a 10x stock phosphate-buffered saline (PBS) to 90 mL dH_2O. Mix and adjust pH to 7.0–7.2. Add 100 µL of TWEEN® 20 and shake into solution.
- Blocking solution: Place 10 mL of PBST in a 15 mL Falcon™ tube. Add 0.1 g of Carnation Instant Milk and vortex for 20 s to dissolve.
- Ammonium chloride (to remove glutaraldehyde): Place 50 mL of dH_2O in a 50 mL Falcon™ tube, add 0.13 g of ammonium chloride and dissolve.
- Hydrogen peroxide (to remove osmium tetroxide): Add 1 mL of 30% (v/v) of H_2O_2 (stored at 4°C) to 5 mL of dH_2O. Wear gloves and use with care.
- Primary antibodies: mAbs for many cellular components may be obtained from Agrisera Inc, or Sigma-Aldrich®. Cell wall epitope recognizing mAbs may be purchased from Plant Probes (Leeds, UK), the Complex Carbohydrate Research Center (Athens, GA, USA), or Biosupplies (Bundoora, Australia); mAbs may be diluted in PBST to the desired concentration.
- Secondary antibody: gold conjugated goat anti-rat IgG (10–15 nm range is best for TEM) may be obtained from Nanoprobes (Yaphank, NY, USA), Sigma-Aldrich®, Agar Scientific, or EMS and is diluted in PBST to the desired concentration.
- Microcentrifuge (Fisher Scientific accuSpin Micro 17).
- Vortex (Vortex Genie2).
- Filter paper ripped in pieces.
- Sheets of Parafilm M®.
- High precision and ultrafine tweezers.
- Disposable pipettes.
- 50 mL beaker.
- Squeeze lab bottle with dH_2O.
- Moisture chamber: Use a plastic box with lid, such as a Tupperware® box, place some sheets of paper, moisture with water, and place a flat bottom glass base (such as a glass Petri dish lid) that fits the box.

- Porcelain multiwell plates.
- Microscope slides.
- Oven with a temperature set at 33°C.

26.4 *Experimental procedures*

26.4.1 *Fixation*

1. Prepare the fixative solution and keep on ice. If the specimen is also to be used for LM sections, glutaraldehyde may be eliminated from the fixative; a low percentage of 0.25% (v/v) has also shown to be successful in the case of brown seaweeds (*Note 2*).
2. Pour 5 mL of the fixative into a 9 cm plastic Petri dish also on ice. Cut portions of the seaweed thallus from the collected specimen and place it immediately in the fixative. Using a razor blade, cut 0.5–2.0 cm of seaweed segments in the fixative (Figure 26.1b). Excision of seaweed parts directly in the fixative allows a faster infiltration of the chemical.
3. Gently collect the seaweed pieces with forceps, place in a screw cap glass scintillation vial containing the fixative (Figure 26.1c), and keep on ice for 30–60 min. For some dense seaweed parts, fixations at 4°C for up to 48 h may be required.
4. If the samples float in the fixative and do not sink, place the vial with a loose cap in a small ice bath in a bell jar connected to a vacuum pump or to the top chamber of a medium filtration unit attached to a sink aspirator (if a vacuum pump is unavailable). (Figure 26.1d, *Note 3*). Once the vacuum is applied, the seaweed pieces typically sink after 2–5 min. Remove the vial and put it back on ice.
5. Remove and discard the fixative with a pipette, wash the samples with ice cold seawater or another buffer for 5 min. Repeat the washing step twice.
6. Postfixation: If using a postfixative such as osmium tetroxide, place the samples in a vial and cover them with the postfixative solution (*Note 4*). Incubate on ice for 2 h, or overnight. Samples can be gently shaken on a laboratory shaker, vacuum infiltrated, or processed in a microwave oven (*Note 5*).
7. Repeat the washing step 5.

26.4.2 *Dehydration*

Most embedding agents are hydrophobic. Therefore, water must be removed from the specimen. Therefore, samples are slowly dehydrated in a series of ethanol or acetone before adding the embedding medium.

Using a pipette to add or remove the solutions, samples are dehydrated in the graded ethanol or acetone series (10% [v/v], 30% [v/v], 50%

[v/v], 70% [v/v], 90% [v/v], and 100% [v/v]) during 30–60 min for each grade, with gentle shaking, either at RT, at 4°C (ice bath or refrigerator) or at −20°C (commercial freezer). For embedding media that may be polymerized with UV light (e.g., LR White), the ethanol dehydration should be carried on at 4°C or −20°C; for epoxy plastics especially of osmicated specimens, RT dehydration is satisfactory. The 70% (v/v) dehydrating stage is an appropriate timing for an overnight incubation (*Note 6*).

26.4.3 Infiltration

After dehydration, the tissue segments are slowly infiltrated in a series of embedding medium dissolved in the dehydrating agent (*Note 7*).

For LR White: Incubate the samples in a series of LR White:ethanol of 1:2, 1:1, 2:1, and 1:0. The first three infiltration steps should be 2–4 h long each, at RT, 4°C or −20°C. The seaweed parts should be lightly shaken during infiltration, placed in pure LR White at the desired temperature, and lightly shaken for a period of up to two days.

For Spurr's medium, prepare the resin beforehand in a plastic cup according to the manufacturer's instructions. Use a similar infiltration series using acetone as the solvent. The final incubation with 100% Spurr's medium is performed overnight, or for one day, at RT.

26.4.4 Polymerization

A variety of specimen molds used for polymerization is commercially available, such as plastic BEEM® capsules, gelatin capsules (Figure 26.1e), and molds for flat embedding made of rubber, Teflon, or flexible plastic (Figure 26.1f).

1. Seaweed parts may be trimmed with a razor blade to fit in the appropriate mold.
2. Add a small amount of the resin of choice (if using LR White and has to be UV polymerized, an accelerator should be added in advance following the manufacturer's instructions). Using forceps, carefully place a piece at the bottom of a BEEM® capsule or at the end of a flat embedding mold in the desired orientation/position, and fill the mold with resin. A dissecting light microscope is often helpful to properly position the seaweed part in the mold.
3. For LR White polymerization, the BEEM® capsule is capped carefully. For flat embedding molds, a piece of Aclar® embedding film should be cut and carefully placed over the molds, making sure that there are no air bubbles between the resin and the film (Figure 26.1f) (*Note 8*). Position a laboratory UV lamp 10–20 cm above the mold and polymerize the sample for 24 h at the desired temperature (*Note 9*).

LR White polymerization has also been proven to work well for brown seaweeds at 60°C for two days (Raimundo et al. 2016, 2017).

4. For Spurr's polymerization, the BEEM® capsules should not be capped or covered. Polymerize the samples at 70°C for 8 h (or at the temperature specified by the manufacturer).
5. After polymerization, the plastic of the BEEM® capsule is cut with a razor blade, releasing the polymerized sample (Figure 26.1g), whereas for gelatin capsules there is no need to remove the gelatin from the sample (Figure 26.1h). For flat embedding molds, the Aclar® film is removed, the mold is gently bent, and the blocks pop out. Blocks can be stored in labeled paper or glassine envelopes at RT for years.

26.4.5 Sectioning

An ultramicrotome is the most convenient instrument to obtain sections ranging in thickness from 500 nm for LM to 50–70 nm for TEM purposes. Depending on the embedding medium chosen, the same block can be conveniently used for correlative studies at the LM and TEM levels. Thick sections may be immunolabeled and imaged with a fluorescence LM, and labeled zones identified and documented. Thin sections are obtained from selected zones and processed for immunogold labeling studies to elucidate a more precise epitope location through electron microscopy. Specific protocols include the following:

1. *Knife manufacture*: For sectioning, a knife that accurately cuts embedded specimens with a specific thickness is a critical tool. Glass or diamond knives can be used to section. Glass knives are relatively cheap and highly effective for all thickness sections, but their durability and shelf life are limited. Water-cooled glass strips and troughs must be assembled and individual knife is quality assessed before each sectioning session. Likewise, the knife maker used for scoring and fracturing the glass to make the knife must be regularly maintained and cleaned. On the other hand, diamond knives are more expensive, but if cared properly, may provide outstanding sections for years. Their cutting edge is extremely durable, whereas glass knives must constantly be replaced, and it is possible to obtain ultrathin sections (50 nm). Information on the various knives, including maintenance and troubleshooting are provided by manufactures, for example, DiATOME (Hatfield, PA, USA).
2. *Preparation of imaging platforms for TEM grids and special tools*: Before sectioning for TEM, grids, that is, the platforms for holding sections (Figure 26.2a) must be prepared. Typically, a 0.25%–1.0% (v/v) layer of Formvar or another transparent plastic is coated on the grid. Grids

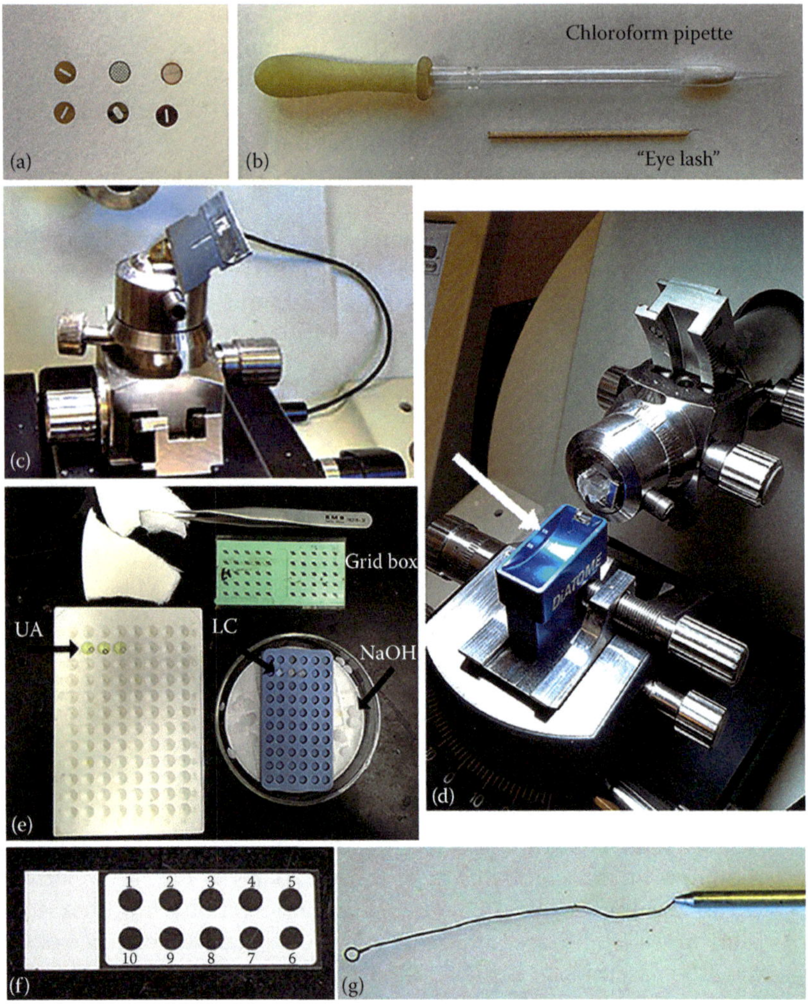

Figure 26.2 (a) Gold, nickel, or copper (first to third column, respectively) grids are used to hold sections for TEM observations. (b) Useful tools for sectioning: chloroform pipette, and eyelash attached to a toothpick. (c) Trimming the block with a razor blade is the first and essential step for a successful sectioning. (d) During the sectioning procedure, sections can be observed floating on the water (arrow). (e) Apparatus used for uranyl acetate/lead citrate staining procedure. A CO_2-free chamber is essential to avoid the formation of lead carbonate deposits. UA, Uranyl acetate drops; LC, Lead citrate chamber; NaOH, Sodium hydroxide pellets. (f) 10 Multiwelled immunoslide is useful when performing immunolabeling with different antibodies. (g) Microbiology loop used to pick sections.

are dried and stored at RT until needed. For routine ultrastructural studies, copper grids are typically used. For immunogold labeling, nickel or gold grids should be used. To help collect the sections onto grids, two valuable tools to have are the *eyelash* and a chloroform vapor-dispensing pipette (Figure 26.2b). The former is easily assembled by affixing an eyelash to a toothpick handle with glue or nail polish. The latter consists of a glass pipette filled with a plug of cotton that is wetted with chloroform. A rubber bulb is placed on the open end of the pipette and through gentle squeezing, *puffs* of chloroform vapor are released from the tip.

3. *Trimming specimen blocks*: To facilitate sectioning of plastic-embedded specimens with a small knife edge (diamond knives cutting edge size range from 0.5–4.0 mm), the block face must be properly trimmed. Using the trimming vice of an ultramicrotome and razor blades (Figure 26.2c), a trapezoid is cut around the specimen with 1.0 mm or less. It is important to trim the sides of the block at a slight angle, as steeper angles result in vibrations and chatter during sectioning.
4. *Sectioning for TEM*: After placing the trimmed block on the microtome and carefully positioning it with respect to the knife edge, 50–70 nm sections are cut and floated onto to the surface of the water that fills the trough behind the knife (Figure 26.2d). The chloroform dispensing pipette is then used to gently release chloroform vapors over the sections (*Note 10*). The sections can be trapped into one area of the trough using the eyelash, collected, and placed on the coated grids. Excess of water should be removed from the grid surface by carefully touching the grid edge with a piece of ripped filter paper without touching the sections. Grids can be stored in a grid storage box indefinitely.

26.4.6 *Uranyl acetate/Lead citrate staining*

For routine ultrastructural work, the sections are stained with conventional uranyl acetate and lead citrate (Venable and Coggeshall 1965, Hayat 1981). Grid staining may not be necessary if using a TEM with energy filtration lenses.

1. Keep the solutions in 15 mL Falcon™ tubes at 4°C for regular use, and before using them make sure to centrifuge for 1 min at 1000 rpm to precipitate unwanted particles.
2. Prepare a CO_2-free chamber: Put a silicone plate in a glass Petri dish, place sodium hydroxide pellets around it, put the number of drops

of lead citrate needed on the wells (one per grid), and close the chamber with the lid (Figure 26.2e, *Note 11*).

3. Put a drop of uranyl acetate on a silicone plate, one drop for each grid, and incubate for 5–10 min (section facing the stain).
4. Wash the grids with a gentle continuous stream of dH_2O, dry with a ripped filter paper piece.
5. Place each grid on a drop of lead citrate, close the chamber, and let it incubate for 1–2 min.
6. Dip the grids in a beaker with the CO_2-free dH_2O, gently, for a few seconds, wash again with a gentle continuous stream of dH_2O, dry with ripped filter paper, and store indefinitely in the grid box until observation.

26.4.7 *Immunogold labeling*

For all the procedures except those requiring the porcelain plates, perform the incubations placing drops of the solutions on Parafilm M® sheets previously placed on the lab bench, one drop for each grid, and incubating the grids on the solutions with the section facing the solution.

1. Spread Parafilm M® sheets on the bench and place drops of H_2O_2 solution to match the number of grids. Float each grid in each drop with section-side (or Formvar-side or dark-side) down onto the solution for 5 min (*Note 12*).
2. Wash the grids with a gentle stream of dH_2O. Blot dry with a ripped filter paper.
3. Float the grids for 20 min on drops of ammonium chloride solution.
4. Repeat washing step 2.
5. Pipette 1 mL aliquots of the blocking solution in Eppendorf tubes and centrifuge at 12,000 rpm for 1 min. Use the supernatant of the blocking solution to incubate the grids for 20–30 min.
6. Repeat washing step 2.
7. Float grids on 20–30 μL of diluted primary mAb in the wells of porcelain plate (*Note 13*).
8. Cover the wells with microscope slides, place the porcelain dish carefully in the moisture chamber, close the lid, and place it carefully in the oven at 33°C for 90 min (this step can also be performed at 4°C overnight).
9. Remove the moisture chamber from the oven; remove the porcelain dish with the grids at RT.
10. Repeat washing step 2.
11. Float the grids on blocking solution for 20–30 min. At this time, also prepare the secondary antibody solutions.

12. Repeat washing step 2.
13. Float the grids on drops of secondary antibody solution in the wells of a new porcelain dish. Cover the wells with microscope slides, place the porcelain dish carefully in the moisture chamber, close the lid, and place it carefully in the oven at 33°C for 90 min.
14. Remove the moisture chamber from the oven, remove the porcelain dish with the grids at RT.
15. Repeat washing step 2.
16. Stain the grids with uranyl acetate/lead citrate (Figure 26.2e). After drying, samples are ready for TEM observation.

26.4.8 Considerations for light microscopy and correlative studies

For examining thick sections (e.g., 500 nm) with LM using the same block of seaweed samples, the following protocol is recommended.

1. Before sectioning, 10- or 12-welled immunoslides should be coated (Figure 26.2f, *Note 14*): Add 25 µL of a 1 mg/mL poly-L-lysine solution to each well of, let it stand for 5 min, and remove the excess with a micropipette. Let it dry at RT. Slides may be kept for several weeks in a refrigerator until needed.
2. Trimming and microtome setup are the same as for TEM sectioning. However, 250–500 nm sections are more common for LM. The eyelash tool is used to corral the sections together. The sections are collected by placing a clean microbiology loop (Figure 26.2g) around the section floating on the water and lifting it up. The section is caught in a film of water in the loop. The loop is gently tapped into a well of the slide where the drop of water with the section is deposited. Clean the microbiology loop regularly by heating it over an alcohol lamp. The other wells of the slide can be filled with as many sections as needed. Using a dissecting microscope, carefully remove the excess of water. Dry the slides at RT or in a 37°C incubator for 1 h. The slides can be stored in a refrigerator for months until use. Alternatively, sections can be dried at 50°C straight after collection.
3. Stain some sections with a reference stain such as toluidine blue (1% [w/v] toluidine blue in 1% [w/v] sodium borate) for 30 s, wash with dH_2O, and dry at 50°C (Figure 26.3a–b, *Note 15*).
4. For immunolabeling procedures, a similar protocol is used as for immunogold labeling, but the secondary antibody used is a goat anti-mouse IgG or goat anti-rat IgG conjugated with a fluorochrome, such as TRITC or FITC (Figure 26.3c–d). In the final washing steps, place 10 µL of PBS in each well, cover with a cover slip 20 × 75 mm, remove the excess of liquid, and seal with nail polish. Slides can be

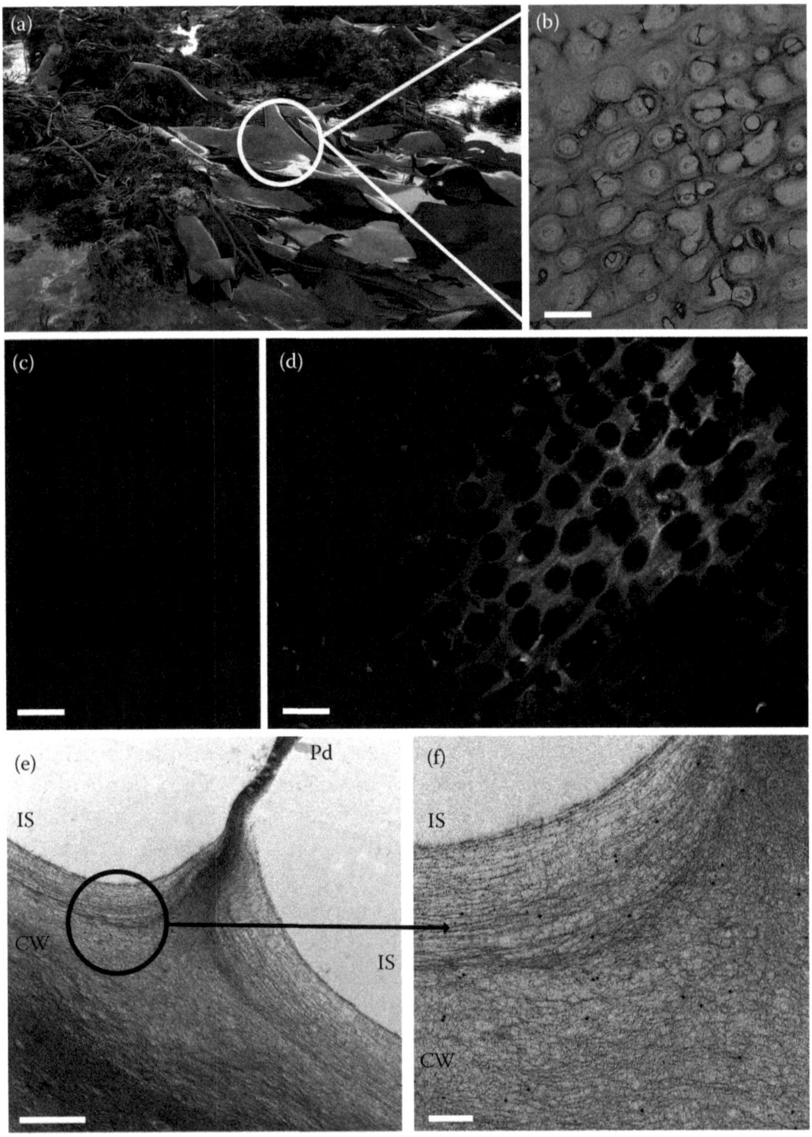

Figure 26.3 (a) Natural habitat of the brown seaweed *Laminaria digitata*. (b) Cross section of *L. digitata* blade stained with toluidine blue. (c) Negative control where no primary mAb is used. (d) Cross section of *L. digitata* blade immunolabeled with LM7 mAb. LM7 is an excellent probe for alginate-containing extracellular matrix in brown seaweed tissues. (e) Transmission electron micrograph of *L. digitata* blade cross section immunogold labeled with LM7. (f) Detail of the gold particles showing the labeling of the cell wall. CW, cell wall; IS, intracellular space; Pd, plasmodesmata. Scale bar: (b–d) = 20 µm, (e) = 1 µm, (f) = 200 nm.

kept in a refrigerator for up to six months. The same tissue areas can them be correlated together with TEM immunogold labeling (Figure 26.3e–f).

26.5 Notes

Variations and refinements in reagents, equipment, and protocols are often necessary for the accommodation of the aforementioned protocols in particular laboratories. The following list includes some common areas where changes may be performed, and comments on details that should be kept in mind while performing the experiments.

Note 1: In some situations, it may be preferable or necessary to perform a fixation in the collection site. A Styrofoam™ box containing the fixative solution, glass scintillation vials, razor blades, plastic Petri dishes, forceps, ice, waste container, and gloves are taken out to the field and fixation is performed *in situ*. The vials with the fixed specimens are kept on the ice, taken to the laboratory as soon as possible, and vacuum infiltrated if needed. Particular care must be taken to avoid spilling fixative into the collection habitat.

Note 2: Glutaraldehyde is known for being an excellent fixative, and it has shown to work well in the preservation of brown seaweed tissues for LM fluorescence observations (Raimundo et al. 2016, 2017), but a low concentration should be used, such as 0.25% (v/v), as glutaraldehyde induces an increase in autofluorescence, commonly observed in many brown seaweed tissues, which will interfere with immunofluorescence observations.

Note 3: During the fixation process, the seaweed pieces should sink to the bottom of the vial. If they keep floating, vacuum infiltration is necessary to remove air trapped in the tissues. The presence of air in the sample will lower the success of the penetration efficacy of the fixative and further embedding procedure.

Note 4: The use of osmium tetroxide as a secondary or postfixative is commonly used to enhance contrast in TEM studies but is not needed for LM.

Note 5: Recently, the use of special microwave ovens (e.g., EMS-9000 precision-pulsed laboratory microwave oven) modified for specimen processing have been used in the processing of specimens for microscopy. The function of these devices is to make the fixation–infiltration–embedding processes faster, with the same results as regular methods (Login and Dvorak 1988, Webster 2007). Currently, there is limited published information regarding their application in the processing of seaweeds.

Note 6: The chemical composition and/or organization of the thallus parts of some seaweeds (especially brown seaweeds) might limit the permeability to the embedding chemicals, therefore longer dehydrating stages may be needed (e.g., 1–4 h each grade). This may be confirmed after sectioning; the seaweed part separating from the embedding agent or the presence of empty holes within the tissue after sectioning and microscopy imaging is an indicator of an incomplete penetration of the embedding medium.

Note 7: It is up to the researcher to use LR White or Spurr's as the embedding medium of choice, as both provide satisfactory results with seaweeds. However, if using LR White, ethanol must be used for the dehydration process and as a solvent for the embedding series, whereas the same steps are performed with acetone if Spurr's medium is the resin of choice.

Note 8: Although Spurr's medium polymerizes in contact with air, oxygen alters the polymerization ability of LR White, therefore the presence of any air bubbles must be avoided by capping the BEEM® or gelatin capsules tightly, and if using flat molds, Aclar® film should cover the resin. If oxygen infiltration becomes a problem, LR White polymerization can be performed in a vacuum desiccator or a chamber where nitrogen gas is pumped.

Note 9: Styrofoam™ blocks are a practical and inexpensive solution to hold the UV lamps under and above the embedding molds.

Note 10: One major problem that occurs when sectioning is that the hydrophobic embedding material wrinkles when floating on water. Wrinkles may be removed by gently releasing vapors of chloroform over the sections using the chloroform-dispensing pipette. The vapor results in the immediate section dewrinkling.

Note 11: During the staining with lead citrate, it is highly recommended that CO_2-free water is used when washing the grids to avoid the formation of lead carbonate deposits on the specimen that will appear as black dots and smudges on the specimen during TEM observations.

Note 12: Incubation with H_2O_2 is only required if the samples were osmicated.

Note 13: Immunogold labeling can only be performed on nickel or gold grids; do not use copper grids for this procedure.

Note 14: Multiwelled immunoslides are very useful because immunolabeling with different mAbs can be performed per slide. Some seaweed species have shown to be difficult to use, as sections detach from the slides during staining/immunolabeling. To avoid such situations, multiwell immunoslides can be coated with poly-L-lysine before used. An alternative that has also proven to be successful with brown seaweeds is the use of noncoated Superfrost™ Plus slides (Fisher Scientific, Pittsburg, PA, USA).

Note 15: When performing immunolabeling procedures, it is important to keep in mind that the labeling will appear on a black background. Therefore, it is useful to have some sections from the same block stained with, for example, toluidine blue, which is an excellent dye for seaweeds to have a reference regarding the localization of the labeling within the tissue/cells.

References

Becker, B., and B. Marin. 2009. Streptophyte algae and the origin of embryophytes. *Annals of Botany* 103 (7):999–1004.

Bold, H. C., and M. J. Wynne. 1978. *Introduction to the Algae: Structure and Reproduction*. Englewood, NJ: Prentice Hall.

Callow, J. A., and M. E. Callow. 2006. The Ulva spore adhesive system. In *Biological Adhesives*, A. M. Smith and J. A. Callow (Eds.), pp. 63–78. Berlin, Germany: Springer.

Cardozo, K. H. M., T. Guaratini, M. P. Barros, V. R. Falcão, A. P. Tonon, N. P. Lopes, S. Campos et al. 2007. Metabolites from algae with economical impact. *Comparative Biochemistry and Physiology, Part C: Toxicology & Pharmacology* 146 (1–2):60–78.

Cock, J. M., L. Sterck, P. Rouzé, D. Scornet, A. E. Allen, G. Amoutzias, V. Anthouard et al. 2010. The *Ectocarpus* genome and the independent evolution of multicellularity in brown algae. *Nature* 465 (7298):617–621.

Collén, J., B. Porcel, W. Carré, S. G. Ball, C. Chaparro, T. Tonon, T. Barbeyron et al. 2013. Genome structure and metabolic features in the red seaweed *Chondrus crispus* shed light on evolution of the Archaeplastida. *Proceedings of the National Academy of Sciences of the United States of America* 110 (13):1–6.

Cunha, L., and A. Grenha. 2016. Sulfated seaweed polysaccharides as multifunctional materials in drug delivery applications. *Marine Drugs* 14 (3):42.

Enquist-Newman, M., A. M. Faust, D. D. Bravo, C. N. Santos, R. M. Raisner, A. Hanel, P. Sarvabhowman et al. 2014. Efficient ethanol production from brown macroalgae sugars by a specific yeast platform. *Nature* 505 (7482):239–243.

Graham, L. E., and L. W. Wilcox. 2000. *Algae*. New York: Prentice Hall.

Hallam, N. D., and S. E. Luff. 1988. The fixation and infiltration of larger brown algae (Phaeophyta) for electron microscopy. *British Phycological Journal* 23 (4):337–346.

Hayat, M. A. 1981. *Principles and Techniques of Electron Microscopy—Biological Applications*. 2nd ed. Vol. 1. Baltimore, MD: University Park Press.

Hervé, C., A. Siméon, M. Jam, A. Cassin, K. L. Johnson, A. A. Salmeán, W. G. T. Willats, M. S. Doblin, A. Bacic, and B. Kloareg. 2016. Arabinogalactan proteins have deep roots in eukaryotes: Identification of genes and epitopes in brown algae and their role in *Fucus serratus* embryo development. *New Phytologist* 209 (4):1428–1441.

Leliaert, F., D. R. Smith, H. Moreau, M. D. Herron, H. Verbruggen, C. F. Delwiche, and O. De Clerck. 2012. Phylogeny and molecular evolution of the green algae. *Critical Reviews in Plant Sciences* 31:1–46.

Lobban, C. S., and P. J. Harrison. 1994. *Seaweed Ecology and Physiology*. Cambridge, UK: Cambridge University Press.

Login, G. R., and A. M. Dvorak. 1988. Microwave fixation provides excellent preservation of tissue, cells and antigens for light and electron microscopy. *The Histochemical Journal* 20 (6):373–387.

Nagasato, C., A. Inoue, M. Mizuno, K. Kanazawa, T. Ojima, K. Okuda, and T. Motomura. 2010. Membrane fusion process and assembly of cell wall during cytokinesis in the brown alga, *Silvetia babingtonii* (Fucales, Phaeophyceae). *Planta* 232 (2):287–298.

Nagasato, C., and T. Motomura. 2002. Ultrastructural study on mitosis and cytokinesis in *Scytosiphon lomentaria* zygotes (Scytosiphonales, Phaeophyceae) by freeze-substitution. *Protoplasma* 219 (3):140–149.

Nagasato, C., N. Kajimura, M. Terauchi, Y. Mineyuki, and T. Motomura. 2014. Electron tomographic analysis of cytokinesis in the brown alga *Silvetia babingtonii* (Fucales, Phaeophyceae). *Protoplasma* 251 (6):1347–1357.

Patterson, D. J. 1999. The diversity of eukaryotes. *The American Naturalist* 154 (S4):S96–S124.

Raimundo, S. C., U. Avci, C. Hopper, S. Pattathil, M. G. Hahn, and Z. A. Popper. 2016. Immunolocalization of cell wall carbohydrate epitopes in seaweeds: Presence of land plant epitopes in *Fucus vesiculosus* L. (Phaeophyceae). *Planta* 243 (2):337–354.

Raimundo, S. C., S. Pattathil, S. Eberhard, M. G. Hahn, and Z. A. Popper. 2017. β-1,3-Glucans are components of brown seaweed (Phaeophyceae) cell walls. *Protoplasma* 254 (2):997–1016.

Torode, T. A., A. Siméon, S. E. Marcus, M. Jam, M.-A. Le Moigne, D. Duffieux, J. P. Knox, and C. Hervé. 2016. Dynamics of cell wall assembly during early embryogenesis in the brown alga *Fucus*. *Journal of Experimental Botany* 67:6089–6100.

Venable, J. H., and R. Coggeshall. 1965. A simplified lead citrate stain for use in electron microscopy. *The Journal of Cell Biology* 25 (2):407–408.

Wargacki, A. J., E. Leonard, M. N. Win, D. D. Regitsky, C. N. S. Santos, P. B. Kim, S. R. Cooper et al. 2012. An engineered microbial platform for direct biofuel production from brown macroalgae. *Science* 335 (6066):308–313.

Webster, P. 2007. Microwave-assisted processing and embedding for transmission electron microscopy. In *Electron Microscopy: Methods and Protocols*, J. Kuo (Ed.), pp. 47–65. Totowa, NJ: Humana Press.

Yoon, H. S., J. D. Hackett, C. Ciniglia, G. Pinto, and D. Bhattacharya. 2004. A molecular timeline for the origin of photosynthetic eukaryotes. *Molecular Biology and Evolution* 21 (5):809–818.

chapter twenty seven

Coralline algae preparation for scanning electron microscopy and optical microscopy

S. Kaleb, G. Alongi, and A. Falace

Contents

27.1 Introduction

Coralline algae (CA) are distributed from tropical regions to the poles, thriving from the intertidal to the lower boundaries of the euphotic zone. Because of the calcium carbonate present in their cell walls, they fill a paramount role as bioconstructors of coral reefs, coralligenous, rhodolith beds, providing three-dimensional habitats and contributing to carbonate deposition in marine ecosystems. CA are object of studies related to global change and are also used in paleoenvironmental reconstruction as biological datalogger (McCoy and Kamenos 2015; Kamenos et al. 2016). The Marine Strategy Framework Directive (MSFD; European Parliament and Council of the European Union 2008; European Commission 2010) has included the assessment of CA distribution and biodiversity within the protocol to evaluate the Good Environmental Status of Coralligenous and Rhodolith beds Habitats (Basso et al. 2016; Piazzi et al. 2015). In particular, the protocol proposed for the monitoring of deep Mediterranean rhodolith beds (Basso et al. 2016) requires time-consuming and expensive laboratory analysis for species identification. In fact, as most coralline species belong to a few genera only, the use of taxonomic ranks higher than species is not useful. Finally, because CA are calcified, the preparation of the material for the microscopy (both optical microscopy and scanning electronic microscopy) can be more demanding than for most other macroalgae.

In the current chapter, we describe how to obtain samples of CA for the observation at scanning electron microscopy (SEM) and optical microscopy (OM) of the anatomical features relevant for taxonomical identification. The protocol applies to herbarium specimens or freshly collected samples. The same procedures can be used to prepare samples for energy dispersive spectrometry (SEM–EDS) in calcium carbonate biomineralization studies.

27.2 State of the art

SEM analyses have been first used in CA biomineralization and morphoanatomical studies starting from the early 1970s (i.e., Bailey and Bisalputra 1970; Alexandersson 1974; Borowitzka et al. 1974; Littler 1976; Woelkerling 1978; Chamberlain 1983). SEM micrographs have been used to reveal the presence of microborers (i.e., Cyanobacteria and endolithic Chlorophyta) in algal tissues (Tribollet and Payri 2001; Ghirardelli 2002).

Some indications on how to prepare samples for CA taxonomical studies are reported in Woelkerling (1988), Irvine and Chamberlain (1994), Bressan and Babbini (2003), and Harvey et al. (2005), but without detailing procedures and protocols. Because of their morphological plasticity, the identification of CA displays some challenges, in particular for the nonspecialists. Moreover, if compared to other macroalgae, the preparation of CA samples for microscopy observation can be more arduous because of the calcium carbonate cell walls. For this reason, it is necessary to employ a straightforward protocol to obtain sections with proper orientation, thus allowing the observation of the more relevant features needed for the identification.

27.3 Materials

27.3.1 Materials for optical microscopy

For the decalcification and stain of CA, different solutions can be used (Woelkerling 1988; Bressan and Babbini 2003).

- Decalcification solutions: (a) 6 M HNO_3 or (b) Perenyi solution (4:3:3 of 10% aqueous NHO_3, 90% aqueous ethanol, 5% aqueous chromic acid, respectively)
- Staining mixtures: (a) 2%–5% aqueous solution of $KMnO_4$ or (b) 5% aqueous solution of aniline blue with a drop of 1% aqueous solution of HCl as fixative
- Ethanol (according to the increasing concentration series: 30%, 60%, 90%, and 100%)
- Medium grade *LR White* resin® (resin + accelerator)
- Clearing agent Eukitt (O. Kindler GmbH & Co, Germany)
- Hand razor blade
- Sledge microtome
- Infrared lamp
- Stoppered small plastic box

27.3.2 *Materials for scanning electron microscopy*

- Hammer (Figure 27.1)
- Chisel (Figure 27.1)
- Single-edged razor blades (Figure 27.1)
- Fine forceps
- Air Hardening Modeling Clay (DAS)
- Permanent marker
- Aluminum stubs (diameter 1 cm)
- Instant adhesive (cyanoacrylate adhesive), acrylic adhesive, or liquid graphite
- Sonicator
- Sputter coater
- Oven

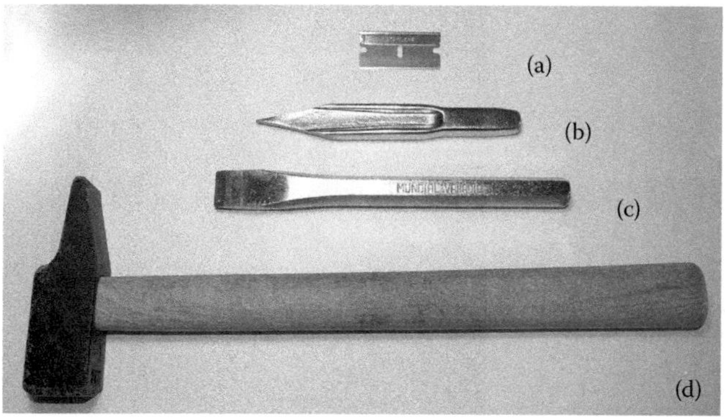

Figure 27.1 Material required for the cut of samples (a–d): Single-edged razor blade, chisels with fine tip, and hammer.

27.4 Experimental procedures

27.4.1 Choice of the portion to cut and orientation of the sample

The anatomical features of CA vary depending on the planes of view. The longitudinal sections/fractures permit to identify the majority of CA genera and allow observing some characteristic such as trichocytes (Figure 27.2), primary and secondary pit connections, cell fusions, calcite crystals disposition in the cell wall (Figure 27.3), or reproductive features (Figure 27.4). The identification of the genus *Sporolithon* requires transverse sections allowing observing the presence of both cell fusions and secondary pit connections (Figure 27.5).

1. Longitudinal view (Figure 27.6a, c):
   a. In crustose and lamellate thalli, the longitudinal/radial section is directed toward the centre of the thallus, perpendicularly to the growth margin (Figure 27.7a and b);
   b. In rhodoliths/maerl or in protuberances, the longitudinal section is directed from the apex to the base (Figure 27.8).

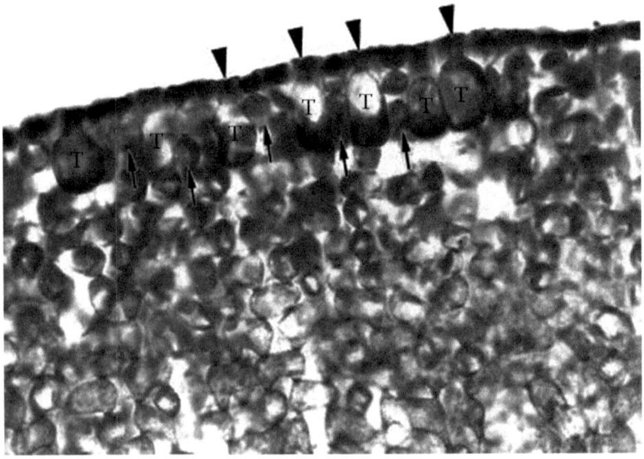

Figure 27.2 Section of the thallus showing solitary and paired trichocytes (T). Note the neck (arrowheads) of trichocyte megacells and the normal vegetative cells (arrows) between the solitary and paired trichocytes.

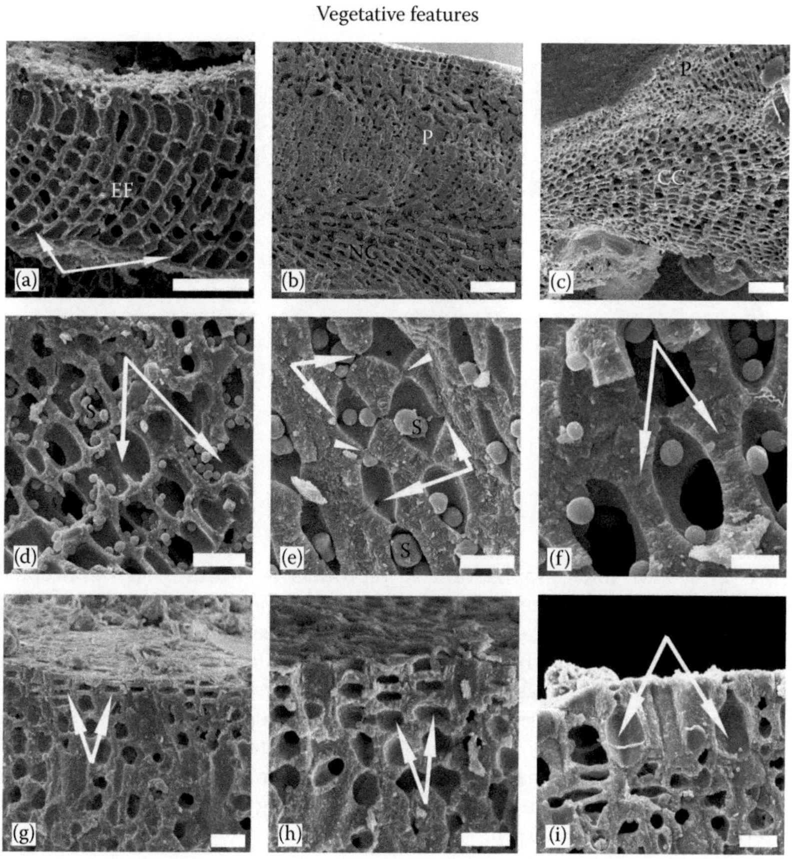

Figure 27.3 SEM micrographs of CA vegetative features. (a) Dimerous thallus: single-layered basal filament (arrows); erect filaments (EF) up to 7 cells long. Scale bar = 50 μm. (b) Monomerous thallus with noncoaxial core region (NC); peripheral region with filaments curving outward (P). Scale bar = 50 μm. (c) Monomerous thallus with a coaxial core region (CC); peripheral region (P). Scale bar = 50 μm. (d) Fusion cells (arrows) with starch grains (S). Scale bar = 20 μm. (e) Cells of adjacent filaments joined by secondary pit connections (arrows). Primary pit connections (arrowheads). Starch grains (S). Scale bar = 10 μm. (f) Calcite crystals of the cell walls. Scale bar = 5 μm. (g) Epithelial cells (arrows). Scale bar = 10 μm. (h) Subepithelial initial cells (arrows). Scale bar = 10 μm. (i) Trichocytes (arrows). Scale bar = 20 μm.

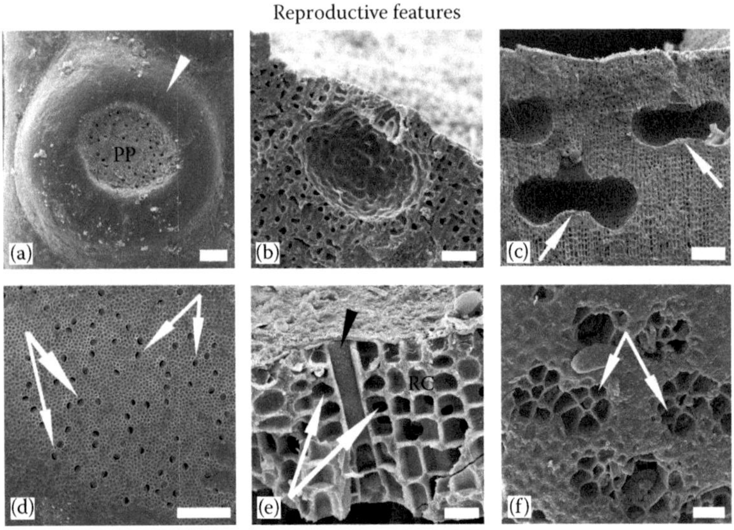

Figure 27.4 SEM micrographs of CA reproductive features. (a) Conceptacle morphology. Note the sunken pore plate (PP) and a conspicuous rim (arrowhead). Scale bar = 100 µm. (b) Vertical section of uniporate conceptacle. Note the conceptacle chamber immersed in the thallus. Scale bar = 20 µm. (c) Vertical section of buried conceptacles. Note the presence of a columella (arrow). Scale bar = 100 µm. (d) Surface view of pores (arrows) in multiporate conceptacles. Scale bar = 100 µm. (e) Vertical section of roof of the multiporate conceptacle. Cells of the roof (RC); cells bordering the pore canal (arrows); pore canal (arrowhead). Scale bar = 10 µm. (f) Surface view of rosette cells (arrows) bordering the pores of multiporate conceptacles. Scale bar = 20 µm.

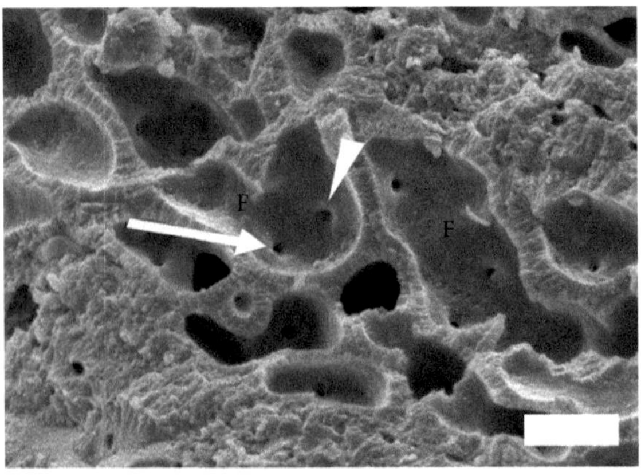

Figure 27.5 Transverse section of *Sporolithon* sp. Cells with both fusions (F) and secondary pit connection (arrowhead). Primary pit connections (arrow) join cells of the same filament. Scale bar = 10 µm.

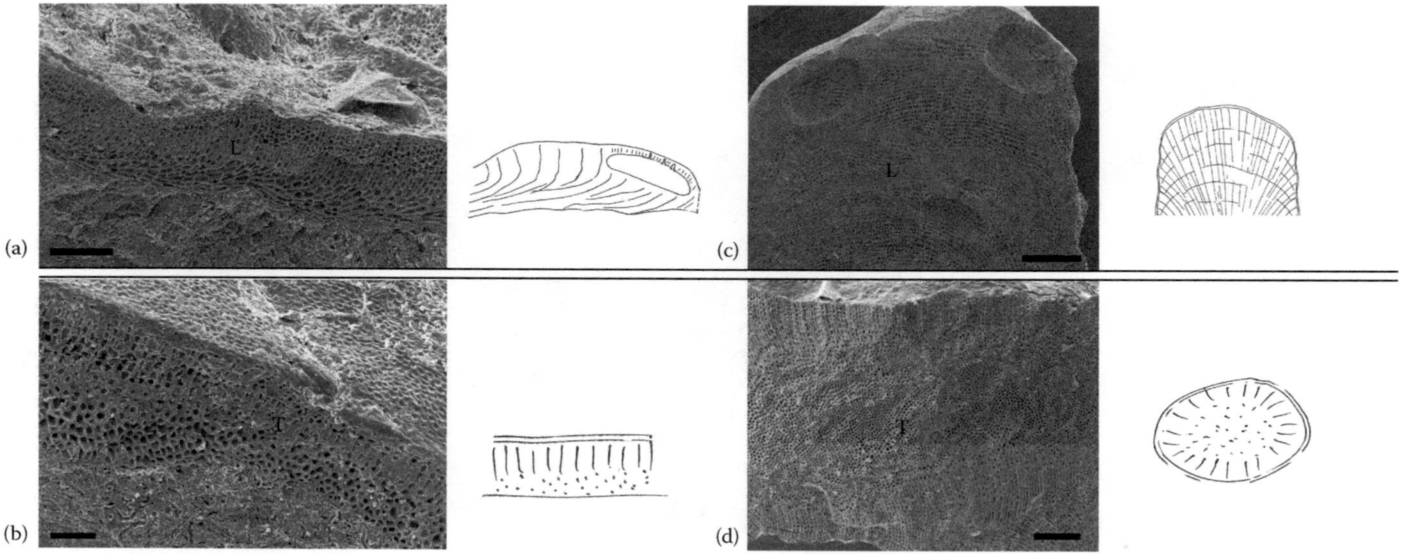

Figure 27.6 (a) Section of crustose thallus in a longitudinal plane of view. (b) Section of crustose thallus in transverse plane of view. Scale bar = 50 μm (a), 30 μm (b). (c) Section of protuberance in longitudinal plane of view. (d) Section of protuberance in the transverse plane of view. Scale bar = 200 μm (c), 50 μm (d).

Figure 27.7 (a) Longitudinally cut perpendicular to margin (arrow) in crustose thallus. (b) Longitudinally cut perpendicular to margin (arrow) in lamellate thallus.

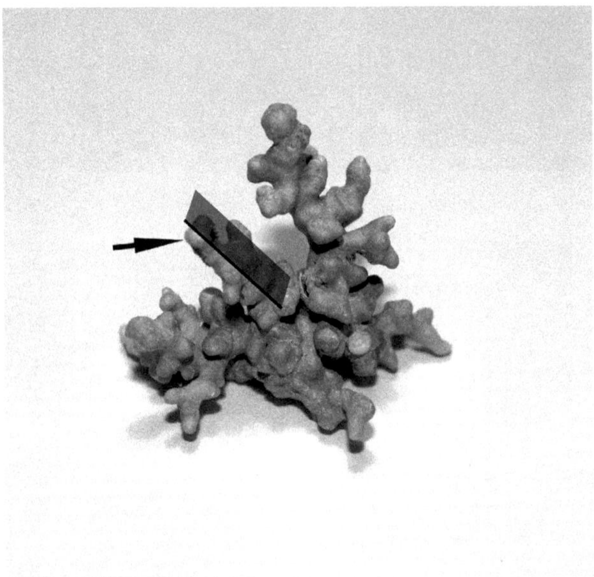

Figure 27.8 Longitudinal section (black line) of protuberance directed from its apex (arrow) to the base.

2. Transverse view (Figure 27.6b, d):
   a. In crustose and lamellate thalli, the transverse section is parallel to the growth margin and perpendicular to the longitudinal/radial section (Figure 27.9a and b);
   b. In rhodoliths/maerl or in protuberances, the transverse section is perpendicular to the growth axis (Figure 27.10).

Figure 27.9 (a) Crustose thallus: transverse section parallel to the growth margin (arrow) and perpendicular to the radial section. (b) Lamellate thallus: Transverse section parallel to the growth margin (arrow) and perpendicular to the radial section.

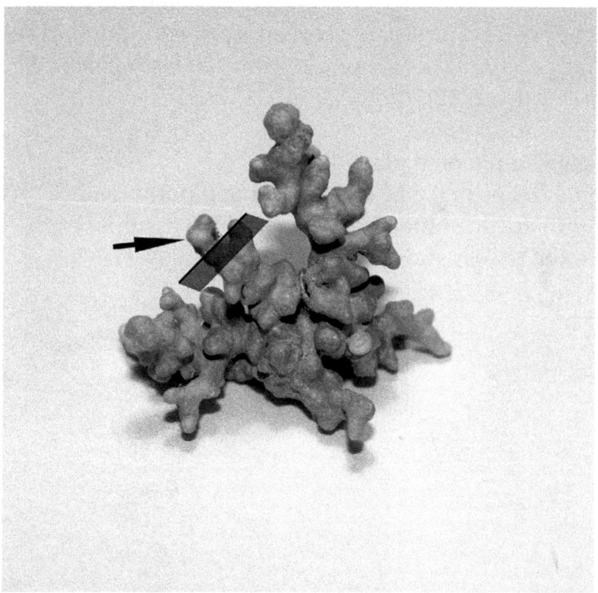

Figure 27.10 Transverse section of protuberance (black line) perpendicular to the growth axis in a rhodolith sample.

3. To observe reproductive structures (conceptacles), the section/fracture must be directed from their surface toward the base. The section must pass through the pore in uniporate conceptacles (Figure 27.4b and c; Figure 27.11a) and through the center of the roof in multiporate conceptacles (Figure 27.11b). Some relevant features can be observed directly from the surface view, without sections (Figure 27.4a, d, and f).

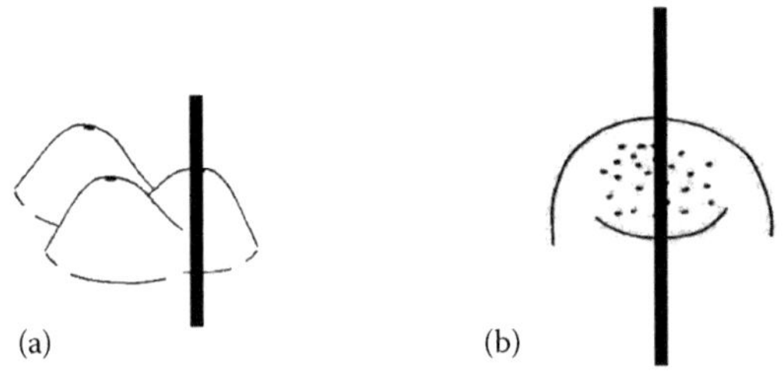

Figure 27.11 For sections of reproductive structures, fractures must be directed from their surface toward the base and pass through the pore in uniporate conceptacles (a) and through the center of the roof in multiporate conceptacles (b).

27.4.2 Optical microscopy

OM is used to observe some of the above-described features and to identify trichocyte position and pattern of spore germination in very thin and small epiphytic thalli of, for example, *Fosliella, Hydrolithon, Melobesia,* and *Pneophyllum* (Figure 27.12).

1. Place the sample on a glass slide
2. Put some drops of 0.6 M HNO_3 with a pipette to decalcify it
3. Stain with aniline blue and a drop of HCl (usually for 3–5 min), then remove the excess stain with drops of water

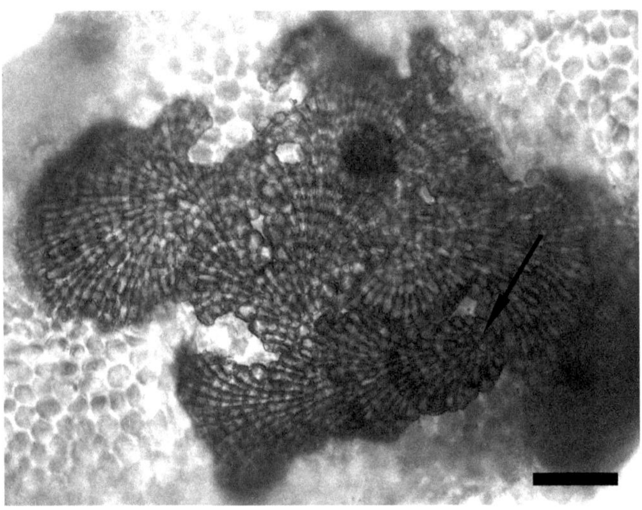

Figure 27.12 Surface view of a thin crust. Note the four-celled element on the original germination disc. Scale bar = 100 μm.

4. Mount either a temporary slide (for a rapid observation) or a permanent one using some drops of Eukitt

For larger specimens:

5. Select the portion to be decalcified under stereomicroscope, following one of these ways:
   a. Break off a small piece of thallus and immerse it in a decalcifying solution (Figure 27.13a and b)
   b. Around the area to be studied, build a silicone rim to form a small bowl in which to pour the decalcifying solution

      The time required for the complete decalcification, recognizable by the end of bubbles production, vary from few minutes to many hours, according to both the size and thickness of the thallus
6. Transfer the decalcified fragments in a small plastic box
7. Stain in 2%–5% aqueous $KMnO_4$ for 30–60 minutes (Figure 27.14); then rinse off the excess of stain with water
8. In the same box, dehydrate through a series of 30%, 60%, 90%, and 100% ethanol (Figure 27.15) at 20/30 min intervals; during these steps, the box should be covered or sealed with parafilm
9. Embed in a medium-grade *LR White* resin according to the following steps:
   a. LR White Resin: Overnight.
   b. Fresh LR White Resin: 2–3 h.
   c. Fresh LR White resin with the accelerator; the embedding solution is made by adding 20 μL of accelerator to 10 mL of LR White resin just before use and mixing well (*Note 1*). The resin sets in 20 min at 20°C.

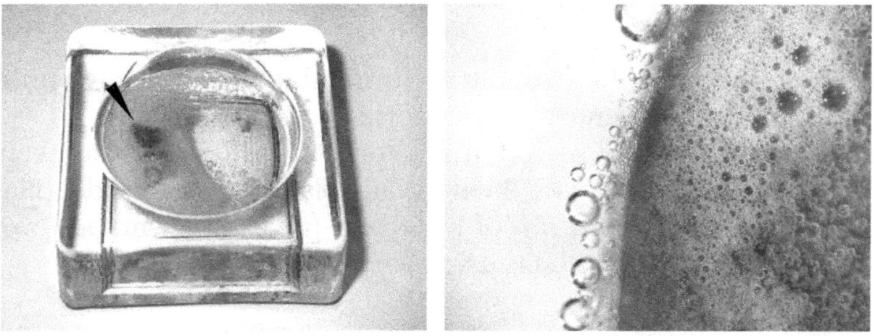

(a) (b)

Figure 27.13 (a) Sample of CA (arrowhead) in the decalcifying solution and (b) Bubbles of CO_2 produced by CA during decalcification process.

Figure 27.14 CA sample (arrowhead) in 2%–5% aqueous $KMnO_4$ staining mixture.

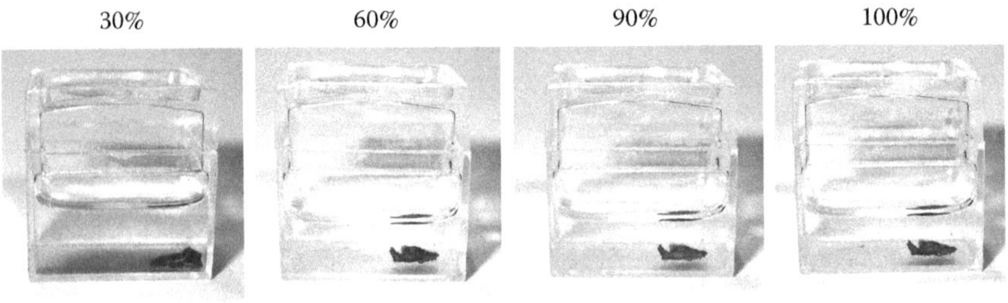

Figure 27.15 CA sample placed in a sealed box for dehydration in a series of 30%, 60%, 90%, and 100% ethanol.

10. Soften the resin blocks under an infrared lamp and trim them with a razor blade to remove surplus resin.
11. Place the resin block on the sledge microtome. Make sections 5–15 μm thick, and place them serially placed on glass slides. Flatten the sections with a drop of histoclear (clearing agent) and mount them on permanent slides using some drops of Eukitt.

For coarse microscopic observations, a hand section with razor blade should be made, after the decalcification; the sections should be stained with aniline blue and a drop of HCl, and rinsed with water to remove the excess of stain.

27.4.3 Scanning electron microscopy

1. Thalli should be air dried for at least 48 h or dried in oven at 50°C for 24 h.
2. Select and detach the portions of thallus to investigate under a stereomicroscope (*Note 2*).
3. Embed the portion of thallus in a small block of air hardening modeling clay (ca. 2 × 2 cm, 3 to 5 mm thick) (*Note 3*). Gently press the fragment until the algal piece is in line with the surface of the clay or slightly below (<1 mm).
4. To trace the fracture line press with a razor blade on the clay in the desired direction (Figure 27.16a–d). The incision line should be 1–2 mm deep.
5. Dry the clay block in the oven at 50°C for 24 h or in air for 48 h.
6. To make the section, place the cutting edge of the razor into the trace line previously prepared and give a sharp hit with the hammer (*Note 4*).
7. After the cut, verify the sectioned portions under a stereomicroscope.
8. Using fine forceps place the cut fragments on aluminum stubs and attach them with a thin and uniform layer of glue (*Note 5*).
9. More than one fragment can be placed on a single stub paying attention that they are distant at least 2–3 mm (Figure 27.17).
10. To remove sediments, diatoms or any residual material formed during cutting, briefly sonicate (ca. 5–10 min) the stubs using distilled water.
11. Dry stubs in oven at 50°C for 24 h.
12. Coat stubs with a mixture of gold and palladium (80% and 20%, respectively) or with carbon, and place them with their support into the chamber of the sputter coater (Figure 27.18). Time for coating (ca. 2–10 min) depends on the model of SEM used.

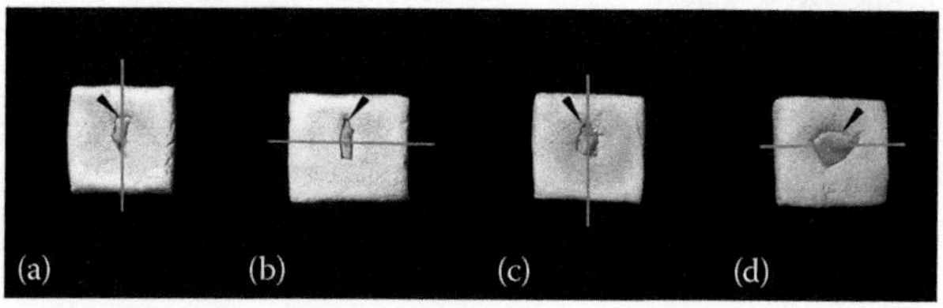

Figure 27.16 Fragments of samples oriented and embedded in the clay. (a) Line of fracture directed from the apex (arrowhead) to the base of protuberance to obtain longitudinal section. (b) Line of fracture perpendicular to the growth axis of protuberance to get a transverse plane of view. (c) Longitudinal section of encrusting thallus. The line of fracture is perpendicular to the margin (arrowhead). (d) Transverse section of lamella. The line of fracture is parallel to the growth margin (arrowhead).

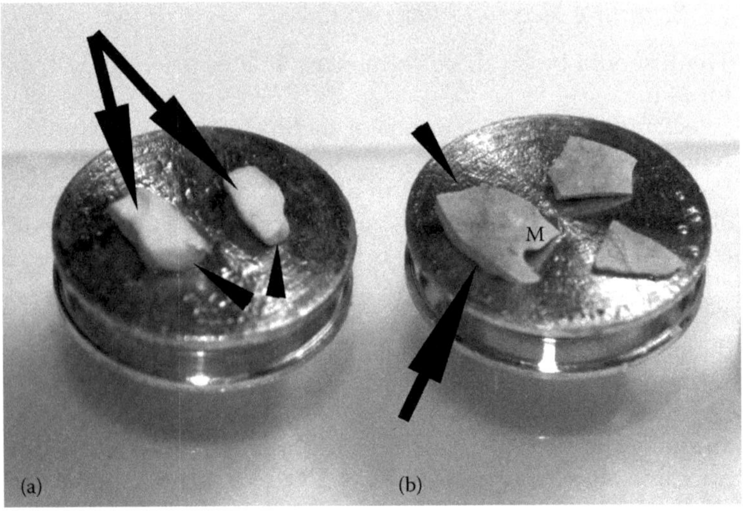

Figure 27.17 Samples mounted on stubs. (a) Protuberance. (b) Lamella. Longitudinal sections are indicated with arrows; note that in the lamellate fragment the longitudinal section is perpendicular to the margin (M). Transverse sections are shown with arrowheads.

Figure 27.18 Stubs with samples collocated in the sputter coater for coating with gold/palladium mixture.

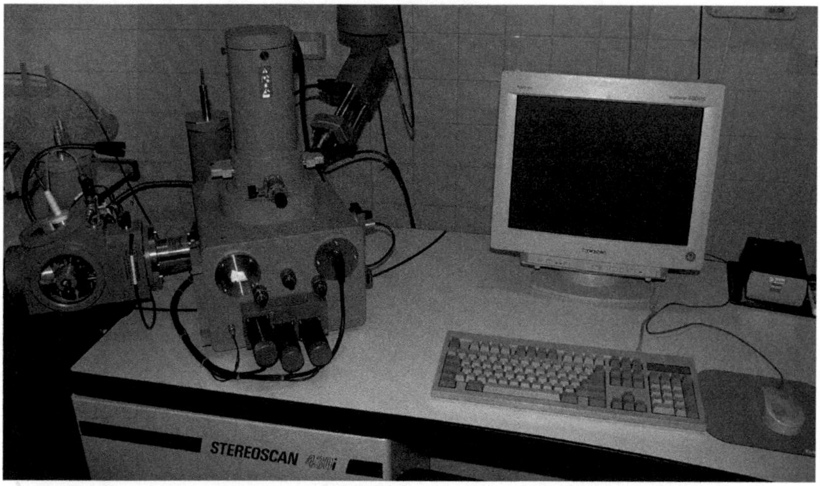

Figure 27.19 SEM (Leica Stereoscan 430i).

13. Samples are thus ready for SEM observations (Figure 27.19). If different planes of a sample need to be examined, the stub should be placed individually into the SEM chamber to allow the tilting of it. Otherwise, if only the upper planes of the samples require observation, several stubs can be placed together in the chamber and viewed consecutively.

27.5 Notes

Note 1: Use the mixture immediately after having added accelerator to the resin. It is very important to seal the box, to exclude any contacts with oxygen and to prevent surface ripples in hardened resin blocks.

Note 2: Usually, the selected portions of CA should include one or more of the following: margins, lamellae, protuberance, and areas with reproductive structures, as shown in Figure 27.20a–d. For taxa with protuberances, the sections must be done on both protuberances and crustose parts. In case of thin crusts, better is to cut them along with the substratum (Figure 27.20a) to avoid breaking of the sample.

Note 3: Orient sample in the clay to obtain the desired section (Figure 27.16). Prepare the clay blocks just before embedding the samples into it. The thickness of the clay should not exceed 5 mm otherwise it could be too hard to be cut.

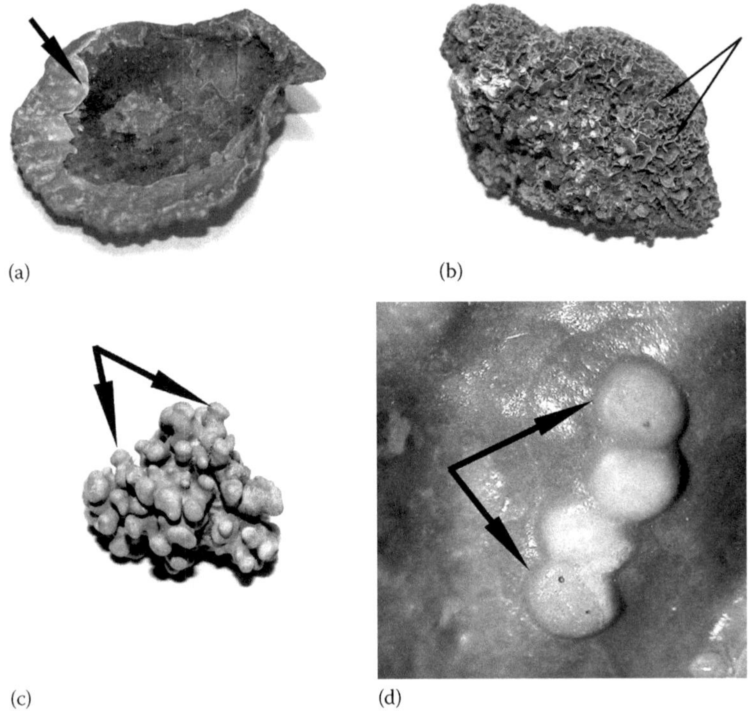

Figure 27.20 (a) Margin of the encrusting thallus (arrow). (b) Thallus with lamellae (arrows). (c) Unattached thallus (rhodolith) with protuberances (arrows). (d) Multiporate conceptacles of *Mesophyllum* sp. (arrows).

Note 4: The razor can be used several times, paying attention that it is not damaged. If damaged razor is used the fracture can be compromised and not useful for observation.

Note 5: As the glue rapidly hardens, determine prior how to orient fragments on the stub. Place the samples with the plane to observe facing up, as shown in Figure 27.17a. If the sample is oriented with the plane to observe is placed lateral (Figure 27.17b), during SEM observations the stub can be tilted also allowing these planes to be viewed. It is a good practice to put several fragments of the same sample on a single stub to have sufficient material to examine. The labeling of the stubs must be done on its bottom side to prevent labels being covered during coating phase.

Acknowledgments

Figure 27.2 has been previously published in Wolf et al., 2015 (Phytotaxa 224(1): 59–71).

References

Alexandersson T. 1974. Carbonate cementation in coralline algal nodules in the Skagerrak, North Sea; biochemical precipitation in undersaturated waters. *J. Sediment. Petrol.* 44: 7–26.

Bailey A., Bisalputra T. 1970. A preliminary account of the application of thin-sectioning, freeze-etching, and scanning electron microscopy to the study of coralline algae. *Phycologia* 9: 83–101.

Basso D., Babbini L., Kaleb S., Bracchi V.A., Falace A. 2016. Monitoring deep Mediterranean rhodolith beds. *Aquatic Conserv: Mar. Freshw. Ecosyst.* 26: 549–561.

Borowitzka M.A., Larkum A.W.D., Nockolds C.E. 1974. A scanning elctron microscope study of the structure and organization of the calcium carbonate deposits of algae. *Phycologia* 13: 195–203.

Bressan G., Babbini L. 2003. Biodiversità marina delle coste Italiane: Corallinales del Mar Mediterraneo: Guida all deteminazione. *Biologia Marina Mediterranea* 10(Suppl. 2): 1–237.

Chamberlain Y.M. 1983. Studies in the Corallinaceae with special reference to Fosliella and Pneophyllum in the British Isles. *Bulletin of the British Museum (Natural History) Botany* 11: 291–463.

European Commission. 2010. Commission decision of 1 September 2010 on criteria and methodological standards on good environmental status of marine waters. *Official Journal of the European Union* L 232/14 .

European Parliament and Council of the European Union. 2008. Directive 2008/56/EC of the European Parliament and of the Council of 17 June 2008 establishing a framework for Community action in the field of marine environmental policy (Marine Strategy Framework Directive). *Official Journal of the European Union* L 164/19.

Ghirardelli L.A. 2002. Endolithic microorganisms in live and dead thalli of coralline red algae (Corallinales, Rhodophyta) in the Northern Adriatic Sea. *Acta Geologica Hispanica* 37(1): 53–60.

Harvey A., Woelkerling W., Farr T., Neill K., Nelson W. 2005. *Coralline Algae of Central New Zealand: An Identification Guide to Common "Crustose" Species.* National Institute of Water & Atmospheric Research Information Series No. 57, Wellington, 145 p.

Irvine L.M., Chamberlain Y.M. 1994. *Seaweeds of the British Isles, Vol. 1, Rhodophyta Part 2B Corallinales, Hildenbrandiales.* HMSO, London. p. 276.

Kamenos N.A., Burdett H.L., Darrenogue N. 2016. Coralline algae as palaeoclimatic proxies. In Riosmena-Rodríguez R., Nelson W., Aguire J. (Eds.), *Rhodolith/Maerl Beds: A Global Perspective.* Switzerland: Springer.

Littler M.M. 1976. Calcification and its role among macroalgae. *Micronesica* 12: 27–41.

McCoy S.J., Kamenos N.A. 2015. Coralline algae (Rhodophyta) in a changing world: Integrating ecological, physiological, and geochemical responses to global change. *J. Phycol.* 51: 6–24.

Piazzi L., Gennaro P., Cecchi E. et al. 2015. Improvement of the ESCA index for the evaluation of ecological quality of coralligenous habitat under the European Framework Directives. *Medit. Mar. Sci.* 16: 419–426. https://doi.org/10.12681/mms.1029

Tribollet A., Payri C. 2001. Bioerosion of the crustose coralline alga Hydrolithon onkodes by microborers in the coral reefs of Moorea, French Polynesia. *Oceanolog Acta* 24: 329–342.

Woelkerling W.J. 1978. *Mastophoropsis canaliculata* (Harvey in Hooker) gen. et comb. nov. (Corallinaceae, Rhodophyta) in southern Australia. *Brit. Phycol. J.* 13: 209–225.

Woelkerling W.J. 1988. *The Coralline Red Algae: An Analysis of the Genera and Subfamilies of Nongeniculate Corallinaceae.* Oxford University Press, New York. p. 268.

Wolf M.A., Maneveldt G.W., Kaleb S., Moro I., Falace A. 2015. Morphological and molecular characterization of *Hydrolithon rupestre* (Corallinaceae, Corallinales, Rhodophyta): First report from the Mediterranean Sea. *Phytotaxa* 224(1): 59–71.

chapter twenty eight

Extraction of high quality RNA from brown algae for transcriptomic analysis

Sandra Heinrich

Contents

28.1 Introduction

Environmental genomics and transcriptomics are powerful approaches to gain novel insights into evolutionary ecology, organism–environment interactions, and ecosystem processes (Vandenkoornhuyse et al. 2010). Main aims of transcriptomics are to catalog all expressed RNA species, to determine the transcriptional structure of genes, and to quantify the changing expression levels of transcripts, for example, during the development or under stressful environmental conditions (Wang et al. 2009). Recent technological advances gave rise to the rapid generation of large-scale sequencing data from nonmodel organisms at a reasonable cost (Ekblom and Galindo 2011). One of the most common applications is transcriptome characterization by high-throughput RNA sequencing, especially for nonmodel species, as it is not limited to detect transcripts that correspond to known genomic sequences (Wang et al. 2009, Deng et al. 2012). However, in nonmodel species, RNA extraction often proves to be difficult, but high-quality RNA is crucial, for example, preparing cDNA libraries or investigating gene expression profiles by RT-PCR (Falcão et al. 2008). The loss in RNA quality, that is, RNA degradation, can be caused by prolonged storage, suboptimal storage conditions, and through cleavage by RNases during handling of RNA samples (Fleige and Pfaffl 2006, Vermeulen et al. 2011). Various studies showed that the quality of data from gene expression analysis using RT-qPCR or microarrays is strongly influenced by the quality of the extracted RNA (Copois et al. 2007, Becker et al. 2010). Johnson et al. (2012) investigated the effects of RNA quality on Illumina sequencing; their results imply that impure RNA hampers cDNA synthesis, RNA of low integrity, in contrast, prevents the assembly of large scaffolds during data analysis. Therefore, assessment of RNA quality before using samples for downstream applications is highly critical to obtain reliable gene expression data (Vermeulen et al. 2011). RNA purity can be determined photometrically using sensitive spectrophotometers, for example, the NanoDrop (peqLab Biotechnologie GmbH, Erlangen, Germany); for standardized RNA, integrity control automated capillary electrophoresis is recommended (Becker et al. 2010).

28.2 State of the art

Within recent years, several transcriptomic studies on macroalgae have been published (Wang et al. 2013, Liang et al. 2014, Heinrich et al. 2015, 2016). Nevertheless, compared to other species groups, that is, green plants, multicellular heterokonts remain highly underrepresented in sequence databases (Pearson et al. 2010). Up to now, only two brown algal genomes were fully sequenced: (1) the genome of *Ectocarpus siliculosus* (Cock et al. 2010) and (2) *Saccharina latissima* (Ye et al. 2015). One reason might be that brown macroalgae exhibit high amounts of polysaccharides (Wang et al. 2005, Varela-Alvarez et al. 2006) and phenolic compounds (Lane et al. 2006, Pearson et al. 2006), which make isolation of high-quality RNA and DNA difficult. Tissues with high levels of polysaccharides yield poor quality RNA as polysaccharides coprecipitate with

RNA in low ionic strength buffers (Falcão et al. 2008). Furthermore, RNA extracted from macroalgae is often contaminated with acidic polysaccharides because of almost identical densities between those molecules (Kim et al. 1997). Commercially available kits using guanidinium salts (e.g., the RNeasy Mini Kit; Qiagen, and Hildesheim) usually fail for brown algae. A more promising approach consists of the utilization of a cetyltrimethylammonium bromide (CTAB)-based buffer and subsequent phenol–chloroform purification followed by selective precipitation with LiCl (Pearson et al. 2006, Dittami et al. 2009). One disadvantage of this method is that lithium salts, even in minute amounts, often interfere with downstream enzymatic applications (e.g., reverse transcription) (Bilgin et al. 2009). Here, a fast and effective approach to extract high-quality RNA from brown algae for transcriptomic analysis is presented. The developed protocol consists of a CTAB extraction followed by the usage of the RNeasy Plant Mini Kit (Qiagen, Hildesheim, Germany), applying a modified protocol based on the company's user manual (Heinrich et al. 2012a). The extraction protocol was tested successfully in several brown algae (e.g., *S. latissima*, *Desmarestia anceps*, and *Fucus serratus*) and in the green alga *Ulva* sp.; thus it might also be effective in the red algae. The isolated RNA was successfully used for downstream applications such as RT-PCR, Microarray hybridizations, 454 sequencing, and Illumina high-throughput RNA sequencing (Heinrich et al. 2012a, b, Heinrich et al. 2016, Iñiguez et al. 2017). A comparison of different RNA extraction methods for brown algae regarding RNA yield, purity, and integrity can be found in Table 28.1.

Table 28.1 Comparison of CTAB-based method, Qiagen RNeasy Plant Mini Kit, and a combined approach of regarding RNA yield, integrity, purity, and working time

	Qiagen RNeasy Plant Mini Kit	CTAB-based method	CTAB + Qiagen RNeasy Plant Mini Kit
Number of samples yielding RNA	3/6	6/6	6/6
RNA yield μg/100 mg sample	0.4 ± 0.2	1.2 ± 0.7	11.8 ± 1.7
RNA integrity	High	High	High
Average OD (260/230)	0.98 ± 0.2	1.41 ± 0.13	2.19 ± 0.04
Average OD (260/280)	1.64 ± 0.14	1.56 ± 0.17	2.2 ± 0.03
Average time to process	1 h	20 h	2.5 h
DNA contamination	No	Yes	No

28.3 Materials

28.3.1 Algal material

Laboratory cultures, and field materials, can be used for RNA extraction. RNA isolation from culture material proves to be easier as it typically shows a low degree of contamination with endo- and epiphytes and often contains lower amounts of polysaccharides and polyphenolic compounds. When using materials collected in the field, collection sites should be well considered. For example, material collected from sites with strong grazing pressure exhibits higher contents of polyphenolic compounds; sporophytes from strong wave-exposed sampling sides usually contain higher amounts of polysaccharides. Both substances might hamper the RNA-extraction process. Furthermore, field material should be carefully cleaned before extracting RNA to minimize contamination. To prevent RNA degradation, samples should be stored at −80°C and processed within four weeks.

28.3.2 Sample homogenization

- Liquid nitrogen
- Liquid nitrogen dewar
- Mortar and pestle (autoclaved)
- Gloves
- Small chemical spoon (autoclaved)
- 2 mL microcentrifuge tubes (PCR-clean)
- RNase Zap (Thermo Fisher Scientific, Waltham, USA)

28.3.3 RNA extraction

28.3.3.1 Reagents

- CTAB buffer (2% cetyltrimethylammonium bromide, 1 M NaCl, 100 mM Tris pH 8, and 50 mM EDTA, pH 8), β-mercaptoethanol, chloroform:isoamyl alcohol (24:1; v/v), ethanol (96%–100%), RNeasy Plant Mini Kit (Qiagen, Hildesheim, Germany), RNase-free DNase Kit (Qiagen, Hildesheim, Germany), and RNase Zap (Thermo Fisher Scientific, Waltham, USA).

28.3.3.2 Equipment

- 2 mL microcentrifuge tubes (PCR-clean)
- Microcentrifuge, for example, Eppendorf 5418R (Eppendorf GmbH, Hamburg, Germany)
- Thermoblock
- Vortexer with a microcentrifuge adapter
- Pipettes 1000, 100, and 10 μL and corresponding tips.

28.3.4 RNA quality control

- Low-volume spectrophotometer: Agilent Bioanalyzer & Agilent RNA 6000 Nano Kit (Agilent, Santa Clara, CA, Model 2100).

28.4 Experimental procedures

28.4.1 Sample homogenization

All experiments are performed under clean conditions; workplaces should be cleaned and thoroughly wiped with RNase Zap (Thermo Fisher Scientific, Waltham, USA) before starting. The sample homogenization is one of the most crucial steps of the RNA-extraction process as insufficiently homogenized samples lead to low RNA extraction yields. For cell disruption, two tissue homogenizers: (1) the Mikro-Dismembrator (U Braun Biotech International, Melsungen, Germany) and (2) the Precellys 24 (Bertin, Montigny-le-Bretonneux, France) have been tested, and manual grinding in liquid nitrogen. The latter method produced the best results.

1. Precool the mortar by pouring liquid nitrogen into it, precool the pestle, and microcentrifuge tubes and chemical spoon by placing them into a liquid nitrogen dewar.
2. Add the sample (at least 500 mg) into the mortar still containing liquid nitrogen and grind with the pestle until a fine powder is produced (*Note 1*).
3. Transfer the homogenized sample into a precooled 2 mL microcentrifuge tube using the precooled chemical spoon. At most 1/4 of the cup should be filled with the homogenized sample. Store the sample at −80°C until further use to avoid RNA degradation.

28.4.2 RNA extraction

RNA extraction should be performed under clean conditions in a fume hood. The workbench should be cleaned and thoroughly wiped with RNase Zap (Thermo Fisher Scientific, Waltham, USA) before starting the RNA extraction. Advice for troubleshooting can be retrieved from Table 28.2.

1. Add 1 mL of CTAB extraction buffer and 20 μL β-mercaptoethanol to the homogenized sample and immediately mix vigorously by vortexing (*Note 2*).
2. Incubate the mixture for 10 min at 45°C in a thermoblock. Mix the samples by inversion 2–3 times during the incubation time (*Note 3*).

Table 28.2 Troubleshooting advice

Problem	Possible reason	Suggestion
Low RNA yield	Too small amounts of biomass	Increase the starting material
	Insufficient homogenization	Increase homogenization time
		Increase incubation time
Low A260/A280	Too high amounts of biomass	Decrease starting material
Low A260/230	Disruption of the organic phase after the chloroform extraction	Do not touch organic phase with the pipet
	Ethanol carry over during the steps RNeasy extraction step 17–20	Carefully remove the column from the collection tube after centrifugation so that the column does not touch the flow through
	Ethanol carry over during the RNeasy extraction elution step 21	Make sure to pipet the water into the center of the membrane; do not touch walls of the column
RNA degraded	Frozen material too old	In general, the material can only be stored for four weeks at −80°C without RNA degradation
	Sample thawed during homogenization	Keep sure to refill liquid nitrogen regularly during homogenization

3. Add 1 mL of chloroform: Isoamyl alcohol (24:1; v/v), vortex vigorously for 10 min, and centrifuge the sample at 12,000 × g for 20 min at 20°C.
4. Collect the aqueous phase (maximal 750 µl) into a clean 2 mL microcentrifuge tube and discard the rest (*Note 4*).
5. Add 0.3 volume of ethanol (96%–100%) and mix gently by inversion.
6. Add 1 mL of chloroform:isoamyl alcohol (24:1; v/v), vortex vigorously for 10 min, and centrifuge the sample at 12,000 × g for 20 min at 20°C.
7. Collect the supernatant (maximal 500 µl) into a clean 2 mL microcentrifuge tube and extract total RNA by using the RNeasy Plant Mini Kit (Qiagen, Hildesheim, Germany) according to this modified form of the manufacturer's protocol for RNA extraction

(https://www.qiagen.com), including on-column DNA digestion. For the following buffers, the terminology of the kit provider is used.

8. Add 500 µL Buffer RLT to the supernatant and vortex vigorously.
9. Transfer 500 µL lysate to a QIAshredder spin column (lilac) placed in a collection tube, and centrifuge for 2 min at full speed. Carefully transfer the flow through to a new microcentrifuge tube and avoid disturbing the pellet.
10. Repeat step 9 with the remaining 500 µL of the supernatant.
11. Add 0.5 volume of ethanol (96%–100%) to the lysate and mix immediately by pipetting.
12. Transfer the sample to a RNeasy spin column (pink) placed in a 2 mL collection tube and centrifuge for 1 min at ≥9000 × g at 20°C. Discard the flow through.
13. Repeat step 12 with the remaining supernatant.
14. Add 350 µL Buffer RW1 to the RNeasy spin column and centrifuge for 1 min at ≥9000 × g at 20°C. Discard the flow through.
15. Add 10 µL DNase I stock solution (RNase-free DNAse Kit) to 70 µL Buffer RDD. Mix by gently inverting the tube.
16. Add the 80 µL DNase I mix directly to the RNeasy spin column membrane and place on the benchtop (20°C–30°C) for 15 min.
17. Add 350 µL Buffer RW1 to the RNeasy spin column and centrifuge for 1 min at ≥9000 × g at 20°C. Discard the flow through.
18. Add 500 µL Buffer RPE to the RNeasy spin column, centrifuge for 1 min at 20°C, and ≥9000 × g and discard the flow through.
19. Add 500 µL Buffer RPE to the RNeasy spin column and centrifuge for 1 min at ≥9000 × g and 20°C.
20. Place the RNeasy spin column into a new 2 mL collection tube and centrifuge at full speed at 20°C for 1 min.
21. Place the RNeasy spin column in a new 1.5 mL collection tube and add 50 µL RNase-free water to the spin column membrane. Centrifuge for 1 min at 9000 × g at 20°C to elute the RNA (*Note 5*).
22. Repeat step 21 using the eluate from step 21.

28.4.3 RNA quality control

28.4.3.1 Measuring RNA concentration and purity

Determine RNA quantity and purity spectrophotometrically by measuring the absorbance at 230, 260, and 280 nm. The purity of the RNA can be estimated by calculating the A260/A280 ratio, which gives a measure for protein contamination. Calculate the A260/A230 ratio to determine whether impurities, for example, polysaccharides and polyphenols, are present. Pure RNA exhibits A260/A280 and A260/A230 ratios ≥1.8.

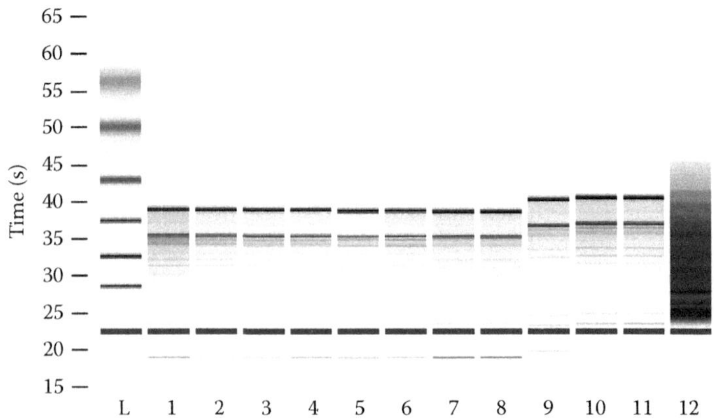

Figure 28.1 Results of a 2100 Bioanalyzer microfluidic electrophoresis chip. The first lane of the gel electrophoresis contains the RNA ladder (L), sample lanes 1–11 contain nondegraded RNA, and sample lane 12 degraded RNA from *Desmarestia anceps*.

28.4.3.2 RNA integrity control

Checking the RNA integrity can be performed using automated capillary electrophoresis or an agarose RNA gel. For downstream applications such as RT-qPCR and RNA sequencing, it is highly recommended to use automated capillary electrophoresis (e.g., the Bioanalyzer microfluidic electrophoresis) because it gives more detailed results on RNA integrity. For total RNA, the RNA 6000 Nano chip can be used. Prepare the RNA 6000 Nano chips according to the manufacturer's protocol (http://rai.unam.mx/manuales/lbg_ARNGuideAgileny.pdf). Figures 28.1 and 28.2 show the results of a Bioanalyzer microfluidic electrophoresis with RNA isolated from the brown algae *Desmarestia anceps*.

28.5 Notes

Note 1: Do not allow the sample to thaw during grinding. Add extra liquid nitrogen to the mortar during the homogenization to avoid thawing. This should be done very slowly to avoid splashing of the material.

Note 2: Do not allow the homogenized material to thaw.

Note 3: Incubation time can be raised up to 30 min if only small amounts of biomass are available or if the RNA extraction yields from previous extractions have been insufficient.

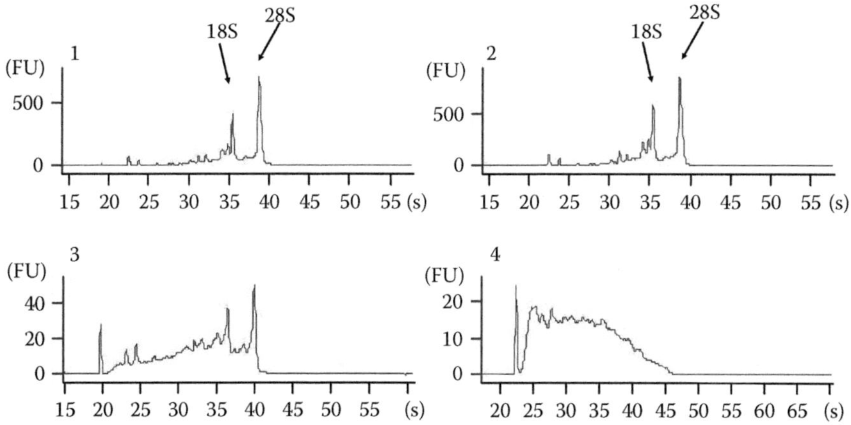

Figure 28.2 Electropherogram of RNA samples of different integrity from *Desmarestia anceps*. Sample 1 and 2: nondegraded RNA, sample 3 and 4 degraded RNA.

Note 4: Be very careful not to disturb the organic phase. Otherwise, contaminations are transferred into the aqueous phase.

Note 5: Pipette the water directly on the membrane. Do not touch the walls of the column to avoid carryover of ethanol.

Acknowledgments

This work was partly funded by the Deutsche Forschungsgemeinschaft (DFG) in the framework of the priority program "Antarctic Research with comparative investigations in Arctic ice areas" by a grant (HE 6734/1-1). Andreas Wagner and Claudia Daniel are acknowledged for technical assistance.

References

Becker, C., A. Hammerle-Fickinger, I. Riedmaier, and M.W. Pfaffl. 2010. mRNA and microRNA quality control for RT-qPCR analysis. *Methods* 50 (4):237–243.

Bilgin, D.D., E.H. DeLucia, and S.J. Clough. 2009. A robust plant RNA isolation method is suitable for Affymetrix GeneChip analysis and quantitative real-time RT-PCR. *Nat. Protoc.* 4 (3):333–340.

Cock, J.M., L. Sterck, P. Rouze et al. 2010. The *Ectocarpus* genome and the independent evolution of multicellularity in brown algae. *Nature* 465 (7298):617–621.

Copois, V., F. Bibeau, C. Bascoul-Mollevi et al. 2007. Impact of RNA degradation on gene expression profiles: Assessment of different methods to reliably determine RNA quality. *J. Biotechnol.* 127 (4):549–559.

Deng, Y., J. Yao, X. Wang, H. Guo, and D. Duan. 2012. Transcriptome sequencing and comparative analysis of *Saccharina japonica* (Laminariales, Phaeophyceae) under blue light induction. *PLoS One* 7 (6):e39704.

Dittami, S.M., D. Scornet, J.L. Petit et al. 2009. Global expression analysis of the brown alga *Ectocarpus siliculosus* (Phaeophyceae) reveals large-scale reprogramming of the transcriptome in response to abiotic stress. *Genome Biol.* 10 (6):51.

Ekblom, R. and J. Galindo. 2011. Applications of next generation sequencing in molecular ecology of non-model organisms. *Heredity* 107 (1):1–15.

Falcão, V.D.R., A.P. Tonon, M.C. Oliveira, and P. Colepicolo. 2008. RNA Isolation method for polysaccharide rich algae: Agar producing *Gracilaria tenuistipitata* (Rhodophyta). *J. Appl. Phycol.* 20 (1):9–12.

Fleige, S. and M.W. Pfaffl. 2006. RNA integrity and the effect on the real-time qRT-PCR performance. *Mol. Asp. Med.* 27 (2):126–139.

Heinrich, S., S. Frickenhaus, G. Glöckner, and K. Valentin. 2012a. A comprehensive cDNA library of temperature stressed *Saccharina latissima* (Phaeophyceae). *Eur. J. Phycol.* 47 (2):83–94.

Heinrich, S., K. Valentin, S. Frickenhaus, U. John, and C. Wiencke. 2012b. Transcriptomic analysis of acclimation to temperature and light stress in *Saccharina latissima* (Phaeophyceae). *PLoS One* 7 (8):e44342.

Heinrich, S., K. Valentin, S. Frickenhaus, and C. Wiencke. 2015. Temperature and light interactively modulate gene expression in *Saccharina latissima* (Phaeophyceae). *J. Phycol.* 51:93–108.

Heinrich, S., K. Valentin, S. Frickenhaus, and C. Wiencke. 2016. Origin matters—Comparative transcriptomics in *Saccharina latissima* (Phaeophyceae). *J. Exp. Mar. Biol. Ecol.* 476:22–30.

Iñiguez, C., S. Heinrich, L. Harms, and F.J.L. Gordillo. 2017. Increased temperature and CO_2 alleviate photoinhibition in *Desmarestia anceps*: From transcriptomics to carbon utilization. *J. Exp. Bot.* 68:3971–3984.

Johnson, M.T.J., E.J. Carpenter, Z. Tian et al. 2012. Evaluating methods for isolating total RNA and predicting the success of sequencing phylogenetically diverse plant transcriptomes. *PLoS One* 7 (11):e50226.

Kim, S.H., Y.K. Lee, C.B. Hong, and I.K. Lee. 1997. A simple method for RNA extraction from marine macroalgae. *Algae* 12:53–56.

Lane, C.E., C. Mayes, L.D. Druehl, and G.W. Saunders. 2006. A multi-gene molecular investigation of the kelp (Laminariales, Phaeophyceae) supports substantial taxonomic re-organization. *J. Phycol.* 42 (2):493–512.

Liang, X., X. Wang, S. Chi et al. 2014. Analysis of *Saccharina japonica* transcriptome using the high-throughput DNA sequencing technique and its vanadium-dependent haloperoxidase gene. *Acta Oceanol. Sin.* 33 (2):27–36.

Pearson, G., G. Hoarau, A. Lago-Leston et al. 2010. An expressed sequence tag analysis of the intertidal brown seaweeds *Fucus serratus* (L.) and *F. vesiculosus* (L.) (Heterokontophyta, Phaeophyceae) in response to abiotic stressors. *Mar. Biotechnol.* 12 (2):195–213.

Pearson, G., A. Lago-Leston, M. Valente, and E. Serrao. 2006. Simple and rapid RNA extraction from freeze-dried tissue of brown algae and seagrasses. *Europ. J. Phycol.* 41 (1):97–104.

Vandenkoornhuyse, P., A. Dufresne, A. Quaiser et al. 2010. Integration of molecular functions at the ecosystemic level: Breakthroughs and future goals of environmental genomics and post-genomics. *Ecol. Lett.* 13 (6):776–791.

Varela-Alvarez, E., N. Andreakis, A. Lago-Leston et al. 2006. Genomic DNA isolation from green and brown algae (Caulerpales and Fucales) for microsatellite library construction. *J. Phycol.* 42 (3):741–745.

Vermeulen, J., K. De Preter, S. Lefever et al. 2011. Measurable impact of RNA quality on gene expression results from quantitative PCR. *Nucleic Acids Res.* 39 (9):e63.

Wang, G.G., Y.H. Li, P. Xia, and D.L. Duan. 2005. A simple method for DNA extraction from sporophyte in the brown alga *Laminaria japonica*. *J. Appl. Phycol.* 17 (1):75–79.

Wang, W.-J., F.-J. Wang, X.-T. Sun, F.-L. Liu, and Z.-R. Liang. 2013. Comparison of transcriptome under red and blue light culture of *Saccharina japonica* (Phaeophyceae). *Planta* 237 (4):1123–1133.

Wang, Z., M. Gerstein, and M. Snyder. 2009. RNA-Seq: A revolutionary tool for transcriptomics. *Nat. Rev. Gen.* 10 (1):57–63.

Ye, N., X. Zhang, M. Miao et al. 2015. *Saccharina* genomes provide novel insight into kelp biology. *Nat. Commun.* 6:6986.

chapter twenty nine

Induction of sexual reproduction in Spirogyra *cultures for laser capture microdissection of gametes and zygotes*

Denis Saint-Marcoux and Jane A. Langdale

Contents

29.1 Introduction

Zygnematophyceae belong to the paraphyletic Charophyte group, which is composed entirely of unicellular and multicellular algae. Together with the land plants, Charophytes form a monophyletic group called *Streptophyta*. Charophytes are of particular interest for evodevo studies because many developmental mechanisms and gene toolkits that allowed land plants to conquer the terrestrial environment must have first evolved in ancient Charophytes. Studying extant Charophytes can thus provide clues about these ancestral developmental toolkits, at least about those that are retained in modern lineages, and can help to understand the evolutionary transition from water to land.

Phylogenetic studies places *Zygnematophyceae* as the sister group to land plants (Wickett et al. 2014). As a consequence, representatives of *Zygnematophyceae*, such as *Spirogyra*, are ideal candidates for evodevo studies. The *Spirogyra* genus is composed of fresh water species that develop an unbranched filamentous body plan. As with all *Zygnematophyceae*, *Spirogyra* has an unusual mode of sexual reproduction in that gametes are not released into the environment. Instead, vegetative cells differentiate cellular protrusions called conjugation tubes that bridge two adjacent cells, either from the same (lateral conjugation) or from two (scalariform conjugation) algal bodies. Gametes differentiate from the vegetative protoplast that initiated the conjugation tube into an amoeboid cell form. In the case of *Spirogyra pratensis*, one of the gametes moves from its gametangium into the other gametangium. After gametic fusion, the resulting zygote stays inside the female gametangium until it is finally released into the surrounding environment on degradation of the gametangium cell wall. One of the most iconic innovations that accompanied the transition of *Streptophyta* to the land was the invention of a multicellular embryo. Growth of the embryo gives rise to a sporophyte that produces spores at maturity, augmenting the dispersal of the products of sexual reproduction. The transition from unicellular zygote to multicellular embryo was a major innovation that contributed to the success of plants during the colonization of land.

To date, very few methods have been developed for working with *Zygnematophyceae* in laboratory conditions, and molecular data are scarce. Vegetative culture can be achieved in fresh-water medium but completing the life cycle in a controlled manner is a challenge, and transformation has only been reported for a single unicellular species, *Penium margaritaceum* (Sørensen et al. 2014). Molecular data are available for a handful of species through the oneKP project and the transcriptome of the vegetative thallus of *S. pratensis* has also been released (Timme and Delwiche 2010; Poel et al. 2016). With a view to understanding the evolution of multicellular embryos, we set up a large transcriptomic survey to compare zygotes from Charophytes (including *Spirogyra*) with embryos from early land plants. Because reproduction occurs through conjugation, *Spirogyra* gametes and zygotes cannot be separated easily from vegetative filaments. As a consequence, we developed a method to microdissect gametes and zygotes from filaments using laser capture microdissection (LCM). In the current chapter, we describe a method for the induction of sexual competence in *Spirogyra*, and also report a protocol for the isolation of gametes and zygotes that are suitable material for RNA extraction. These protocols have been used to obtain gamete- and zygote-specific transcriptomes after RNA amplification and sequencing on an Illumina platform (unpublished data).

29.2 *State of the art*

Vegetative growth of *Spirogyra* is easily established in controlled conditions. Filaments can be cultivated in a variety of fresh-water growth media, either based on soil solutions or on synthetic components, for example, Woods Hole medium (Nichols 1973; Drummond et al. 2005), Closteridium medium, Reichart's medium, artificial pond water (Ichimura 1971; Nagata 1973; Ikegaya et al. 2012), Guillard's medium (Andersen et al. 2005; Poel et al. 2016), or Pringsheim's soil–water medium (Starr 1964; Simons et al. 1984). Growth medium is often supplemented with vitamins, notably thiamine, cyanocobalamin, and biotin.

Gametogenesis and sexual competency have previously proven difficult to induce in laboratory cultures of *Spirogyra*. Over the past 60 years, growth stage (Czurda 1933), light intensity (Allen 1958), nitrogen depletion (Grote 1977; Simons et al. 1984; Drummond et al. 2005), temperature and light, CO_2, sucrose availability, growth media dilution (Simons et al. 1984), and growth on solid medium (Hoshaw et al. 1985; Ikegaya et al. 2012) have all been tested for the ability to induce sexual reproduction in lab cultures. Of these factors, nitrogen depletion seemed to have the most reliable effect, and indeed nitrogen depletion is a potent trigger for the induction of gametogenesis in the unicellular alga *Chlamydomonas reinhardtii* (Sager and Granick 1954). A recent study confirmed the importance of nutrient ratio, and notably nitrogen quantity, for gametogenesis induction in *Spirogyra* but no single determinant could be clearly identified (Zwirn et al. 2013).

Since the mid-1990s, LCM has been used for microdissecting cell types out of tissues (Emmert-Buck et al. 1996). Generally, LCM involves capturing cells from tissue sections that are visualized using an optical microscope. Modern iterations of the technology use glass slides covered with a plastic membrane onto which sections are attached. A laser cuts out the desired portion of the section, which is then collected in vessels (tubes or tube caps) for further treatments. LCM can be used on fresh frozen sections, fixed and embedded sections, or whole structures that are either fixed or alive, providing the organism or tissue is not too thick. Often, tissues are fixed and dehydrated to ease the processing of the samples before and during laser capture. For a comprehensive description of the use and applications of this technology, see Espina et al. 2006. A minute amount of material is typically collected during LCM. As a consequence, downstream analysis methods must be adapted, and in the case of molecular biology and sequencing, an amplification step is often required before further processing. LCM has been successfully applied to the study of gene expression in animal cells (Emmert-Buck et al. 1996), plant cells (Nakazono et al. 2003), and fungi (Gomez and Harrison 2009).

Recently, LCM has been used with macroalgae to capture specific cell types of *Ectocarpus siliculosus* sporophytic thallus (Saint-Marcoux et al. 2015). In this case, algal filaments were grown directly on the LCM slide immersed in cultivation medium. Growing *Spirogyra* filaments attached to a LCM slide was not an option, however, because although fine cut pieces of *Spirogyra* thallus initially attached themselves to the LCM slide they quickly became detached. As a consequence, filaments had to be grown freely in cultivation medium and then attached to a slide before processing for LCM. *Ectocarpus* filaments were fixed in a 100% acetone solution before LCM (Saint-Marcoux et al. 2015). Acetone is a precipitative fixative that coagulates the cell content by essentially denaturing the proteins. This type of fixative is favored in LCM experiments because cross-linking fixatives create chemical bonds between cellular components and tend to render downstream molecular extraction difficult. However, acetone fixation is harsh and *Spirogyra* cells were damaged to the point that filaments could not be used for laser capture (not shown). As a consequence, a gentler fixative composed of ethanol and acetic acid was developed and successfully used with sexually induced *Spirogyra* filaments.

29.3 Materials

29.3.1 Equipment

29.3.1.1 Algae cultivation

- Temperature and light-controlled growth chamber
- Laminar flowhood
- 500 µL sterile cell culture plastic flask with vented cap
- Glass Pasteur pipettes
- 1 L glass bottles
- 50 mL sterile plastic tubes
- Horizontal microscope slide holder
- Plastic sandwich boxes to contain 200–500 mL
- 0.2 µm Millipore filter and large plastic syringe
- Standard lab equipment (gloves, etc.)

29.3.1.2 Laser microdissection

- LCM unit equipped with a 40x magnification objective; here a Carl Zeiss PALM MicroBeam with an AxioVert 200 microscope and a CryLas UV laser, assisted by the PalmRobo 4.5 software were used
- UV cross-linker or a flowhood with UV light
- 37°C oven or chamber
- Nuclease-free 1.0 polyethylene naphthalate (PEN) membrane slides (Carl Zeiss Microscopy, #415190-9081-000)

- 500 μL plastic tubes with adhesive tube caps (Carl Zeiss Microscopy, #415190-9201-000)
- Soft paint brush

29.3.2 Reagents

- Bold modified basal freshwater solution 50X (SIGMA #B5282); on arrival, aliquot stock in 50 mL plastic tubes and freeze at –20°C
- Vitamins:
  - Kao and Michayluk vitamin solution 100X (SIGMA #K3129)

  Or
  - Biotin powder (SIGMA #B4639)
  - Cyanocobalamin (SIGMA #C3607)
  - Thiamine (SIGMA #T1270)
- Pure water
- Ethanol 100%
- Acetic acid 100%
- ProtectRNA™ RNase Inhibitor 500X concentrate (SIGMA #R7397)
- RNaseZap™ RNase Decontamination solution (Invitrogen #AM9780)
- Poly-L-lysine solution 0.1% (w/v) in H_2O (SIGMA #P8920)
- RNase-free water
- Silica gel

29.4 Experimental procedures

29.4.1 Cultivation of Spirogyra *and induction of sexual reproduction*

Cultures are grown in Bold's basal medium (BSM) supplemented with vitamins. Vitamins can be obtained as a combined solution or prepared individually in the lab. Induction of sexual reproduction is carried out in 1/10th dilution BSM. Vitamin concentration is kept undiluted.

29.4.1.1 Media preparation

1. Prepare BSM by adding 784 mL pure H_2O and 16 mL of Bold-modified basal freshwater stock solution 50X in a 1 L glass bottle
2. Adjust the pH solution to 7.0 with 0.1 or 1 M KOH
3. Autoclave and keep at room temperature until further use
4. Prepare BSM 1/10th by adding 798.4 mL pure H_2O and 1.6 mL of Bold-modified basal freshwater stock solution 50X in a 1 L glass bottle

If not using the commercial combined solution of vitamins

5. Prepare a vitamin stock solution (Table 29.1), filter through a 0.2 μm filter, aliquot and freeze at –20°C for storage, or keep at 4°C for routine use

Table 29.1 Vitamin solution recipe

	MW (g mol^{-1})	For 500 mL	In stock	Final
Biotin	244.1	2.5 mg	20 μM	2 nM
Cyanocobalamin	1355.37	2.75 mg	4 μM	0.4 nM
Thiamine	337.27	500 mg	3 mM	0.3 μM

29.4.1.2 Vegetative growth

1. Using a flame, close the tip of a glass Pasteur pipette by melting the glass and then carefully curve the tip to form a hook.
2. Sterilize the curved pipette using ethanol.
3. Pour 50 mL of BSM per culture flask and add 0.5 mL of Kao and Michayluk vitamin solution or add 5 μL of lab-made vitamin solution.
4. Using the curved Pasteur pipette, grab a few filaments of *Spirogyra* from the old culture and plunge them in the new medium. If the culture is old and filaments are fragmented, bits of thalli can be collected using a sterile pipette.
5. Place the flasks in the cultivation chamber at 22°C, a light radiance of 150 μE m^{-2} s^{-1}, and a photoperiod of 14:10 h of a light:dark cycle. No agitation is required. When starting from a few filaments, a saturated culture is obtained within 2–3 weeks. Cultures can survive for several months in these conditions. Figure 29.1a and c shows filaments in vegetative culture.

29.4.1.3 Sexual reproduction induction

1. Before attempting to induce sexual reproduction, it is important to stimulate vegetative growth by changing the vegetative growth medium every 4–5 days for a minimum of three times (*Note 1*).
2. Then transfer a full pipette hook of filaments in a culture flask containing 50 mL of BSM 1/10th supplemented with 0.5 mL of Kao and Michayluk or 5 μL of lab-made vitamin solution.
3. Place the flask back in the culture chamber. From the sixth day in culture, start to look for the first signs of gametogenesis using a dissecting microscope: cells within the filaments should start to enlarge and protoplasts start to condense soon after (Figure 29.1d). In our conditions, the maximum peak of fertilization and zygote formation was attained after 8–9 days in these culture conditions (Figure 29.1b, e).

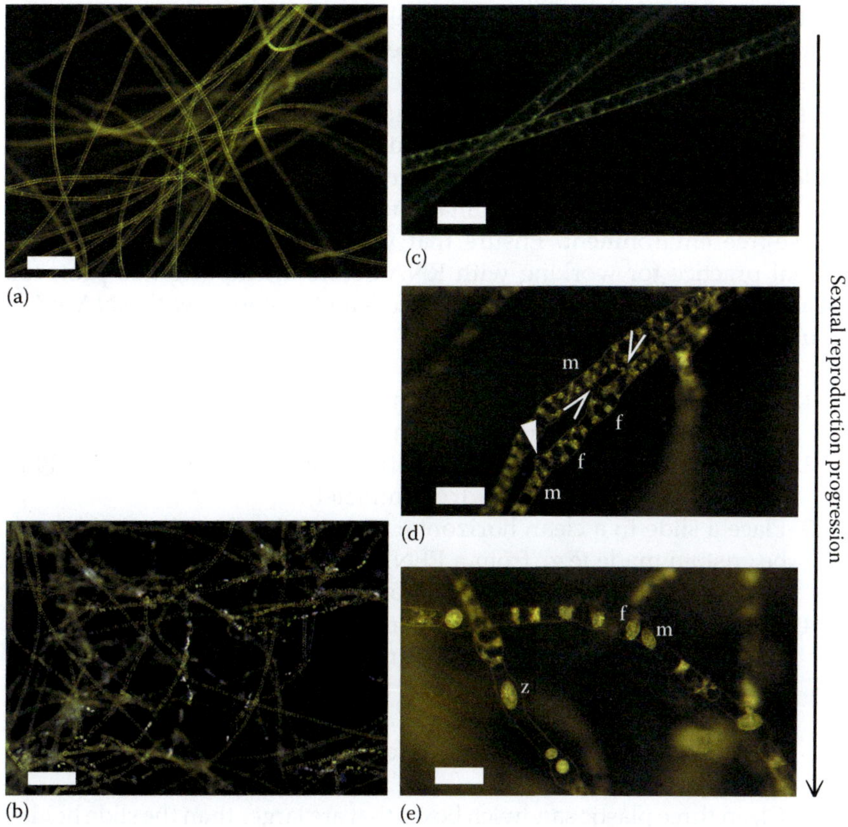

Figure 29.1 Induction of sexual reproduction in *S. pratensis*. *S. pratensis* (UTEX LB 928 strain) vegetative thallus (a) and after sexual reproduction (b). (c–e) close-up and progression of sexual reproduction: vegetative stage (c); intermediate stage where male and female gametes can be distinguished and conjugation tubes are forming (d); sexually mature filaments containing both gametes and zygotes (e) m: male; f: female; z: zygote; filled and hollow arrowheads point respectively to a lateral conjugation tube and a scalariform conjugation tube. Scale bars a and b: 500 μm; c–e: 100 μm.

29.4.2 Fixation of thallus before laser capture microdissection

Fixation is carried out using a more aqueous version of the traditional acetic acid:ethanol fixative, that is supplemented with RNase inhibitor. Together with its fixative properties, this solution allows fixed material to be stained, which helps to visualize cell structures during LCM. Fixation is carried out on the LCM slides because the algal filaments are too fragile to be manipulated afterward. Slides are covered with a solution of poly-L-lysine to help filaments stick to the surface.

NEVER touch the PEN membrane directly with anything but a brush (*Note* 2). The plastic membrane is extremely fragile.

FROM this step onward, all manipulations must be carried out in an RNase-free environment. Ensure that you have mastered good experimental practice for working with RNA before attempting this protocol. Wear gloves at all times. Clean all surfaces and materials with RNAseZap solution. Use only RNase-free water.

1. Place the PEN membrane slides under UV rays for 30 min. This helps to make the plastic surface more hydrophilic.
2. Pour ~1 mL of poly-L-lysine solution onto the surface of the PEN membrane slides and let it dry completely.
3. Place a slide in a clean horizontal slide holder. The slide holder can be custom made (e.g., from a PEN membrane slide plastic box). It is important that the slide can be moved while keeping it horizontal.
4. Prepare the following three solutions:
   - BSM 1/10th cultivation medium
   - 25% ethanol, 10% acetic acid, and 1X SIGMA RNase inhibitor mixed with RNase-free H_2O
   - 25% ethanol, 1X SIGMA RNase inhibitor mixed with RNase-free H_2O
5. Clean three plastic sandwich boxes that are larger than the slide holder.
6. Place the slide holder in a plastic box then pour the BSM 1/10th medium up to the limit of the surface of the slide. The solution must barely cover the slide.
7. Using a dissecting microscope, select fragments of thalli where reproduction and gametogenesis took place. Transfer pieces of fragments to the surface of the membrane slide using a clean brush and then spread the filaments equally over the slide (*Note* 2).
8. Very gently transfer the slide holder into a second plastic box (keeping the slide as horizontal as possible) and then pour the 25% ethanol, 10% acetic acid, and 1X SIGMA RNase inhibitor solution up to the limit of the slide. Again, make sure that the solution barely covers the slide. Note at this step that you will lose some pieces of thallus as they will be washed off of the slide. Incubate for 10 min.
9. Transfer the slide holder and the slide into a third plastic box and pour the 25% ethanol, 1X SIGMA RNase inhibitor solution as abovementioned. Incubate for 10 s.

10. Remove and transfer the slide holder and the slide to 37°C to dry (~30 min).
11. Once dried, the slides can be kept at room temperature in a box containing silica gel for a few days before LCM.

29.4.3 Laser capture microdissection

LCM procedures can vary significantly from one unit to another. You should be trained before attempting LCM. In our case, cells were collected in adhesive tube caps and we applied the following settings:

- *Cutting energy*: 45%–60%
- *Cutting speed*: 5%
- *Cutting focus*: 35%
- *LPC delta*: 25
- *LPC focus delta*: 0

A rule of thumb is to lower the laser energy as much as possible, while maintaining a good cutting performance. Cut as many gametes or zygotes as necessary depending on the downstream application. We found that approximately 200 captures were enough to ensure appropriate RNA amplification with our protocol. Male and female gametes differ in size and can be identified as such: female gametangia are enlarged compared to male gametangia. Zygotes that are contained in the female gametangium are linked through a conjugation tube to an empty male gametangium. Figure 29.2a shows filaments fixed on a LCM slide and Figure 29.2b is a close-up to a zygote. Figure 29.2c to k shows capturing of both gametes and zygotes.

29.5 Notes

Note 1: Gametogenesis induction is at best unreliable. We found that having very healthy cultures was crucial, and thus it is important to subculture the algae every 4–5 days (several times) before attempting induction. Counterintuitively, we also found that gametogenesis works best when only a moderate number of filaments are transferred to BSM 1/10th; do not try to saturate the culture from the onset. *Spirogyra* can be easily cultured on agar plates. In this case, we found that gametogenesis occurred more easily, even on a full BSM. Induction can thus be attempted on 1% agar plates containing BSM 1/10th.

Note 2: Transferring *Spirogyra* filaments to LCM slides is a tedious task and it is very important not to touch the membrane with a hard

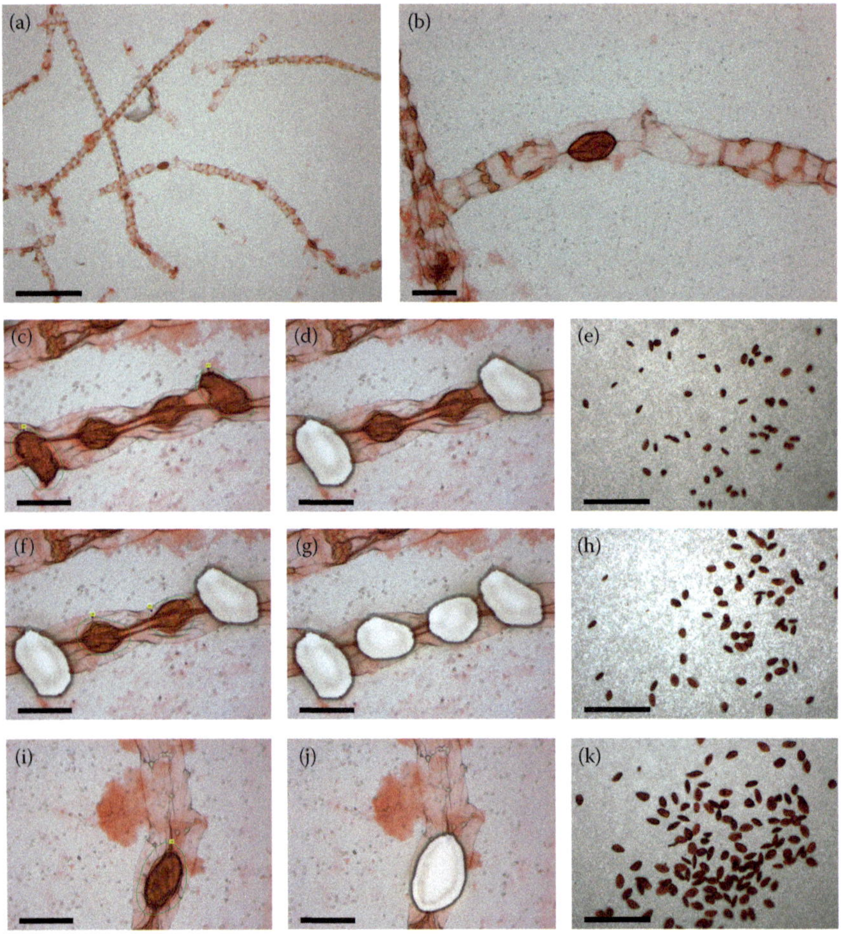

Figure 29.2 Laser capture microdissection of *S. pratensis* gametes and zygotes. (a, b) *S. pratensis* thallus fragments attached to an LCM slide before microdissection. Zygotes are visible. (c–k) Capture of male gametes (c–e), female gametes (f–h), and zygotes (i–k) showing the LCM slide (c, d, f, g, i, j) and the adhesive cap containing the captured cells (e, h, k). Scale bars a, e, h, k: 300 μm; b: 25 μm; c, d, f, g, i, j: 50 μm.

object to avoid tearing. A soft paint brush is ideal to manipulate the thallus fragments over the slide surface and it can be used to disentangle clumps, thus ensuring that the cellular structure of the filaments is visible for capture. Only a thin film of liquid BSM is needed over the surface of the slide when spreading the filaments. If there is too much liquid, filaments will float free from the slide.

Acknowledgment

This work was funded by an ERC Advanced Investigator Grant (EDIP) to JAL. We are grateful to Julie Bull for technical support.

References

Allen, M. A. 1958. *The Biology of a Species Complex in Spirogyra*. Bloomington, IA: Indiana University.

Andersen, R. A., J. A. Berges, P. J. Harrison, and M. M. Watanabe. 2005. Recipes for freshwater and saltwater media. In *Algal Culturing Techniques*, edited by R. A. Andersen, pp. 429–538. London, UK: Elsevier Academic Press.

Czurda, V. 1933. Experimentelle analyse der kopulationsauslösenden Bedingungen bei Mikroorganismen—1. Untersuchungen an Algen (Spirogyra, Zygnema Und Hyalotheca). *Beihefte zum Botanischen Centralblatt* 51: 711–761.

Drummond, C. S., J. Hall, K. G. Karol, C. F. Delwiche, and R. M. McCourt. 2005. Phylogeny of *Spirogyra and Sirogonium* (Zygnematophyceae) based on rbcL sequence data. *Journal of Phycology* 41 (5): 1055–1064. doi:10.1111/j.1529-8817.2005.00130.x.

Emmert-Buck, M. R., R. F. Bonner, P. D. Smith, R. F. Chuaqui, Z. Zhuang, S. R. Goldstein, R. A. Weiss, and L. A. Liotta. 1996. Laser capture microdissection. *Science* 274 (5289): 998–1001. doi:10.1126/science.274.5289.998.

Espina, V., J. D. Wulfkuhle, V. S. Calvert, A. VanMeter, W. Zhou, G. Coukos, D. H. Geho, E. F. Petricoin, and L. A. Liotta. 2006. Laser-capture microdissection. *Nature Protocols* 1 (2): 586–603. doi:10.1038/nprot.2006.85.

Gomez, S. K., and M. J. Harrison. 2009. Laser microdissection and its application to analyze gene expression in *Arbuscular Mycorrhizal* symbiosis. *Pest Management Science* 65 (5): 504–511. doi:10.1002/ps.1715.

Grote, M. 1977. Über die auslösung der generativen fortpflanzung unter kontrollierten bedingungen bei der grünalge *Spirogyra Majuscula. Zeitschrift Für Pflanzenphysiologie* 83 (2): 95–107. doi:10.1016/S0044-328X(77)80064-0.

Hoshaw, R. W., J.-C. Wang, R. McCourt, and H. M. Hull. 1985. Ploidal changes in clonal cultures of *Spirogyra Communis* and implications for species definition. *ResearchGate* 72 (7): 1005–1011.

Ichimura, T. 1971. Sexual cell division and conjugation-papilla formation in sexual reproduction of *Closterium Strigosum*. In *Proceedings of the 7th International Seaweed Symposium*, edited by K. Nishizawa, pp. 208–214, Tokyo, Japan: University of Tokyo Press.

Ikegaya, H., T. Nakase, K. Iwata, H. Tsuchida, S. Sonobe, and T. Shimmen. 2012. Studies on conjugation of *Spirogyra* using monoclonal culture. *Journal of Plant Research* 125 (3): 457. doi:10.1007/s10265-011-0457-3.

Nagata, Y. 1973. Rhizoid differentiation in *Spirogyra* I. Basic features of rhizoid formation. *Plant and Cell Physiology* 14 (3): 531–541.

Nakazono, M., F. Qiu, L. A. Borsuk, and P. S. Schnable. 2003. Laser-capture microdissection, a tool for the global analysis of gene expression in specific plant cell types: Identification of genes expressed differentially in epidermal cells or vascular tissues of maize. *The Plant Cell Online* 15 (3): 583–596. doi:10.1105/tpc.008102.

Nichols, H. W. 1973. Growth media—Freshwater. In *Handbook of Phycological Methods: Culture Methods and Growth Measurements*, edited by J. R. Stein, pp. 7–24, Cambridge, UK: Cambridge University Press.

Poel, B. V., E. D. Cooper, D. Van Der Straeten, C. Chang, and C. F. Delwiche. 2016. Transcriptome profiling of the green alga *Spirogyra* pratensis (Charophyta) suggests an ancestral role for ethylene in cell wall metabolism, photosynthesis, and abiotic stress responses. *Plant Physiology* 172 (1): 533–545. doi:10.1104/pp.16.00299.

Sager, R., and S. Granick. 1954. Nutritional control of sexuality in *Chlamydomonas reinhardi*. *The Journal of General Physiology* 37 (6): 729–742.

Saint-Marcoux, D., B. Billoud, J. A. Langdale, and B. Charrier. 2015. Laser capture microdissection in *Ectocarpus Siliculosus*: The pathway to cell-specific transcriptomics in brown algae. *Frontiers in Plant Science* 6: 54. doi:10.3389/fpls.2015.00054.

Simons, J., A. P. Van Beem, and P. J. R. De Vries. 1984. Induction of conjugation and spore formation in species of *Spirogyra* (Chlorophyceae, Zygnematales). *Acta Botanica Neerlandica* 33 (3): 323–334. doi:10.1111/j.1438-8677.1984.tb01823.x.

Sørensen, I., Z. Fei, A. Andreas, W. G. T. Willats, D. S. Domozych, and J. K. C. Rose. 2014. Stable transformation and reverse genetic analysis of *Penium Margaritaceum*: A platform for studies of *Charophyte Green Algae*, the immediate ancestors of land plants. *The Plant Journal* 77 (3): 339–351. doi:10.1111/tpj.12375.

Starr, R. J. 1964. The culture collection of algae at Indiana University. *American Journal of Botany* 51: 1013–1044.

Timme, R. E., and C. F. Delwiche. 2010. Uncovering the evolutionary origin of plant molecular processes: Comparison of *Coleochaete* (Coleochaetales) and *Spirogyra* (Zygnematales) transcriptomes. *BMC Plant Biology* 10 (1): 96. doi:10.1186/1471-2229-10-96.

Wickett, N. J., S. Mirarab, N. Nguyen, T. Warnow, E. Carpenter, N. Matasci, S. Ayyampalayam et al. 2014. Phylotranscriptomic analysis of the origin and early diversification of land plants. *Proceedings of the National Academy of Sciences* 111 (45): E4859–E4868. doi:10.1073/pnas.1323926111.

Zwirn, M., C. Chen, B. Uher, and M. Schagerl. 2013. Induction of sexual reproduction in *Spirogyra* clones—Does an universal trigger exist? *Fottea* 13 (1): 77–85. doi:10.5507/fot.2013.007.

chapter thirty

Cloning and expression strategies for the postgenomic analysis of brown algae

Agnès Groisillier

Contents

30.1 Introduction

The marine environment is highly complex and contains the vast majority of known and unknown biodiversity. It is also the last frontier to understand the control of the global climate and hides a wealth of biological resources still to be tapped for food, health, and energy. Marine macroalgae are thus photosynthetic eukaryotes that provide relevant and significant biological models to gain insight into the deep evolution of fundamental biological processes, which are at the same time complementary to those obtained by the study of other models such as plants or animals. Seaweeds that grow in the intertidal zone live in a constant stressful environment. They undergo the disadvantages of terrestrial habitats combined with disadvantages of aquatic environments (desiccation, freezing, osmotic stress, UV radiation, and variable temperatures), to which the mechanical stresses imposed by waves are added. In this intertidal zone, macroalgae, particularly brown algae, have a remarkable ability to adapt and acclimate to many different adverse conditions. Hence, these organisms should possess key processes that have evolved to cope with these frequent changes in environmental conditions.

Brown algae are a large group of marine seaweeds including almost 1800 species with a characteristic olive-green to dark-brown color derived from fucoxanthin (Wei et al. 2013). They are multicellular and have evolved independently from animals and land plants. They belong to the Stramenopiles (Heterokonta), a phylum that diverged from the Archaeplastida and Opisthokonta 1 billion years ago (Yoon et al. 2004). Within the Stramenopiles, brown algae are a relatively recent lineage that emerged about 200 million years ago (Brown and Sorhannus 2010; Silberfeld et al. 2010). Little is known about their biology, as they were mostly studied for their taxonomic and ecological diversity. So, knowledge about primary metabolic processes is essential for the understanding of the physiology and ecology of seaweeds.

Global genome sequencing efforts have generated large numbers of new genes, the biological or biochemical functions of which remain unknown. The complete sequences of more than 24,000 genomes are currently available in the public databases (https://www.ncbi.nlm.nih.gov/genome/).

However, up to now, only three draft genomes of brown algae have been decoded: *Ectocarpus siliculosus* (*E. siliculosus*), order Ectocarpales (Cock et al. 2010); *Saccharina japonica*, order Laminariales (Ye et al. 2015); and *Cladosiphon okamuranus* S-strain, order Chordariales (Nishitsuji et al. 2016). In newly sequenced genomes, genes are annotated on the basis of sequence similarity to other proteins that have already been characterized (Altschul et al. 1997). One major challenge is to assign biological function and to elucidate the mechanism of action of such genes. This challenge involves techniques to elucidate the structure and function of the gene products, interactions between proteins, and/or global protein changes. Overexpression of recombinant proteins is one of the possible solutions to answer these biological questions.

30.2 State of the art

Overexpression of a recombinant protein is essential for its structural and biochemical characterization, and the aim is to obtain as much correctly folded protein as possible. Most recombinant protein expression is achieved in heterologous hosts, in which low yields can be the direct result of protein toxicity to the host cell, or incorrect folding and subsequent insolubility of the protein. *Escherichia coli* (*E. coli*) is often used as a heterologous expression host as expression tests can be carried out relatively quickly and economically, allowing various constructs and conditions to be efficiently screened. Heterologous expression of algal genes is notoriously difficult and has presented a bottleneck in projects aiming at characterizing enzymes of macroalgal origins (Kurland and Gallant 1996; Groisillier et al. 2010). Thus far, there are only a few examples of successful expression of brown algal proteins in heterologous systems.

The publication of the first genome sequence of a brown alga (Cock et al. 2010) paved the way for experimental analyses of some of the unique metabolic pathways featured by these organisms. Among those that play pivotal roles in the physiology of these seaweeds, mannitol metabolism is of great importance due to its implication in carbon storage and in stress response. Several metabolic pathways for biosynthesis and degradation have been suggested, and a mannitol cycle (reviewed by Iwamoto and Shiraiwa 2005) has been proposed in a number of organisms, including micro- and macroalgae. This cycle, branching off from glycolysis at the level of fructose-6-phosphate, comprises a mannitol-1-phosphate dehydrogenase (M1PDH, EC1.1.1.17), a mannitol-1-phosphatase (M1Pase, EC3.1.3.22), a mannitol-2-dehydrogenase (M2DH), and a hexokinase. In 2011, for the first time, the activity of a recombinant M1PDH of *E. siliculosus* in *E. coli* cell–free extracts was shown (Rousvoal et al. 2011).

The purification and characterization of the recombinant catalytic domain of EsM1PDH1 provides the first insights on the biochemical behavior of such enzymes (Bonin et al. 2015). These latter results complement the characterization of a new type of phosphatase in *Ectocarpus*, which is specific for M1P (Groisillier et al. 2014). Such observations pave the way for future analysis to gain a better understanding about the physiological role, functioning, and regulation of mannitol synthesis in brown algae, notably by assessing the relationships between the structure and function of relevant enzymes and by altering mannitol synthesis *in vivo* when protocol(s) for targeted genetic modification of *Ectocarpus* will be available. In addition, a gene encoding a M2DH was recently characterized from the kelp *S. japonica*, but the purified recombinant protein was not active (Shao et al. 2014). Other original metabolic processes have been studied because of recombinant protein expression in *E. siliculosus*. In brown algae, Mannuronan C5-epimerases (ManC5-Es) catalyze the remodeling of alginate, a major cell-wall component that is involved in many biological functions. In 2016, the heterologous production and activity characterization of two ManC5-Es were successfully performed. The ManC5-E from the brown alga *S. japonica* was expressed because of Sf9 insect cells system (Inoue et al. 2016), and ManC5-E from *E. siliculosus* was purified in an active form after refolding on affinity columns (Fischl et al. 2016). Functional characterization of glutathione transferases (GSTs, EC 2.5.1.18) has been established by measuring the enzymatic activity of recombinant microsomal EsGSTs in *Saccharomyces cerevisiae* cell-free extracts (de Franco et al. 2009). In the same way, the first recombinant CYP74-related enzyme of oxylipin biosynthesis (Toporkova et al. 2017) and the first recombinant Guanosine DiPhosphate (GDP)-mannose dehydrogenase providing the precursor for the alginate polymer (Tenhaken et al. 2011) of *E. siliculosus* were purified and characterized. The only X-ray crystallographic structure of an *E. siliculosus* protein concerns a type III polyketide synthase (Meslet-Cladière et al. 2013). The current study has provided phylogenetic, biochemical, and structural evidence that support a primary role of PolyKetide Synthase 1 (PKS1) in the condensation of malonyl-CoA to produce phloroglucinol, the direct precursor of phlorotannins unique to brown algae. In 2011, the activity of a type III PKS *Sargassum binderi* PolyKetide Synthase (SbPKS) from a brown seaweed, *Sargassum binderi*, has been characterized but only in *E. coli* cell–free extracts (Baharum et al. 2011).

30.3 Materials

The protocol described below is based on techniques used for the purification of recombinant M1Pase and M1PDH from *E. siliculosus* (Groisillier et al. 2014; Bonin et al. 2015).

30.3.1 *Strains and vector*

30.3.1.1 E. coli

- *DH5α*: F-Φ80dlacZΔM15Δ(lacZYA-argF) U169 recA1 endA1 hsdR17 (r_k−, m_k+) phoA supE44 λ-thi-1 gyrA96 relA1.
- *BL21 (DE3)*: fhuA2 [lon] ompT gal (λ DE3) [dcm] ΔhsdS λ DE3 = λ sBamHIo ΔEcoRI-B int::(lacI::PlacUV5::T7 gene1) i21 Δnin5.

30.3.1.2 Vector

- *pFO4 vector*: AmpR, T7lac promoter hexa-histidine tail at the N-terminal of recombinant (a vector modified from pET15b, Novagen) (Groisillier et al. 2010) (*Note 1*).

30.3.2 *General stock solutions*

- Ultrapure (purified) water (such as MilliQ water, Millipore).
- 5 M NaCl, filter on 0.2 μm and store at room temperature.
- 1 M Tris–HCl, pH 7.5, filter on 0.2 μm and store at room temperature.
- 5 M Imidazole, filter on 0.2 μm and store at room temperature.
- 6 M Urea, filter on 0.2 μm and store at room temperature.
- 1 M Isopropyl β-D-1-thiogalactopyranoside (IPTG) in water, filter-sterilize and store in 1 mL aliquots at −20°C.
- 100 mg mL^{-1} ampicillin in water, filter-sterilize and store in 1 mL aliquots at −20°C.
- 0% (v/v) glycerol, autoclave, and store at room temperature.
- Super Optimal broth with Catabolite repression (SOC) buffer (Merck).
- *LB medium*: 1% Tryptone, 0.5% yeast extract, 200 mM NaCl, sterilize by autoclaving.
- *LB2x*: LB medium prepared in 0.5 volume.
- *LB-agar plates*: LB medium + 1.5% agar, add antibiotic after autoclave; pour 60-mm Petri dishes.
- ZYP5052 medium, prepared according to Studier (2005) and supplemented with 200 μg mL^{-1} of ampicillin (final concentration).

30.3.3 *PCR, cloning and transformation procedures*

- Primers were ordered from Eurogentec.
- Gene of interest was synthetized by GeneArtR Gene Synthesis (Life Technologies).
- *Bam*HI HF, *Eco*RI HF, and 10× Cutsmart Buffer (New England Biolabs).
- Calf intestinal alkaline phosphatase (CIP) (New England Biolabs).
- T4 DNA Ligase and 10× ligase buffer (New England Biolabs).
- DNA ladder.

- *Pfu* polymerase and 10× Pfu buffer (Promega).
- dNTP (10 mM).
- PCR master mix (Promega).
- Plasmid prep kit.
- PCR clean-up kit.
- *Buffers for nucleotide electrophoresis*: Commonly available formulations.

30.3.4 Protein expression and purification

- *Lysis buffer*: Tris–HCl 50 mM, pH 7.5; 250 mM NaCl; 1 mM Ethylenediaminetetraacetic acid (EDTA); 1 mg mL^{-1} lysozyme; 0.1 mg mL^{-1} DNAse (final concentrations)
- His Microspin columns (GE Healthcare, USA) or equivalent
- *Complete, EDTA-free (Roche)*: One tablet for 20 mL expression cells suspended in loading buffer
- *Loading (wash) buffer*: 20 mM Tris–HCl, pH 7.5, 200 mM NaCl, 10 mM imidazole (final concentrations)
- *Elution buffer*: 20 mM Tris–HCl, pH 7.5, 200 mM NaCl, 500 mM imidazole (final concentrations)
- *Gel filtration buffer*: 20 mM Tris–HCl, pH 7.5, 200 mM NaCl (final concentrations)
- 12% Criterion precast Bis-Tris gels (Biorad) or equivalent
- *Buffers for protein gel electrophoresis* Sodium Dodecyl Sulfate PolyAcrylamide Gel Electrophoresis (SDS–PAGE): commonly available formulations
- Anti-His antibody Horseradish Peroxidase (HRP)-conjugated
- TransBlot Turbo transfer pack 0.2 μ, nitrocellulose membrane mini (Biorad) or equivalent
- Enhanced ChemiLuminescent (ECL) Western Blotting Substrate
- *TBS-tween buffer*: 20 mM Tris–HCl pH 7.6, 0.8% NaCl, 0.1% Tween-20 (final concentrations)

30.3.5 Equipment

- Water bath set at 42°C
- Incubator set at 37°C
- Shaking incubator at 20°C or 37°C
- Autoclave
- 96-well thermocycler with heated lid
- OD600 absorbance reader
- Nanodrop spectrophotometer
- Sonicator
- French press
- ÄKTA Avant system (GE Healthcare, USA) or equivalent

- HisPrep™ FF 16/10 column (GE Healthcare, USA) or equivalent
- HiLoad™ Superdex 200 column (GE Healthcare, USA) or equivalent
- DNA and protein gel electrophoresis systems
- TransBlot Turbo system (Biorad) or equivalent

30.3.6 Software

- Hyper 32 (Informer Technologies, model: hyper32)
- Microsoft Excel or similar spreadsheet
- UNICORN software associated with ÄKTA Avant

30.4 Experimental procedures

The current chapter does not include description of standard techniques and gel electrophoresis, the use of equipment, and commercial kits. The reader should refer to texts, manuals, and protocols provided by suppliers for instructions.

30.4.1 Bioinformatics analysis

- Basic Local Alignment Search Tool (BLAST) analyses in National Center for Biotechnology Information (NCBI) to predicted related proteins (https://blast.ncbi.nlm.nih.gov/).
- SignalP 3.0 server to predict signal peptides (Bendtsen et al. 2004).
- TransMembrane Helices based on a hidden Markov Model (TMHMM) servers to predict transmembrane domains (Krogh et al. 2001).
- The crucial choices of the N- and C-terminal boundaries of each module were refined using hydrophobic cluster analysis (Callebaut et al. 1997).

30.4.2 Cloning

Gene of interest was synthetized and codon-optimized for expression in *E. coli* by GeneArtR Gene Synthesis, Life Technologies (USA) (Bonin et al. 2015) or searched in the genome sequence or the GenBank Expressed Sequence Tag (EST) library of *E. siliculosus* (Groisillier et al. 2014) (*Note 2*).

30.4.2.1 Amplification of gene of interest

Primers are designed to contain restriction sites contained in multicloning site of pFO4 vector:

- *Forward primer*: 5′-GGGGGG*GATCC*X^n-3′ (*Bam*HI restriction site in italic).
- *Reverse primer*: 5′-CCCCCC*GAATTC*X^n-3′ (*Eco*RI restriction site in italic).

- X^n corresponds to gene primers chosen to have a Tm of 70°C–72°C (*Note 3*).
- Prepare the PCR mix, on ice:

Pfu polymerase	0.5 μL (3 u μL^{-1})
10× buffer	5 μL
dNTP	2 μL (10 mM)
Primer forward	0.5 μL (10 μmol μL^{-1})
Primer reverse	0.5 μL (10 μmol μL^{-1})
Gene DNA	1 μL (10 ng μL^{-1})
Sterile water	qsp 50 μL

- *Program the thermocycler as follows*: 5 min at 94°C, followed by 30 cycles of 30 s at 94°C, 60 s at 50°C, 90 s at 72°C, then 10 min at 72°C. The duration of the extension step depended on the length (nucleotides) gene fragment, approximately 60 s kb^{-1}.
- Put 5 μL of PCR result on 1% agarose gel for analysis (size and amount).

30.4.2.2 *PCR fragment purification*

When proper amplification of the desired fragment has occurred, purify the PCR product using PCR clean-up kit. Finally, add 30 μL of sterile water to suspend PCR fragment.

30.4.2.3 *PCR fragment digestion*

- Digest the PCR fragment with reaction mix as follows:

PCR fragment	25 μL
*Eco*RI HF (20 u μL^{-1})	0.5 μL
*Bam*HI HF (20 u μL^{-1})	0.5 μL
10× Cutsmart Buffer	3 μL
Sterile water	qsp 30 μL

- Incubate 2 h at 37°C.
- Purify the linearized vector using PCR clean-up kit, elute in 30 μL sterile water.

30.4.2.4 *pFO4 vector preparation*

- Digest the vector with reaction mix as follows:

Vector DNA	1 μg
*Eco*RI HF (20 u μL^{-1})	1 μL
*Bam*HI HF (20 u μL^{-1})	1 μL
10× Cutsmart Buffer	2 μL
Sterile water	qsp 20 μL

- Incubate 2 h at 37°C.
- Add 1 μL of CIP and then incubate 1 h at 37°C.
- Purify the linearized vector using PCR clean-up kit. Finally, add 100 μL sterile water to get a vector concentration of 10 ng μL^{-1}.

30.4.2.5 Ligation step

- Prepare the ligation mix:

Linearized vector (2.5 ng μL^{-1})	2 μL
Digested PCR fragment	5 μL
10× ligation buffer	0.5 μL
T4 DNA ligase	0.5 μL
Sterile MilliQ water	qsp 10 μL

- To optimize ligation, incubate at room temperature for 1 h and then at 4°C O/N.

30.4.2.6 *Transformation in* E. coli *DH5α*

- Prepare LB-agar with antibiotic plates at room temperature.
- Thaw competent cells on ice.
- Mix 50 μL DH5α competent cells with ligation mix in 10 mL sterile culture tube.
- Incubate 30 min on ice.
- Heat shock exactly 45 s at 42°C in water bath.
- Put back on ice 5 min.
- Add 100 μL of SOC medium.
- Incubate 45 min at 37°C with agitation.
- Plate 2 × 80 μL on selective plates then incubate at 37°C O/N.

30.4.2.7 PCR screening

PCR screening was performed directly on the DH5α bacterial colonies to verify the clones with inserts of the expected size, by using PCR primers that annealed upstream (5′-GCAGCAGCCACCATCACCATCACC-3′) and downstream (5′-CCTTTCGGGCTTTGTTAGCAGCCGG-3′) from the insertion site of pFO4 vector.

- PCR screening Mix:
  10 μL Master mix
  0.2 μL primer pFO4 forward
  0.2 μL primer pFO4 reverse
  With a sterile plastic tip, pick up a single well-isolated colony and inoculate it in a PCR reaction tube. Repeat this step for each colony to be screened.

- *PCR program*: (cf. Section 30.4.2.1).
- Put 10 μL on 1% agarose gel for size analysis.

30.4.2.8 Plasmid extraction and glycerol stock

- Positive DH5α clones were cultivated in 2 mL of LB medium + ampicillin (100 μg mL^{-1}) at 37°C O/N. One milliliter was used to stock clones in glycerol at −80°C, and plasmids were purified from the other mL.
- *Glycerol stocks*:
  500 μL glycerol 100%
  500 μL LB 2×
  1 μL ampicillin (100 mg mL^{-1})
  1 mL culture
- Plasmid extraction was carried out using Plasmid Prep kit. Plasmids were eluted with 50 μL H_2O sterile. The integrity of sequences was verified by sequencing.

30.4.2.9 Transformation of expression strain, PCR screening, and glycerol stock

Steps in Section 30.4.2.6 and 30.4.2.7, and glycerol stock were repeated with *E. coli* BL21 (DE3) cells and recombinant pFO4 vector obtained at step in Section 30.4.2.8 (*Note 1*).

30.4.3 Small-scale test expression

BL21 (DE3) positive clones (between 4 and 8) were tested for the expression of the desired protein (*Note 4*).

30.4.3.1 Cultivation was performed in two phases

First, transformed colonies were grown at 37°C overnight in 1 mL LB medium containing 100 μg mL^{-1} ampicillin; then, cultures were diluted 1:100 in 2 mL autoinductible ZYP5052 medium containing 200 μg mL^{-1} ampicillin and subjected to further incubation at 20°C, 200 rpm, during 72 h.

30.4.3.2 Lysis of cells and detection of proteins

- For solubility assay, cell pellets from small-scale expression cultures were resuspended in 500 μL of lysis buffer and incubated at 18°C for 1 h with light agitation and the soluble and insoluble fractions separated by centrifugation (14,000 g, 20 min, 4°C).
- Insoluble pellets were resuspended in 200 μL of lysis buffer supplemented with urea 6 M, incubated at 4°C for 1 h with light agitation and the insoluble protein fraction and the pellet separated by centrifugation (14,000 g, 20 min, 4°C).
- Samples from soluble and insoluble fractions were separated by 12% SDS–PAGE.

Presence of the protein of interest is not always easy to observe in SDS–PAGE, due to the presence of *E. coli* proteins. Two methods can be used to confirm the presence of the recombinant protein:

- First, soluble fractions could be purified using His Microspin columns according to the protocol recommended by the supplier. The results were also analyzed by 12% SDS–PAGE.
- Second, a Western blot can be carried out. For this, transfer from gel onto ready-to-use 0.2 μm nitrocellulose membrane was done by using a TransBlot Turbo system under the condition specified by the manufacturer. Monoclonal antipolyhistidine peroxidase conjugate antibody that specifically recognizes his-tagged fusion proteins was used at a final concentration of 1/10,000.

30.4.4 Large-scale expression and purification

30.4.4.1 Expression culture

- Expression tests were performed as described in Section 30.4.3.1 but adjusted to a 1 L culture (*Note 5*).

30.4.4.2 Affinity chromatography

- To purify the recombinant proteins, bacteria cells were harvested by centrifugation at 4000 rpm for 20 min and then resuspended in 5 to 20 mL of loading buffer (depending of the pellet volume) before lysis with a French press. The lysate was centrifuged for 1 h at 14,000 rpm, and the cell-free supernatant was filtered through 0.2 μm filter before loading onto a HisPrep column.
- Proteins were eluted with a linear increasing gradient (0%–100%) of elution buffer within minimum 10 column volumes, and fractions of 1 mL were collected (Figure 30.1a).
- Presence of the recombinant protein was verified by 12% SDS–PAGE (Figure 30.1a).

30.4.4.3 Size-exclusion chromatography

- To improve the purification of the recombinant protein, fractions containing the recombinant tagged protein were injected onto a HiLoad™ Superdex 200 column previously equilibrated with gel filtration buffer. Fractions were then eluted with the same buffer at a flow rate of 1 mL min^{-1} (Figure 30.1b).

30.4.4.4 Analysis of the chromatography

After each step of purification, protein fractions were analyzed by SDS–PAGE and by Western blot if necessary (Figure 30.1b).

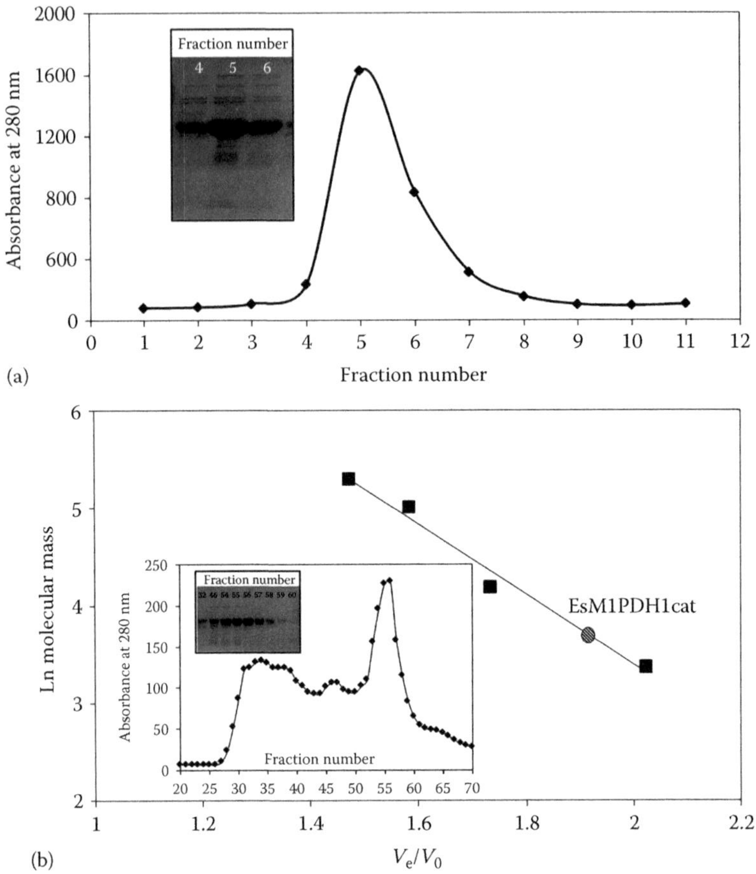

Figure 30.1 Purification (a) and estimation of molecular mass (b) of the recombinant His-tagged EsM1PDH1cat domain. (a) Proteins were purified by Ni^{2+} affinity chromatography, and eluted fractions were analyzed by SDS–PAGE (inset); (b) proteins were resolved by gel filtration onto a Superdex 200 HiLoad™ column (inset), and the following proteins were used as standards for column calibration: carbonic anhydrase (29 kDa), albumin (66 kDa), alcohol dehydrogenase (150 kDa), b-amylase (200 kDa); V_e, elution volume; V_0, void volume. (From Bonin, P. et al., *Phytochemistry*, 117, 509–520, 2015.)

30.5 Notes

Note 1: The *E. coli* system is the first-choice host for the initial screening of the recombinant protein expression, because these cells can be readily manipulated, are cultured inexpensively, and grow rapidly. In recent years, numerous new strains, vectors, and tags have been developed to overcome the limitations of this system, which include codon bias, inclusion body formation, toxicity, protein inactivity, mRNA instability, and lack of posttranslational modification (Khow and Suntrarachun 2012; Gopal and Kumar 2013; Rosano and Ceccarelli 2014).

Note 2: The few example of expression of proteins in brown algae does not make it possible to know whether gene synthesis and optimization in *E. coli* is the best solution to obtain a soluble protein, but now, relatively cheap commercial gene synthesis is often a preferred option, providing an outsourced service that affords the benefit of codon-optimized cDNAs that can be custom cloned into the required expression vectors by the supplier (Quax et al. 2015).

Note 3: This protocol uses restriction sites dependent on the multiple site cloning present on the expression vectors, but other cloning techniques may be commonly used as Gateway Technology, Flexi R Vector Systems, and Creator™ DNA Cloning System (Festa et al. 2013). The primers are always chosen with the same Tm (70°C–72°C) to simplify the amplification steps in the case of multiple clonings performed in parallel (Groisillier et al. 2010).

Note 4: Figure 30.2 shows the importance of examining several clones for the same expression strain transformed by a recombinant plasmid.

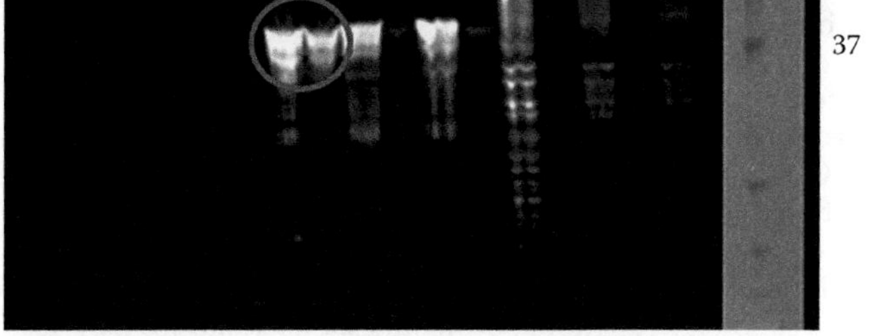

Figure 30.2 Western blot analysis of insoluble and soluble fractions of three clones of *E. coli* cell-free extracts containing catalytic domain of M1PDH (M1Short NC, without induction), M1PDH (M1Short N), and full M1PDH (M1full NC) of *E. siliculosus*. *M*: molecular weight marker in kDa. (From Bonin, P., Étude du métabolisme du mannitol chez l'algue brune modèle *Ectocarpus siliculosus*: caractérisation de l'enzyme clé mannitol-1-phosphate déshydrogénase. *Thèse de doctorat de Biologie*. Université Pierre et Marie Curie, Paris, France, 2014.)

Note 5: Culture condition of recombinant *E. coli* may also increase gene expression levels and the solubility of expressed proteins. The solubility of recombinant proteins is increased by prolonged induction at low temperatures with decreased amounts of IPTG. Sometimes addition of some chemicals in the culture medium also increases the expression of the gene-of-interest because these chemicals induce chaperone expression (Gopal and Kumar 2013).

References

Altschul, S.F., Madden, T.L., Schaffer, A.A., Zhang, J., Zhang, Z., Miller, W., Lipman, D.J. 1997. Gapped BLAST and PSI-BLAST: A new generation of protein database search programs. *Nucleic Acids Res* 25: 3389–3402.

Baharum, H., Morita, H., Tomitsuka, A., Lee, F.-C., Ng, K.-Y., Rahim, R.A., Abe, I., Ho, C.-L. 2011. Molecular cloning, modeling, and site-directed mutagenesis of Type III polyketide synthase from *Sargassum binderi* (Phaeophyta). *Mar Biotechnol* 13: 845–856.

Bendtsen, J.D., Nielsen, H., von Heijne, G., Brunak, S. 2004. Improved prediction of signal peptides: SignalP 3.0. *J Mol Biol* 340: 783–795.

Bonin, P. 2014. Étude du métabolisme du mannitol chez l'algue brune modèle *Ectocarpus siliculosus*: Caractérisation de l'enzyme clé mannitol-1-phosphate déshydrogénase. *Thèse de doctorat de Biologie*. Université Pierre et Marie Curie, Paris, France.

Bonin, P., Groisillier, A., Raimbault, A., Guibert, A., Boyen, C., Tonon, T. 2015. Molecular and biochemical characterization of mannitol-1-phosphate dehydrogenase from the model brown alga *Ectocarpus* sp. *Phytochem* 117: 509–520.

Brown, J.W., Sorhannus, U. 2010. A molecular genetic timescale for the diversification of autotrophic stramenopiles (Ochrophyta): Substantive underestimation of putative fossil ages. *PLoS One* 5: e12759.

Callebaut, I., Labesse, G., Durand, P., Poupon, A., Canard, L., Chomilier, J., Henrissat, B., Mornon, J.P. 1997. Deciphering protein sequence information through hydrophobic cluster analysis (HCA): Current status and perspectives. *Cell Mol Life Sci* 53: 621–645.

Cock, J.M., Sterck, L., Rouzé, P. et al. 2010. The *Ectocarpus* genome and the independent evolution of multicellularity in brown algae. *Nature* 465: 617–621.

de Franco, P.O., Rousvoal, S., Tonon, T., Boyen, C. 2009. Whole genome survey of the glutathione transferase family in the brown algal model. *Mar Genomics* 1: 135–148.

Festa, F., Steel, J., Bian, X., Labaer, J. 2013. High-throughput cloning and expression library creation for functional proteomics. *Proteomics* 13: 1381–1399.

Fischl, R., Bertelsen, K., Gaillard, F., Coelho, S., Michel, G., Klinger, M., Boyen, C., Czjzek, M., Hervé, C. 2016. The cell-wall active mannuronan C5-epimerases in the model brown alga *Ectocarpus*: From gene context to recombinant protein. *Glycobiology* 26: 973–983.

Gopal, J., Kumar, A. 2013. Strategies for the production of recombinant protein in *Escherichia coli*. *Protein J* 32: 419–425.

Groisillier, A., Hervé, C., Jeudy, A., Rebuffet, E., Pluchon, P.F., Chevolot, Y., Flament, D. et al. 2010. MARINE-EXPRESS: Taking advantage of high throughput cloning and expression strategies for the post-genomic analysis of marine organisms. *Microb Cell Fact* 9: 45.

Groisillier, A., Shao, Z., Michel, G., Goulitquer, S., Bonin, P., Krahulec, S., Nidetzky, B., Duan, D., Boyen, C., Tonon, T. 2014. Mannitol metabolism in brown algae involves a new phosphatase family. *J Exp Bot* 65: 559–570.

Inoue, A., Satoh, A., Morishita, M., Tokunaga, Y., Miyakawa, T., Tanokura, M., Ojima, T. 2016. Functional heterologous expression and characterization of mannuronan C5-epimerase from the brown alga *Saccharina japonica*. *Algal Res* 16: 282–291.

Iwamoto, K., Shiraiwa, Y. 2005. Salt-regulated mannitol metabolism in algae. *Mar Biotechnol* 7: 407–415.

Khow, O., Suntrarachun, S. 2012. Strategies for production of active eukaryotic proteins in bacterial expression system. *Asian Pac J Trop Biomed* 2: 159–162.

Krogh, A., Larsson, B., von Heijne, G., Sonnhammer, E.L. 2001. Predicting transmembrane protein topology with a hidden Markov model: application to complete genomes. *J Mol Biol* 305: 567–580.

Kurland, C., Gallant, J. 1996. Errors of heterologous protein expression. *Curr Opin Biotechnol* 7: 489–493.

Meslet-Cladière, L., Delage, L., Leroux, C.J.-J., Goulitquer, S., Leblanc, C., Creis, E., Ar Gall, E., Stiger-Pouvreau, V., Czjzek, M., Potin, P. 2013. Structure/function analysis of a Type III polyketide synthase in the brown alga *Ectocarpus siliculosus* reveals a biochemical pathway in phlorotannin monomer biosynthesis. *The Plant Cell* 25: 3089–3103.

Nishitsuji, K., Arimoto, A., Iwai, K., Sudo, Y., Hisata, K., Fujie, M., Arakaki, N. et al. 2016. A draft genome of the brown alga, *Cladosiphon okamuranus*, S-strain: A platform for future studies of "mozuku" biology. *DNA Res* 23: 561–570.

Quax, T.E., Claassens, N.J., Söll, D., van der Oost, J. 2015. Codon bias as a means to fine-tune gene expression. *Mol Cell* 59: 149–161.

Rosano, G.L., Ceccarelli, E.A. 2014. Recombinant protein expression in *Escherichia coli*: Advances and challenges. *Front Microbiol* 5: 172.

Rousvoal, S., Groisillier, A., Dittami, S.M., Michel, G., Boyen, C., Tonon, T. 2011. Mannitol-1-phosphate dehydrogenase activity in *Ectocarpus siliculosus*, a key role for mannitol synthesis in brown algae. *Planta* 233: 261–273.

Shao, Z., Zhang, P., Li, Q., Wang, X., Duan, D. 2014. Characterization of mannitol-2-dehydrogenase in *Saccharina japonica*: Evidence for a new polyol-specific longchain dehydrogenases/reductase. *PLoS One* 9: e97935.

Silberfeld, T., Leigh, J.W., Verbruggen, H., Cruaud, C., de Reviers, B., Rousseau, F. 2010. A multi-locus time-calibrated phylogeny of the brown algae (Heterokonta, Ochrophyta, Phaeophyceae): Investigating the evolutionary nature of the "brown algal crown radiation". *Mol Phylogenetics Evol* 56: 659–674.

Studier, F.W. 2005. Protein production by auto-induction in highdensity shaking cultures. *Protein Expr Purif* 41: 207–234.

Tenhaken, R., Voglas, E., Cock, J.M., Neu, V., Huber, C.G. 2011. Characterization of GDP-mannose dehydrogenase from the brown alga *Ectocarpus siliculosus* providing the precursor for the alginate polymer. *J Biol Chem* 286: 16707–16715.

Toporkova, Y.Y., Fatykhova, V.S., Gogolev, Y.V., Khairutdinov, B.I., Mukhtarova, L.S., Grechkin, A.N. 2017. Epoxyalcohol synthase of *Ectocarpus siliculosus*. First CYP74-related enzyme of oxylipin biosynthesis in brown algae. *Biochimica et Biophysica Acta* 1862: 167–175.

Wei, N., Quarterman, J., Jin, Y. 2013. Marine macroalgae: An untapped resource for producing fuels and chemicals. *Trends Biotechnol* 31: 70–77.

Ye, N., Zhang, X., Miao, M., Fan, X., Zheng, Y., Xu, D., Wang, J. et al. 2015. *Saccharina* genomes provide novel insight into kelp biology. *Nat Commun* 6: 6986.

Yoon, H.S., Hackett, J.D., Ciniglia, C., Pinto, G., Bhattacharya, D. 2004. A molecular timeline for the origin of photosynthetic eukaryotes. *Mol Biol Evol* 21: 809–818.

chapter thirty one

Polyethylene glycol-mediated transformation in the green macroalga Ulva mutabilis *(Chlorophyta)*

A forward genetics approach

Jens Boesger, Michiel Kwantes, and Thomas Wichard

Contents

31.1 Introduction

The number of research groups looking for opportunities to transform algae has increased steadily with the increasing interest in algae as novel plant model organisms and potential next-generation fuel producers (Walker et al. 2005, Hallmann 2007, Mikami 2013). Genetic engineering programs aim to manipulate algal metabolism, but a stable transformation is still a prerequisite and often the limiting factor to achieving the next level of skills in macroalgal research.

Vector transformation is the process by which exogenous DNA is transferred into a host organism. Methods based on polyethylene glycol (PEG)-mediated DNA transfer, particle bombardment, electroporation, and gene transfer using *Agrobacterium* are common approaches to create transgenic cell lines, which were also all applied to green algae (Kathiresan et al. 2009, Oertel et al. 2015). Transformation can be divided into genetic (stable) and transient transformations, which has recently been more frequently reported for macroalgae (Mikami 2013). On transformation, genes introduced by genetic recombination are maintained in the genome through generations of cells, whereas in transient transformation, rapid loss of introduced foreign genes is usually observed. At this moment, the critical step for genetic transformation is to generate a plasmid vector containing all the genetic elements for expression, including promoter, enhancer, and transcription termination sequences. The vector involves the manipulation of selection marker expression cassettes (usually a gene conferring resistance to an antibiotic) for either selection of the plasmid in bacteria during the cloning steps or gene transfer in the host organism. Recent genome data of several macroalgae now support the application of recombinant techniques for these species (Cock et al. 2010, Collén et al. 2013), because they provide a rich resource of DNA elements, such as, for example, promoters that can be chosen for transgenic expression.

Macroalgae very often reveal a strong resistance to antibiotics, so that an appropriate selection of antibiotic markers is hard to find. At the same time, morphogenesis of macroalgae (i.e., *Ulva*) depends on its associated bacteria (Wichard 2015). Therefore, those essential growth-promoting and/or morphogenesis-inducing bacteria must be equally resistant to the chosen antibiotics such as the transgenic *Ulva* (Oertel et al. 2015).

Establishing the genetic-transformation system requires five main parameters (Mikami 2013): (1) easily accessible target cells (e.g., *Ulva* gametes); (2) an efficient gene transfer system; (3) an efficient expression system for foreign genes, considering the codon usage of the host; (4) an integration (and targeting) system to integrate the foreign gene into the genome; and (5) a selection system for transformed lines that does not interfere with the growth of the bacteria associated to *Ulva*. In fact, a stable transformation system will be accomplished by the proof of stable DNA integration into the algal genome by PCR techniques or Southern blots. A stable transformation system in hand allows us to apply several important techniques in forward and reverse genetics; for example, insertional mutagenesis, promoter trapping, gene silencing, or gene editing. In the current chapter, the PEG-mediated DNA transfer into *Ulva*'s gametes is described for stable transformation facilitating a forward genetic approach (Figure 31.1).

31.2 State of the art

Hallmann (2007), in his review *Algal Transgenics and Biotechnology*, highlighted several critical questions that have to be addressed when transgenic algae of new model systems are to be produced: (1) *Are there any close relatives that have been transformed before?* (2) *Which DNA constructs might be the most suitable for selection of my transgenic algae?* (3) *Which promoters should work and result in a high level of expression?* (4) *How should the DNA be transferred into the cells?* [and] (5) *How can expression of a given gene be monitored?* (Hallmann 2007). Answers to these questions could be given for *Ulva*, but not for other macroalgae, such as *Ectocarpus siliculosus* or *Porphyra* species, which are still under investigation. In any case, considerable technical improvements have been made, for example, regarding transient transformation in red seaweeds (Mikami 2014) and stable transformation in *Ulva* (Oertel et al. 2015). Oertel et al. (2015) adopted a protocol for introducing DNA into *U. mutabilis* cells that was originally developed for the high-frequency transformation of yeast protoplasts (Hinnen et al. 1978). In contrast to yeast, *Ulva* gametes or zoids can be used directly as receiver cells, because they are devoid of a rigid cell wall. After settling and adhesion, germ cells resorb their flagella and form a protoplast.

Indeed, the first successful stable transformation of macroalgae was achieved by PEG-mediated direct DNA transfer into germ cells of *U. mutabilis* in Oertel's laboratory (University of Regensburg, Regensburg, Germany) to target green fluorescent protein expression in the cell nucleus during early cell divisions of the germling and to generate developmental mutants showing an atypical morphogenesis (Oertel et al. 2015). Transformation vectors possess (Figure 31.2) the chromosomal *RBCS* gene from *U. mutabilis* with the bleomycin resistance gene (*ble*) as a dominant selection marker gene. Transformation frequencies are in average 10^4 in 10^7 cells (Oertel et al. 2015). A similar approach revealed the expression of mitochondrial targeting green fluorescent protein upon PEG transformation in *U. partita* (Suzuki et al. 2016). The insertional mutagenesis approach is currently evolving to a reliable genetic tool kit along with the genome project of *U. mutabilis* (Wichard et al. 2015).

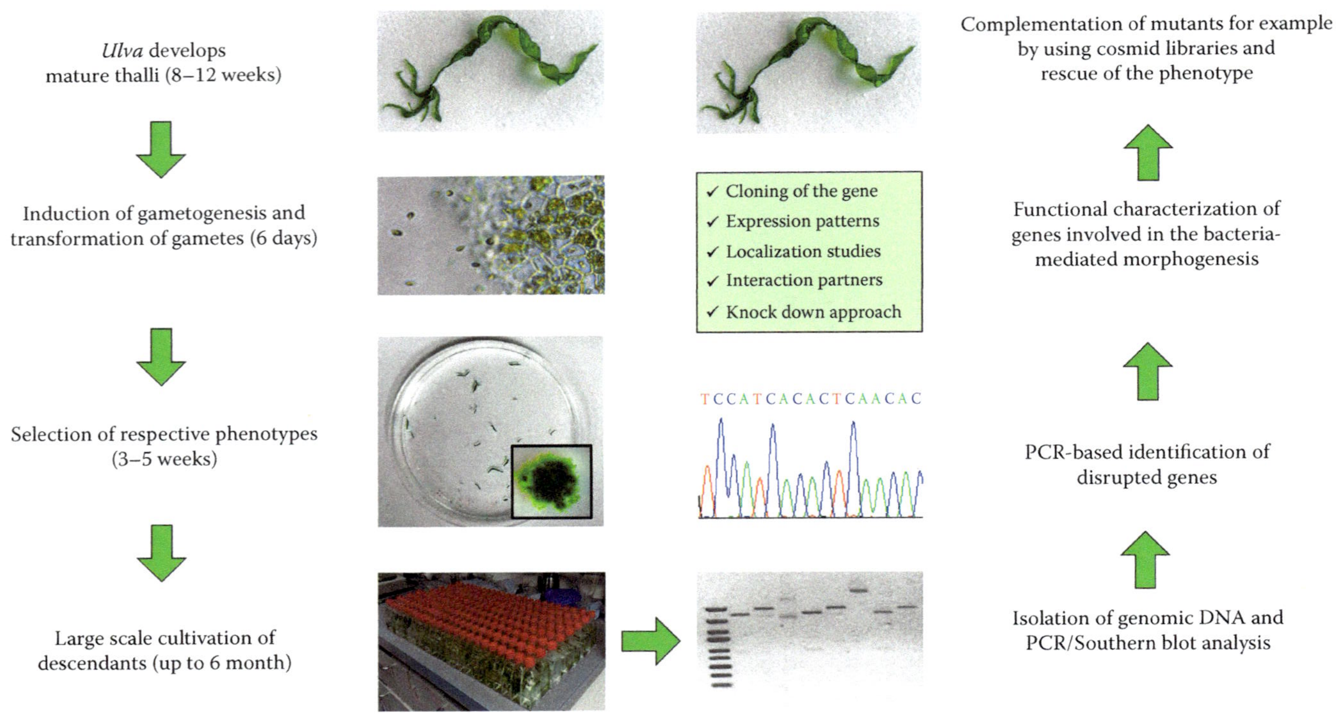

Figure 31.1 Workflow of the forward genetic approach. The selection criteria and screenings for the phenotype of interest are the bottleneck before the (multiple) insertion sites of the phleomycin-resistant cassette can be determined.

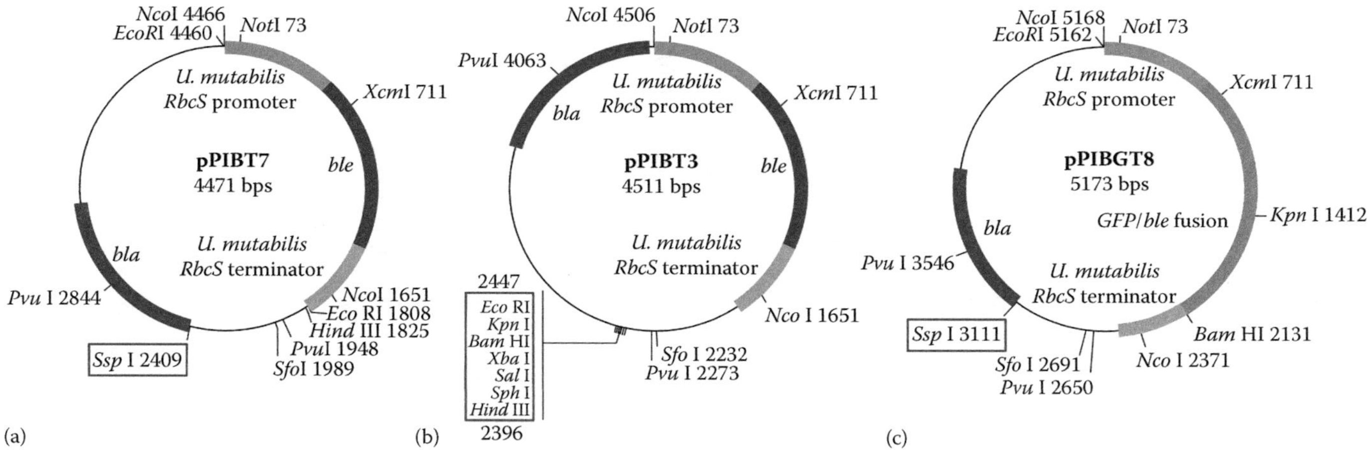

Figure 31.2 Vector plasmids for the transformation of *Ulva mutabilis* (Oertel et al. 2015). (a–c) Codon adapted phleomycin-resistant cassettes: *RBCS*. Ribulose-1.5-bisphosphate carboxylase/oxygenase small subunit; *ble*. phleomycin resistant gene; *bla*. ß-lactamase. Suitable restriction site marked with the frame.

31.3 Materials

31.3.1 Equipment, consumables, and reagents

31.3.1.1 Equipment and consumables

- Bead mill, for example, Tissuelyser (Qiagen, Hilden, Germany)
- 2–3 mm diameter steel beads
- Erlenmeyer flasks (2 L)
- Nylon sieve (mesh size 30 µm)
- Micro centrifuge tubes (1.5 mL, 2 mL, and for PCR)
- Centrifuge tubes (15 and 50 mL)
- Membrane filters (0.2 µm)
- Disposable syringes (sterile; 10/20/50 mL range)
- Pipettes and tips (2/10/100/1000 µL range)
- Petri dishes (diameter 145 mm) or culture flasks (surface area 182 cm^2, capacity 550 mL)
- Pasteur pipettes with long, thin opening
- Light source (e.g., a mobile fluorescent lamp)
- Common PCR thermal cycler
- Aluminum foil and paper tissues

31.3.1.2 Reagents

- *Ulva* culture medium (UCM, Stratmann et al. 1996)
- *Ulva* transformation buffer 1 (UTB1, Table 31.1)
- *Ulva* transformation buffer 2 (UTB2, Table 31.2)
- Phleomycin (InvivoGen, San Diego CA, USA; stock solution: 10 mg mL^{-1})
- Bacterial growth factors of *Ulva* or stationary cultures of *Maribacter* sp. MS6 and *Roseovarius* sp. MS2 (Spoerner et al. 2012, Grueneberg et al. 2016)
- Vector plasmids for the transformation of *U. mutabilis*, for example, pPIBT7 (EU176859.1), pPIBT3 (EU176858.1), pPIBGT8 (EU196041.1) (Figure 31.2; Oertel et al. 2015)
- *Oligonucleotides*: Ph-2–5′-TTAGTCCTGCTCCTCGTCCACGAAGTG-3′; RbcS_F1—GGCTTTCCAGGAGGGCAGTC; Ubi1-F1–5′-TACGGAGGC TGTTGCACACG-3′; Ubi1_R2—AGCAGCATGACAGACCATCTTGTG
- PrimeSTAR GXL DNA Polymerase (Takara Bio Inc, USA); alternative Pfu-Turbo DNA Polymerase (Agilent, USA)
- Dimethyl sulfoxide
- Silica-based DNA purification kit, for example, QIAquick (Qiagen, Hilden, Germany)
- Genomic DNA purification kit, for example, GenElute Plant Genomic DNA Miniprep Kit (Sigma, St. Louis, MO, USA)
- Suitable restriction endonucleases

Table 31.1 Composition of *Ulva* transformation buffer 1 (UTB1) (*Note 1*)

Composition	Final concentration	Quantity to add (for 50 mL final)
Sorbitol	1 M	9.1085 g
Tris–HCl pH 7.5	10 mM	500 μL of a stock solution at 1 M
$CaCl_2$	10 mM	55.49 mg
UCM	10% (v/v)	5 mL

Table 31.2 Composition of *Ulva* transformation buffer 2 (UTB2) (*Note 1*)

Composition	Final concentration	Quantity to add (for 50 mL final)
PEG 4000	40% (w/v)	20 g
Tris–HCl pH 7.5	10 mM	500 μL of a stock solution at 1 M
$CaCl_2$	10 mM	55.49 mg

31.3.2 Ulva *culture medium*

A specific UCM was developed that allows *Ulva* and its associated bacteria to develop sufficiently (see Chapter 9 by Califano and Wichard 2018, Stratmann et al. 1996).

31.4 Experimental procedures

31.4.1 Appropriate vector preparation for transformation of U. mutabilis

1. Linearize 20 μg of a transformation vector with a suitable single cutting restriction endonuclease.
2. Perform a final purification of the linearized transformation vector by using a silica-based DNA purification kit, for example, QIAquick (Qiagen, Hilden, Germany), and elute in the smallest volume.
3. Determine the DNA concentration of the linearized and purified transformation vector.

31.4.2 Preparation of gametes of U. mutabilis *for transformation*

As *Ulva* gametogenesis is regulated by a light-triggered and endogenous circadian clock, releasing of gametes from mature gametangia should take place in the morning of the third day after the induction of gametogenesis according to Kessler et al. (2018) (please see Chapters 8 and 9).

Gametes must be used directly as they are depleted of a cell wall for a few hours only. Therefore, the transformation procedure must be performed not later than 4 h after the hatching of gametes. Moreover, zoids from parthenosporophytes lack also a rigid cell wall, making them suitable for this method as well (see *Note* 2).

1. Induce the gametogenesis as described by Stratmann et al. (1996) and in Chapter 8.
2. In the morning of the third day after induction, decant the tissue material through a nylon sieve and transfer all thalli-fragments into fresh UCM to wash out the swarming inhibitor *via* exchanging the media (see also Chapter 9).
3. Place a light source at one site of the cultivation vessel. After 10–40 min, released gametes accumulate at the brightest spot facing the light source.
4. Collect the motile gametes in centrifuge tubes (15 mL) and place these again with the tip orientated to the light source for additional accumulation and concentration. The movement of gametes to the light can be enhanced by covering the centrifuge tubes with aluminum foil up to the end of the tip of the tube. After accumulation, the gametes can be carefully collected by using Pasteur pipettes with a long thin tip. Transfer the accumulated gametes carefully into a micro centrifuge tube (1.5 mL).
5. *Optional*: Due to the overall research interest, it might be necessary to work with axenic cultures, which can be prepared from gametes as described in Chapter 9 (*Note 3*).
6. Wash the gametes after centrifugation (2 min, 13,500 × g, 20°C) by resuspending in 1 mL UTB1.
7. Determine the cell density of the gametes or zoids (Chapter 9) and dilute cells with UTB1 to 10^6–10^8 cells/100 µL with UTB1 (Table 31.1).

31.4.3 *Transformation of* U. mutabilis *gametes*

1. Add 2–5 µg of linearized and purified plasmid DNA to the 100 µL cell suspension and incubate for 10 min at room temperature.
2. Proceed with 100 µL cell suspension without purified plasmid DNA analogous to the next steps for control (*Note 4*).
3. Dilute the sample with 1 mL UTB2 (Table 31.2, *Note 5*) and incubate for an additional 30 min.
4. Centrifuge the transformed gametes for 10 min at 13,500 × g at 20°C and discard the supernatant immediately.
5. Resuspend the transformed gametes carefully in 500 µL UCM (*Note 6*).

31.4.4 Cultivation of transgenic Ulva lines and selection by phleomycin treatment

1. Inoculate 20 mL of fresh UCM in a Petri dish (145 mm diameter) with either 250 µL of resuspended transformed gametes or nontransformed gametes as a negative control, respectively (Figure 31.3).
2. Incubate the cells in darkness at 18°C until the next morning to allow the expression of the phleomycin resistance gene and attachment of gametes to the surface of the Petri dish (or culture flask).
3. Add phleomycin from the stock solution to a final concentration of 5 µg mL^{-1} (*Note 7*).
4. A prerequisite when working with already purified (axenic) gametes: Add 0.5 µL from each stationary culture of *Maribacter* sp. MS6 and *Roseovarius* sp. MS2 (*Note 8*).
5. Cultivate the transgenic lines and the control for two to three weeks in a 7 h light/17 h dark cycle at 18°C without changing the media, including phleomycin.

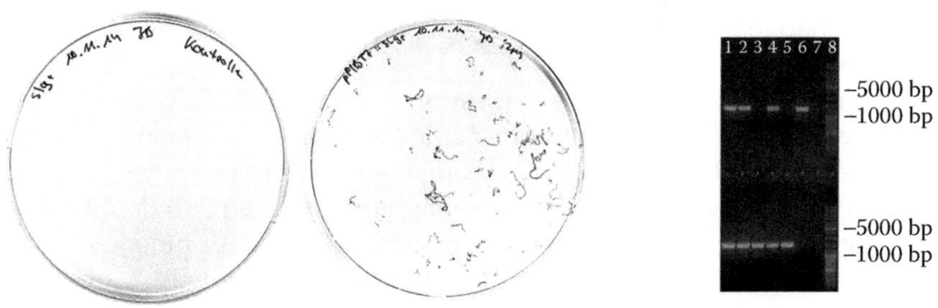

(a) Control (b) Transformants (c)

Figure 31.3 Transformants of the *Ulva mutabilis* mutant (slender) derived from gametes. (a) The negative control (=transformation experiment without DNA) did not show any surviving germling or contaminant. (b) *U. mutabilis* gametes were transformed with the linearized vector plasmid pPIBT7. The picture was taken by inverted microscopy in the third week after transformation. The transformation was conducted under the conditions described. (c) Agarose gel photograph showing PCR genotyping results. The upper part of the gel was loaded with 5 µL PCR product from the reaction, using primers Ph-2 and RbcS_F1 (amplicon length 1376 bp), and the lower part product from the control reaction with primers UBQ1_F1 and Ubi_R2 (amplicon length 1601 bp). PCR products, lanes 1–4: gDNA from independently selected transformed *Ulva* lines; lane 5: gDNA from untransformed *Ulva*; lane 6: 0.5 pg *Ssp*I-linearized pPIBT7; lane 7: water. Lane 8 contains 5 µL of molecular marker.

31.4.5 *Determination of the transformation efficiency and further single-strain cultivation*

Regarding plasticity in morphogenesis of *U. mutabilis*, random insertion of transformation vectors very often reveals a variety of different morphotypes in transgenic lines with different developmental paces and states. Different growth rates are also furthered by differences in the sensitivity of the individual insertional mutants against phleomycin, caused by multiple insertions of the resistance gene.

The first green transgenic lines usually become visible within the third week after transformation. It also includes thalli that are free-floating in the media. Early separation of larger seedlings in single culture flasks or multiwell plates induces faster growth of the remaining smaller seedling. Further cultivation of the transgenic strains does not require a continuance of the antibiotic treatment due to the stable insertion of the resistance gene. Moreover, persistent treatment of phleomycin can cause additional random mutations in the individual transgenic lines. Due to various insertion events into the genome and the disrupted genes, different growth rates of the transformants should be expected. Novel morphotypes of the thallus can be regularly observed five weeks after the transformation (Figure 31.4).

1. Exchange the media containing phleomycin for fresh UCM without phleomycin.
2. Preferably use a nylon sieve (mesh size 30 μm) to avoid a loss of free-floating or very tiny transgenic seedlings when exchanging the media supplemented with phleomycin to UCM.
3. Count the number of seedlings that have survived in a defined area by using a calibrated inverted microscope or with binoculars. No seedling should develop on the control plates (Figure 31.3).
4. Depending on the size and developmental status of the transgenic seedlings, transfer each seedling to single culture flasks or multiwell plates for further cultivation. Frequent changes of media (every week) expedite sporulation of the transgenic strains, which is necessary to produce larger populations of the transformants.

31.4.6 *Confirmation of stable transformation*

The most straightforward methods to verify whether the *Ulva* germlings that survived the phleomycin selection also carry the inserted vector are Southern blots and PCR analyses. To optimize these techniques for *Ulva*, information about genome size and GC-content would be beneficial because these parameters influence the efficiency of both approaches.

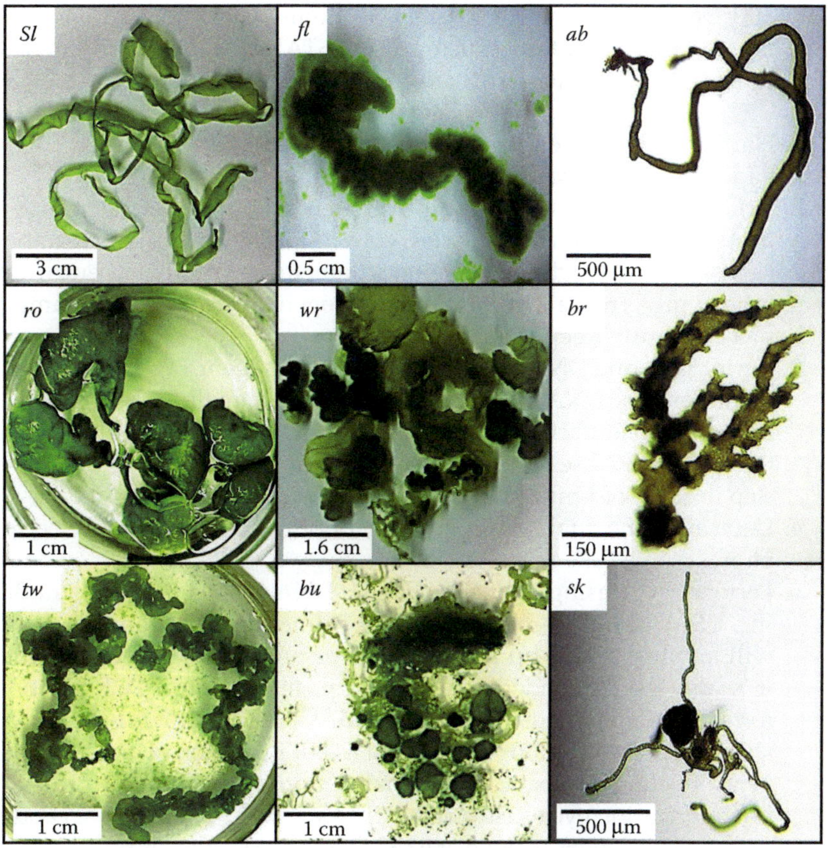

Figure 31.4 Mutants with atypical thallus formation on polyethylene glycol (PEG)-mediated transformation of gametes of *Ulva mutabilis*. Cultures of *Ulva* are three to five weeks old. Sl, slender (control strain); fl, filamentous branched; ab, additional blade branch; ro, rosette-shaped; wr, wrinkled; br, branched; tw, twisted; bu, bubble; sk, skinny.

However, currently, there is no clear consensus on the exact genome size or the GC-content of the *Ulva* genome (Freshwater et al. 1990, Kapraun 2007). In addition, high methylation of the genome (Pakhomov et al. 1968) and the presence of secondary metabolites, such as sulfated polysaccharides (Haug 1976), further affect the amenability of *Ulva* to standard molecular techniques. Therefore, protocols and kits that work effectively for other organisms often need considerable testing before they can be applied successfully to *Ulva*. Here, we used PCR analysis to verify transgene insertion in *Ulva* primary transformants (Figure 31.3). We were most successful in the purification of genomic DNA using the GenElute Plant

Genomic DNA Miniprep Kit (Sigma, St. Louis, USA). Regarding PCR, the best results were obtained using PrimeSTAR GXL DNA polymerase (Takara Bio Inc, USA) (alternative: *Pfu*-Turbo DNA polymerase).

1. Harvest algal material and squeeze it thoroughly between tissue paper to remove excess culture medium.
2. Grind the tissue, under liquid nitrogen, to a fine powder using a pestle and mortar. Alternatively, collect up to 100 mg tissue together with two steel beads (approximate diameter 2–3 mm) in a centrifuge tube (2 mL), freeze it using liquid nitrogen, and disrupt the sample in a bead mill. Keep the samples frozen.
3. Purify genomic DNA from the disrupted tissue using the GenElute Plant Genomic DNA Miniprep Kit (Sigma, St. Louis, MO, USA), according to the manufacturer's instructions. It is recommended to include the optional RNaseA digestion and perform only a single final elution step in 50 µL of 5 mM Tris/HCl (pH 8.5) solution heated to 65°C.
4. Determine the quality and quantity of purified DNA by gel electrophoresis on a 0.8% agarose gel.
5. Perform PCR using PrimeSTAR GXL DNA Polymerase (Takara Bio Inc, USA). The PCR reactions, using primer pair Ph-2 and RbcS_F1, will amplify a 1376 bp region of the bleomycin resistance cassette. It is also recommended to include a positive control PCR reaction for each sample by using primers pair Ubi1-F1 and Ubi1-R2, which should result in the amplification of a 1601 bp region of an *Ulva UBQ* gene. Cycling protocol: 10 s denaturation at 98°C followed by 30 polymerization cycles (10 s at 98°C, 10 s at 62°C, 105 s at 68°C) and a final 1 min elongation step at 68°C (Table 31.3).
6. Analyze the PCR products length on a 1.0% (w/v) agarose gel.

Table 31.3 PCR reaction mixture composition for optimized conditions

Component	Amount for a 50 µL reaction volume	Final concentration
5× PrimeSTAR GXL Buffer	10 µL	1×
dNTP Mixture (2.5 mM each)	4 µL	0.2 mM each
Primer FW (10 µM)	1 µL	200 nM
Primer RV (10 µM)	1 µL	200 nM
DNA template	1–100 ng	–
PrimeSTAR GXL DNA Polymerase (1.25 U/µL)	0.75 µL	0.94 U/50 µL
Dimethyl sulfoxide	2.5 µL	5%
Water	To 50 µL	–

31.5 Notes

Note 1: Before the use of any transformation buffer, sterilize the buffer by passing through a membrane filter (0.2 μm) using a sterile disposable syringe.

Note 2: *Ulva* zoids from parthenosporophytes were transformed by the same procedure as gametes. However, a partial purification of these cells by their movement to the light could not be applied, because zoids are not phototactic. Therefore, zoids were used directly after hatching from mature sporangia. For further applications, cells must be processed within 30 min after the sporangia have discharged the zoids and before they start settling or aggregating.

Note 3: Development and morphogenesis in a variety of green macroalgae are strongly dependent on the appropriate microbiome in the environment. Therefore, it is necessary for certain scientific issues to work further with axenic algal cultures or in the presence of a defined microbiome. A smart way to separate gametes from their accompanying bacteria can be performed by the horizontal movement of phototactic gametes through glass capillary pipettes (for a detailed protocol, see Chapter 9 by Califano and Wichard 2018). The generation of axenic insertional mutants also requires highly purified DNA, and, therefore, additional purification steps by isopropanol or ethanol precipitation might be necessary.

Note 4: Increasing concentrations of the transformation vector not only result in better transformation efficiencies but also may cause multiple insertion events accompanied by problems in further characterization, including the identification of the insertional mutants.

Note 5: Because of the high viscosity of PEG 4000 and the UTB2, careful mixing is recommended through repeated aspiration with a microliter pipette with a wide-open tip.

Note 6: Based on experience, gametes do not precipitate in the highly concentrated UTB2 solution at the tip of the 1.5 mL micro centrifuge tubes after centrifugation. Gametes usually stick around the wall of the microcentrifuge tubes and are only visible as a light-green halo. Therefore, resuspend the transformed gametes carefully by repeated aspiration with a microliter pipette with a wide-open tip or gently shaking the tube in 500 μL UCM after discarding the supernatant.

Note 7: The concentration of phleomycin can be increased up to 50 μg mL^{-1}.

Note 8: A cross-kingdom cross-talk between macroalgae and bacteria controls a variety of algal morphogenic processes, such development, settlement, and growth. For that reason, it is essential to inoculate each medium with the appropriate marine bacteria.

At least two bacterial strains are necessary for complete morphogenesis of *Ulva*. *Roseovarius* sp. MS2 induces cell division, whereas the *Maribacter* sp. MS6 promotes a viable basal stem cell and primary rhizoid cells in addition to the induction of the proper cell wall formation. Moreover, the MS2 species exhibits a specific chemotactic affinity to the rhizoid cells of *U. mutabilis* (Kessler et al. 2018b) and seems to cooperate with MS6 and alga by chemical communication. Both bacterial strains need to be resistant to phleomycin. Alternatively, it is also possible to supplement 20 mL UCM with 100 μL of the sterile-filtered supernatant of stationary cultures of both MS6 and MS2 (Spoerner et al. 2012). The supernatants contain sufficient amounts of the essential bacterial growth factors for *Ulva*. Alternatively, the partly purified bacterial factors could be used (Spoerner et al. 2012). The morphogenetic substances are added repetitively to the growth medium of axenic cultures of *Ulva*.

Acknowledgments

This work is supported by the Deutsche Forschungsgemeinschaft (DFG) through the Collaborative Research Centre 1127 "Chemical Mediators in complex Biosystems" (T.B., M.K., and T.W.).

References

Califano, G., and T. Wichard. 2018. Preparation of axenic cultures in *Ulva* (Chlorophyta). In *Protocols for Macroalgae Research*, B. Charrier, T. Wichard, C.R.K. Reddy (Eds.), Chapter 9. Boca Raton, FL: CRC Press, Taylor & Francis Group.

Cock, J.M., L. Sterck, P. Rouze et al. 2010. The *Ectocarpus* genome and the independent evolution of multicellularity in brown algae. *Nature* 465: 617–621.

Collén, J., B. Porcel, W. Carré et al. 2013. Genome structure and metabolic features in the red seaweed *Chondrus crispus* shed light on evolution of the archaeplastida. *Proc. Natl. Acad. Sci.* 110: 5247–5252.

Freshwater, D.W., J.A. Dutcher, D.F. Kapraun, and R.K. Sizemore. 1990. Variation in nuclear DNA base composition (mol % G + C) in three orders of marine green algae. *Hydrobiologia* 204: 167–172.

Grueneberg, J., A. Engelen, R. Costa, and T. Wichard. 2016. Macroalgal morphogenesis induced by waterborne compounds and bacteria in coastal seawater. *PLoS One* 11: e0146307–e07.

Hallmann, A. 2007. Algal transgenics and biotechnology. *Transgenetic Plant J.* 1: 81–98.

Haug, A. 1976. Influence of borate and calcium on gel formation of a sulfated polysaccharide from *Ulva lactuca*. *Acta Chem. Scandinaviaca B* 30: 562–566.

Hinnen, A., J.B. Hicks, and G.R. Fink. 1978. Transformation of yeast. *Proc. Natl. Acad. Sci.* 75: 1929–1933.

Kapraun, D.F. 2007. Nuclear DNA content estimates in green algal lineages: Chlorophyta and Streptophyta. *Ann. Bot.* 99: 677–701.

Kathiresan, S., A. Chandrashekar, G.A. Ravishankar, and R. Sarada. 2009. *Agrobacterium*-mediated transformation in the green alga *Haematococcus pluvialis* (chlorophyceae, volvocales). *J. Phycol.* 45: 642–649.

Kessler, R.W., T. Alsufyani, and T. Wichard. 2018. Purification of sporulation and swarming inhibitors from *Ulva*: Application in algal life cycle controlling. In *Protocols for Macroalgae Research*, B. Charrier, T. Wichard, C.R.K. Reddy (Eds.), Chapter 8. Boca Raton, FL: CRC Press, Taylor & Francis Group.

Mikami, K. 2014. A technical breakthrough close at hand: Feasible approaches toward establishing a gene-targeting genetic transformation system in seaweeds. *Front. Plant Sci.* 5: 498.

Mikami, K. 2013. Current advances in seaweed transformation. In *Biochemistry, Genetics and Molecular Biology an Integrated View of the Molecular Recognition and Toxinology—From Analytical Procedures to Biomedical Applications*, G.R. Baptista (Ed.), Chapter 13. Rijeka, Croatia, InTech.

Oertel, W., T. Wichard, and A. Weissgerber. 2015. Transformation of *Ulva mutabilis* (chlorophyta) by vector plasmids integrating into the genome. *J. Phycol.* 51: 963–979.

Pakhomov, M.V., G.N. Zaitseva, and A.N. Belozerskii. 1968. Presence of 5-methylcytosine and 6-methylaminopyrine in composition of DNA of certain algae. *Dok. Akad. Nauk SSSR* 182: 712–714.

Spoerner, M., T. Wichard, T. Bachhuber, J. Stratmann, and W. Oertel. 2012. Growth and thallus morphogenesis of *Ulva mutabilis* (Chlorophyta) depends on a combination of two bacterial species excreting regulatory factors. *J. Phycol.* 48: 1433–1447.

Stratmann, J., G. Paputsoglu, and W. Oertel. 1996. Differentiation of *Ulva mutabilis* (chlorophyta) gametangia and gamete release are controlled by extracellular inhibitors. *J. Phycol.* 32: 1009–1021.

Suzuki, R., S. Ota, T. Yamazaki et al. 2016. Morphological changes of giant mitochondria in the unicellular to multicellular phase during parthenogenesis of *Ulva partita* (Ulvophyceae) revealed by expression of mitochondrial targeting GFP and PEG transformation. *Phycol. Res.* 64: 176–184.

Walker, T.L., C. Collet, and S. Purton. 2005. Algal transgenics in the genomic era. *J. Phycol.* 41: 1077–1093.

Wichard, T. 2015. Exploring bacteria-induced growth and morphogenesis in the green macroalga order ulvales (chlorophyta). *Front. Plant Sci.* 6: 86.

Wichard, T., B. Charrier, F. Mineur et al. 2015. The green seaweed *Ulva*: A model system to study morphogenesis. *Front. Plant Sci.* 6: 72.

Index

Note: Page numbers followed by f and t refer to figures and tables respectively.

L

M

N

O

V

W

Z